CONTENTS

Acknowledgements

We are grateful to those listed for permission to reprint copyright material.

Figure 1.5 from *Rival States, Rival Firms: Competition for World Markets*, John Stopford and Susan Strange with John S. Henley, Figure 1.6, p.22. Reprinted with permission of Cambridge University Press.

Figure 3.1 from *Global Politics: Globalization and the Nation-State*, Anthony G. McGrew, Paul G. Lewis *et al.* (eds) (1992), Figure 1.7, p. 13, Polity Press, Oxford. Reprinted with permission of Blackwell Publishers.

Figure 4.1 Reprinted from *Between MITI and the Market: Japanese Industrial Policy for High Technology*, Daniel I. Okimoto, Figure 1.5, p. 51, with the permission of the publishers, Stanford University Press. © 1989 by the Board of Trustees of the Lealand Stanford Junior University.

Figure 5.2 from The challenge of new technologies, C. Freeman, © OECD 1987, *Interdependence and Cooperation in Tomorrow's World, a symposium marking the 25th Anniversary of the OECD.* Reproduced with permission of the OECD.

Figure 5.7 from The geography of international business telecommunications: the role of leased networks, John V. Langdale, *Annals of the Association of American Geographers,* 1989, Vol. 79, Figure 1. Reprinted with permission of Blackwell Publishers.

Figure 6.2 Reprinted by permission of Harvard Business School Press. From *The Product Life Cycle and International Trade*, Louis T. Wells Jr., Boston, MA, 1972, p. 15. Copyright © 1972 by the President and Fellows of Harvard College, all rights reserved.

Table 6.1 from *Multinational Enterprises and the Global Economy*, 1993, John H. Dunning, Table 4.2, pp. 82–83. Reprinted by permission of Addison-Wesley Longman Ltd.

Figure 7.12 from The role of collaborative integration in industrial organization: observations from the Canadian aerospace industry, Malcolm Anderson, *Economic Geography*, 1995, Vol. 71, Figure 1, p. 60.

Figure 12.6 from Telecommunications and the globalization of financial services, B. Warf, *Professional Geographer*, 1989, Vol. 41, Figure 5. Reprinted by permission of Blackwell Publishers.

Figure 12.8 from The internationalization of Japanese commercial banking, in M. J. Taylor and N. J. Thrift (eds) (1986) *Multinationals and the Restructuring of the World Economy*, chapter 7, M. Fujika and K. Ishigaki, Table 7.6, Croom Helm, London.

Table 12.5 from *Human Resources and Corporate Strategy: Technological Change in Banks and Insurance Companies*, O. Bertrand and T. Noyelle, Table 4.1. Reproduced with permission of the OECD.

Preface

Preoccupation with the 'global' has become one of the emblematic features of our time. Almost all national political leaders invoke the mantra of globalization to justify their particular economic and social policies. Two examples, both taken from annual budget statements made by national finance ministers, make the point in graphic terms:

> . . . we have no choice but to open and to compete in the *world* market to survive and prosper. We can grow faster by taking advantage of *global* markets and advanced technology . . . our openness exposes us inevitably to the fluctuation of *global* business and demand cycles . . . Adapting nimbly to changes in the environment and staying relevant to *global* demand remains fundamental to Singapore's survival.
> (Dr Richard Hu, Singapore Finance Minister, 10 July 1997, emphasis added)

> the central purpose of this Budget is to ensure that Britain is equipped to rise to the challenge of the new and fast-changing *global* economy . . . The impact of the *global* market in goods and services, and of rapidly advancing technology, is now being felt in every home and every community in our country . . . This new *global* economy driven by skills creativity, and adaptability offers a historic opportunity . . . In a *global* economy, long-term investment will come to those countries that demonstrate stability in their monetary and fiscal policies, and in their trading relationships . . .
> (Gordon Brown, UK Chancellor of the Exchequer, 7 July 1997, emphasis added)

The electronic media are the instant window on 'global' events or, more accurately, on events occurring at a local scale which are then projected globally. At the same time, it is virtually impossible to read any (reputable) newspaper without finding some reference to the phenomemon of globalization. Not surprisingly, then, there is a huge and fast-growing academic and popular literature on the 'global'. A good deal of that literature is stronger on rhetoric and hype than on reality. On the one hand, ambitious extrapolations are made, often from a small number of cases, to paint a scenario of unstoppable global forces leading to an ultimately homogenized world in which local differences will be virtually eradicated. On the other hand, there are those who argue that, really, nothing has changed very much; that the globalization story *is* little more than hype.

In line with its two predecessors, this third edition of *Global Shift* takes a more balanced view of the processes that are transforming the world economy, although there is no doubt at all that major transformations are, indeed, occurring. In the six years since the second edition appeared, not only have there been massive changes in the world economy and in its constituent parts but also in the debates about the nature and meaning of these changes. This edition attempts to capture both these sets of developments through a judicious mix of conceptual and empirical materials drawn from a very wide variety of academic disciplines and sources: particularly in economic geography, economics, sociology, international business, international political economy. Each chapter has been totally revised and, in several cases, completely rewritten. The statistical data on trends in the global economy are the latest available as of early 1997. Although the broad four-part structure of the second edition is retained, significant changes have been made in content and emphasis. The book remains unique in its particular combination of empirical and conceptual material, and, especially, in the inclusion of detailed sectoral case studies of key global industries, as well as in its concern with the broader social impacts of change, especially on employment.

Perhaps most significantly – and appropriately given my own background – I emphasize to a greater extent than in previous editions the geographical unevenness of the economic, political and technological processes which, together, create global shifts in economic activity. Despite the undoubted revolutionary effects of new technologies which shrink time and space, it is highly misleading to talk glibly of 'the death of distance'. Just as we have certainly not seen the 'end of history' neither have we seen – or are we going to see – the 'end of geography'. Place and space remain fundamental to the operation of all forms of human organization. All economic activities are physically grounded in specific locations; the geographical clustering of economic activities is the normal state of affairs. What is changing, however, is the scale and complexity of the structures within which such activities are embedded. Between the 'global' and the 'local' there is a continuum of inter-related geographical scales through which the processes of economic transformation are mediated. These issues are addressed in a totally new first chapter which outlines the major dimensions of the current globalization debate and then puts in place the fundamental building blocks of my own approach. I emphasize, in particular, the significance of production chains and the major ways in which they are being configured and reconfigured both organizationally and geographically.

The basic approach of this book is that globalization is a complex set of *processes* which operate very unevenly in both time and space. What we see 'on the ground', at various geographical scales, is primarily the outcome of the interaction between two major sets of institutions – transnational corporations and states – set within the context of a volatile technological environment. Through a complex and dynamic set of interactions, these constitute the primary generators of global economic transformation. Transnational corporations, through their geographically extensive operations, and states, through their trade, foreign investment and industry policies (both singly and in such regional groupings as the EU or NAFTA) shape and reshape the global economic map. At the same time, revolutionary changes in the technologies of transport and communication and in the processes of production have facilitated the internationalization and globalization of production of both manufactured goods and of services. Few parts of the world are unaffected as TNCs restructure their operations, as national governments attempt to build or preserve their own economies, and as the pace of technological change quickens and its nature changes. Such changes are both geographically and sectorally uneven. It is for that reason that we examine not only the general processes of change but also their specific expression in different places and in different industries (both manufacturing and services). But the process is not simply 'top down', from the global to the local. Conditions at a range of geographical scales, themselves, play an active role in mediating and influencing the precise operation of 'global' forces.

In this edition, I have expanded the treatment of each of these three major forces in specific ways. In discussing the continuing significance of the state in the world economy, more emphasis is placed upon the nature of states as containers of distinctive institutions and practices, as regulators, and as competitors, as well as their increasing propensity to participate in regional economic groups (Chapter 3). A separate chapter (Chapter 4) provides a range of case examples of state policies drawn from both the older industrialized economies of Europe, the United States and Japan, and from newly industrializing economies in east and southeast Asia (including China) and Latin America. In the two chapters on transnational corporations, I have added a new section which deals with the myth of the 'placeless' TNC (in Chapter 6) and substantially expanded the treatment of business networks in Chapter 7 to incorporate some discussion of business networks in east Asia. The dynamic interaction between TNCs and states – both conflictual and collaborative – receives greater, and more focused, attention in a new chapter (Chapter 8), which incorporates the material on the costs and benefits of TNCs which appeared in Chapter 12 of the second edition. The chapter on technology (Chapter 5) has been expanded to incorporate a fuller discussion of evolutionary technological change and, especially, of alternative forms of 'after-Fordist' production systems. Explicit attention is also given to the propensity for technological innovation and learning to occur in localized innovative milieux or technology districts. To make space for these major changes, I have collapsed what were formerly two separate empirical chapters on the changing global map of production, trade and investment into a single chapter (Chapter 2). This not only saves space but it also, I believe, creates a more clearly focused and integrated treatment. Finally, I have replaced what was a very brief speculative 'epilogue' in the second edition by a more substantial, though restricted, discussion (Chapter 14) of some issues of global governance. In particular, I outline some of the currently politically sensitive questions surrounding the relationships between international trade, labour standards and the environment.

The intention of the book remains the same as the earlier editions. It is designed as a book which crosses disciplinary boundaries, both in terms of its subject-matter and also in its approach. Precisely how it is used will vary according to the background of the reader. To some it will be seen as an introductory-level book; to others it will be regarded as an upper-level undergraduate or a postgraduate text. The notes for further reading at the end of each chapter and the extensive bibliography should facilitate its use at these different levels.

As always, I owe enormous intellectual (and social) debts to colleagues and friends across the world whose stimulating and constructively critical comments have done so much to help (though, of course, none of them is to blame for the inevitable weaknesses that remain). Such debts are especially great when the subject-matter of the book is global. The challenge of presenting seminars and lectures on the topic of *Global Shift* to colleagues and students in departments of geography, economics, sociology, politics and international business in Europe, the United States, Canada, Mexico, Australia, Hong Kong, Singapore, has been hugely beneficial to me in sharpening up my ideas and in helping me to see the world (both literally and figuratively) from different perspectives.

It is impossible to name every individual who has directly or indirectly helped in the process of writing this book. However, I would like to thank, in particular, the following people. In my home department at the University of Manchester, I have been immensely fortunate to be associated with a great group of colleagues and friends, notably Jamie Peck and Adam Tickell, together with our dynamic cluster of irreverent graduate students. Among my recent graduate students, I particularly want to thank Henry Yeung, now of the National University of Singapore, for teaching me about Asian business (and for keeping me on my toes and cheering me up even on email). Also at the University of Manchester, Nick Scarle has again done a miraculous job with the illustrations for the book, turning scrappy sketches and masses of statistical data into elegant diagrams. I also benefit greatly from involvement with such social science and business school colleagues as Jeff Henderson, Richard Whitley, Diane Elson and Huw Beynon, while Adam Holden and Steve Quilley helped in updating some of the case study material in Part III. Elsewhere in the UK, I owe a much greater debt than he realizes to Roger Lee, who takes more care in commenting on papers and drafts than anybody I know (and who also has excellent musical taste). I also wish to thank Yoshihiro Miyamachi for stimulating collaboration on Japanese trading companies, and Andy Leyshon and others for their helpful comments on the outline for this edition. Within Europe, I was very fortunate to be involved in a European Science Foundation project on regional and urban restructuring and I would especially express my gratitude to colleagues on the RURE Programme: Anders Malmberg (who also made invaluable comments on my plans for this edition), Sture Öberg, Arie Shachar, Ash Amin, Ray Hudson, Eike Schamp and Nigel Thrift. In North America, Meric Gertler deserves special thanks for being such a stimulating and caring colleague, as do Erica Schoenberger and Amy Glasmeier. I would also like to thank Gary Gereffi for his very helpful comments. Finally, in the last two years I have been very fortunate in being involved with a new venture, the Duxx Graduate School of Business Leadership in Monterrey, Mexico and I would like to thank Carlo Brumat and his colleagues, as well as the first two generations of students there, for providing such a stimulating and friendly environment. Also in Monterrey, Jesús Treviño has been extremely helpful in obtaining up-to-date materials. Both publishers of the book have continued to be extremely supportive. I particularly want to thank Paul Chapman and Marianne Lagrange of Paul Chapman Publishing, London, and Peter Wissoker of Guilford Publications, New York. Ultimately, of course, I come back to where it all really begins and ends – with my family: Valerie, Michael and Christopher. They are what it is really all about. Their continuing love and healthy scepticism of academic pretentiousness, including their scatalogical translation of the book's title, really does make it all worth while. The Bridgewater Hall also helps.

Peter Dicken
Manchester and Singapore, July 1997

CHAPTER 1

A new geo-economy

Something is happening out there

The notion that something fundamental is happening, or indeed has happened, in the world economy is now generally accepted. As we look around us, all we seem to see is the confusion of change, the acceleration of uncertainty; feelings currently intensified by our proximity to the new millennium with all its promises – and threats – of epochal change. Television news reports and specials, press headlines and the like constantly remind us of our uneasy present and precarious future. Turbulence and change are, of course, nothing new in human affairs. But there is no doubt that the world economy, and its constituent parts, are being buffeted by extremely volatile forces. To the individual citizen the most obvious indicators of change are those which impinge most directly on his or her daily activities – making a living and consuming the necessities and luxuries of life. To those currently employed in a job, what matters is job security and the wages or salary received or anticipated. To those seeking a job, what matters is availability. On both counts, the situation seems to have become increasingly uncertain. In the industrialized countries, in particular, there is a real fear that the dual (and connected forces) of technological change and geographical shifts in the location of manufacturing and service activities are transforming the employment scene in adverse ways for many people, notably less educated and less skilled blue-collar workers although there has also been very considerable job volatility among white-collar workers as well. As consumers, the most obvious indicator of change is the vast increase in the number of products whose origins lie on the other side of the world but which are now either literally on our doorsteps (through superstores) or metaphorically in our homes (through the all-pervasive TV commercial). Whatever our particular position, however, we cannot fail to be aware that what is happening in our own back-yards is largely the product of forces operating at a much larger geographical scale.

The immediacy and longer-term impact of these major forces of change are enormously enhanced by the growing interconnections between all parts of the world. The most significant development in the world economy during the past few decades has been the *increasing internationalization – and, arguably, the increasing globalization – of economic activities*. The internationalization of economic activities is nothing new. Some commodities have had an international character for centuries; an obvious example being the long-established trading patterns in spices and other exotic goods. Such internationalization was much enhanced by the spread of industrialization from the eighteenth century onwards in Europe. Nevertheless until very recently the production process itself 'was primarily organized *within* national economies or parts of them. International trade . . . developed primarily as an exchange of raw materials and foodstuffs . . . [with] . . . products manufactured and finished in single national economies . . . *In terms of production, plant, firm and industry were essentially national phenomena* (Hobsbawm, 1979, p. 313, emphasis added).

1

The nature of the world economy has changed dramatically, however, especially since the 1950s. National boundaries no longer act as 'watertight' containers of the production process. Rather, they are more like sieves through which extensive leakage occurs. The implications are far reaching. Each one of us is now more fully involved in a global economic system than were our parents and grandparents. Few, if any, industries now have much 'natural protection' from international competition whereas in the past, of course, geographical distance created a strong insulating effect. Today, in contrast, fewer and fewer industries are oriented towards local, regional or even national markets. A growing number of economic activities have meaning only in a global context. Thus, whereas a hundred or more years ago only rare and exotic products and some basic raw materials were involved in truly international trade, today virtually everything one can think of is involved in long-distance movement. And because of the increasingly complex ways in which production is organized across national boundaries, rather than contained within them, the actual origin of individual products may be very difficult to ascertain.

Something of this increased global diversity of production can be gleaned simply from examining the labels on products. Many labels are geographically misleading, however, particularly in the case of products consisting of a large number of individual components, each of which may have been made in different countries. Generally, the labels signify the country of the final (assembly) stage of production. But where are such products really made? Under such conditions what is a 'British' car, an 'American' computer, a 'Dutch' television or a 'German' camera? In today's global economy, some products can be regarded as having been made almost everywhere – or nowhere – such is the geographical complexity of some production processes.

What these developments imply is the emergence of a *new global division of labour* which reflects a change in the geographical pattern of specialization at the global scale. Originally, as defined by the eighteenth-century political economist Adam Smith, the 'division of labour' referred simply to the specialization of workers in different parts of the production process. It had no explicitly geographical connotations at all. But quite early in the evolution of industrial economies the division of labour took on a geographical dimension. Some areas came to specialize in particular types of economic activity. Within the rapidly evolving industrial nations of Europe and the United States regional specialization – in iron and steel, shipbuilding, textiles, engineering and so on – became a characteristic feature. At the global scale the broad division of labour was between the industrial countries on the one hand, producing manufactured goods, and the non-industrialized countries on the other, whose major international function was to supply raw materials and agricultural products to the industrial nations and to act as a market for some manufactured goods. Such geographical specialization – structured around a *core*, a *semi-periphery* and a *periphery* – formed the underlying basis of much of the world's trade for many years.

This relatively simple pattern (although it was never quite as simple as the description above suggests) no longer applies. During the past few decades trade flows have become far more complex. The straightforward exchange between core and peripheral areas, based upon a broad division of labour, is being transformed into a highly complex, kaleidoscopic structure involving the *fragmentation* of many production processes and their *geographical relocation* on a global scale in ways which slice through national boundaries. In addition, we have seen the emergence of new centres of industrial production in the newly industrializing economies (NIEs). Both old and new industries are involved in this re-sorting of the global jigsaw puzzle in ways which also reflect the development of technologies of transport and communications, of corporate organization and of the production process. The technology of production itself is undergoing substantial and far-reaching change as the emphasis on large-scale, mass-production, assembly-line techniques is shifting to a more flexible production technology. And just as we can identify a new international division of labour in production so, too, we can identify a 'new international financial system', based on rapidly emerging twenty-four hour global transactions concentrated primarily in the three major financial centres of New York, London and Tokyo.

A 'new' geo-economy? The globalization debate

So, something is undoubtedly happening 'out there'. But precisely what that 'something' might be – and whether it really represents something new – is a subject of enormous controversy amongst academics, politicians, popular writers and journalists alike.[1] Box 1.1 provides a sample of quotations to reflect a spectrum of opinions. The first two quotations – by Cohen and Zysman and by Drucker – argue that fundamental, possibly irreversible, change has occurred in the world economy and that the result is a deep-seated shift in its structure and operations. But Reich and Ohmae go much further and assert that, in effect, we now live in a borderless world in which the 'national' is no longer relevant. It is this latter view which has become especially pervasive. 'Globalization' is the new economic (as well as political and cultural) order. We live, it is asserted, in a global*ized* world in which nation-states are no longer significant actors or meaningful economic units; in which consumer tastes and cultures are homogenized and satisfied through the provision of standardized global products created by global corporations with no allegiance to place or community. The global is, thus, claimed to be the natural order of affairs in today's technologically driven world in which time-space has been compressed, the 'end of geography' has arrived and everywhere is becoming the same.

Although the notion of a *globalized* world has become pervasive there are strong opponents who argue, in effect, that globalization is a mirage. According to this view – represented in Box 1.1 by the quotations from Gordon, Glyn and Sutcliffe and Hirst and Thompson – the 'newness' of the current situation has been grossly exaggerated. The world economy, it is argued, was actually more open and more integrated in the half century prior to World War One (1870–1913) in which there was

> unprecedented international integration. An open regulatory framework prevailed; short- and long-term capital movements were unsupervised, the transfer of profits was unhampered; the gold standard was at its height and encompassed almost all the major industrial countries by the period's close and most smaller agrarian nations . . . ; citizenship was freely granted to immigrants; and direct political influence over the allocation of resources was limited . . . Under these conditions, markets linked a growing share of world resources and output; exports outgrew domestic output in the core capitalist countries . . . and the migration of labour was unprecedented.
>
> (Kozul-Wright, 1995, pp. 139–40)

So, on the one hand, we have the view that we do, indeed, live in a new – *globalized* – world economy in which our lives are dominated by global forces. On the other hand, we have the view that not all that much has changed; that we still inhabit an *international*, rather than a globalized, world economy in which national forces

The world economy is changing in fundamental ways. The changes add up to a basic transition, a structural shift in international markets and in the production base of advanced countries. It will change how production is organized, where it occurs, and who plays what role in the process.

(Cohen and Zysman, 1987, p. 79)

The talk today is of the 'changing world economy'. I wish to argue that the world economy is not 'changing'; it has already changed – in its foundations and in its structure – and in all probability the change is irreversible.

(Drucker, 1986, p. 768)

We are living through a transformation that will rearrange the politics and economics of the coming century. There will be no *national* products or technologies, no national corporations, no national industries. There will no longer be national economies, at least as we have come to understand that concept . . . As almost every factor of production – money, technology, factories, and equipment – moves effortlessly across borders, the very idea of an American economy is becoming meaningless, as are the notions of an American corporation, American capital, American products, and American technology. A similar transformation is affecting every other nation, some faster and more profoundly than others; witness Europe, hurtling toward economic union.

(Reich, 1991, pp. 3, 8)

Today's global economy is genuinely borderless. Information, capital, and innovation flow all over the world at top speed, enabled by technology and fueled by consumers' desires for access to the best and least expensive products.

(Ohmae, 1995a, inside front cover)

Globalization seems to be as much an overstatement as it is an ideology and an analytical concept.

(Ruigrok and van Tulder, 1993, p. 22)

I would argue that we have *not* witnessed movement toward an increasingly 'open' international economy, with productive capital buzzing around the globe, but that we have moved rapidly toward an increasingly 'closed' economy for productive investment . . . The international economy . . . has witnessed *declining* rather than *increasing* mobility of productive capital . . . the role of the State has grown substantially since the early 1970s; state policies have become increasingly decisive on the international front, not more futile.

(Gordon, 1988, p. 63)

The system has . . . become more integrated or globalized in many respects . . . Nonetheless what has resulted is still very far from a globally integrated economy . . . In short, the world economy is considerably more globalized than 50 years ago; but much less so than is theoretically possible. In many ways it is less globalized than 100 years ago. The widespread view that the present degree of globalization is in some way new and unprecedented is, therefore, false.

(Glyn and Sutcliffe, 1992, p. 91)

We do not have a fully globalized economy, we do have an international economy and national policy responses to it.

(Hirst and Thompson, 1992, p. 394)

Box 1.1   Some alternative views in the globalization debate remain highly significant. The truth, it seems to me, lies in neither of these two polarized positions. Although in quantitative terms the world economy was perhaps at least as integrated economically before 1913 as it is today – in some respects, even more so – the nature of that integration was *qualitatively* very different (UNCTAD, 1993b, p.113):

- International economic integration before 1913 – and, in fact, until only about three decades ago – was essentially *shallow integration* manifested largely through arm's length *trade* in goods and services between independent firms and through international movements of portfolio capital.
- Today, we live in a world in which *deep integration*, organized primarily by transnational corporations (TNCs), is becoming increasingly pervasive.'"Deep" integration extends to the level of the *production* of goods and services and, in addition, increases visible and invisible trade. Linkages between national economies are therefore increasingly influenced by the cross-border value adding activities within . . . TNCs and within networks established by TNCs' (UNCTAD, 1993b, p. 113).

However, although there are undoubtedly global*izing* forces at work we do not have a fully global*ized* world economy. Globalization tendencies can be at work without this resulting in the all-encompassing end-state – the globalized economy – in which all unevenness and difference are ironed out, market forces are rampant and uncontrollable, and the nation-state merely passive and supine.[2] The position taken in this book is that globalization is a complex of inter-related *processes*, rather than an end-state. Such tendencies are highly uneven in time and space. In taking such a process-oriented approach it is important to distinguish between processes of *internationalization* and processes of *globalization*:

- *Internationalization processes* involve the simple extension of economic activities across national boundaries. It is, essentially, a *quantitative* process which leads to a more extensive geographical pattern of economic activity.
- *Globalization processes* are *qualitatively* different from internationalization processes. They involve not merely the geographical extension of economic activity across national boundaries but also – and more importantly – the *functional integration* of such internationally dispersed activities.

Both processes – internationalization and globalization – coexist. In some cases, what we are seeing is no more than the continuation of long-established international dispersion of activities. In others, however, we are undoubtedly seeing an increasing dispersion and integration of activities across national boundaries. The pervasive internationalization, and growing globalization, of economic life ensure that changes originating in one part of the world are rapidly diffused to others. We live in a world of increasing complexity, interconnectedness and volatility; a world in which the lives and livelihoods of every one of us are bound up with processes operating at a global scale.

However, although we are often led to believe that the world is becoming increasingly homogenized economically (and perhaps even culturally) with the use of such labels as 'global village', 'global marketplace' or 'global factory', we need to treat such all-embracing claims with some caution. The 'globalization' tag is too often applied very loosely and indiscriminately to imply a totally pervasive set of forces and changes with uniform effects on countries, regions and localities. There are, indeed, powerful forces of globalization at work – they are the central focus of this book – but we need to adopt a sensitive and discriminating approach to get beneath the hype and to lay bare the reality. Change does not occur everywhere in the same way and at the same rate; the processes of globalization are not geographically uniform. The particular character of individual countries, of regions and of localities interacts with the larger-scale general processes of change to produce quite specific outcomes. Reality is far more complex and messy than many of the grander themes and explanations tend to suggest.

A new geo-economy: unravelling the complexity

We *are* witnessing the emergence of a new geo-economy which is qualitatively different from the past but in which both processes of internationalization and globalization and of shallow and deep integration continue to coexist. However, they do so in ways which are highly uneven in space, in time and across economic sectors. Very few industries are truly and completely global although many display some globalizing tendencies. The question is: how can we begin to unravel the dynamic, kaleidoscopic complexity of this geo-economy?

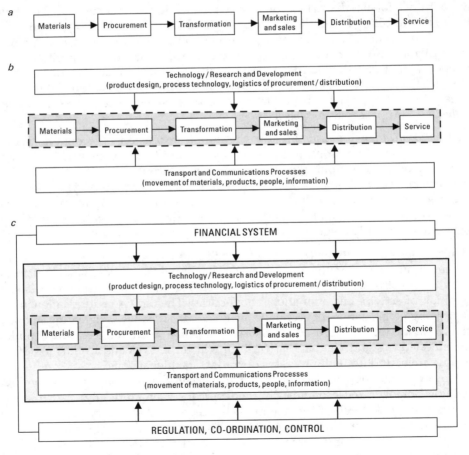

Figure 1.1   The basic production chain
Note: Labour inputs are involved in each element and are, therefore, not shown

The conventional unit of analysis in studies of the world economy is the nation-state. Virtually all the statistical data on production, trade, investment and the like are aggregated into national 'boxes'. Such a level of aggregation is less and less useful, given the nature of the changes occurring in the organization of economic activity. This is not to imply that the national level is unimportant. On the contrary, one of the major themes of this book is that nation-states continue to be key players in the contemporary global economy (see, for example, Chapter 3). In any case, we shall have to rely heavily on national level data to explore the changing maps of production trade and investment (as in Chapter 2, for example). But, as we noted earlier, national boundaries no longer 'contain' production processes in the way they once did. Such processes slice through national boundaries and transcend them in a bewildering array of relationships that operate at different geographical and organizational scales. We need to be able to get both below and above the national scale to understand what is going on.

Production chains: a basic building block

One especially useful conceptual point of entry is the *production chain*[3] which can be defined as

a transactionally linked sequence of functions in which each stage adds value to the process of production of goods or services.

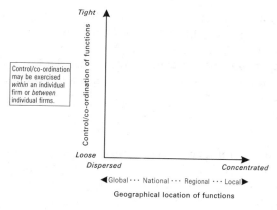

Figure 1.2   The two primary dimensions of production chains
Source: Based, in part, on material in Porter (1986, Chapter 1)

Figure 1.1 illustrates the basic elements of a hypothetical production chain. Figure 1.1(a) shows its most basic form. Figure 1.1(b) demonstrates that each of the individual elements in the production chain and their transactional links depend upon various kinds of technological inputs. Finally, Figure 1.1(c) indicates that each production chain is both embedded within a financial system, which provides the necessary investment and operating capital (notably credit), and also has to be co-ordinated, regulated and controlled. They key point is that many production chains are increasingly global in their geographical extent.

Two aspects of production chains are especially important from our point of view (Figure 1.2):

- their *co-ordination and regulation*
- their *geographical configuration.*

Co-ordination and regulation of production chains

Production chains are co-ordinated and regulated at two levels. First and foremost, they are co-ordinated by business firms, through the multifarious forms of intra- and

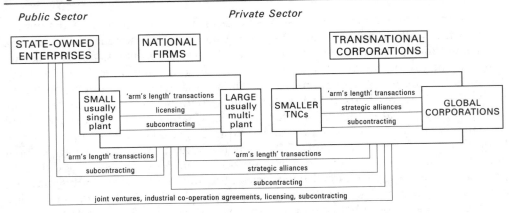

Figure 1.3   Major organizational segments within an economic system

interorganizational relationships that make up an economic system. As Figure 1.3 shows, economies are made up of different types of business organization – transnational and domestic, large and small, public and private – in varying combinations and inter-relationships. The firms in each of the segments shown in Figure 1.3 operate over widely varying geographical ranges and perform rather different roles in the economic system.

One of the major themes of this book is that it is increasingly the *transnational corporation* (TNC) which plays the key role in co-ordinating production chains and, therefore, in shaping the new geo-economy (see Chapter 7).[4] However, we need to use a broad definition of the TNC – one which goes beyond the conventional definition based upon levels of ownership of internationally based assets – to capture the diversity and complexity of transnational networks. Thus, a TNC will be defined as

> a firm which has the power to co-ordinate and control operations in more than one country, even if it does not own them.

This definition implies that it is not essential for a firm to *own* productive assets in different countries in order to be able to control how such assets are used. TNCs generally do own such assets but they are also typically involved in a spider's web of collaborative relationships with other legally independent firms across the globe.

At one extreme, each function in a specific production chain may be performed by individual, independent, firms so that the links in the chain consist of a series of *externalized transactions*. In other words, the transactions are organized through 'the market'. At the other extreme, the whole chain may be performed within a single firm as a *vertically integrated* system. In this case, the links in the chain consist of a series of *internalized transactions*. Here, transactions are organized 'hierarchically' through the firm's internal organizational structure.[5] In fact, this dichotomy, between externalized, market-governed transactions and internalized, hierarchically governed transactions, grossly simplifies the richness of the regulatory mechanisms in the contemporary economy. The boundary between internalization and externalization is continually shifting as firms make decisions about which functions to perform 'in-house' and which to 'out-source' to other firms. What we have in reality, therefore, is a spectrum of different forms of co-ordination which consist of networks of inter-relationships within and between firms structured by different degrees of power and influence. Such networks increasingly consist of a mix of intrafirm and interfirm structures. These networks are dynamic and in a continuous state of flux. However, there will invariably be a primary co-ordinator driving any particular production chain or network.

In this regard, Gereffi (1994, p. 97) makes a useful distinction between two types of 'driver':

- *Producer-driven chains*:

  refer to those industries in which transnational corporations (TNCs) or other large integrated industrial enterprises play the central role in controlling the production system (including its backward and forward linkages). This is most characteristic of capital- and technology-intensive industries like automobiles, computers, aircraft, and electrical machinery . . . What distinguishes 'producer-driven' production systems is the control exercised by the administrative headquarters of the TNCs.

- *Buyer-driven chains*:

  'refer to those industries in which large retailers, brand-named merchandisers, and trading companies play the pivotal role in setting up decentralized production networks in a variety of exporting countries.' (It is important to emphasize that, in terms of the definition introduced above, such firms are also TNCs.)

Figure 1.4 illustrates the major features of producer-driven and buyer-driven production chains.

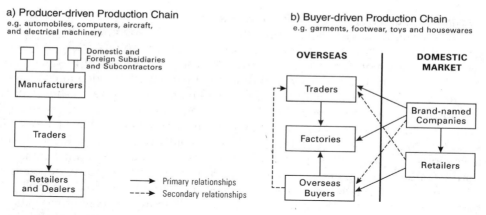

Figure 1.4 Producer-driven and buyer-driven production chains
Source: Based on Gereffi (1994, Figure 5.1)

The second level at which production chains are regulated is that of the *state*. Contrary to those who argue that the state is either dead or dying as a viable force in the contemporary global economy the position taken here is that the state remains a fundamentally significant influence. All the elements in the production chain are regulated within some kind of political structure whose basic unit is the nation-state but which also includes such supranational institutions as the International Monetary Fund or the World Trade Organization, as well as regional economic groupings such as the European Union or the North American Free Trade Agreement. All markets are socially constructed. Even supposedly 'deregulated' markets are still subject to some kind of political regulation. All states operate a battery of economic policies whose objective is to enhance national welfare. However, the particular policy orientation

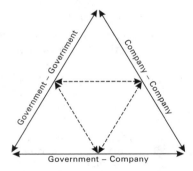

Figure 1.5   The triangular nexus of relationships between states and firms
Source: Stopford and Strange (1991, Figure 1.6)

and policy mix varies according to the political, social and cultural complexion of the individual state. Hence, just as there is great diversity in TNC behaviour so, too, states vary in their behaviour depending upon their position along the ideological spectrum.

Consequently, all business organizations – even the most global TNC – have to operate within national and international regulatory systems. They have to conform to national business legislation. It is true, of course, that TNCs attempt to take advantage of national differences in regulatory regimes while states attempt to minimize such 'regulatory arbitrage'. The result is a very complex situation in which firms and states are engaged in various kinds of power play; what Stopford and Strange call a *triangular nexus* of interactions comprising firm–firm, state–state and firm–state relationships (Figure 1.5). In other words, the new geo-economy is essentially being structured and restructured not by the actions of either firms or states alone but by complex, dynamic interactions between the two sets of institutions (see Chapter 8).

Geographical configuration of production chains

Where each element in a production chain is located geographically and how the geographical structure of the different parts of the firm are physically linked is the second fundamental dimension of Figure 1.2. Just as we can identify a spectrum of organizational arrangements for co-ordinating a particular production chain so, too, we can identify a geographical spectrum of possibilities. As Figure 1.2 shows, production functions may be *geographically dispersed* at one end of the spectrum or *geographically concentrated* at the other along a continuum from the global through to the local scale. One obvious influence on the geographical configuration of production chains is technological – primarily the technologies of transport and communications which transform the meaning of geographical distance. In general, therefore, there has been a tendency for the geographical extensiveness of virtually all production chains to increase. However, different types of production chain may be configured geographically in very different ways.

Even in a globalizing world, all economic activities are geographically localized

'The end of geography'; 'the death of distance'. These two phrases resonate, either explicitly or implicitly, throughout much of the globalization literature. According to this view, dramatic developments in the technologies of transport and communication have made capital – and the firms controlling it – 'hyper-mobile', freed from the 'tyranny of distance' and no longer tied to 'place'. In other words, it implies that economic activity is becoming 'deterritorialized'. The sociologist Manuel Castells argues that the forces of globalization, especially those driven by the new information technologies, are replacing this 'space of places' with a 'space of flows'.[6] Anything can be located anywhere and, if that does not work out, can be moved somewhere else with ease (see the Reich quotation in Box 1.1). Seductive as such ideas might be, a moment's thought will show just how misleading they are. Although transport and communications technologies have indeed been revolutionized (see Chapter 5) both geographical distance and, especially, *place* remain fundamental. Every component in the production chain, every firm, every economic activity is, quite literally, 'grounded' in specific locations. Such grounding is both physical, in the form of sunk costs,[7] and less tangible in the form of localized social relationships.

Geographical clustering of economic activities is the norm

Not only does every economic activity have to be located somewhere; more significantly, there is also a very strong propensity for economic activities to form *localized geographical clusters* or *agglomerations*. In fact, the geographical concentration of economic activities, at a local or subnational scale, is the norm not the exception. The pervasiveness and the significance of geographical clustering has recently been recognized – and has come to occupy a central position – in the writings of some leading economists and management theorists, notably Paul Krugman, Michael Porter and Kenichi Ohmae.[8] However, economic geographers and location theorists have been pointing to the pervasiveness of this phenomenon of geographical concentration for decades.[9]

The reasons for the origins of specific geographical clusters are highly contingent and often shrouded in the mists of history. As Myrdal (1958, p. 26) pointed out many years ago: 'Within broad limits the power of attraction today of a center has its origin mainly in the historical accident that something once started there, and not in a number of other places where it could equally well or better have started, and that the start met with success.' Whatever the specific reason for the initiation of a localized economic cluster its subsequent growth and development tends to be based upon two sets of agglomerative forces:

- *Traded interdependencies*. Geographical proximity between firms performing different – but linked – functions in the production chain may reduce the transaction costs involved and make possible a higher intensity of interfirm transactions between neighbouring firms. In fact, it does not always follow that firms located close to each other are actually linked together through such transactions. Firms may be geographically proximate but functionally unrelated. They will, however, benefit from the second category of agglomerative forces –
- *untraded intredependencies*. These are the less tangible benefits derived from geographical clustering, both economic – such as the development of an appropriate pool of labour – and sociocultural. Amin and Thrift (1994) emphasize this sociocultural basis of agglomeration, arguing that it facilitates three particular processes: 1) face-to-face contact; 2) social and cultural interaction – 'to act as places of sociability, of gathering information, establishing coalitions, monitoring and maintaining trust, and developing rules of behaviour' (p. 13); and 3) enhancement of knowledge and innovation – 'centres are needed to develop, test, and track innovations, to provide a critical mass of knowledgeable people and structures, and socio-institutional networks, in order to identify new gaps in the market, new uses for and definitions of technology, and rapid responses to changes in demand patterns' (p. 13). More broadly, large-scale urban agglomerations make possible the supply of a whole range of other facilities which would not be possible under geographically dispersed circumstances.

In a whole variety of ways, therefore, once established a localized economic cluster or agglomeration will tend to grow through a process of cumulative, self-reinforcing development:

> Provided that markets continue to grow, the leading locale is now likely to be subject to a many-sided process of developmental self-transformation in which the agglomeration effects . . . will be greatly amplified. Thus, there is apt to be a deepening and widening of the social division of labor leading to economic diversification and increased industrial synergies in the local area. Concomitantly, new labor skills are likely to emerge, and the general rounding out of local labor markets will occur. The industrial atmosphere of the locale will tend to thicken, and the business community may well begin to take on identifiable cultural attributes marked by distinctive conventions and routines. Information exchanges and learning effects are liable to become increasingly densely-textured, with a corresponding sharpening of the stimuli to technological and commercial innovation . . . And as these processes move forward, a complex regional economic system will start to materialize and – at least for a time – to evolve forward on the basis of a deepening stock of external economies of scale and scope.
>
> (Scott, 1995, p. 56)

The cumulative nature of these processes of localized economic development emphasizes the significance of historical trajectory. It has become common to use terminology from evolutionary economics[10] to describe the process as being *path-dependent*. Thus, a region's (or a nation's) economy becomes 'locked in' to a pattern which is strongly influenced by its particular history. This may be either a source of continued strength or, if it embodies too much organizational rigidity, a source of weakness. However, even for 'successful' regions, such path dependency does not imply the absolute inevitability of continued success. As Scott (1995, p. 57) points out,

> the notion of path-dependence also implies the existence of critical branching points, representing conjunctures where the regional economy may move in any one of a number of different possible directions (though once it has moved, its future is then to that degree committed) . . . the onward march of development in economically successful regions is always in practice subject to eventual cessation or reversal, not only because there *are* usually limits to the continued appropriation of external economies, but also because radical shifts in markets, technologies, skills, and so on, can undermine any given regional configuration of production. Indeed, the very existence of lock-in effects means that regions, as they develop and grow, will eventually find it difficult to adapt to certain kinds of external shocks.

A central argument, then, is that *place* matters; that 'territorialization' remains a significant component in the organization of economic activity.

Scales of activity; scales of analysis

The geo-economy, therefore, can be pictured as a geographically uneven, highly complex and dynamic web of production chains, economic spaces and places connected together through threads of flows. But the spatial *scale* at which these processes operate is, itself, variable. So, too, is the meaning which different scales have for different actors within the global economic system. The tendency is to collapse the scale dimension to just two: the global and the local and much has been written about the *global–local tension* at the interface between the two. Firms, states, local communities, it is argued, are each faced with the problem of resolving that tension.

There is no doubt that this is a real problem. However, it is not always the case that the terms 'global' and, especially, 'local', mean the same thing in different

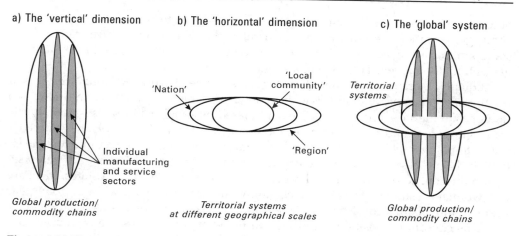

Figure 1.6 The interconnecting dimensions of a globalizing economy
Source: Based, in part, on Humbert (1994, Figure 1)

contexts. In the international business literature, for example, the term 'local' generally refers to the national, or even the larger regional, scale (i.e. at the level of Europe, Asia, North America). But for most people, 'local' refers to a very much smaller spatial scale: that of the local community in which they live. However, it is a mistake to focus only on the two extremes of the scale – the global and the local – at which economic activities occur. It is more realistic to think in terms of inter-related scales of activity and of analysis: for example, the local, the national, the regional (i.e. supranational) and the global. These have meaning both as activity spaces in which economic and political actors operate and also as analytical categories which more accurately capture some of the complexity of the real world.

However, we need to bear in mind that the scales are not independent entities. Figure 1.6 captures the major dimensions of these relationships. Individual industries (production/commodity chains) can be regarded as vertically organized structures which operate across increasingly extensive geographical scales. Cutting across these vertical structures are the territorially defined political-economic systems which, again, are manifested at different geographical scales. It is at the points of intersection of these dimensions in 'real' geographical space where specific outcomes occur, where the problems of existing within a globalizing economy – whether as a business firm, a government, a local community or as an individual – have to be resolved.

Aims and organization of the book

The title *Global Shift* tries to capture the idea that both absolute and relative geographical changes have been, and are, occurring in the world economy; that there has been a progressive shift in the centre of gravity of the world system and a fundamental redrawing of the global economic map. My basic aim in this book is to explore the kinds of processes and issues outlined in this introductory chapter and to identify their major impacts on the economic well-being of communities occupying different positions within the global economic system. Its underlying theme is that, while there are indeed globalizing processes at work in transforming the world economy into what might reasonably be called a new geo-economy, such processes – and their outcomes – are far more diverse than we are generally led to believe. So, when we talk about globalization we must always remember that it is a set of *tendencies* and not some kind of achieved condition. These tendencies are both geographically and organizationally uneven. There is not a single predetermined trajectory.

Equally, there is not a single transformative force at work. Although the role of the transnational corporation (TNC) is given considerable prominence in this book I do not take the view that the TNC has rendered all other institutions – notably the state – impotent and irrelevant as economic actors. On the contrary, the state continues to play a highly significant role. It seems to me to be more accurate to conceptualize the process as one of a complex interaction between TNCs and states set within the context of a volatile technological environment. Firms and states, then, are the two major shapers of the global economy and are embedded within a triangular nexus of interactions consisting of firm–firm, state–state and firm–state relationships.

This conceptual position is reflected in the way the chapters are organized. But we must realize that there are several alternative ways of organizing the treatment of such a broad and complex subject. The processes involved are tightly interconnected and mutually interact with one another in intricate ways. We are not dealing with a linear process in which each element can be dealt with one at a time without regard for the others. Unfortunately, the constraints of language necessitate such a linear treatment for something which is not linear at all. As the American playwright, Arthur Miller (1987, p. 144) acutely observes: 'the only art in which simultaneity . . . [is] . . . really possible . . . [is] . . . music. Words . . . [can] . . . not make chords; they ha[v]e to be uttered in a line, one after the other'.

The book is organized into four distinct, but closely related, parts:

- Part I consists of a single chapter (Chapter 2) which describes the changing global map of production, trade and investment. Most of the data employed in that chapter are at the national scale – both because that is the scale at which most data are collected and published and also because of the importance of the state as an economic institution. However, we shall not lose sight of the fact that economic activity actually takes place at a localized scale.
- Part II is the explanatory core of the book. It is concerned with the *processes* of transformation and is built around the three elements of TNCs, states and technology. Chapters 3 and 4 focus on the state as the 'traditional' building block of the international economy. Chapter 5 analyses the nature and significance of technological change. Chapters 6 and 7 look at transnational corporations as networks of relationships. Chapter 8 examines the nature and dynamics of TNC–state interactions.
- Part III explores in detail the globalization process in specific sectors. The precise form of the internationalization of production, the form of global production systems and the manner in which global shifts have occurred vary substantially from one sector to another. Chapters 9–12, therefore, consist of case studies of individual sectors – textiles and clothing, automobiles, electronics and services – chosen to demonstrate some of this diversity of experience. Global shifts are particularly evident in these sectors and have major implications for both the older and newer industrialized nations.
- Part IV is concerned with the stresses and strains created by global shift and with the problems facing national, regional and local economies. It begins with a brief connective summary of the trends and processes identified in the first three parts of

the book. Chapter 13 then examines the problems of adjustment arising from global shifts in economic activity. The interconnected problems facing economies occupying different positions in the global economy are explored, respectively, for the older industrialized countries, the newly industrializing countries and the less industrialized countries. The primary focus is on employment. The question of 'where will the jobs come from?' is one of the most crucial contemporary problems in all parts of the world. Finally, Chapter 14 concludes by addressing some of the major issues of governance in a globalizing economy.

Notes for further reading

1. The literature on this topic is huge and seemingly growing by the week. Major proponents of the globalization position include Reich (1991), Barnet and Cavanagh (1994), Ohmae (1985a,b; 1990; 1995). The underlying concept may be traced back to McLuhan's (1960) notion of the 'global village' but was explicitly introduced into the management literature in Levitt's (1983) paper on the globalization of markets. Strong counter-views are voiced by Gordon (1988), Glyn and Sutcliffe (1992), Hirst and Thompson (1992; 1996).
2. This argument is developed more fully in Dicken, Peck and Tickell (1997).
3. The 'chain' metaphor has been used in a number of different disciplinary contexts with slightly varying terminology. In the economic geography literature, for example, Walker (1988) uses the concept of the *filière* ('the connecting filament among technologically related activities') in his analysis of the geographical organization of production systems. Storper (1992) uses the concept of the *commodity chain* as the basis for his analysis of technology districts in a global context. In the business literature Porter uses the term *value chain* while Johnston and Lawrence (1988) use the more conventional economic term, *value-added* chain. In the 'world systems' literature in international sociology the term *commodity chain* is generally used (Hopkins and Wallerstein, 1986). In this latter respect, the most extensive and systematic development of the *'global' commodity chain* (GCC) concept has been produced by Gereffi (1994).
4. The term 'transnational' corporation is preferred rather than 'multinational' corporation simply because it is the more generic term. The 'multinational' label implies operations in a substantial number of countries whereas 'transnational' simply implies operations in at least two countries, including the firm's home country. In effect, all multinational corporations are transnational corporations but not all transnational corporations are *multi*national. Cowling and Sugden (1987) explicitly emphasize the importance of 'strategic co-ordination' as the distinguishing characteristic of a TNC which they define as 'the means of coordinating production from one centre of strategic decision-making when this coordination takes a firm across national boundaries' (p. 60).
5. This 'markets and hierarchies' view of the governance of economic transactions was developed by Williamson (1975). As a concept, it derives from the work of Ronald Coase (1937), who addressed the fundamental question of why multifunction firms exist at all.
6. Castells's arguments were developed initially in his book *The Informational City* (1989) and have been elaborated more recently in *The Rise of the Network Society. Volume 1* (1996).
7. Clark (1994), Clark and Wrigley (1995), Schoenberger (1997) explore the nature and significance of sunk costs in corporate decision-making and corporate restructuring in specifically spatial contexts.
8. See, for example, Porter (1990), Ohmae (1995a), Krugman (1991; 1995; 1996). Martin and Sunley (1996) provide a very useful comparison of Krugman's writings and those in the 'new industrial geography'.
9. Dicken and Lloyd (1990, Chapters 5 and 6) review the traditional economic-geographical and location-theoretic approaches to spatial concentration. The newer generation of economic geographical literature is represented in, for example, Storper and Walker (1989), Amin and Robins (1990), Malmberg and Solvell (1995), Scott (1988a,b; 1995), Malmberg (1996) Storper (1995; 1997).
10. See, for example, Hodgson (1993), de la Mothe and Paquet (1996).

PART I

Patterns of global shift

INTRODUCTION

A brief historical perspective

Our primary concern in this book is with the transformation of the world economy during the past few decades but, especially, since the 1960s. This is when, with the benefit of hindsight, we can see that substantial qualitative as well as quantitative economic changes began to emerge. However, it is impossible to understand the current situation without at least an outline knowledge of what has gone before. We need a baseline against which to measure current trends.

Evolution of a global economy

Although the development of a global economy was greatly accelerated by the process of industrialization beginning in the second half of the eighteenth century, its basis was established much earlier. This global system was a capitalist system which originated and developed in quite specific geographical locations.[1] The beginnings of a world economy were evident first in the expansion of trade during the period from 1450 to 1640, a period which Wallerstein (1979) has labelled the 'long sixteenth century'. Prior to, and even during, this period the pattern of trade was bimodal:

- On the one hand, there was *local trade* – short distance, mainly concerned with basic necessities – the kind of trade focused upon the medieval market towns.
- On the other hand there was the much smaller volume of *long-distance trade* in luxury goods and rare items for a very tiny fraction of the population: the spice and fine-cloth trade, for example, together with other exotic goods from distant parts of the world.

During the late fifteenth and early sixteenth centuries, however, the geographical extent of trade increased dramatically as a result of the expansion of a small number of European maritime nations which came to form the core of an evolving world economy. By the middle of the seventeenth century economic leadership was centred on northwest Europe. The development of a world trading system over a period of several centuries resulted in a tripartite geographical structure of core, semi-periphery and periphery. It also laid the foundations for a process which was to have even more far-reaching effects: *industrialization*. In turn, industrialization greatly accelerated the expansion of world trade and further transformed its character. As the nineteenth century progressed the nature and geographical pattern of world trade changed to one in which the core (initially Britain) exported manufactured goods throughout the world and imported raw materials, especially from the colonies. Exports of textiles led the way, followed, in the second half of the nineteenth century, by heavy manufactured goods, such as iron and steel and also coal. In effect, a *new international division of labour* – a new pattern of geographical specialization – had emerged.

Progressively, the roles of core and periphery in the geographical division of labour became increasingly clearly defined:

- Industrial production became pre-eminently a core activity, a process reinforced by the political process of colonialization. Industrial goods were both traded between core nations and also exported to the periphery.
- Conversely, the periphery's role was a dual one. First, it supplied the core with primary commodities – raw materials for transformation into manufactured products in the core; food stuffs to help feed the industrial nations. Secondly, it purchased manufactured goods from the core, particularly capital goods in the form of machinery and equipment.

The relative fortunes of the core countries themselves waxed and waned. Most notably, the United States and Germany emerged to overtake the previously undisputed leader, Britain. By 1913 the United States was producing 36 per cent of total world industrial output while Britain's share had fallen to 14 per cent.

Of course, the process of development of the world economy is not a continuous, uninterrupted sequence of events. By its very nature it is a discontinuous process; periods of rapid growth of production and trade and geographical expansion are punctuated by periods of stagnation and recession. Such business cycles vary enormously in their frequency, duration and intensity. In recent years much attention has been focused upon the notion of long waves of economic development of roughly fifty years' duration, which are associated with a fundamental restructuring of economic activity at a global scale (see Chapter 5).

The emergence of transnational corporations

The evolution of a global economy based on increasingly extensive flows of international trade, and structured around the broad framework of a core and periphery/semi-periphery, also involved the growth and spread of international investment. It involved, in other words, the early development of the transnational corporation, the institution which has subsequently come to dominate the global economy[2]. The development of companies with interests and activities located outside their home country was part and parcel of the early development of an international economy. A number of chartered trading companies emerged in Europe from the fifteenth century onwards as an important part of the early evolution of the world economic system. Companies such as the East India Company, the Hudson's Bay Company and others created vast trading empires on the world scale. But, despite their worldwide extent, their main *raison d'être* was trade and exchange rather than production. In fact the first firms to engage in production outside their home country did not emerge until the second half of the nineteenth century, and then only hesitantly at first. But by the eve of the First World War there was substantial overseas manufacturing production by United States, British and continental European companies.

The growth of United States transnational activity and manufacturing in the late nineteenth and early twentieth centuries reflected the country's emergence as the world's major industrial nation. But in 1914 the major source of overseas investment was still the United Kingdom. In fact, the geographical spread of UK overseas manufacturing investment was considerably broader than that of US firms or of those from continental Europe. The UK pattern, of course, strongly reflected the nation's imperial position. Although US and UK transnational manufacturing investment was similar in several respects – both were investing heavily in food, chemicals and engineering industries – there were some differences which persisted for a long time. In particular, US transnational investments, from a very early stage, leaned towards the newer, more technologically sophisticated products in both producer and consumer goods. A good deal of UK transnational investment was in textiles. The early overseas manufacturing investments by continental European firms also displayed a distinctive industrial complexion; much of their investment was in chemicals and in electrical machinery. Between the First and Second World Wars, transnational manufacturing investment grew considerably, with especially rapid increases in the foreign network of United States TNCs. By 1939 the United States had become the major source of international investment in manufacturing as well as the world's leading industrial nation.

Post-1945: the shaping of a new global economic system

These broad contours of the global economic map persisted until the Second World War (1939–45). Global production and trade were dominated by the old-established core economies of northwest Europe and the United States. Manufacturing production remained strongly concentrated in this industrialized core. Of world manufacturing production, 71 per cent was concentrated in just four countries and almost 90 per cent in only eleven countries. Japan produced only 3.5 per cent of the world total. The group of core industrial countries sold two-thirds of its manufactured exports to the periphery and absorbed four-fifths of the periphery's primary products (League of Nations, 1945). A clear international division of labour was apparent. International direct investment by the rapidly developing transnational corporations was also dominated by firms in these leading core nations and was most strongly concentrated in the developing countries which, on the eve of the Second World War, were host to two-thirds of total foreign direct investment.

This relatively stable and long-established structure was shattered by the Second World War, which devastated the global economy, creating 'a great dividing line' (Scammell, 1980, p. 2) or 'one of the great punctuation marks in human history' (Stubbs and Underhill, 1994, p. 145). The vast majority of the world's industrial capacity (outside North America) was destroyed and had to be rebuilt. At the same time, new technologies (including what were to become the new information technologies) were created and many industrial technologies were refined and improved in the process of waging war. Hence, the world economic system which emerged after 1945 was in many ways a new beginning. It reflected both the new political realities of the postwar period – particularly the sharp division between East and West – and also the harsh economic and social experiences of the 1930s. The kinds of international economic institutions devised in the aftermath of war grew out of both these factors.[3]

The major political division of the world after 1945 was essentially that between the West (led by the United States) and the East (the Soviet-dominated nations of eastern Europe). Outside these two major power blocs was the so-called 'Third' World, a highly heterogeneous – but generally impoverished – group of nations, many of them still at that time under colonial domination. The Third World was far from immune from the East–West confrontation. Both major powers (the United States and the USSR) made strenuous efforts to extend their spheres of influence, a process which had considerable implications for the subsequent pattern of global economic change. The Soviet bloc drew clear boundaries around itself and its eastern

European satellites and created its own economic system, quite separate from the capitalist market economies of the West, at least initially. In the West the kind of economic order built after 1945 reflected the economic and political domination of the United States. Alone of all the major industrial nations, the United States emerged from the war strengthened rather than weakened. It had both the economic and technological capacity and also the political will to lead the way in building a new order.

The institutional basis of this new order came into being formally at an international conference at Bretton Woods, New Hampshire, in 1944. It resulted in the creation of two international financial institutions: the International Monetary Fund (IMF) and the International Bank for Reconstruction and Development (later renamed the World Bank).[4] The primary objective of the Bretton Woods system was to stabilize and regulate international financial transactions between nations on the basis of fixed currency exchange rates in which the US dollar played the central role. In this way it was hoped to provide the necessary financial 'lubricant' for a reconstructed world economy. The other major pillar of postwar international economic order was to be that of free trade. The view that the 'beggar-my-neighbour' protectionist policies of the 1930s should not be allowed to recur after the war was reflected in the establishment of another international institution in 1947: the General Agreement on Tariffs and Trade (GATT, whose purpose was to reduce tariff barriers and to prohibit other types of trade discrimination – see Chapter 3). Together, this triad of international bodies formed the international institutional framework in which the rebuilt world economy evolved.

However, although the world economy was rebuilt anew after the devastation of the Second World War such rebuilding did not take place on completely fresh, unbroken ground. The postwar 'economic architecture' did, indeed, point to a new world economic order but it was an order containing many traces of what had gone before. As pointed out in Chapter 1, an internationally integrated economy – at least in a shallow form – had begun to emerge even before 1913. But even by the eve of the Second World War, the geography of the world economy was still essentially based upon the simple international division of labour that had existed for decades. It is from this historical baseline, therefore, that recent global shifts in economic activity will be examined in the next chapter. As we will demonstrate, today's world is far more complex than it was even a few decades ago. The half century between the end of the Second World War in 1945 and the late 1990s has witnessed a truly fundamental transformation of the world economy.

Notes for further reading

1. For a superb historical survey of the evolution of the global economy over the *longue durée*, see Braudel's magisterial three-volume work (1984). Bairoch (1982; 1993) presents a detailed historical survey of international industrialization from 1750, while Kitson and Michie (1995) focus specifically on the period since 1870.
2. A broad survey of the evolution of international direct investment from the late nineteenth century is provided by Dunning (1983; 1993, Chapter 5). See also Kozul-Wright (1995).
3. Excellent accounts of the politics of international economic relations in the postwar period are provided by Kennedy (1987), McGrew and Lewis (1992), Stubbs and Underhill (1994). MacBean and Snowden (1981) and various chapters in Stubbs and Underhill (1994) provide details of the major international financial and trade institutions which have underpinned the operation of the postwar world economy.

4. The International Monetary Fund's primary purpose was to encourage international monetary co-operation among nations through a set of rules for world payments and currencies. Each member nation contributes to the fund (a quota) and voting rights are proportional to the size of a nation's quota. A major function of the IMF has been to aid member states in temporary balance of payments difficulties. A country can obtain foreign exchange from the IMF in return for its own currency which is deposited with the IMF. A condition of such aid is IMF supervision or advice on the necessary corrective policies. The World Bank's role, as its full name suggests, is to facilitate development through capital investment. Its initial focus was Europe in the immediate postwar period. Subsequently, its attention shifted to the less developed economies.

CHAPTER 2

The global economic map: trends in production, trade and investment

The roller-coaster: aggregate trends in global economic activity

The path of economic change is like a roller-coaster. Sometimes the ride is relatively gentle with just minor ups and downs; at other times the ride is truly stomach-wrenching, with steep upward gradients being separated by vertiginous descents to what seem like bottomless depths. The experience of the world economy during the past few decades clearly bears this out. The years immediately following 1945 were ones of basic reconstruction of war-damaged economies throughout the world. It was to be expected that there would be considerable growth of production and trade during the 1950s as the world economy caught up after the deep recession of the 1930s and after the war itself. At the time it was felt that growth rates would then slacken in the 1960s. This did not happen. Instead, rates of economic growth reached unprecedented levels:

- Between 1948 and 1953 world trade increased at an average annual rate of 6.7 per cent.
- Between 1958 and 1963 the rate had risen to 7.4 per cent.
- Between 1963 and 1968 it had accelerated further to 8.6 per cent.

Such growth rates were unprecedented. However, more important than rates of growth was the fact that *trade increased more rapidly than production*, a clear indicator of the *increased internationalization* of economic activities and of the greater *interconnectedness* which have come to characterize the world economy.[1] As Figure 2.1 shows, by 1994 total world exports were more than fourteen times greater than in 1950 whereas total world output was a little over five times greater than in 1950.

The period between the early 1950s and the early 1970s was one of almost continual growth in world production and trade, with only minor and short-lived interruptions. People began to believe that the roller-coaster days were over. Growth, in the western industrialized economies at least, came to be an expectation rather than a hope, even though such growth was extremely variable from one country to another and between different parts of the same country. As Webber and Rigby (1996, p. 6) point out, 'the golden age was only partly golden: it was more golden in some places than others, for some people than others'. But then, in the early 1970s, the sky fell in. The long boom suddenly went 'bust', the 'golden age' of growth became tarnished though, again, more so for some places rather than others. As Figure 2.2 shows growth rates of both production and trade declined with each successive decade.

24

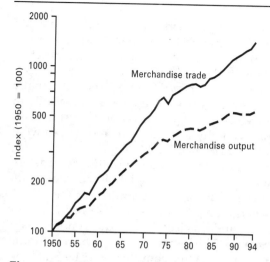

Figure 2.1 The growing interconnectedness of the world economy: the widening gap between trade and production
Source: Based on WTO (1995, *International Trade, 1995*, Chart II.1)

Throughout the 1980s and 1990s annual rates of growth were very variable indeed. The roller-coaster had come back with a vengeance.

The most obvious explanation, at the time, seemed to be the decision by the Organization of Petroleum Exporting Countries (OPEC) in 1973 to quadruple oil prices. Certainly this was a massive shock to the system but, with hindsight, it is clear that this was merely the final piece in a jigsaw puzzle which had been taking shape since the late 1960s. From about 1968 a number of important changes had been occurring which ultimately came together to throw the world economy into reverse. Commodity prices, other than oil, had been rising steeply. Labour costs in all the industrialized nations had begun to accelerate as the level of wage settlements rose. The international monetary system created at Bretton Woods in 1944 became increasingly unstable as national currencies became more and more out of line with the fixed exchange rates. In 1971 the United States moved to a floating exchange rate, other countries followed and the Bretton Woods system was no more.

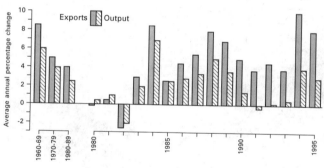

Figure 2.2 The roller-coaster of world merchandise production and trade, 1960–95
Source: Based on GATT (1990, *International Trade, 1989–1900*, Vol. II, Chart I.1); WTO (1996, *Annual Report, 1996: Volume I*, Chart II.1)

Initially, some regarded the problems of the 1970s as being the result of 'an unusual bunching of unfortunate circumstances'. But, as Eric Hobsbawm (1994, p. 286) acutely observes, 'any historian who puts major changes in the configuration of the world economy down to bad luck and avoidable accidents should think again'. As the recession deepened during the second half of the 1970s and into the early 1980s emphasis shifted towards the view that the changes were the result of more deep-seated and fundamental processes. A good deal of attention came to be focused on what is, in fact, a very old idea: that economic activity proceeds in a series of long waves each of which lasts for approximately fifty years.[2] A separate, though not entirely unrelated, strand of explanation, was that of the *regulationist* school which explained the specific downturn in the 1970s in terms of the breakdown of the former dominant regime of accumulation and its associated mode of social regulation – what became universally known as *Fordism*.[3]

During the 1980s, rates of growth were extremely variable, ranging from the negative growth rates of 1982 through to two years (1984 and 1988) when growth of world merchandise trade reached the levels of the 1960s once again. Overall, growth – albeit uneven growth – reappeared. But then, in the early 1990s, recession occurred again. As Figure 2.2 shows, world production was static in 1990 and 1993 and actually declined in 1991. In 1994 and 1995, however, strong growth reappeared, especially in exports. Nevertheless, the unevenness and volatility of world economic growth reflects the continuing difficulties of a world economy struggling to rediscover what had appeared to be, in the 1960s, a virtuous circle of growth.

These aggregate figures on international production and international trade are important in their own right as indicators of a changing global economy and, especially, of its increased interdependencies. But aggregate figures at the global scale mask very significant geographical variations. Changes in the production of goods and services and the trade in products and services have varied enormously from place to place on the world economic map. The geography of global economic change is exceptionally complex. In trying to unravel this complexity let us first look at production.

Global shifts in production of goods and services

Manufacturing production

Although there are virtually no countries in the world which do not have at least some manufacturing activity, the map of world manufacturing industry (Figure 2.3) shows enormous geographical unevenness. The overwhelming majority of production is

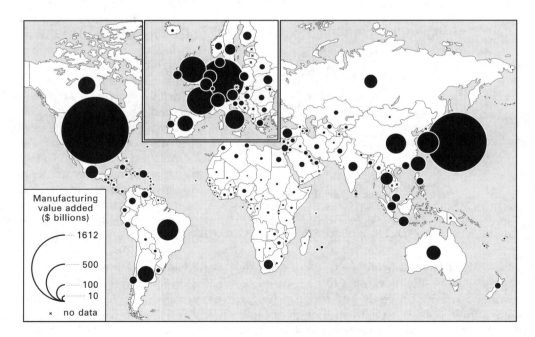

Figure 2.3 The map of world manufacturing industry, 1994
Source: Based on UNIDO (1996, Statistical Annex)

Table 2.1   The world 'league table' of manufacturing production, 1994

Rank	Country	Manufacturing value-added (US$ million)	Percentage of world total
1	United States	1,611,763	26.9
2	Japan	1,257,761	21.0
3	Germany	692,191	11.6
4	France	268,611	4.5
5	United Kingdom	243,653	4.1
6	South Korea	159,172	2.7
7	Brazil	154,425	2.6
8	China	139,031	2.3
9	Italy	128,486	2.2
10	Canada	100,322	1.7
11	Argentina	88,366	1.5
12	Spain	81,196	1.4
13	Taiwan	73,295	1.2
14	Australia	64,417	1.1
15	Switzerland	60,111	1.0
	Total		85.8

Source: Based on data in UNIDO (1996).

concentrated in a relatively small number of countries. Four-fifths of world manufacturing production is located in North America, western Europe and Japan. These features can be seen more clearly in Table 2.1, which is the upper section of the world 'league table' of manufacturing production. The fifteen countries listed produce 86 per cent of total world manufactured output. However, a mere three countries – the United States, Japan and Germany – account for 60 per cent of the total. Only five or six of the countries listed in Table 2.1 can be regarded as developing countries. It is abundantly clear from both Figure 2.3 and Table 2.1, therefore, that manufacturing production is highly concentrated globally. The vast majority of developing countries have only a very small manufacturing base; the 'manufacturing tail' of the world economy is very long indeed.

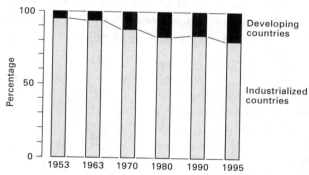

Figure 2.4   The changing distribution of world manufacturing production between industrialized and developing countries, 1953–95
Source: Based on UNIDO (1986, World industry: a statistical review, 1985, *Industry and Development*, Vol. 18, Figure 1); UNIDO data base

However, if we compare the current situation with that of only forty years ago we can see that some very substantial global shifts have occurred. Figure 2.4 shows that between 1953 and 1995 the industrialized economies' share of world manufacturing output declined from 95 to 80 per cent while that of the developing economies quadrupled, albeit from a very low base level, to 20 per cent.[4] Within these two broad categories, however, there has been wide variation in manufacturing growth.

Table 2.2   Changing manufacturing fortunes of leading industrialized countries, 1963–94

	Rank			Share of world manufacturing value-added (%)			Average annual rates of change (%)			
	1963	1987	1994	1963	1987	1994	1960/70	1970/81	1980/87	1990/94
United States	1	1	1	40.3	24.0	26.9	5.3	2.9	3.9	3.20
Japan	5	2	2	5.5	19.4	21.0	13.6	6.5	6.7	0.03
Germany*	2	3	3	9.7	10.1	11.6	5.4	2.1	1.0	−1.10
France	4	4	4	6.3	5.4	4.5	7.8	3.2	−0.5	−0.30
United Kingdom	3	5	5	6.5	3.3	4.1	3.3	−0.5	1.3	0.50
Italy	6	6	9	3.4	2.2	2.2	8.0	3.7	0.9	1.10
Canada	7	7	10	3.0	1.8	1.7	6.8	3.2	3.6	1.20

Notes:
1990/94 rates of change are based upon constant 1990 dollars.
*For 1994, the data refer to the reunified Germany.
Source: Based on data in UNIDO, *Industrial Development: Global Report*, various issues; World Bank, *World Development Report*, various issues.

Shifts between the older industrialized economies

The continued dominance of the core industrial nations in aggregate terms masks some very important changes in the manufacturing fortunes of individual nations. Table 2.2 summarizes these changes for the seven leading manufacturing countries between 1963 and 1994. Three major trends are apparent:

- *The substantial decline in the United States' relative share of world manufacturing production.* In 1963 the United States produced 40 per cent of world output; in 1994 its share had declined to 27 per cent. Even so, the United States remains the world's largest manufacturing producer and has sustained higher annual rates of growth than western Europe.
- *The uneven manufacturing performance of the western European economies.* As a group, western European countries have had far lower growth rates since the 1970s, with some countries showing negative growth rates (e.g. the United Kingdom between 1970 and 1981, France in both 1980–87 and 1990–94 and Germany between 1990 and 1994). The United Kingdom also moved down the league table, from third in 1963 to fifth today. The German experience is, of course, complicated by the reunification of the former West and East Germany in 1991. Even so, the former 'miracle economy' of West Germany did not perform quite as convincingly in the 1980s as the earlier experience suggested it would.
- *The rise of Japan.* By far the most spectacular manufacturing performance of all the major industrialized nations in the postwar period – at least until the early 1990s – has been that of Japan. In 1963 Japan ranked fifth in the world league table with 5.5 per cent of the world manufacturing total. In 1994 it ranked second with a share of 21 per cent. During the 1960s, manufacturing growth in Japan averaged 13.6 per cent, two and a half times greater than that of the United States and four times greater than the United Kingdom. Even though Japan's growth rates fell to half of that level in succeeding periods, such growth continued to be very much greater than that of all the other established industrialized countries until the end of the 1980s when Japan's 'bubble economy' burst.

Thus, within the core industrial nations there has been a pronounced shift in the centre of gravity of manufacturing production. Although the United States retained its leadership, its dominance was much reduced and its position increasingly challenged by the spectacular manufacturing growth of Japan.

The 'transitional economies' of Europe and the former USSR

Since 1989, there has been a further significant development in the changing geography of manufacturing output. The political collapse of the USSR-led group of countries and, indeed, of the USSR itself has produced a group of so-called 'transitional economies': former centrally planned economies now in various stages of transition to a capitalist market economy. These countries have fared badly in terms of their manufacturing sectors. In 1985, for example, the USSR accounted for almost 10 per cent of world manufacturing output. In 1995, its share had declined to 1.5 per cent. The countries of central and eastern Europe also experienced a substantial – though far less spectacular – decline. In 1985 they produced 3.5 per cent of the world total; in 1995 they produced only 1.6 per cent (UNIDO, 1996, p. 39).

Shifts within the developing market economies

We saw earlier (Figure 2.4) that the developing countries as a whole increased their share of world manufacturing output from 5 per cent in 1953 to 20 per cent in 1994. However, the developing economies constitute an extremely varied group. For the vast majority, manufacturing remains relatively unimportant, as Figure 2.3 revealed. In fact, most of the rapid growth in manufacturing production has occurred in what the World Bank terms 'the middle-income group' of developing countries. But there are pronounced geographical variations in this process as Figure 2.5 shows.

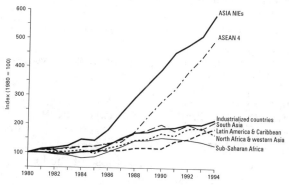

Figure 2.5   Differences in manufacturing growth rates between country groups, 1980–94
Source: Based on data in UNIDO (1996, Figures 3, 6, 8, 10, 12, 13, 15)

Although growth fluctuated over the period (the roller-coaster effect at a regional scale), the rates of manufacturing growth were highest of all in east and southeast Asia and lowest in sub-Saharan Africa. Growth rates in Latin America and in south Asia (both important manufacturing locations in absolute terms) were between these two extremes but with variability in manufacturing growth being much greater in Latin America. In fact only a relatively small number of developing countries – the so-called newly industrializing economies (NIEs) – can be regarded as significant centres of manufacturing production on a world scale.

Table 2.3 shows that although the relative importance of each individual country (as a proportion of total world manufacturing output) is small, they have grown extremely rapidly. Most striking, however, is the fact that the gap between the NIEs of east and southeast Asia, on the one hand, and those of Latin America, south Asia and southern Europe on the other has widened enormously. In 1980, Brazil, Argentina, Mexico and Spain were all more important centres of manufacturing production than any of the east and southeast Asian economies apart from China. By 1994, this situation had changed dramatically. Throughout the 1960s, 1970s and 1980s, the four 'tiger' economies of South Korea, Taiwan, Singapore and Hong Kong dominated the picture. More recently, as Table 2.3 shows, other east and southeast Asian countries – notably

Table 2.3   Growth of manufacturing production in newly industrializing economies, 1963–94

	Share of world manufacturing output (%)			Average annual change (%)				Employment in manufacturing (000s)	
	1963	1980	1994	1960/70	1970/81	1980/87	1990/94*	1980	1994
East and southeast Asia									
South Korea	0.1	0.6	2.7	17.6	15.6	10.6	8.2	2015	2936
Taiwan	0.1	0.5	1.2	16.3	13.5	7.5	4.7	1997	2170
Hong Kong	0.1	0.2	0.2	–	10.1	–	–0.2	937	496
Singapore	0.1	0.1	0.3	13.0	9.7	3.3	8.5	287	366
Malaysia	–	0.1	0.3	–	11.1	6.3	15.8	456	1210
Thailand	–	0.3	0.8	–	–	–	13.1	742	1720
Indonesia	–	0.1	0.5	–	–	–	10.7	964	3801
Philippines	–	0.2	0.2	–	–	–	0.7	949	1029
China	–	2.9	2.3	–	–	–	20.6	24390	61931
Southern Asia									
India	–	0.4	0.4	4.7	5.0	8.3	3.0	6992	8382
Latin America									
Brazil	1.6	2.4	2.6	–	8.7	1.2	2.7	5562	4688
Argentina	–	0.8	1.5	5.6	0.7	0.0	10.0	1346	982
Mexico	1.0	1.4	0.8	10.1	7.1	0.0	2.4	2417	1993
Southern Europe									
Spain	0.9	1.7	1.4	–	6.0	0.4	0.3	2383	1907
Portugal	0.2	0.2	0.2	8.9	4.5	–	–1.5	680	443
Greece	0.2	0.2	0.2	10.2	5.5	0.0	–1.2	378	312

Note:
*1990/94 rates of change are based upon constant 1990 dollars.
Source: Based on data in OECD (1979, Table 1); UNIDO, *Industrial Development: Global Report*, various issues; World Bank, *World Development Report*, various issues.

Malaysia, Thailand, Indonesia and China – have grown extremely rapidly. These east and southeast Asian manufacturing growth rates should be contrasted with those of the leading developed market economies discussed earlier (Table 2.2). It is especially notable that in the east and southeast Asian NIEs manufacturing growth rates remained at a high level throughout the 1970s and 1980s whereas those of the leading industrialized economies fell to half or less of their 1960s' levels.

In terms of manufacturing production, therefore, it is clear that there has been a considerable acceleration of growth in the global periphery. But it is also clear that such manufacturing growth is very unevenly distributed. A small group of developing countries has begun to make a real impact on the world manufacturing scene, adding further to global shifts in the manufacturing system. Although the core industrialized countries continue to dominate, manufacturing production is no longer exclusively a core activity. Whereas in the first quarter of this century 95 per cent of world manufacturing production was concentrated into only ten countries, by 1994 some twenty-five countries were responsible for the same proportion of world output.

Services production

To focus only on manufacturing production in exploring the transformation of the world economy is to tell only a part of the story. In all the industrialized economies, the services sector accounts for a larger share of gross domestic product (GDP) – and of employment – than manufacturing, as Table 2.4 shows. Perhaps more surprisingly, Table 2.4 also shows that the service sector is immensely important in

Table 2.4   The contribution of the service sector to gross domestic product

Country group	Percentage of GDP		
	1965	1987	1994
Low income	30	32	38
Lower middle income	50	46	48
Upper middle income	42	50	55
High income	54	61	64

Source: Based on data in World Bank, *World Development Report*, various issues.

virtually all national economies. Contrary to much popular opinion, a large service sector is not confined to the more developed economies although it is certainly true that the relative importance of services does vary more or less directly with a country's income level. What does differ is the kind of service activity involved.

Much of the service sector in both developing and industrialized countries is either low-skill, low-technology private service activity (including wholesale and retail activity) or activities within the public sector (e.g. health, education, welfare services, etc.). The service sector is exceptionally diverse. From the perspective of this book, the most important services are what might be termed *circulation* activities, notably the whole range of business services (commercial and financial) which are crucial components of production chains. In effect, such services 'lubricate' the wheels of production and trade. Though distinctive in their characteristics they should not be regarded as totally separate from the rest of the economic system. Indeed, as will become apparent at various points throughout this book, the relationship between manufacturing and services is strongly symbiotic. It is the business services which tend to be much more strongly developed in the industrialized countries and far less developed in countries lower down the development scale.

The changing fabric of trade

Shifts in the global network of trade in manufactures

A distinctive feature of the postwar period was the very rapid growth of manufacturing trade (see Figure 2.1). Indeed, manufacturing has come to account for an increasingly greater proportion of total exports in both developed and developing market economies. In the developed market economies, manufactured exports increased from 70 per cent of the total in 1960 to 77 per cent in 1988. However, the shift was particularly strong in the developing market economies, where manufacturing counted for only 20 per cent of total exports in 1960 but for as much as 47 per cent by 1988. In fact, by the end of the 1970s – for the very first time – the value of manufactured products exported from the developing market economies exceeded that of food and raw materials. After 1973 exports of manufactured goods from the Third World grew at twice the rate of exports of raw materials. Without doubt the old international division of labour had been displaced. But as in the case of manufacturing production, it is a small number of newly industrializing economies which accounts for the most important changes among developing countries. For example, for the group of east and southeast Asian NIEs as a whole manufactures accounted for 49 per cent of total exports in 1980. By 1994, this share had increased to 78 per cent.

The geography of trade in manufactures is far more complicated than that of production, simply because trade consists of *flows* between areas. Theoretically every nation can trade with every other although, in fact, trade flows tend to be channelled

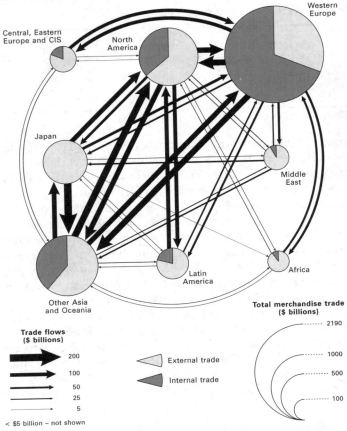

Figure 2.6 The network of world merchandise trade, 1995
Source: Based on data in WTO (1996, *Annual Report, 1996: Volume II*, Tables A2)

into certain dominant routes. Figure 2.6 provides a broad-brush picture of the global network of merchandise trade in 1995. As in the case of production, the global pattern of trade in manufactures has changed a great deal – in some respects quite dramatically – during the past four decades. In fact, it is through the lens of trade that global shifts in the world manufacturing system can be seen most clearly. The evolving pattern of trade in manufactures reflects the emergence of a globalizing system of manufacturing.

Viewed in isolation international trade figures can be misleading. For example, other things being equal, international trade will be far more important relatively speaking for a small economy than for a larger one (an obvious example would be to contrast the United States with Singapore or Hong Kong). Similarly, although in general there is a relationship between manufacturing production and trade, the relationship is not exact. Some countries which are very significant as producers are less significant as exporters (again, the United States is an example as are Brazil and India). Figure 2.7 shows the global map of merchandise trade in 1995. The upper map shows the pattern of manufactured exports by country and reinforces the points made earlier about the unevenness of world manufacturing activity. The lower part of Figure 2.7 shows the net difference between exports and imports and reveals the marked contrast between the large manufacturing trade surplus of Japan and Germany and the massive trade deficit of the United States. Table 2.5 is the 'league table' of trade by country and should be compared with Table 2.1. Such a comparison reveals some interesting differences. First, the origins of manufactured exports are less concentrated than those of production. Secondly, no single nation dominates exports in manufactures to the extent that the United States dominates world manufacturing production. The United States was the most important exporter of manufactured goods in 1995, with 12.4 per cent of the total (compared with its 26.9 per cent of manufacturing output). In fact several manufacturing nations occupy very different positions in the league tables of manufacturing production and manufacturing exports. For example, Hong Kong and Singapore both appear in the top fifteen exporting countries but not in the top fifteen manufacturing producers. Conversely, Brazil and Argentina, both high-ranking producers of manufactures, do not appear among the leading exporters of manufacturers.

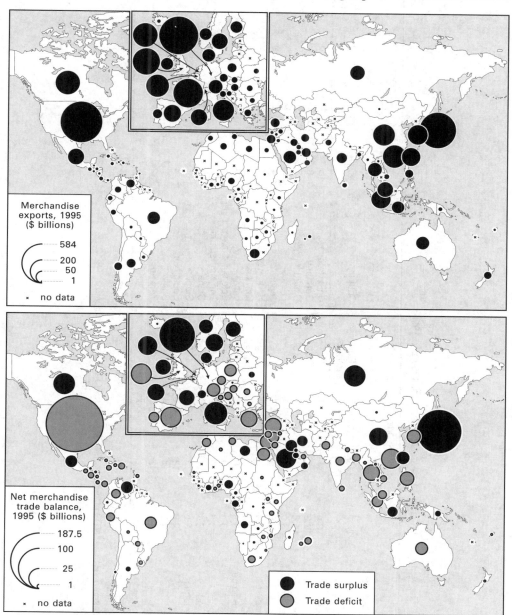

Figure 2.7 The map of world merchandise trade, 1995
Source: Based on data in WTO (1996, *Annual Report, 1996: Volume II*, Tables A3, A4)

Table 2.5   The world 'league table' of manufacturing trade, 1963 and 1995

Exports

Country	1995 (%)	Rank 1995	1963 (%)	Rank 1963
United States	12.4	1	17.4	1
Germany	12.3	2	15.6	2
Japan	11.6	3	6.1	5
France	6.0	4	7.0	4
Italy	5.6	5	4.7	6
United Kingdom	5.4	6	11.4	3
Hong Kong	4.4	7	0.9	15
Belgium–Luxembourg	3.5	8	4.3	7
China	3.4	9	–	–
Canada	3.3	10	2.6	12
South Korea	3.1	11	0.0	–
The Netherlands	3.0	12	3.3	9
Taiwan	2.9	13	0.2	–
Singapore	2.7	14	0.4	–
Switzerland	2.1	15	2.7	11
Total	81.7		76.7	

Imports

Country	1995 (%)	Rank 1995	1963 (%)	Rank 1963	Balance (1995) ($ billion)
United States	16.0	1	8.6	1	–158
Germany	8.5	2	6.2	2	+122
United Kingdom	5.6	3	4.6	6	–14
France	5.5	4	4.7	5	+10
Japan	4.7	5	1.9	13	+244
Hong Kong	4.4	6	0.9	–	–3
Canada	3.6	7	5.0	4	–16
Italy	3.5	8	4.1	8	+71
Belgium–Luxembourg	3.0	9	3.4	9	+12
The Netherlands	3.0	10	4.3	7	–3
China	2.7	11	–	–	+21
Singapore	2.7	12	0.7	–	–4
South Korea	2.4	13	0.3	–	+25
Spain	2.1	14	1.2	–	–11
Taiwan	2.0	15	0.2	–	+29
Total	69.7		46.1		

Note:
– data unavailable.
Source: Based on data in WTO (1996, *Annual Report, 1996: Volume II*, Tables II.11, II.12).

Continued dominance of the developed market economies

Three-quarters of all world merchandise exports are generated by the core economies of the global system. Roughly 60 per cent of this trade is conducted within the core itself as trade between the developed market economies themselves. Expressed in a different way, approximately half of total world trade in manufactures is intracore trade. In fact, during the 1950s and the early 1960s it was largely the dynamic nature of trade between the industrialized economies which drove the global manufacturing system. Since then, however, the dynamic has shifted.

As Table 2.5 shows, the United States is the leading exporter of manufactures (with 12.4 per cent of the world total), followed by Germany (12.3 per cent) and Japan (11.6 per cent). But their individual export performance differed substantially. Between 1963 and 1995, the United States' share fell from 17 per cent. The share of the pioneer manufacturing exporter, the United Kingdom, declined dramatically from 11.4 per cent of the world total in 1963 to around 5 per cent in 1995. The 'star performer' among the developed market economies was undoubtedly Japan: in 1963 it generated just over 6 per cent of total world manufactured exports; by 1995 its share had doubled. The other side of the coin is, of course, imports and the resulting balance between exports and imports for individual countries. As Table 2.5 reveals there was, again, very significant change between 1963 and 1995. Most strikingly, the United States' share of world manufactured imports almost doubled: to 16 per cent of the total. In 1995, it had a negative trade balance in aggregate of $158 billion. Conversely, Japan remained far less significant as an importer. Even though Japan's share of world manufactured imports grew from 1.9 per cent in 1963 to 4.7 per cent in 1995 this compared very unfavourably with its 11.6 per cent share of world manufactured exports. Herein lies one of the major sources of political tension in the world economy. While the United States had a manufacturing trade deficit of $158 billion in 1995, Japan had a massive surplus of $244 billion.

One very marked feature of the manufacturing trade network of the industrialized countries is its strong regional pattern. This is especially apparent in western Europe, the world's largest trading region. Roughly 68 per cent of total exports generated by western European countries takes the form of intraregional trade. Some 42 per cent of Japan's manufactured exports goes to other Asian countries whilst, in the case of North America, Canada's immensely close ties to the US market is reflected in the fact that almost 90 per cent of Canada's exports of manufactures goes to the United States market. However, the manufacturing trade network in North America is highly asymmetrical; only 24 per cent of the United States' manufactured exports go to Canada.

Acceleration of exports from the newly industrializing economies

We saw in our discussion of global shifts in manufacturing production that a relatively small number of developing countries – the NIEs – have emerged in the past few decades as new centres of production. This new feature of the global economic map is even more evident when we consider trade in manufactured products. Figure 2.8 and Table 2.6 show that, as in the case of production, it is the east and southeast Asian countries which dominate the picture with consistently high export growth rates. In the global reorganization of manufacturing trade the increased importance of east and southeast Asia as an exporter of manufactures is unique in its magnitude. As Table 2.6 shows, the eight countries listed in the top half of the table increased their collective

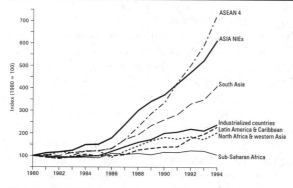

Figure 2.8   Differences in growth rates of manufactured exports between country groups, 1980–94
Source: Based on data in UNIDO (1996, Figures 3, 6, 8, 10, 12, 13, 15)

share of total world manufactured exports from a mere 1.5 per cent in 1963 to almost 20 per cent in 1995.

So, it is especially in their role as exporters that the east and southeast Asian NIEs are particularly significant. As centres of manufacturing production their share of the world total is fairly modest (a glance back to Table 2.3 shows that, in 1994, the four 'first generation' NIEs accounted for just 4.4 per cent of world output; the eight Asian economies (including China) for 8.5 per cent of the total). The first generation of Asian NIEs – South Korea, Taiwan, Hong Kong and Singapore – alone generated 13 per cent of world manufactured exports. As the final two columns of Table 2.6 show, in 1980 three of the four economies already had an export profile that was almost totally dominated by manufacturing. By 1994, this was also becoming the case for the others. In some cases the transformation has been nothing short of spectacular. For example, in 1980, less than 20 per cent of Malaysia's exports were of manufacturers; by 1994 the figure was 84 per cent. Indonesia provides an even more striking experience. In 1980 a mere 2.3 per cent of the country's exports were of manufactures; in 1994 more than half was in that category.

Table 2.6   Growth of manufactured exports from newly industrializing countries, 1963–95

Country	Share of world total per cent		Average annual growth rate (%)		Manufactures as % of total exports	
	1963	1995	1970/80	1980/94	1980	1994
East and southeast Asia						
South Korea	0.01	3.1	22.7	12.3	89.4	92.7
Taiwan	0.20	2.9	16.5	10.0	88.0	93.7
Hong Kong	0.80	4.4	9.9	15.8	88.3	93.0
Singapore	0.40	2.7	–	12.7	43.0	82.1
Malaysia	0.10	1.5	3.3	12.6	18.7	83.6
Thailand	0.00	1.1	8.9	15.5	25.0	71.8
Indonesia	0.00	0.8	6.5	6.7	2.3	51.6
China	–	3.4	8.7	11.5	47.7	82.5
Subtotal	1.51	19.9				
Southern Asia						
India	0.80	0.6	5.9	7.0	51.4	75.6
Latin America						
Brazil	0.10	0.8	8.6	5.2	37.2	54.5
Argentina	–	–	8.9	3.2	23.1	33.7
Mexico	0.20	1.5	5.5	5.4	10.2	77.4
Southern Europe						
Spain	0.30	1.8	12.6	7.4	71.8	76.9
Portugal	0.30	0.4	1.5	10.6	70.3	72.2
Greece	–	–	11.7	5.3	46.5	50.6

Source: Based on data in GATT, *International Trade*, various issues; World Bank, *World Development Report*, various issues; WTO (1996, *Annual Report, 1996: Volume II*, Table II.11).

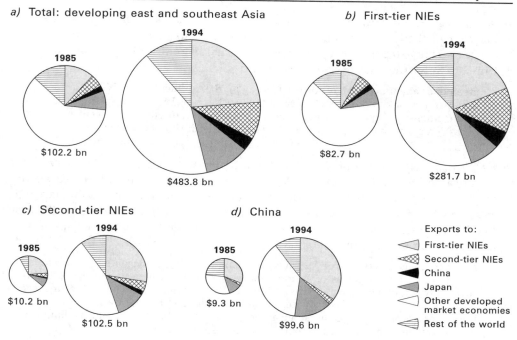

a) Total: developing east and southeast Asia

b) First-tier NIEs

c) Second-tier NIEs

d) China

Exports to:
First-tier NIEs
Second-tier NIEs
China
Japan
Other developed market economies
Rest of the world

Figure 2.9 The trade network of east and southeast Asian NIEs
Source: Based on data in UNCTAD (1996a, Table 24)

The basis of these transformations was the incredibly high annual growth rates that were sustained over a very long period of time as the middle columns of Table 2.6 show. In comparison, the manufacturing export performance of India and the Latin American and southern European countries was relatively modest. As a result, the gap between the east and southeast Asian NIEs and other developing countries has widened substantially. By any criterion, the first generation of Asian NIEs are now, in effect, industrial*ized* – rather than industrializing – countries. Both Singapore and Hong Kong now have per capita GNP higher than a number of industrialized countries, including the United Kingdom, Italy and Canada. Taiwan and South Korea are on the threshold of joining the World Bank's 'high-income' category. Apart from South Korea, these are all relatively small countries in terms of population. However, the newly emerging group of east and southeast Asian NIEs are very much larger. China, of course, is in a size league of its own with its population of 1.2 billions. But Indonesia, too, has a population larger than any industrialized country other than the United States. The geography of the NIEs' manufacturing trade has undergone some very significant changes during the past decade as Figure 2.9 shows. The east and southeast Asian NIEs are divided into three categories: the 'first-tier NIEs' (Hong Kong, Korea, Singapore, Taiwan), the 'second-tier NIEs' (Indonesia, Malaysia, Philippines, Thailand) and China. The changing export structure of these three groups reflects the intraregional dynamics of the east and southeast Asian region and, especially, the transition from lower-skilled, labour-intensive manufactured products to more technologically sophisticated products by the first-tier NIEs. The major elements of change shown in Figure 2.9 are:

- *the relatively minor importance of Japan as an export destination for these NIEs.* In 1985, 8 per cent of their total manufactured exports went to Japan; in 1994 this had increased to only 10 per cent. The Japanese market is significantly more important for China than for the other Asian NIEs. In fact

  > Japan does not appear to have been a leading market for the first-tier NIEs during their initial stages of development . . . Similarly, in the late 1980s, only a small share of labour-intensive exports (10 per cent or less . . .) from ASEAN-4 went to Japan. However, such exports appear to have risen rapidly during the present decade.
  >
  > (UNCTAD, 1996a, pp. 86, 88)

- *the changing significance of the other developed market economies* (essentially North America and western Europe). In 1985, 60 per cent of total manufactured exports from the Asian NIEs went to the developed market economies, excluding Japan. In 1994, this had fallen to 43 per cent. The decline was especially marked for the first-tier NIEs, less so for the second tier while the developed market economies increased in importance for China. However,

  > all three areas . . . run large surpluses in manufacturing trade with the developed market economy countries of Europe and America. For China and ASEAN-4, this surplus reflects their rising exports of labour-intensive items . . . The picture is more complex for the first-tier NIEs. These countries still have a large trade surplus with Europe and North America in labour-intensive products, but this surplus is diminishing as they develop and shift to more sophisticated products.
  >
  > (UNCTAD, 1996a, p. 38)

- *the increased significance of the 'developing' east and southeast Asian market.* In 1985, less than one-fifth of the Asian NIEs' manufactured exports were sold within the region itself. In 1994, the share had increased to almost two-fifths. The increase was especially marked for the first-tier NIEs but the same trend applied to all three NIE groups:

  > The evidence thus suggests that there is a rapid process of integration taking place within the smaller countries of East Asia, a process in which even China is becoming involved to some extent. Trade among the countries of East Asia (excluding Japan) has been rising much faster than trade with the outside world . . . Although Japan has a huge trade surplus with the rest of East Asia, the latter countries on average now import more manufactures from each other than they do from Japan.
  >
  > (UNCTAD, 1996a, p. 90)

NIEs' import penetration of developed country markets

This discussion of the very high geographical concentration of manufactured exports from the east and southeast Asian NIEs leads us to a broader consideration of the extent of import penetration of developed country markets. At the global scale one of the most controversial aspects of the spectacular export growth of NIEs in general, and of the Asian NIEs in particular, has been their alleged growing penetration of the domestic markets of the developed market economies as well as their success in competing with industrialized nations elsewhere. In Chapter 13 we shall look at the economic and social implications of such increased penetration, particularly for employment in the developed market economies. At this point our concern is with the scale and nature of such import penetration.

Figure 2.10 shows the marked changes which occurred between 1963 and 1993 in the trade between the NIEs and the developed market economies. It reveals the

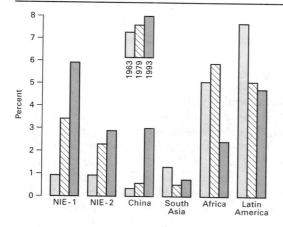

Figure 2.10 Shares of NIEs in manufactured imports to developed countries, 1963–93
Source: Based on data in UNCTAD (1996a, Chart 5)

increasing difference in penetrating developed country markets between the east and southeast Asian NIEs and those from other parts of the world, especially Latin America.

> In this respect, both the first-tier and the second-tier . . . [Asian] . . . NIEs performed much more successfully than most other developing countries over the past three decades. While their shares in total OECD imports rose steadily from 1962 to 1993, those of all other developing regions have fallen . . . The situation since 1979 has become even more dramatic: from 1979 to 1993 their combined share rose from 5.7 per cent to 8.8 per cent, whereas for Latin America the share declined further, from 5.1 per cent to 4.7 per cent.
>
> (UNCTAD, 1996a, p. 123)

The 1996 UNCTAD report goes on to show that not only has the overall penetration of developed country markets by east and southeast Asian NIEs increased far more than that of other NIEs but also that they are far more concentrated in those goods for which demand is growing most rapidly:

> The East Asian economies have been singularly successful in increasing their exports of products that are growing in importance in international trade. In 1990 about three-quarters of their exports were in goods for which the share in total OECD imports had been expanding over the previous three decades. In contrast, the proportion for Latin America was only 38 per cent, and only 24 per cent if Mexico is excluded. There was, however, some difference between the first-tier and second-tier NIEs . . . the second-tier NIEs, in this respect, resemble more closely non-Asian NIEs, such as Mexico, Tunisia, Morocco and Turkey . . . [T]he East Asian economies have been far more successful than other developing economies in entering markets for products with high income elasticities of demand. While 9 of the 10 leading exports from the first-tier NIEs to OECD in 1993 were income-elastic, there were only 2 in Latin America.
>
> (UNCTAD, 1996a, pp. 124, 126)

In general, a key characteristics of NIEs' exports has been their selective nature. They tend to be in certain sectors rather in others. Although their overall penetration of developed country markets is far from overwhelming (around 20 per cent of total imports) they do tend to be far higher in specific industries. The industries involved have tended to be those which are especially sensitive to such global shifts in production and trade and we shall look at some of these in detail in the case studies in Part III of this book. However, we should not forget the other side of the import penetration coin. The east and southeast Asian NIEs are not just export generators. They are, as demonstrated in Figure 2.9, increasingly important as *markets* for imports. In fact, they make up one of the fastest-growing markets in the world. In 1984, the east and southeast Asian NIEs constituted less than 10 per cent of the world market (in terms of import values); in 1994 this had increased to 17 per cent. In comparison, the combined market size of North America, the European Union and Japan fell from 52 per cent to 47 per cent. Hence, we must lose the habit of seeing the east and southeast

Asian NIEs as merely the generators of cheap exports. Not only are the first-tier NIEs producing increasingly sophisticated goods but also they, and their neighbours, are now major global markets.

The network of international trade in services

The growth of international manufacturing activity and of international trade in manufactured goods can only occur with the parallel development of circulation activities within the production chain, notably the whole variety of commercial, financial and business services. These are, in themselves, becoming increasingly internationalized. In effect, 'the production – and trade – of goods and services are becoming increasingly inter-linked' (GATT, 1989, p. 23). Although many service activities are not 'tradable' – that is, they have to be provided to customers on a face-to-face basis – the growth of international trade in commercial services is accelerating rapidly. In the 1970s such trade grew more slowly than manufacturing trade but during the 1980s the position was reversed. This was especially true of telecommunications, financial services, management, advertising, professional and technical services as we shall see in Chapter 12.

Table 2.7 is the commercial services equivalent of Table 2.5. A comparison of the two tables reveals a general similarity but with some significant differences. The top fifteen exporters of services account for 73 per cent of the world total (a lower level of concentration than for manufacturing exports) and for 69 per cent of imports. The United States is the leader in both cases. However, both the United Kingdom and France are more important as exporters of commercial services than of manufactured goods, while Japan and Germany are less important as exporters of commercial services than they are as manufacturing exporters. Equally significant is the substantial increase in ranking of the leading east and southeast Asian NIEs as exporters of commercial services, a clear indication that these economies are now substantially more than mere assemblers of manufactured goods. Indeed, although at a broad regional scale, western Europe is dominant in total commercial services accounting for 47 per cent of the world total in 1994, followed by Asia (20 per cent) and North America (18 per cent), in terms of *rates of growth* the most dynamic region was Asia:

> Asia's exports and imports of commercial services continued to grow at a significantly faster rate than those of North America or Western Europe. Although Asia's share in world commercial services exports and imports is lower than in world merchandise trade, the value of its services exports have exceeded those of North America since 1993 and its services imports are about 80 per cent larger than those of North America. Japan accounts for about one-third of Asia's services imports and for a quarter of Asia's services exports . . . In both 1994 and 1995, Asia was the region with the most dynamic traders in commercial services. At least four traders – Hong Kong, the Republic of Korea, Singapore and Thailand recorded an expansion of their commercial services trade in excess of 20 per cent.
>
> (WTO, 1995, pp. 11, 12)

It is important, once again, to reiterate the intimate functional relationship which exists between the internationalization of services and the internationalization of manufacturing production and trade. 'Increases in merchandise trade stimulate the expansion of such trade-related services as shipping, port services and merchandise insurance . . . also . . . the causation can run the other way . . . advances in these traditional services, as well as the availability of new services, can stimulate merchandise trade' (GATT, 1989, p. 40). But the relationship is not simply one between

Table 2.7 The world 'league table' of commercial services trade, 1994

	Exports				Imports			
Country	1994 (%)	Rank (1994)	Rank (1970)	Country	1994 (%)	Rank (1994)	Rank (1970)	Balance ($ billion)
United States	17.2	1	1	United States	11.9	1	1	+53.0
France	8.6	2	3	Japan	10.1	2	5	–48.7
United Kingdom	5.8	3	2	Germany	9.5	3	2	–44.8
Italy	5.7	4	5	France	6.7	4	4	+19.2
Japan	5.5	5	6	Italy	5.5	5	6	+1.3
Germany	5.3	6	4	United Kgdm	4.9	6	3	+9.1
The Netherlands	4.1	7	6	The Netherlands	3.9	7	8	+1.5
Belgium–Luxembourg	3.5	8	10	Belgium–Luxembourg	3.2	8	9	+2.5
Spain	3.3	9	8	Canada	2.7	9	7	–8.9
Hong Kong	3.0	10	–	Austria	2.0	10	18	+7.6
Austria	2.8	11	15	Taiwan	2.0	11	–	–7.5
Singapore	2.2	12	22	South Korea	1.9	12	32	–1.4
Switzerland	2.2	13	12	Spain	1.8	13	16	+15.3
South Korea	1.8	14	27	Hong Kong	1.7	14	–	+13.3
Canada	1.8	15	9	Thailand	1.5	15	–	
Total	72.8				69.3			

Source: Based on data in WTO (1996, *Annual Report, 1996: Volume II*, Table I.7).

services trade and merchandise trade. Development of some kinds of services may also stimulate the growth of other services. These relationships become particularly significant components of the internationalization of business activity and, especially, the increasing pursuit of globalization strategies by business firms. These are all key issues to be explored throughout this book.

The changing global map of foreign direct investment

The changing global maps of production and of trade, described in the previous sections of this chapter, are important indicators of the increasing internationalization of economic activity. A third, intimately related, indicator is the growth in the scale and complexity of international investment. Here, it is *foreign direct investment* (FDI), rather than portfolio investment, which is especially important.[5] In this section we examine global trends in foreign direct investment over the past few decades. FDI statistics constitute the most comprehensive single indicator of the activities of transnational corporations and of the growth of international production. However, FDI is only one measure – albeit a very important one – of TNC activity. It does not capture the increasingly diverse ways in which firms engage in international operations, for example, through various kinds of collaborative ventures and alliances, or through their co-ordination and control of production chain transactions. We will look at these issues in Chapter 7.

Aggregate trends in foreign direct investment

Although there was very considerable growth and spread of foreign direct investment during the first half of the present century, that was as nothing compared with its spectacular acceleration and spread since the end of the Second World War. The

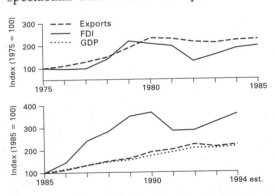

Figure 2.11   Growth of foreign direct investment compared with trade and production, 1975–94
Source: Based on material supplied by UNCTAD

postwar surge of FDI was an integral part of the 'golden age' of economic growth of the 1960s. In fact, during the 1960s, FDI grew at twice the rate of global gross national product and 40 per cent faster than world exports. Figure 2.11 shows that during the 1970s and into the first half of the 1980s the trend lines of both FDI and exports ran more or less in parallel. Then, from 1985 to 1990 the rate of growth of FDI and of exports and GDP diverged in an unprecedented manner. Between 1986 and 1990 FDI outflows grew at an average annual rate of 28 per cent and cumulative FDI stocks at a rate of 20 per cent a year compared with a growth rate of world exports of 14 per cent. One calculation[6] suggests that FDI during the 1980s grew more than four times faster than world GNP. The early 1990s' recession reduced the FDI growth rates significantly but by the mid-1990s the upward trend had resumed:

> Following the end of the FDI recession in 1993, investment *inflows* rose by 9 per cent in 1994 . . . and by another 40 per cent . . . in 1995 . . . to reach a record of $315 billion . . .

Investment *outflows* also hit new highs in 1995 . . . an increase of 38 per cent over 1994 . . . In 1995, FDI growth was substantially higher than that of exports of goods and non-factor services (18 per cent), world output (2.4 per cent) and gross domestic capital formation (5.3 per cent).

(UNCTAD, 1996b, p. 3)

Of course, these trends in the growth of FDI, trade and production are not independent of one another. The common element is the transnational corporation. UNCTAD (1996b) calculates that there are some 39,000 parent-company TNCs controlling around 265,000 foreign affiliates. This is almost certainly an underestimate. However, this minuscule proportion of the total number of business firms in the world is responsible for a highly disproportionate share of global production and trade. UNCTAD estimates that TNCs account for around two-thirds of world exports of goods and services. However, not only do TNCs generate a large proportion of world trade in a general sense but also a significant share of that trade is *intrafirm trade*. In other words, it is trade which takes place inside the boundaries of the firm – although across national boundaries – as transactions between different parts of the same firm. Unlike the kind of trade assumed in international trade theory – and in the trade statistics collected by national and international agencies – intrafirm trade does not take place on an 'arm's-length' basis. It is, therefore, not subject to external market prices but to the internal decisions of TNCs.

In effect,

> international trade in manufactured goods looks less and less like the trade of basic economic models in which buyers and sellers interact freely with one another (in reasonably competitive markets) to establish the volume and prices of traded goods. It is increasingly managed by . . . [transnational] . . . corporations as part of their systems of international production and distribution.

(Helleiner and Lavergne, 1979, p. 307)

Unfortunately, there are no comprehensive and reliable statistics on intrafirm trade. The 'ball park' figure is that approximately one-third of total world trade is intrafirm although, again, that could well be a substantial underestimate. For example, roughly four-fifths of the United Kingdom's manufactured exports are flows of intrafirm trade either within UK enterprises with overseas operations or within foreign-controlled firms with operations in the United Kingdom.

Geographical origins of FDI: the old and the new

Foreign direct investment originates overwhelmingly from the developed market economies. Figure 2.12 and Table 2.8 illustrate FDI trends among the leading developed country sources. The most significant feature of Figure 2.12 is the difference in relative growth rates for individual source nations and the consequent shifts in their relative shares of world foreign direct investment. For most of its history, world foreign direct investment has been overwhelmingly dominated by TNCs from the United States, the United Kingdom and one or two continental European countries. From the 1950s to the mid-1970s, US firms accounted for between 40 and 50 per cent of the world total. In 1960, US and UK TNCs made up two-thirds of the world total. But although TNCs from both countries have continued to invest heavily overseas other countries' outward investment has increased more rapidly, as Figure 2.12 shows. By 1985 the combined US–UK share of the world total had fallen to around half. Conversely, the German share of the total increased from 1.2 to 8.8 per cent while

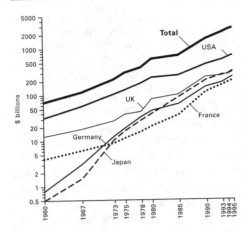

Figure 2.12   Growth of foreign direct investment, 1960–95
Source: Based on data in UNCTAD, *World Investment Report* (various issues)

Japan's share had grown even more sharply, from 0.7 to 6.5 per cent. From being a very minor player in terms of foreign direct investment in 1960, and not especially important in 1975, Japan had surged up the league table to fourth place by 1985. 'Le défi Americain' was being replaced by 'le défi Japponais'.[7]

As shown above, 1985 marked a major acceleration in the growth of world FDI to unprecedented levels. In that acceleration, Japan was undoubtedly the leading player. Japanese outward direct investment grew from $44 billion in 1985 to $306 billion by 1995 (even though there was a pronounced slackening in the rate of growth in the early 1990s). Comparison with the UK underlines the dramatic change in Japan's global position since 1960. In that year, Japan's share of the world FDI total – 0.7 per cent – was a minuscule one-twenty-sixth of that of the UK (18.3 per cent). In 1995, Japan accounted for 11.2 per cent of the world FDI total, virtually the same as the United Kingdom. In contrast to Japan, Germany's share of the world FDI total has remained virtually unchanged since 1985 at just under 9 per cent. Of course, it must be emphasized that these are all shares of a much larger total. Foreign direct investment from virtually all developed economies is now very much larger in absolute terms than ever before.

Apart from the rapid growth of Japanese FDI, the other – more recent and, as yet, more embryonic – development is the emergence of TNCs from developing countries. In 1960, 99 per cent of world FDI came from the developed economies. Fifteen years later, in 1985, the developing countries' share was only around 3 per cent. By 1995, however, this had more than doubled so that around 8 per cent of world FDI originated from developing countries (Figure 2.13). As yet, only a small number of

Table 2.8   World foreign direct investment: changing relative importance of leading source countries, 1960–95

Country of origin	Per cent of world total of outward direct investment stock				
	1960	1975	1985	1990	1995
United States	47.1	44.0	36.6	25.8	25.9
United Kingdom	18.3	13.1	14.6	13.7	11.7
Japan	0.7	5.7	6.5	12.2	11.2
Germany	1.2	6.5	8.8	9.0	8.6
France	6.1	3.8	5.4	6.5	7.4
Subtotal	73.4	73.1	71.9	67.2	64.8
The Netherlands	10.3	7.1	7.0	6.5	5.8
Canada	3.7	3.7	6.0	4.7	4.0
Switzerland	3.4	8.0	3.1	4.0	4.0
Italy	1.6	1.2	2.3	3.3	3.2
Sweden	0.6	1.7	1.8	3.0	2.3
Total	93.0	94.8	92.1	88.7	85.1

Source: Based on data in UNCTAD, *World Investment Report*, various issues.

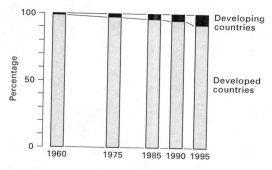

Figure 2.13   The increasing share of foreign direct investment from developing countries, 1960–95
Source: Based on data in UNCTAD, *World Investment Report* (various issues)

developing countries is involved.[8] Four-fifths of all developing country FDI originates from just seven countries. Six of those seven are Asian NIEs as Figure 2.14 shows. However, though modest in scale, this is undoubtedly the harbinger of an important new development. Again, therefore, we see clear signs that the relatively simplistic division of the global economy has disappeared. The world's population of TNCs is not only growing very rapidly but also there has been a marked increase in the geographical diversity of its origins in ways which cut across the old international division of labour.

Geographical destinations of FDI: increasing interpenetration

As well as relative shifts in the geographical origins of foreign direct investment, equally significant, but very different, shifts have been occurring in its geographical destinations. The geographical structure of FDI has become far more complex in recent years, a further indication of increased interconnectedness within the global economy. Figure 2.15 maps the overall pattern of the origins and destinations of FDI stocks for 1995. Despite some changes in recent years – and contrary to much popular

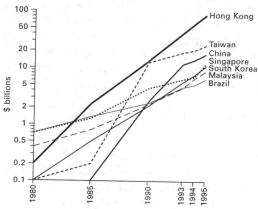

Figure 2.14   Foreign direct investment from NIEs, 1980–95
Source: Based on data in UNCTAD (1995, Annex Table 4; 1996b, Annex Table 4)

thinking – the geographical destinations of foreign direct investment are still strongly concentrated in the developed market economies. Indeed, this developed country bias has increased rather than decreased. In 1938, 66 per cent of the world's foreign direct investment was located in the developing countries; by 1995 this share had fallen to only 26 per cent. Thus, the developed market economies are not only the dominant source of transnational investment, with 92 per cent of the total, but also the dominant destination. Three-quarters of all FDI in the world in the mid-1990s was located in the developed market economies.

Changes in the relative position of developed countries as FDI destinations

Within that highly concentrated structure the detailed distribution of FDI has changed very considerably. During the 1950s and 1960s it was reasonable to distinguish between those countries that were primarily sources of FDI and those that were primarily destinations for FDI. Among the developed economies – with the notable exception of Japan – that distinction no longer applies. As Figure 2.15 shows, virtually

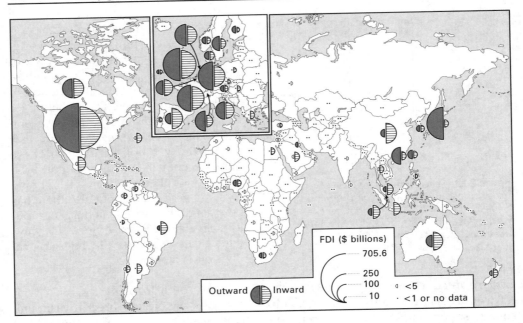

Figure 2.15   The map of foreign direct investment, 1995
Source: Based on data in UNCTAD (1996b, Annex Tables 3, 4)

all developed economies have substantial outward *and* inward direct investment. If we measure this as a simple FDI ratio (outward FDI/inward FDI) then a value of 1.0 would signify an exact balance between outward and inward investment. With just four exceptions, all the countries of western Europe and North America had FDI ratios within the range 0.8 and 1.5 in 1995. The exceptions were Germany (1.75), Sweden (1.9), Finland (2.2) and Switzerland (2.5). What these patterns imply, in fact, is a high degree of *cross-investment* between the major developed market economies: each is investing in each other's home territory.

Within this general increase in interpenetration of FDI between developed economies, three distinctive features stand out:

- *The dramatic change in the position of the United States as a host country for FDI.* Although the United States has attracted FDI for many decades, such inward investment was a tiny fraction of the country's outward direct investment. As we have seen, the United States in 1960 generated almost half of world outward FDI. Even in 1975, the United States' FDI ratio was 4.5: that is, its outward investment was four-and-a-half times greater than its inward investment. In comparison, the ratio for western Europe was 1.2 (outward and inward investment close to balance). The United States' share of world inward FDI stock increased from 11 per cent in 1975 to 21 per cent in 1995. For every leading investing country, the United States became significantly more important as firms from Europe, Japan and, more recently, some east Asian NIEs reoriented the geographical focus of their overseas direct investments.
- *The continuation of Europe's attractiveness to inward investment.* Western Europe's share of total inward investment fell quite substantially between 1975 and 1985

(from 41 per cent of the world total to 33 per cent). But by the mid-1990s its share had risen again to 41 per cent. Much of this resurgence is explained by the European Community's programme to complete the Single European Market (see Chapter 3). Non-European companies made major efforts to gain a direct presence in the single market spurred on by the fear of a potential 'fortress Europe'. Both statistical and anecdotal evidence suggest a major upsurge in direct investment in Europe by both United States and, especially, Japanese and other Asian companies. At the same time, there continues to be a very high level of intra-European FDI. For all the major European countries (excluding the United Kingdom), more than half of their FDI outflows are to other European countries. In most cases, this regional orientation has actually increased. For example, in 1985, 49 per cent of Italy's outward FDI was in Europe. In 1994 this share had grown to 73 per cent. The comparable figures for Germany were 44 per cent and 61 per cent; for The Netherlands 40 per cent and 55 per cent, for France 58 per cent and 63 per cent. For the United Kingdom, in contrast, the figures were 28 per cent (in 1987) and only 38 per cent in 1994. The transformation of the political situation in central and eastern Europe is also leading to the growth of FDI in these economies although, as yet, the volumes are relatively small. In 1995, only 1.2 per cent of world inward FDI was located in central and eastern Europe. However, this represented a ten-fold increase over the position in 1990.

- *The persisting asymmetry of Japan's FDI position.* While the US direct investment position has been transformed from one of being overwhelmingly a home country for FDI to one in which the ratio of outward to inward investment is almost in balance the same certainly cannot be said of Japan. While Japanese outward investment has grown spectacularly, there has been only very limited growth of inward investment. Indeed, the outward/inward ratio actually increased – from an already exceptionally high level of 10.7 in 1975 to 17.2 in 1995. Whereas in 1995 Japan accounted for 11.2 per cent of total world outward direct investment, it was the host to only 0.7 per cent of world inward investment. Along with trade frictions, this huge imbalance in the Japanese direct investment account continues to cause major concern among businesses and policy-makers in the west.

The concentration of FDI among developing countries

As we have seen, the developing countries as a whole are host to only one-quarter of world FDI. Within that relatively small share there is a very high level of concentration

Table 2.9   The concentration of foreign direct investment in developing countries, 1995

Country	FDI stock ($ million)	% of developing country total
China	128,959	18.6
Mexico	61,322	8.8
Singapore	55,491	8.0
Indonesia	50,755	7.3
Brazil	49,530	7.1
Malaysia	38,453	5.6
Argentina	26,801	3.9
Saudi Arabia	26,510	3.8
Hong Kong	21,769	3.1
Thailand	16,775	2.4
Total, 10 developing countries	476,365	68.7

Source: Based on data in UNCTAD (1996b, Annex Table 3).

in a small number of countries. Figure 2.15 shows that inward FDI is minuscule in the majority of developing countries. There is a clear regional dimension to this. Africa's share of the developing countries' total had declined to a mere 8.6 per cent by 1995. Latin America's share had fallen to 33 per cent. In contrast, the Asian share had increased more than two-and-a-half times, from 21 per cent in 1975 to 58 per cent in 1995. Table 2.9 shows that ten countries account for more than two-thirds of all FDI in developing countries. Six of these are in Asia, including by far the largest host country – China. Indeed, the extent to which FDI in China has grown since the early 1980s is nothing short of spectacular.

The relative importance of inward FDI to host countries

Statistics on the scale of FDI are important in showing us the pronounced geographical variations in destinations. But they tell us nothing about how important such investment is to an individual host economy. In fact, such importance varies enormously from one country to another. One measure of relative domestic importance of inward FDI is to compare it to a country's gross domestic product (GDP). Table 2.10 shows that for developed countries as a whole, inward FDI constituted around 9 per cent of GDP in 1994. For western Europe, the share was much higher at 13 per cent. Within Europe, FDI was most significant as a share of GDP in Belgium–Luxembourg (32 per cent), The Netherlands (28 per cent), Spain (25 per cent), Greece (24 per cent) and the United Kingdom (21 per cent). In North America, there was a pronounced contrast between Canada (where FDI contributed 19 per cent of GDP) and the United States (7.5 per cent). Not surprisingly we find the lowest contribution of all in the case of Japan (0.4 per cent).

Similar variation in the relative significance of foreign penetration is evident among developing countries. In Latin America foreign firms are especially important to the economies of Mexico (14 per cent of GDP) and Chile (19 per cent). Within Asia the range is especially wide. Foreign firms are dominant in the economy of Singapore (73 per cent of GDP), extremely significant in Malaysia (46 per cent), Indonesia (27 per cent) and Hong Kong (21 per cent), significant in China (18 per cent of GDP). But in

Table 2.10   Inward foreign direct investment as a share of gross domestic product, 1994

	Share of GDP (%)		Share of GDP (%)
Developed countries	8.6	*Developing countries*	12.5
Belgium–Luxembourg	31.7	Singapore	72.8
The Netherlands	27.7	Malaysia	46.2
Spain	25.0	Indonesia	26.5
Greece	23.5	Hong Kong	20.5
United Kingdom	20.9	Chile	19.2
Canada	19.2	China	17.9
Denmark	12.6	Mexico	14.4
France	10.7	Thailand	10.1
Ireland	10.3	Philippines	8.3
Sweden	9.7	Argentina	8.1
United States	7.5	Brazil	8.0
Germany	6.8	Taiwan	6.6
Portugal	6.6	Pakistan	6.0
Finland	5.9	South Korea	3.3
Italy	5.9	India	0.9
Japan	0.4	Bangladesh	0.7

Source: Based on data in UNCTAD (1996b, Annex Table 6).

Taiwan, inward FDI contributed only 6.6 per cent of GDP and a mere 3.3 per cent in South Korea. Clearly, the direct role of TNCs in the 'economic miracle' of the east and southeast Asian countries varies enormously.

Sectoral tendencies in foreign direct investment

Just as foreign direct investment as a whole is unevenly distributed geographically so, too, it tends to be concentrated rather more in some types of economic activity than in others. But important changes have been occurring as Figure 2.16 reveals. At this broad level of aggregation the two major trends have been, first, the relative decline in the proportion of outward FDI in the primary sector and, secondly, the corresponding increase in importance of the service sector. Within each of these broad categories, however, important transformations have also been taking place.[9]

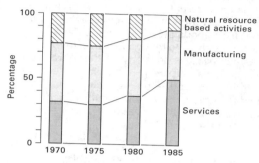

Figure 2.16 The changing sectoral distribution of foreign direct investment *Source:* Based on data in UNCTAD (1993b, Table III.1)

Historically, the major proportion of FDI was concentrated in the *natural resource-based sectors*: the mining of geographically localized minerals and ores, the operation of large-scale plantations for the production of commercial foodstuffs for export, and the like. 'For historical, technical and financial reasons, FDI has featured disproportionately in the development of the primary sector. Few firms in developing countries have the resources and know-how to conduct large-scale exploration or exploit the commercial potential of deposits' (UNCTAD, 1993b, p. 65). TNCs are still heavily involved in such commodity sectors, the most obvious example being the giant petroleum companies. In some extractive activities, national governments have acquired control of their natural resource operations but, even so, TNCs often remain in control of the commodities' marketing channels.

Although foreign direct investment in the extractive industries remains extremely important, the emphasis has certainly shifted. Until the early 1980s, the bulk of FDI was in *manufacturing* industry and was an integral part of the rapid growth and internationalization of economic activity which, as we have seen, helped to transform the world economy. Three broad types of manufacturing industry have an especially large TNC involvement:

- *Technologically more advanced sectors* – for example pharmaceuticals, computers, scientific instruments, electronics, synthetic fibres.
- *Large-volume, medium-technology consumer goods industries* – for example motor vehicles, tyres, televisions, refrigerators.
- *Mass-production consumer goods industries supplying branded products* – for example cigarettes, soft drinks, toilet preparations, breakfast cereals.

These are the industries in which a high level of technological expertise and resource (both human and financial) are required, in which demand is strongly income elastic and for which the extensive operations of TNCs are especially suitable. In fact, there has been a 'gradual move of FDI away from labour-intensive, low-cost, low-skill manufacturing and towards more capital-, knowldge- and skill-intensive industries' (UNCTAD, 1993b, p. 71).

It is in the *service industries* that the most significant relative change has occurred. However, the internationalization of services through foreign direct investment has happened more recently than the more general structural changes which have affected virtually all economies:

> The main home and host countries became predominantly service economies some time ago, whereas the surge of services FDI is relatively recent. Clearly, the FDI changes happened with a lag. The nature of that lag reflects the different sectoral roles of trade and FDI as conduits of international integration. Where goods are concerned, by and large trade has preceded FDI as the main way of delivering them to foreign markets. In the case of services, this sequence was not really feasible, because so many of them are non-tradable. As a result, FDI was the only means to participate in international service transactions. But in many important service industries, and in both developed and developing countries, FDI was initially prohibited for strategic, political or cultural reasons. Consequently, the rise of FDI in services had to wait for the liberalization of major service industries to catch up with the domestic processes of structural change.
> (UNCTAD, 1993b, p. 62)

FDI in services tends to be concentrated in certain key sectors, notably:

- *financial services* (banking, insurance, accounting)
- *trade-related services* (wholesaling, marketing, distribution)
- *telecommunication services*
- *business services* (consulting, advertising, hotels, transportation, construction)
- *some consumer services* (retailing, fast food):

> Within the services sector, finance- and trade-related activities account for two-thirds of the FDI stock for developed countries and the majority of the stock in many host developing countries. This is not surprising, given that financial TNCs (banks, insurance companies) and trading companies (for example, Japanese *sogo shosha*) are among the most prominent TSCs . . . [transnational service companies] . . . , that manufacturing and petroleum companies have invested heavily in wholesale and marketing affiliates and that all kinds of TNCs tend to establish their own finance-related foreign affiliates. In addition, other service industries in which TNCs are prominent are either small (such as advertising), or rely heavily on non-equity forms of investment (such as the hotel and restaurant industry), or both (accounting and business consultancy).
> (UNCTAD, 1993b, p. 78)

A closer look: some case examples

Quite clearly, TNC activity as a whole tends to be unevenly distributed, both geographically and sectorally. However, there are significant differences between individual source nations. In this section we look, very briefly, at the broad trends over the past decade in the geography and sectoral composition of FDI from four developed countries (the United States, the United Kingdom, Germany, Japan) and from some east Asian NIEs (Hong Kong, Taiwan, South Korea, China). The aim is to provide no more than a brief 'thumb-nail' sketch of each case. The reasons for doing this are twofold. First, it helps to add some individual detail to the broad picture painted in this chapter so far. Secondly, one of the themes of this book is that TNCs are not all the same. Although, as profit-seeking enterprises operating within a capitalist market system they do, indeed, share some common characteristics, they are far from being homogeneous. There are many reasons for such differences, as we shall see in later chapters. One very important reason is, without doubt, the influence of the TNC's *home country* environment and its political, social, cultural and economic characteristics.

United States foreign direct investment

In 1986, *The Economist* published an article headed 'American multinationals: the urge to go home' and asserted that American firms were 'turning inward'. In a relative sense, as Figure 2.17 shows, there was some subsequent evidence for this slackening in growth of overseas investment. Although United States FDI grew by 13 per cent in 1986 and by 21 per cent in 1987 its subsequent rates of growth have been more modest. Even so, in six of the nine years shown, the growth rate was still in double figures. United States FDI has indeed continued to grow (Figure 2.17). Although its share of the world foreign investment total has fallen it should be remembered that the world total itself has expanded dramatically. In absolute, if not in relative terms, United States TNCs are more significant today than they were in the past. They constitute the largest and most extensive network of international production facilities

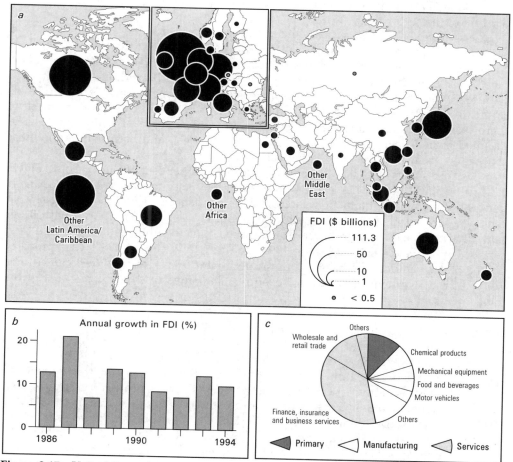

Figure 2.17   United States foreign direct investment
Source: Based on data in OECD (1996a)

in the world. The – not unreasonable – attention being given to the rapid growth of TNCs and of FDI from other countries (particularly Japan) should not let us forget this fact. Reports of the decline of United States FDI, like the reported death of Mark Twain, have been greatly exaggerated.

The *sectoral* distribution of United States FDI has changed substantially in recent years. In 1985, 27 per cent was in the primary sector (predominantly in oil), 41 per cent in manufacturing and 31 per cent in services. In manufacturing, five industries (food, beverages, tobacco; chemicals; mechanical equipment; electric and electronic equipment; motor vehicles) accounted for three-quarters of the total. In services, nine-tenths was in finance, insurance and business services and in wholesale and retail trade. By 1994, the picture had changed in some important respects. The share of the primary sector had fallen to 12 per cent, that of manufacturing to 35 per cent while the service sector had grown to 53 per cent. Within manufacturing, the major increases were in food, drink and tobacco and in chemicals set against a major decline in the share of mechnical equipment. The shares of FDI in motor vehicles and in electric and electronic equipment increased only slightly. The biggest relative shift was in the services sector where FDI in finance, insurance and business services had grown from 54 to 69 per cent of the sector total.

Geographically, 50 per cent of United States FDI was located in Europe in 1994. More than one-third of the European total was in one country: the United Kingdom. Germany, the second most important European host country for United States FDI, had 13 per cent. The trend between the two countries has been different. Compared with 1985, the United Kingdom share increased; the German share decreased. Even so, the two continue to host half of all United States FDI in Europe. In the Americas, there was a marked contrast between Canada (whose share of United States FDI declined from 20 per cent in 1985 to 12 per cent in 1994) and Latin America, whose share grew from 12 per cent to 18 per cent.

East and southeast Asia contained only around 14 per cent of United States FDI in 1994 (up from 11 per cent in 1985). Within Asia, the major foci in 1994 were Japan (43 per cent of the regional total compared with 38 per cent in 1985), Hong Kong (15 per cent compared with 13 per cent) and Singapore (12 per cent compared with 8 per cent). It is notable that South Korea, Taiwan and China together contained only 11 per cent of United States FDI in Asia in 1994, up from 7 per cent in 1985. The spread of United States FDI outside Europe, Latin America and Asia was very thin indeed. The whole of the continent of Africa, for example, contained less than 1 per cent of the total!

United Kingdom foreign direct investment

United Kingdom FDI shares at least one common feature with that of the United States: both have experienced a relative decline in their share of world direct investment in the last two or three decades. However, the year-by-year trend has been rather different, as Figure 2.18 shows. The United Kingdom's growth pattern has been far more volatile than that of the United States with very high growth in both 1988 and 1989, a decline in 1990, slight recovery in 1991, a return to double-digit growth in 1992 and 1993 and then more modest growth in 1994. At the broad level of aggregation, the *sectoral* structure of United Kingdom FDI changed only slightly between 1987 and 1994 (Figure 2.18). The share of the primary sector fell slightly, that of the manufacturing sector increased slightly. The share of the total in services was around 40 per cent

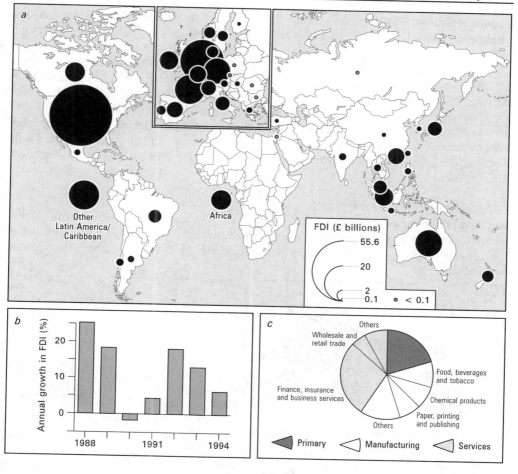

Figure 2.18 United Kingdom foreign direct investment
Source: Based on data in OECD (1996a)

at both dates. But large-scale changes occurred within both the manufacturing and services categories. Within manufacturing, the overwhelming emphasis is on three industries: food, drink and tobacco (24 per cent), chemicals (21 per cent), paper, printing and publishing (20 per cent – up from only 9 per cent in 1987). The biggest change was the dramatic fall in the share of electric and electronic equipment: from 11 per cent to less than 1 per cent. Within services, predictably, it was finance, insurance and business services which grew fastest to account for 67 per cent of the sector total in 1994 (compared with 35 per cent in 1987).

United Kingdom FDI has always displayed a particularly extensive *geographical* spread. Historically, that was clearly related to the extensiveness of its colonial possessions. But the geographical emphasis has certainly changed with a much greater emphasis on Europe. However, it is noticeable that the United Kingdom's FDI in Africa is considerably greater than that of the United States. There is also still important United Kingdom FDI in Australia although it is relatively less important than it used to be. The major shift has been towards Europe where almost 40 per cent of

United Kingdom FDI is now located (compared with 28 per cent in 1987). Within Europe, almost 40 per cent is in The Netherlands, 17 per cent in France and 12 per cent in Germany.

Outside Europe, United Kingdom FDI is heavily oriented towards the United States (32 per cent of the world total) but Canada's significance as a host country has declined. East and southeast Asia is far less important to the United Kingdom than to the United States. Only 9 per cent of United Kingdom FDI is in that most dynamic of all world regions. Sixteen per cent of the regional total is in Japan (unchanged from 1987), but the major focus otherwise is on the former colonies of Singapore, Hong Kong and Malaysia. Together, these account for 65 per cent of all United Kingdom FDI in Asia.

German foreign direct investment

German FDI grew especially rapidly during the 1960s and 1970s. Since the mid-1980s, however, its relative growth has been less pronounced (Figure 2.19). Only in three of the nine years between 1985 and 1994 did German FDI grow by more than single

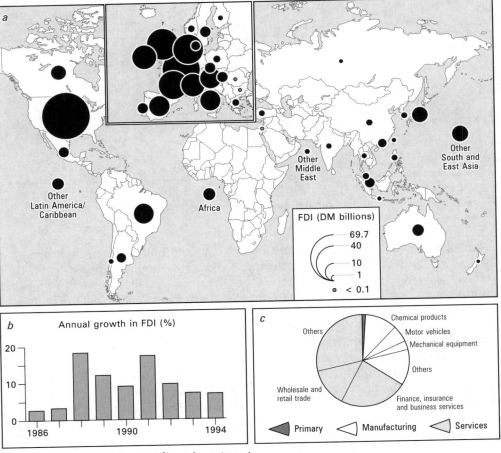

Figure 2.19   German foreign direct investment
Source: Based on data in OECD (1996a)

figures. In *sectoral* terms, German FDI shows a number of distinctive features. FDI in the primary sector is virtually non-existent whereas the oil industry's significance in the United States and the United Kingdom is very marked. The significance of manufacturing halved between 1985 and 1994, from 60 per cent to 33 per cent. Within manufacturing, again, the German profile is distinctive. The chemicals industry dominates the sector (34 per cent) but motor vehicles (18 per cent) are also very significant. There is far more of a 'high-tech' flavour about German FDI in manufacturing. In the service sector, too, the pattern is different from both the United States and the United Kingdom. Finance, insurance and business services are certainly significant but they constitute only 36 per cent of the sector total, half of that of the United States and the United Kingdom. Conversely, FDI in wholesale and retail trade is far more important in the German case than in the others.

The *geographical* distribution of German FDI has become overwhelmingly oriented towards Europe. In 1994, almost two-thirds of the world total was located in Europe (up from 44 per cent in 1985). Although the largest concentration is in Belgium–Luxembourg, German FDI in Europe is rather more evenly spread than that of the United Kingdom. In addition, however, German firms have a far stronger orientation towards central and eastern European countries. As the emphasis on Europe has increased the relative importance of North America has declined; from 33 per cent in 1985 to 23 per cent in 1994. The Latin American focus is also rather less significant than it was (down from 9 per cent to 6 per cent) but almost 60 per cent of the Latin American FDI is in Brazil. East and southeast Asia hosts only 7 per cent of German FDI, a remarkably low figure. Of this, almost one-third is located in Japan with a further 11 per cent in Singapore. Otherwise, German FDI in Asia is very thinly spread indeed.

Japanese foreign direct investment

Until relatively recently, FDI from Japan was extremely small. However, it has grown with great speed since the 1980s in particular. From 1985 to 1990, annual rates of Japanese FDI growth were between 22 and 36 per cent, by far the highest of any country. Only after 1991 was there a marked slowdown but, even so, Japanese FDI stock still grew at around 10 per cent a year up to 1994 (Figure 2.20). The *sectoral* trajectory of Japanese FDI has been quite different from that of the other industrialized countries. In the early stages (the 1960s) the emphasis was overwhelmingly on FDI in the natural resources sector; a reflection of Japan's almost total lack of indigenous industrial raw materials. By 1995, however, the dominant sector was services (67 per cent) – far higher than that of the United States, the United Kingdom and Germany. In fact, contrary to much popular opinion, manufacturing has never been the dominant component of Japanese FDI. Within manufacturing, there is a strong focus upon electric and electronic equipment, motor vehicles and chemicals. The services sector is also distinctive. FDI in distribution services has been especially significant.

Between 1985 and 1994, several major changes occurred in the *geography* of Japanese FDI. The relative importance of east and southeast Asia declined (from 23 per cent to 16 per cent of the total) as did Latin America (from 19 per cent to 12 per cent). Conversely, both North America and Europe increased very substantially in importance. The North American share increased from 32 per cent to 44 per cent; the European share from 13 to 19 per cent. Japanese FDI in Europe is especially heavily

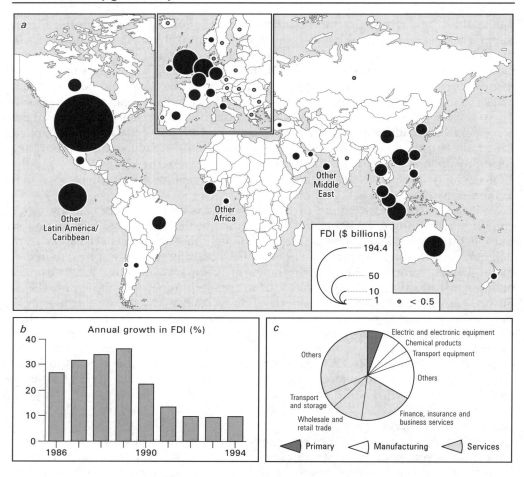

Figure 2.20   Japanese foreign direct investment
Source: Based on data in OECD (1996a)

biased geographically towards the United Kingdom, which contains almost 40 per cent of all Japanese FDI in Europe, followed by The Netherlands. In east and southeast Asia, the major changes have been a huge increase in the relative importance of China as a host country (up from 2 per cent in 1985 to 12 per cent in 1994) as well as – more modestly – Hong Kong (15 to 18 per cent) and Thailand (4 per cent to 9 per cent). On the other hand, Indonesia's relative importance has waned (from 43 per cent in 1985 to 22 per cent in 1994).

A new wave: the emergence of FDI from newly industrializing economies

As shown earlier, by 1995 around 8 per cent of world FDI originated from developing countries, primarily from NIEs in east and southeast Asia. Contrary to much of the conventional literature on the so-called 'Third World multinationals' the sectoral and geographical distribution of FDI from Asian NIEs is very diverse. In particular, the idea that such FDI is predominantly located in developing countries is far from being universally true. The following brief cases provide at least a hint of such diversity:

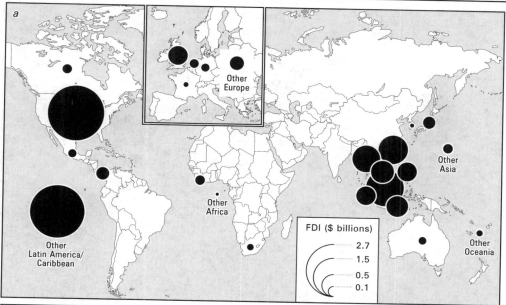

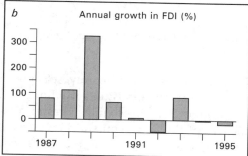

Figure 2.21   Taiwanese foreign direct investment
Source: Based on data in Ministry of Economic Affairs, ROC (1996, *Statistics on Outward Investment, 1996,* Table 11)

- *Taiwan.* Taiwanese FDI was initiated as long ago as 1959 (Chen, 1986) when a cement plant was established in Malaysia. It was a further three years before the second recorded overseas direct investment by a Taiwanese firm. From the 1960s, there was steady growth of FDI from Taiwan but it was not until the mid-1980s that the growth of such investment really began to accelerate. From the early stages one of the most striking features has been the substantial presence in the United States. In 1985 more than half of all Taiwanese FDI was located in the United States. This is certainly not the kind of pattern predicted by the conventional models. Just over one-third was located in Asia and a mere 2 per cent in Europe. Figure 2.21 shows the geographical distribution of Taiwanese FDI in 1995. Although the United States remained very significant, its relative share had halved to 26 per cent of the total. In comparison, the Taiwanese involvement in Europe, though still small, had more than doubled to 5.5 per cent of the total. Around 40 per cent of the total is located in Asia with a particularly strong focus on Malaysia, Hong Kong and Thailand. The missing piece, of course, is China. There is massive Taiwanese investment in China but it is not shown in the official statistics. Much of it flows through Hong Kong.

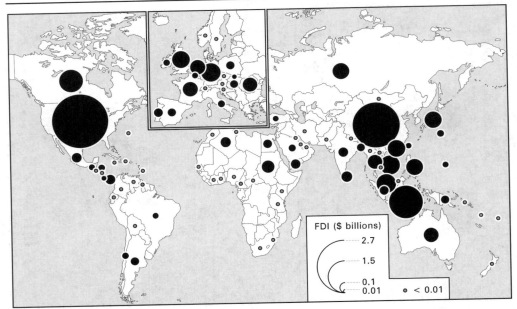

Figure 2.22   South Korean foreign direct investment
Source: Based on data in Bank of Korea (1996, *Overseas Direct Investment Statistics Yearbook,*
1996, Table 4)

- *South Korea.* Korean outward FDI began in the late 1960s with investments in
  Indonesia to procure timber for the Korean plywood industry (Euh and Min, 1986).
  The first overseas manufacturing plant, to produce food seasonings, was also located in Indonesia in the early 1970s. By the late 1970s, Korean outward investment
  was picking up speed and, like Taiwanese FDI, accelerated during the 1980s. The
  two share some similar features in the geography of their investment. Like Taiwan,
  South Korean FDI, from at least the late 1980s, has had a strong orientation towards
  North America. In 1985, 32 per cent of the total was located there compared with 22
  per cent in Asia and 11 per cent in Europe. However, by 1995, the pattern had
  changed somewhat as Figure 2.22 shows. The Asian share of the total had increased
  to 41 per cent and the European share had grown to 14 per cent. Conversely, the
  North American share was unchanged. However, during the mid-1990s, in particular, the huge Korean conglomerate firms (*chaebol*) were making massive investments in Europe and the United States.
- *Hong Kong.* As Yeung shows in his detailed analyses, Hong Kong FDI is both very
  large in quantitative terms but also very strongly oriented towards the Asian region
  – far more so than either Taiwan or South Korea:

  > More than two-thirds of Hong Kong FDI outflows have gone to East, South
  > and Southeast Asia, with another third to the USA. Within the Asia–Pacific
  > region, a large proportion of Hong Kong FDI has been invested in China, in particu-
  > lar in the Guangdong Province. In June 1991, up to US$16 billion of investment in the
  > Guangdong Province (four-fifths of the total FDI in Guangdong) came from Hong
  > Kong.
  >
  > (Yeung, 1994b, p. 1942)

A substantial proportion of that investment almost certainly comes originally from other countries (notably Taiwan) and is channelled through Hong Kong. 'By the early 1990s, there were as many as 3 million workers in Guangdong employed directly or indirectly by Hong Kong manufacturers' (Yeung, 1995b, p. 327). Within ASEAN, Hong Kong FDI is especially targeted towards the two largest economies: Indonesia and Thailand.

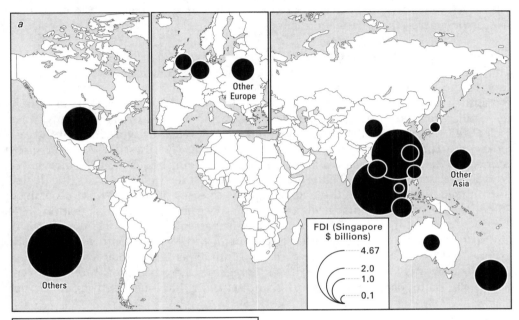

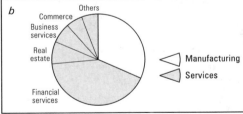

Figure 2.23 Singaporean foreign direct investment
Source: Based on data in Republic of Singapore Department of Statistics (1996, *Singapore's Investment Abroad, 1990–1993*, Table A21)

- *Singapore.* In common with the other Asian NIEs, FDI from Singapore grew quite rapidly during the early 1980s. The 1985 recession slowed down the rate of growth but in the early 1990s, in particular, considerable acceleration occurred. In 1991, outward FDI from Singapore increased by 10.2 per cent, in 1992 by 20.6 per cent, and in 1993 by 25.5 per cent (Singapore Department of Statistics, 1996, p. 17). One-third of Singapore's outward FDI is in the manufacturing sector and 42 per cent in the financial sector. Geographically, as Figure 2.23 shows, Singaporean FDI is very strongly oriented towards other east and southeast Asian countries. More than half goes to Asia, notably to Malaysia (41 per cent of the Asian total), Hong Kong (35 per cent). In comparison, only 8 per cent of Singapore's FDI goes to the United States and 7 per cent to Europe. Within Europe, Belgium is the first-ranking country followed by the United Kingdom.

Changing the lens: the macro-, micro- and mesoscale geography of the world economy

Our analytical lens throughout this chapter has been that of the *national* unit. This is primarily because the data we need to explore economic change at the global scale are only available at that level of spatial aggregation. But although it may be difficult to obtain comprehensive data at other spatial scales we can at least conceive of other highly relevant scales at which economic activities operate. Here, we look at three scales: the macro-, the micro- and the mesoscales.

The macroscale: a global triad

The Japanese management writer, Kenichi Ohmae (1985), coined the term *global triad* to capture the macroscale trends in the world economy. In Ohmae's view, the world economy is now essentially organized around a tripolar, macroregional structure whose three pillars are North America, Europe and east and southeast Asia. Certainly, if we look at the statistical data on production, trade and foreign direct investment the case looks pretty convincing as Figures 2.24 and 2.25 show. In 1994, these three macroregions contained 87 per cent of total world manufacturing output and generated 80 per cent of world merchandise exports (Figure 2.24). Over the past ten to fifteen years, the degree of economic concentration in these three regions increased. In 1980, the triad contained 76 per cent of world manufacturing output (nine percentage points less than in 1994) and 71 per cent of merchandise trade (again, nine percentage points less than in 1994). Not surprisingly, in vew of our earlier discussion, it was east and southeast Asia which experienced the most marked increase. In 1980, only 16 per cent of total world manufacturing output was located there. In 1994, the region's share had grown to 30 per cent whilst its share of total world merchandise trade had increased from 17 per

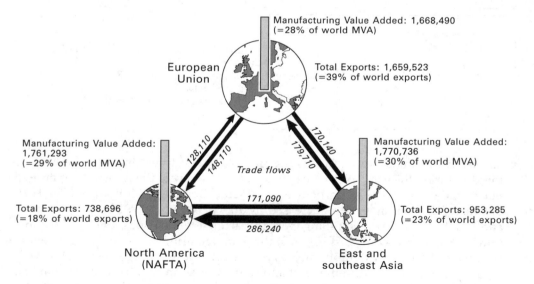

Figure 2.24   The global triad – concentration of manufacturing production and merchandise trade
Note: All figures are in millions of dollars
Source: Based on data in UNIDO (1996, Statistical Annex); WTO (1995, *International Trade, 1995*, Table A2)

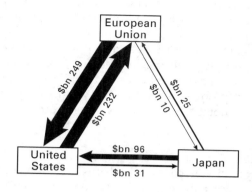

Figure 2.25 The global triad – concentration of foreign direct investment
Source: Based on UNCTAD (1995, Figure I.6)

cent to 23 per cent. The triad is even more dominant in terms of foreign direct investment. Figure 2.25 shows that more than 90 per cent of the world stock of FDI originated in the three regions in 1993. It also reveals very clearly the assymetrical nature of FDI flows between the three macroregions. The biggest flows by far involve the United States.

These trends imply that the global triad is, in effect, 'sucking in' more and more of the world's productive activity, trade and direct investment. The triad sits astride the global economy like a modern three-legged Colossus. It constitutes the world's 'mega-markets'. Whether it is more than a statistical

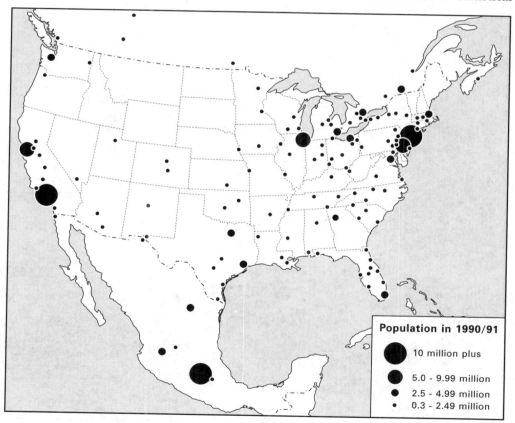

Figure 2.26 Major urban centres in North America and Mexico
Source: Based on United Nations Centre for Human Settlement (1996, *An Urbanizing World: Global Report on Human Settlements, 1996*, Map 2.1, 2.2)

artifact, however, is subject to debate. Two of the three pillars – Europe and North America – are both more formally organized politically into regional trading blocs (the European Union and the North American Free Trade Agreement, respectively). The third – and most dynamic of the pillars of the triad – east and southeast Asia has no formal political organization although, as we shall see in Chapter 3, there are various possibilities being proposed. But if the triad does represent a functional reality (actual or potential) with internally oriented production and trade systems it poses major problems for those parts of the world – notably the least developed countries – which are not integrated into the system.[10]

The microscale: a 'fine-grained mosaic'

The global triad is one view of our complex reality. In so far as both business decision-makers and politicians believe in its existence it will continue to have a major influence on decisions that affect all our lives and livelihoods, regardless of where we actually live. But, of course, as individuals we do not think of ourselves, first and

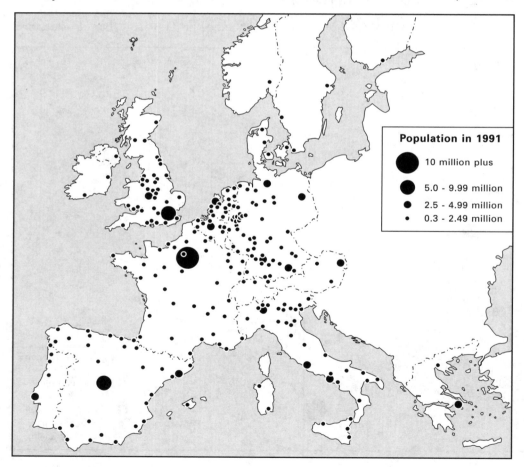

Figure 2.27   Major urban centres in western Europe
Source: Based on United Nations Centre for Human Settlement (1996, *An Urbanizing World: Global Report on Human Settlements, 1996*, Map 2.3)

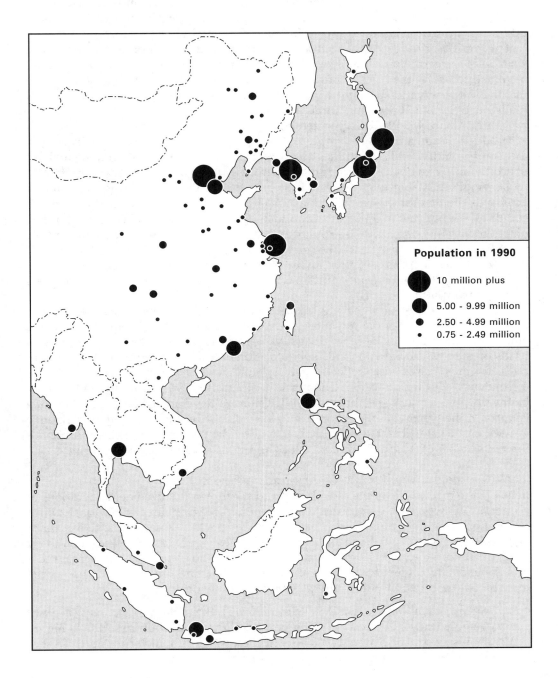

Figure 2.28   Major urban centres in Pacific Asia
Source: Based on United Nations Centre for Human Settlement (1996, *An Urbanizing World:*
Global Report on Human Settlements, 1996, Map 2.5)

foremost, as living in one or other parts of the global triad. We live in specific localized communites – cities, towns, villages. Indeed, as pointed out in Chapter 1, all economic activity is ultimately localized in specific places. Several observers have pointed out that if you observe the earth from a very high altitude and look at the 'economic surface' of the earth you certainly do not see the kinds of national economic boxes we have had to use as the basis of our discussion in this chapter. Particularly if you are making the observation at night what you see are distinctive clusters marked out by the lights of localized agglomerations of activity.

Unfortunately, data disaggregated in this way at the world scale, showing details of production and trade, are simply not available. We have to resort to surrogate measures or individual case studies. But it is vital to stress this most fundamental fact of economic life: the place-specific and clustered nature of most economic activity. The most widely available microscale indicator of the localized clustering of economic activity is the map of the world's cities which can be found in any good atlas. In all countries, virtually all manufacturing and business service activity is located in urban places of varying sizes. In some countries, economic life is dominated by just one or perhaps two major cities (often, though not always, including the political capital); in other countries there is a 'flatter' urban hierarchy and a wider spread of activity. Figures 2.26–2.28 show the major urban centres within each of the three macroregions of the global triad.

It is these cities, and their associated local regions, which contain a nation's economic activity, not some statistical box. Within any individual country, there will almost certainly be considerable diversity between cities/local regions not only in terms of their particular economic specializations but also in terms of their growth rates. In most cases, this reflects their specific historical trajectory – the 'path-dependency' idea introduced in Chapter 1. In others, however, such differentials may be the outcome of very specific political decisions to develop one particular part of a country rather than another. For example, China's recent spectacular economic growth has been articulated quite deliberately by the Chinese government around a limited number of Special Economic Zones and Coastal Cities (see Chapter 4).

The mesoscale: transborder clusters and corridors

Between the macroscale of the global triad and the highly localized agglomerations of economic activity lies a mesoscale of economic-geographic organization which crosses, or sometimes aligns with, national boundaries. In some cases, this scale of organization is actually defined and created by the existence of the political boundary itself. In others it develops in spite of such boundaries and simply extends across them in a functionally organized manner. Three examples, one drawn from each of the three global triad regions, illustrate this mesoscale phenomenon:

● *Europe's major economic growth axis.* Within Europe, the pattern of economic activity is extremely uneven, both within and between individual countries. But as Figure 2.29 shows, we can also identify a distinctive 'growth axis' which runs north west to south east across the core area, cutting across national boundaries:

> The most advanced areas of Europe and most of Europe's major international cities lie on or near an axis extending from the north-west of London through Germany to Northern Italy . . . Along this axis lie two major foci: in the north-west are found the historic capitals of Europe's major colonial powers (Paris, London and Randstad-Holland) whose interconnectedness the Channel Tunnel will reinforce; in the south-

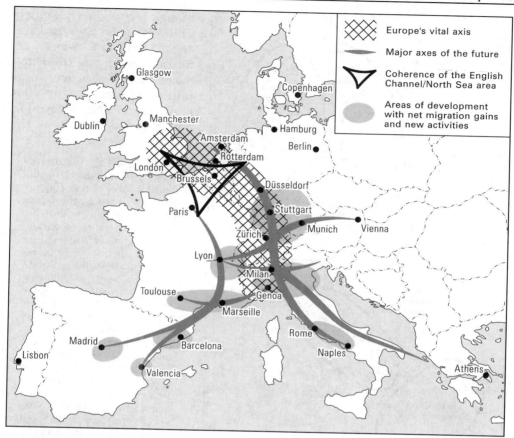

Figure 2.29 Europe's major growth axis
Source: Based on Dunford and Kafkalas (1992, Figure 1.4)

east are cities and regions whose faster recent economic growth has pulled the axis' centre of gravity to the south-east. A parallel axis extends from Paris to the Mediterranean – and a south-western extension stretches down to the major cities in Iberia. With the collapse of communism another parallel axis may emerge in the east extending from Hamburg to Berlin, Leipzig, Prague and Vienna. As distance from Europe's major axes and strong vertical and lateral communications infrastructures increases one moves increasingly into orbital areas of underdevelopment concentrated in the South, East and, as Europe's centre of gravity moves eastwards and Atlantic Europe declines, in the West.

(Dunford and Kafkalas, 1992, pp. 25, 27)

- *Emerging urban corridors in Pacific Asia.* The European 'growth axis' has evolved in a part of the world which has been strongly industrialized for a very long time and also in a region which has had a strong political manifestation through the European Community/European Union. However, we can see a similar phenomenon, at least in embryonic form, developing in Pacific Asia. As Yeung and Lo (1996, p. 37) note, 'several economic hubs have emerged in the region that have essentially taken advantage of a certain complementarity, particularly of labour supply, across national boundaries'. Such 'growth triangles' include the Singapore–Batam–Johor

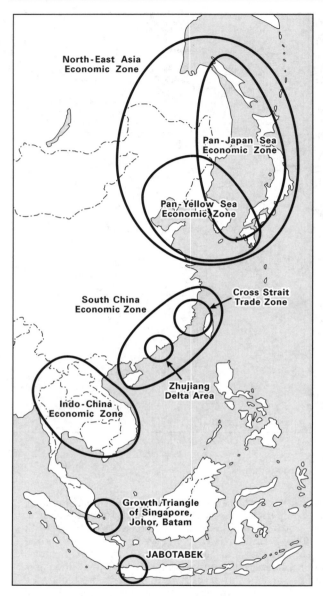

North-East Asia
Economic Zone

Pan-Japan Sea
Economic Zone

Pan-Yellow Sea
Economic Zone

South China
Economic Zone

Cross Strait
Trade Zone

Zhujiang
Delta Area

Indo-China
Economic Zone

Growth Triangle
of Singapore,
Johor, Batam

JABOTABEK

Figure 2.30   Emerging urban corridors in Pacific Asia
Source: Based on Yeung and Lo (1996, Figure 2.8)

triangle and the southern China–Hong Kong–Taiwan triangle, both of which are focused upon a distinctive major city. But some Asian urban scholars argue that much larger urban corridors are becoming evident, as Figure 2.30 suggests:

> Each of these . . . [urban corridors] . . . epitomizes the highest level of interconnectedness among the cities they encompass within the delineated limits as part of the regional restructured economy . . . It has to be recognised that these urban corridors are at varying stages of formation, some exhibiting but incipient development whereas others are quite advanced in form and connectivity among the mega-cities. The best illustration of a mature urban corridor is . . . an inverted S-shaped 1,500 km urban belt from Beijing to Tokyo via Pyongyang and Seoul . . . [which] . . . connects 77 cities of over 200,000 inhabitants each. More than 97 million urban dwellers live in this urban corridor, which, in fact, links four separate megalopolises in four countries in one.
>
> (Yeung and Lo, 1996, pp. 39, 41)

- *The United States–Mexico border zone.* The two previous examples illustrate the development of mesoscale regions which cut across national boundaries and create 'transnational regions'. But there are other cases where the form of economic and urban development is actually defined and created by the existence of a border between countries. Where there is a very marked differential between two adjacent countries – for example, in taxation rates or production costs – there is often a strong incentive for development to occur on one side of the border to take advantages of benefits on the other side. One of the best examples of this is the United

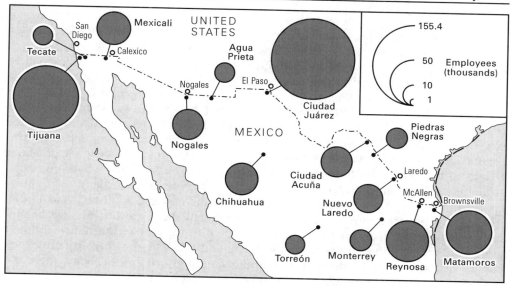

Figure 2.31　Employment in the major *maquiladora* centres
Source: Instituto Nacional de Estadistica, Geografia e Informatica (1996), *Industria Maquiladora de Exportacion)*

States–Mexico border which constitutes the sharpest interface between an extremely wealthy economy and a much poorer developing economy. Far from being just a line on the map, the US–Mexican border is defined in the starkest of physical terms by a whole string of towns and concentrations of manufacturing activity along its entire length. These are not simply spontaneous clusters of activity, however. They owe their existence to political decisions taken initially in 1965 when the Mexican government instituted its Border Industrialization Program to help alleviate the severe economic and social problems of the northern border towns. Its focus was the 'in-bond' assembly (*maquiladora*) plant which was allowed to import materials and components from the United States free of duty provided the end-products were then exported.[11] For United States manufacturers, the attraction of using the *maquiladoras* lay in their very cheap labour and the less stringent environmental controls south of the border.

The result was spectacular growth in both employment and population in towns on both sides of the border but especially on the Mexican side. Figure 2.31 shows the current scale of employment in the major *maquiladora* centres. In 1996, a total of 643,500 workers was employed in these plants compared with 310,000 in 1988. The associated growth of the main towns has been very substantial. In 1950, around 390,000 people lived in the six major Mexican cities along the border. In 1990, this had increased to 2,814,000.

Conclusion: a multipolar, kaleidoscopic global economy

As we move into the new millennium, the world map of production, trade and investment is vastly more complicated than that of only forty or fifty years ago. Although there are clear elements of continuity, dramatic changes have occurred. The overall trajectory of world economic growth has become increasingly volatile; the 'golden age' of the 1950s and 1960s is now seen to have been an atypical section on the roller-coaster. Since those supposedly halcyon days ended in the early- mid-1970s (and we must remember that large parts of the world did not benefit from the boom era) the pattern has been one of short-lived surges in economic growth punctuated by periods of down-turn or even recession. In other words, it has been a return to business as usual.

Within this uneven trajectory, however, there has been a substantial re-organization of the global geography of economic activity. Although the world economic system continues to be dominated by a small group of core economies, manufacturing production – and particularly trade – is no longer almost exclusively a core-region activity as it had been for the previous two hundred years. Although a handful of core economies still dominates international trade flows, the most spectacular growth rates – apart from that of Japan – have been achieved by the east and southeast Asian NIEs. Although foreign direct investment is still dominated by the core economies we are now seeing very substantial growth in such investment from some of the NIEs. Although most transnational investment originates from, and flows to, the core economies, the pattern of the investment flows has become increasingly complex. Today, there is a great deal of *interpenetration* of investment between national economies.

From the viewpoint of North America and western Europe, therefore, the empirical evidence provided in this chapter demonstrates that the centre of gravity of the world economic system has begun to shift towards the Pacific. A number of formerly peripheral countries in that region now challenge the core of the world's manufacturing system. Certainly, in terms of rates of growth, the east and southeast Asian NIEs have continued to perform substantially better than the developed market economies. Whereas the manufacturing production and trade of the developed market economies have grown at much reduced rates that of the East and South East Asian NIEs has continued at high levels. The fact remains, however, that the actual extent of global shifts in economic activity is extremely uneven. Only a small number of developing countries have experienced substantial economic growth; a good many are in deep financial difficulty whilst others are at, or even beyond, the margins of survival. Thus, although we can indeed think in terms of a new international division of labour, its extent is far more limited than is sometimes claimed.

What is clear is that the relatively simple international division of labour organized around the three components of core, semi-periphery and periphery no longer exists. It has been replaced by a far more complex, multiscale, structure. The global economy is not made up of a set of neatly packaged national boxes. It is much closer to what Storper and Walker (1984, p. 37) called 'a mosaic of unevenness in a continuous state of flux'. That mosaic is, however, made up of processes which operate – and are manifested – at different, but inter-related spatial scales. The constraints of the available data on production, trade and investment artificially restrict our vision to the national scale. But just because we cannot so easily see the other scales at which economic activity operates does not mean that they do not exist. Not only do they exist but they are, in many ways, a more significant scale for analysis.

Notes for further reading

1. The uniqueness of this pattern is clear when the longer period from 1870 to the present is considered. As Figure 1.1 in Kitson and Michie (1995) shows, the general experience was that world output invariably grew faster than world trade.

2. The nature of such long waves and, especially, the role of technological change, is discussed in Chapter 5. In the specific case of the downturn in economic activity of the mid-1970s, Webber and Rigby (1996, p. 11) argue that 'to the extent there has occurred a slowdown, it was inherent in the dynamics of the golden age. That is, the forms of technical change that sustained the real-wage rises of the 1950s and 1960s also diminished the profitability of production; our present disarray is caused by the continuing lack of profits in production . . . The end of the period of relatively rapid growth was the product not of mistakes of policy, nor of slower productivity growth, much less of intensified competition from imports, but of the fact that it was a period of rapid growth . . . to the extent there has occurred a slowdown, it was inherent in the dynamics of the golden age'. Hobsbawm (1994) takes a broadly similar view.

3. The *regulationist* approach to the explanation of the transition from one dominant form of economic organization to another originated in France during the 1970s. Not only has it spawned an immense literature but also there is considerable divergence between different regulationist 'schools'. 'Two key concepts are "regime of accumulation" and "mode of regulation". The regime of accumulation refers to a "set of regularities at the level of the whole economy, enabling a more or less coherent process of capital accumulation" . . . It includes norms pertaining to the organization of production and work (the labour process), relationships and forms of exchange between branches of the economy, common rules of industrial and commercial management, principles of income sharing between wages, profits and taxes, norms of consumption and patterns of demand in the marketplace, and other aspects of the macro-economy. The mode of regulation . . . refers to "the institutional ensemble (laws, agreements, etc) and the complex of cultural habits and norms which secures capitalist reproduction as such. It consists of a set of formal or informal 'rules' that codify the main social relationships" . . . It therefore refers to institutions and conventions which "regulate" and reproduce a given accumulation regime through application across a wide range of areas, including the law, state policy, political practices, industrial codes, governance philosophies, rules of negotiation and bargaining, cultures of consumption and social expectations' (Amin, 1994, p. 8). Within the regulationist interpretation, the period 1945 to the early 1970s is usually labelled *Fordism*: a system of intensive accumulation based upon the kinds of mass production of standardized products epitomized by the Ford automobile company and the mode of social regulation based upon Keynesian principles of macroeconomic management by the state. For excellent reviews of the *regulationist* approach, see Dunford (1990), Tickell and Peck (1992), Amin (1994).

4. Prior to 1989, it was necessary to identify a separate category within the industrialized country group – the centrally planned economies of the USSR and of central and eastern Europe. With the collapse of that political system all these economies are now engaged in the transition to market economies (see Chapter 3).

5. 'Direct' investment is defined as the investment by one firm in another with the intention of gaining a degree of control over that firm's operations. 'International' or 'foreign' direct investment is simply direct investment which occurs across national boundaries, that is, when a firm from one country buys a controlling investment in a firm in another country or where a firm sets up a branch or subsidiary operation in another country. 'Portfolio' investment refers to the situation in which firms purchase stocks/shares in other companies purely for financial reasons; that is, like any other investor, they build up portfolios of company shares. But, unlike direct investment, such investments are not made to gain control.

6. Julius (1990).

7. The term 'le défi Americain' was coined by the French writer, Jean-Jacques Servan-Schreiber in 1968. He made the infamous – but wrong – prediction that 'fifteen years from now it is quite possible that the world's third greatest industrial power just after the United States and Russia, will not be Europe, but American industry in Europe' (p. 3). With the relative decline of the United States and the rise of Japan, some have used a similar analogy regarding the growth of Japanese industry abroad.

8. Recent contributions to the relatively small, but growing, literature on FDI and TNCs from developing countries include Ramstetter (1993), Yeung (1994a; 1994b; 1995a; 1995b), Zhan (1995).

9. UNCTAD (1993b, Chapter III) contains a very useful analysis of sectoral trends in foreign direct investment. See also Daniels (1993, Chapter 3), for a discussion of services.
10. This issue will be addressed in Part IV. Stallings (1995, Chapter 11) provides a useful discussion.
11. Sklair (1989) provides a very thorough discussion of the *maquiladora* programme.

PART II

Processes of global shift

Traditional explanations and the need for a new approach

In Part I we explored some of the major empirical trends in today's rapidly evolving global economy: the global shifts in production and trade in goods and services and in international investment. These broad patterns represent some of the major indicators of the internationalization and globalization of economic activity which are the central focus of this book. Part I was entirely descriptive. But what are the major *processes* which create the internationalization and globalization of economic activities? How can we begin to explain the patterns? The following six chapters, which make up Part II of this book, address this question of *explanation*. But before plunging into the detailed arguments of these chapters, we need to be aware of some of the more important traditional explanations of international production, trade and investment which have been, and continue to be in some cases, influential.

Many of the roots of modern explanations of international trade and production are traceable back to the ideas of the 'classical' economists of the late eighteenth and early nineteenth centuries. The most influential figures were Adam Smith, whose *Inquiry into the Nature and Causes of the Wealth of Nations* was published in 1776, and David Ricardo, whose book *On the Principles of Political Economy and Taxation* appeared in 1817. Smith developed the concept of the *division of labour* as a key process in economic development. He also put forward the idea that an economy operates harmoniously without any overt control but through the guidance of the 'invisible hand' of the market. Ricardo introduced what is still probably the most fundamental concept not only in the theory of international trade but also in the trading policies of many nations: the *principle of comparative advantage*. The framework of classical economics, with its assumptions regarding the unfettered operation of markets and the nature of economic decision-making, also spawned *theories of the location of economic activity*. Indeed, the pioneer work in location theory (concerned with agriculture) was written by J.H. von Thünen in 1826, only a few years after Ricardo and very much in the same mould. From our viewpoint the most important of the early location theorists was Alfred Weber, whose work was published in Germany in 1909 and translated into English as the *Theory of the Location of Industries* in 1929. Although both trade theory and location theory have been greatly refined, extended and criticized over the years the influence of the pioneers on much current thinking remains strong, especially in the case of trade theory.

International trade theory

The most basic concept in the whole of international trade theory is the *principle of comparative advantage*, first introduced by Ricardo in 1817. This states that a country (or any geographical area) should specialize in producing and exporting those products in which it has a comparative, or relative cost, advantage compared with other countries and should import those goods in which it has a comparative disadvantage. Out of such specialization, it is argued, will accrue greater benefit for all.[1]

The principle of comparative advantage is at the heart of traditional attempts to explain geographical differences in production and trade. But, in itself, it says nothing about *why* such differences occur. The pioneering work in this respect was carried out by two Swedish economists: Eli Heckscher, writing in 1919, and Bertil Ohlin, whose *Interregional and International Trade*, published in 1933, refined and extended Heckscher's work. As a result their approach is referred to collectively as the Heckscher–Ohlin (H–O) theory of trade. As with the principle of comparative advantage the essence of the H–O theory is very simple. All products require a combination of different factors of production – natural resources (land and raw materials), a labour supply, capital in the form of money to buy the materials and machinery, technology and so on. Different products vary in the precise combination in which these factors are used. Some industries, for example clothing, use a great deal of labour and are labour-intensive industries. Others, such as chemicals, employ relatively few workers but a vast amount of capital equipment and can be described as capital intensive. Some industries occupy huge areas of land whilst others operate on tiny sites. Factors of production, of course, are very unevenly distributed geographically, particularly at an international scale. Some nations have an abundance of natural resources but a sparse population and, therefore, a limited labour force. Others have huge potential labour supplies but little capital, and so on.

The H–O approach was based upon such variations in the *factor endowments* of countries. Thus, a country (or region) will export those goods which use intensively the factors of production with which it is best endowed and import those goods which incorporate those factors of production with which it is poorly endowed. Although the H–O theory recognizes the existence of a number of production factors, for simplicity the theory was worked out using only two: capital and labour. In these terms, countries with an abundant supply of labour should export labour-intensive products while capital-abundant countries should export capital-intensive products.

There is a good deal of general substance in the H–O idea. Much of the spectacular growth of manufactured exports from the NIEs has been in those industries which are, indeed, labour intensive and, therefore, reflect that particular source of comparative advantage. But, as we shall show, it is not quite as simple as this in the real world.[2] The factor endowment theory of international trade attributes trade flows simply to differences in factor endowments. Yet much of world trade in manufactures, as we saw in Chapter 2, actually occurs between countries with broadly *similar* factor endowments (the developed market economies).

Four particular assumptions of traditional trade theory are especially important:

- *Mobility of factors of production.* Factors are assumed to be fixed and immobile geographically. An area has a particular endowment of production factors and this endowment forms the basis of the area's comparative advantage.
- *Transport costs.* Such costs are assumed to be zero. In other words, despite its explicit concern with trade between areas, trade theory, in common with much other economic theory, remains curiously spaceless.
- *Technology.* Technology is assumed to be a given and also geographically constant.

- *Economies of scale.* The assumption of perfect competition embodied in traditional trade theory denies the existence of economies of scale whether these are internal economies (reduction of unit costs with increasing size of a firm) or external economies (reduction of unit costs derived from the scale of an entire industry or from more general externalities).

In fact, none of these assumptions can be maintained in a study of the real world. Apart from land, factors of production are not geographically immobile and fixed in their location, although there are differences in the degree of mobility of different factors. In general, for example, capital is far more mobile geographically than labour, while skilled labour tends to be more mobile than unskilled labour. Such differential factor mobility is an important element in the shifting global pattern of economic activity. Geographical distance imposes a cost on movement, whether of materials, finished products, people or less tangible things such as knowledge and information. Each of these differs in its degree of sensitivity to geographical distance. Within manufacturing industry itself, some materials and products are extremely costly to move in relation to their value whilst others are, relatively speaking, cheap to move. Either way, any explanation of the geographical distribution of economic activity at the global scale must incorporate the role of distance and transport costs.

Likewise, technology is an enormously important influence on production (through its impact on the production function) and, therefore, on trade. Technology both originates and spreads/is adopted in a highly uneven geographical manner. Technological advantage is created, not given. Finally, economies of scale are exceptionally important in creating differential advantage. In particular, it is increasingly accepted that external economies based upon the clustering of activities are especially significant, both at the national and the subnational scales. Recent developments in so-called 'new' international trade theory explicitly incorporate technology and economies of scale and relate them to a drive for 'strategic' trade policy.

Classical location theory

Historically, it has been the location theorists (a generally maverick band of scholars anxious to break out of the 'spaceless' world of economic theory) who have explored some of these issues, especially those of factor mobility, transport costs and external economies of scale.[3] However, most location theorists have been more concerned with incorporating space into economic theory than with attempting to explain the actual location of economic activities. In addition, location theory is not as extensively developed as trade theory. But, as in the latter case, there are elements of location theory which are useful in helping us to understand the world about us. There are two broad bodies of location theory. One focuses on the *costs of production* as determining the location of industry; the other is concerned with the size and shape of a firm's *market area*. Attempts to integrate the two approaches have been largely unsuccessful although both approaches share a central interest in the role of geographical distance in shaping the location pattern of industry.[4]

The pioneer theorist of industrial location theory in general, and of the least-cost approach in particular, was Alfred Weber. Almost all subsequent work in industrial location theory has consisted of attempts to refine or elaborate upon Weber's approach. Weber's concern was to identify the optimal location for an individual firm (operating just one plant), given certain assumptions. On the basis of these assumptions, Weber envisaged the location of industry as being determined by two sets of factors:

- *primary or 'general' factors*: transport costs and labour costs
- *secondary or 'local' factors*: forces of agglomeration or deglomeration.

Weber regarded transport costs as the initial determinant of industrial location. The primary aim was to minimize transport costs incurred in gathering together the necessary materials and transporting the finished product to the market. The nature of the materials used – whether they were weight-losing or weight-gaining in the process of production, whether they were geographically localized in their distribution or available more or less everywhere – was a most important consideration. Heavy materials, or those whose weight was reduced in the process of manufacture, would tend to pull the location of production to the source of the material. Manufacturing processes which added to the weight or bulk of the materials would tend to be located at the market. Thus, Weber's basic argument was that a producer would locate primarily at the point of minimum total transport costs. But he recognized that the best location might well differ from this if other locational forces were more powerful.

The major cause of such deviation, he suggested, would be the existence of geographical differences in labour costs. However, any savings in labour costs would have to offset more than the additional transport costs which would inevitably be incurred in locating at other than the minimum-transport-cost point. Weber's emphasis on the importance of transport costs both for products and for factors of production added a significant dimension to the picture painted by the international trade theorists. At first sight it would appear that Weber's approach is most relevant for those economic activities in which transport costs are especially important. But this is a shrinking category as a result of technological changes in the transport media themselves and in the process of production (Chapter 5). As transport costs and raw material sources have come to exert less of a locational influence, Weber's emphasis on the locational attractions of labour locations has become increasingly relevant, particularly at the global scale.

Weber adopted a similar line of reasoning to determine the influence of his third location factor – agglomeration (external economies of scale). As we discussed in Chapter 1, the spatial concentration of producers in a single location may well generate additional economies, particularly where producers are linked together in a functional manner. For example, in some industries closeness to suppliers and customers may be desirable or even obligatory. More generally, a cluster of economic activities makes possible the provision of a variety of services which might not be feasible for a single, isolated firm. Such external economies might also include an appropriate pool of labour. In this respect, labour and agglomeration factors may coincide in creating a strong locational pull. But in Weber's analysis, the location of production would shift from the minimum-transport-cost point only if the savings at the agglomeration location were greater than the additional transport costs which would be incurred. In fact, this tends to downplay the significance of agglomeration or spatial clustering. It could well be argued that, today, such clustering is *the* most important single factor in helping to explain the geography of economic activity.

Early attempts to explain international direct investment

In comparison with theories of international trade and industrial location, early theoretical explanations of international direct investment were far less satisfactory.[5] For a considerable time, international direct investment was regarded as being simply one variant of international capital theory. In its most basic form this body of theory states that firms will place their investments in those locations where the financial return on the investment is greatest. There is, of course, much truth in this. But the early explanations of direct investment tended to equate such returns with geographical differences in interest rates on the financial investment. This was a reasonable assumption for portfolio investment but not for the kind of direct investment embodied in the transnational corporation. As Dunning (1973, p. 299) pointed out, 'unlike movements in portfolio capital . . . direct investment . . . involve[s] the transmission of other factor inputs than money capital, viz. entrepreneurship, technology and management expertise, and is as likely to be affected by the relative profitability of the use of these resources in different countries as that of money capital'. It was not until the early 1960s that a satisfactory body of theory concerned explicitly with foreign direct investment and the transnational corporation began to emerge (see Chapter 6).

The need for a new approach to explaining global shift

In unmodified form the traditional explanations of patterns of trade and of the location of production and investment are clearly inadequate to explain the complexities of the modern global economy. Nevertheless, they should not all be rejected out of hand. In particular, elements of both trade and location theory remain important to our understanding. Comparative advantage based upon different factor endowments *is* a most important element. Geographical distance reflected in movement costs *does* influence the spatial pattern of economic activities at the global scale. Differential labour costs and economies of spatial agglomeration *are* highly significant influences on global shifts in economic activity.

But although we should continue to utilize some of the insights of these bodies of theory, we have to recognize their limitations. For example, both trade and location theory have tended to be *static* whereas we live in a *volatile* and dynamic global system in which technological change is rapid and endemic. Both trade and location theories have tended to assume very simple economic-geographical relationships and very simple decision-making processes, whereas the real world is made up of complex interactions between firms of widely varying sizes, although increasingly dominated by transnational corporations. It is made up, too, of national governments and multinational political groupings each of which pursues its own, often widely differing, industrial and trading policies. These create important 'discontinuities' at national and regional boundaries and add greatly to the complexity of the global economic system. It is these organizational and institutional forces which are creating global shifts in economic activity and redrawing the global economic map. Comparative advantage is not simply 'given'; it is created and recreated by human action.

The major problem we face in explaining global shifts in economic activity and in trying to understand the processes of internationalization and globalization is that we are dealing with very complex and highly interconnected processes. The overall logic of Part II is that the globalization of economic activity is the outcome of four sets of processes:

- the strategies of firms, notably transnational corporations
- the strategies of states
- the complex and dynamic interaction between firms and states
- technological change.

In fact, there is no unique point of entry in trying to understand these processes; no unique sequence in which each of them should be discussed because they each form part of an inter-related whole. The six chapters in Part II attempt to capture both the individual nature of these processes and also the interactions between them. We choose to begin with the state (Chapters 3 and 4) as the traditional building block of the international economy. Contrary to much popular opinion, the state continues to play a highly significant role in the global economy, particularly as a container of distinctive institutional and cultural practices and as a regulator of economic transactions. We then examine the nature of technological change in Chapter 5, seeing it as a set of transformative processes which have greatly changed the context within which states pursue their national policies and within which business firms pursue their commercial strategies. Such firms – in the form of transnational corporations – are the focus of Chapters 6 and 7. Emphasis is placed particularly on the reasons for the growth of TNCs, the continuing significance of place to their operations, and on the web of networked relationships within which such firms operate their production chains, both organizationally and geographically. Finally, in Chapter 8 we address the question of the relationships – both conflictual and collaborative – between firms and states.

Notes for further reading

1. All the standard texts on international economics contain explanations of the principle of comparative advantage as do most texts on international business. See, for example, Winters (1985), Root (1990), Krugman and Obstfeld (1994).
2. See Dunning (1973) for a summary of criticisms of international trade theory.
3. Dicken and Lloyd (1990) provide a detailed account of location theories.
4. Several writers have pointed to the close link which exists between trade theory and location theory and urged their integration into a single body of theory. In 1933 Ohlin asserted that the theory of international trade is only a part of a general localisation theory. Following Ohlin's lead, some twenty years later Isard (1956, p. 207) stated that 'location and trade are as the two sides of the same coin'. He argued (p. 53) that 'location cannot be explained unless at the same time trade is accounted for and . . . trade cannot be explained without the simultaneous determination of locations'. Despite these and other pleas for integration between the two bodies of theory, however, relatively little progress has been made other than at the most general and abstract level.
5. General reviews of the early attempts to explain international direct investment are provided by Dunning (1973), Pitelis and Sugden (1991).

CHAPTER 3

'The state is dead . . . long live the state'

'Contested territory': the state in a globalizing economy[1]

For some three hundred years, from its emergence in the mid-seventeenth century, the state – in the specific form of the *nation-state* – was rightly regarded as the dominant actor in international economic relationships. The governance[2] of the international economy was synonymous with govern*ment* by nation-states as they exerted their legitimized control over their sovereign territories. Historically, the state was the primary regulator of its national economic system. The world economy, quite reasonably, could be conceptualized as a set of interlocking *national* economies. Trade and investment in the world economy were literally 'inter-national'.

The major theme of this book, of course, is that the world is changing as economic activities have become not just increasingly internationalized but are also globalizing. The degree of interdependence and interconnection within the world economy has increased dramatically. The same can be said of the international political system which can best be depicted as a 'web of interdependencies' as shown in Figure 3.1, involving national governments, international organizations (such as the United Nations, the International Monetary Fund, World Bank, World Trade Organization) and transnational organizations (non-governmental bodies, such as transnational corporations, international trades unions, environmental groups, welfare organizations and the like).

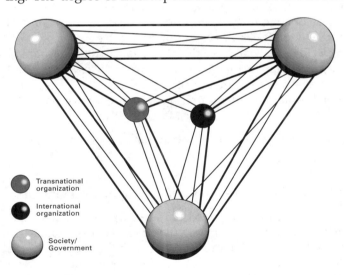

Transnational organization

International organization

Society/ Government

Figure 3.1 The web of interdependencies in the global political system
Source: McGrew (1992, Figure 1.7)

Such a framework emphasizes both the *permeability* of the nation-state and also the *polycentric* nature of the global political system, 'with states as merely one level in a complex system of overlapping and often competing agencies of governance' (Hirst and Thompson, 1996, p. 183). Nation-states thus exist within a world system of differential power relationships which operate at different geographical scales. Such a view of the changed position of the state in the international system – especially the economic system – has led some writers to question the significance of the nation-state as an important force in the world economy. Thirty years ago, Kindleberger (1969, p. 207) bluntly asserted that 'the nation state is just about through as an economic unit'. Boyer and Drache (1996, p. 1) claim that 'globalization is redefining the role of the nation-state as an effective manager of the national economy'. On the other hand, Porter's (1990, p. 19) opinion is that 'while globalization of competition might appear to make the nation less important, instead it seems to make it more so'. More explicitly in relation to Kindleberger's apocalyptic view, Wade (1996) observes, in Mark Twain style, that 'reports of the death of the national economy are greatly exaggerated'.

It is these latter views which underpin the discussion in this chapter rather than that of Kindleberger. While recognizing that the position of the nation-state is being redefined, we reject the view that it is no longer a major player. While some of the state's capabilities are, indeed, being reduced and while there may well be a process of 'hollowing out' of the national state, the process is not a simple one of uniform decline on all fronts.[3] The nation-state remains a most significant force in shaping the world economy. It has, whether explicitly or implicitly, played an extremely important role in the process of industrialization in all countries; *all* governments intervene to varying degrees in the operation of the market and, therefore, help to shape different parts of the global economic map.

The chapter is organized in the following sequence:

- First, we explore the major characteristics of the state as an economic actor, focusing on its roles as 'container' of distinctive institutions and practices, as 'regulator' of economic activities and, more contentiously, as a 'competition state'.
- Second, we examine three specific policy areas of direct relevance to the reshaping of the global economy: trade, foreign investment and industry policies.
- Third, we change the scale of analysis from that of the nation-state to the regional bloc, recognizing that regional economic integration has become a very powerful force in the global economy.

In Chapter 4 we illustrate these general aspects of state activity by looking at specific national cases.

States as containers, states as regulators, states as competitors

Today, there are some 170 nation-states in the world (Figure 3.2). The number has increased progressively in the past forty years, first with the process of decolonization by the former imperial powers of Europe and, more recently, with the breakup of the old USSR, which was replaced not only by a new Russian Federation but also by a number of newly independent states after 1989. The breakup of the former Yugoslavia also resulted in the emergence (or re-emergence) of new states in the early 1990s. The most important distinguishing characteristic of the nation-state is its *territorial* basis. The state's sovereignty over its territory is enshrined in international law. The state has legitimized power over what happens within its own sovereign territory. How that power is exercised, of course, varies according to the particular political system in place. But it is a maxim of international law that how a state organizes its internal affairs is entirely its own business as long as it does not attempt to extend its territorial

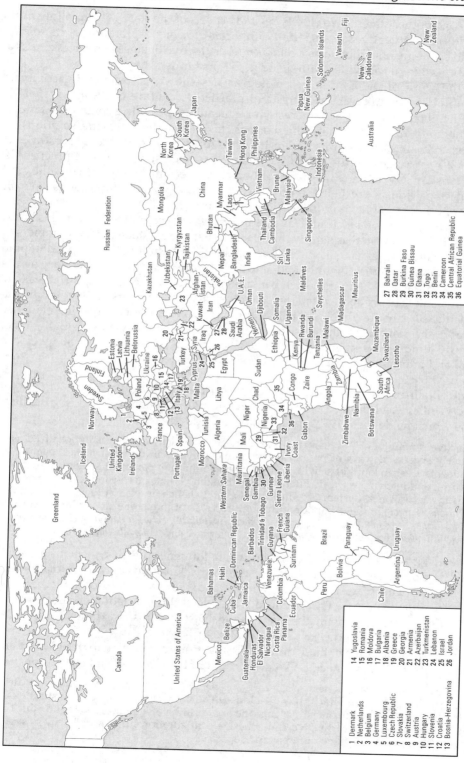

Figure 3.2 A world of nation states

influence over other states. Two closely related aspects of the state, which follow from this territorial prerogative, are especially important for our exploration of its role in the world economy: its function as a *container* of distinctive business practices and as a *regulator* of economic activities within and across its borders.[4]

As the work of the economic sociologists has shown, all economic activity is *embedded* in broader cultural structures and practices:[5] 'Culture is a learned, shared, compelling, interrelated set of symbols whose meanings provide a set of orientations for members of a society. These orientations, taken together, provide solutions to problems that all societies must solve if they are to remain viable' (Terpstra and David, 1991, p. 6). One of the primary containers of such structures and practices is the nation-state. Over time, 'the national homeland became a cultural container . . . where national ideals are being reproduced in schooling, the mass media and all manner of other social institutions' (Taylor, 1994, pp. 155, 156).

The whole question of 'culture' and its meanings is beyond the scope of this book. The point being made here is simply this: the fact that nation-states act as containers of distinctive cultures means that 'ways of doing things' – including how they attempt to *regulate* economic activities within and across their jurisdictions – tend to vary across national boundaries.[6] Of course, such containers are not water-tight; cultural leakage is common and is being accelerated by technological developments in the communications media (see Chapter 5). Nevertheless, there is a good deal of compelling evidence to support the notion of the persistence of national distinctiveness – though not necessarily uniqueness – in structures and practices which help to shape local, national and international patterns of economic activity.

National business systems; national competitive 'diamonds'
Richard Whitley is a leading advocate of the concept of nationally distinctive business systems which he defines (1992a, p. 13, emphasis added) as

> distinctive configurations of hierarchy-market relations which become institutionalized as relatively successful ways of organizing economic activities in different institutional environments. Certain kinds of activities are coordinated through particular sorts of authority structures and interconnected in different ways through various quasi-contractual arrangements in each business system . . . They develop and change in relation to *dominant social institutions*, especially those important during processes of industrialization. *The coherence and stability of these institutions, together with their dissimilarity between nation-states, determine the extent to which business systems are distinctive, integrated and nationally differentiated.*

The major social-instutional variables which influence business systems are summarized in Box 3.1. Whitley divides these into 'background' and 'proximate' variables, the particular assemblage of which tends to vary considerably between different national contexts. The extent to which such differentiation occurs, together with its effect in creating distinctive national business systems, is explored by Whitley in both the European and the east Asian contexts.

We will pick up some of these differences at various points both in this chapter and in later chapters. Here, we need only take account of the more general conclusions:

> The distinctiveness of business systems . . . depends on the integrated and separate nature of the contexts in which they developed. The more that major social institutions such as the political and financial systems, the organization of labour markets and educational institutions, form distinctive and cohesive configurations, the more business systems in those societies will be different and separate.
>
> (Whitley, 1992a, p. 14)

Proximate social institutions
- Business dependence on a strong and cohesive state
- State commitment to industrial development and risk-sharing
- A capital market or credit-based financial system
- A unitary or a dual education and training system
- The strength of skill-based labour unions
- The significance of publicly certified skills and professional expertise

Background social institutions
- The degree, and basis, of trust between non-kin
- The extent of commitment and loyalty to non-family collectivities
- The importance of individual identities, rights and commitments
- The extent of the depersonalization and formalization of authority relations
- The degree of differentiation of authority roles
- The reciprocity, distance and scope of authority relations

Source: Based on material in Whitley (1992b, pp. 20, 27).

Box 3.1   The social-institutional bases of business systems

Thus, while authority relations and structures vary significantly between Japan, South Korea, Taiwan and Hong Kong, those in most Western firms are typically derived from, and share a common reliance on, legal-rational norms and bases of legitimacy. Similarly, the capital-market-based financial system and reliance on 'professional' modes of skill development and organization which are shared by most 'Anglo-Saxon' societies, especially the USA and UK, ensure that business systems in these societies have certain characteristics in common, such as a strong finance function and preference for internalizing risk in the absence of close bank–firm connections, which are less apparent in many continental European business systems.

(Whitley, 1992b, pp. 36–37)

Thus, Whitley does not claim that there is a straightforward one-to-one relationship between a particular kind of business system and a particular nation-state but he does claim that there are clear national/cultural tendencies which tend to differentiate between individual, or groups of, countries. One of his most significant findings is that the east Asian economies are far from being homogeneous, a view at variance with much of the popular understanding of the 'Asian miracle'.

Recognizing that countries continue to differ as 'containers' and 'regulators' of distinctive structures and practices is important in emphasizing that we do not live in a homogenized world and, therefore, that we need to develop our theories accordingly. But there is a more instrumentalist interpretation of the significance of national differences: that they determine the extent to which a country is *economically competitive*. The principal advocate of this view is Michael Porter. The theme of his book *The Competitive Advantage of Nations* (1990, p. 19) is that

Competitive advantage is created and sustained through a highly localized process. Differences in national economic structures, values, cultures, institutions, and histories contribute profoundly to competitive success. The role of the home nation seems to be as strong as or stronger than ever. While globalization of competition might appear to make the nation less important, instead it seems to make it more so. With fewer impediments to trade to shelter uncompetitive domestic firms and industries, the home nation takes on growing significance because it is the source of the skills and technology that underpin competitive advantage.

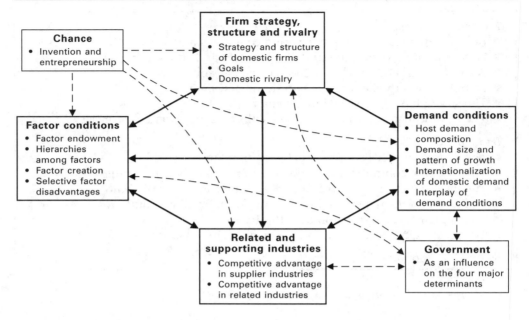

Figure 3.3 National competitive advantage: the Porter 'diamond'
Source: Based on material in Porter (1990, Chapter 3)

In other words, in Porter's view, the particular combination of conditions within nation-states has an enormous influence on the competitive strengths of the firms located there.

Porter conceives of this situation as a 'diamond': an interconnected system of four major determinants (Figure 3.3). Connecting all the components (other than 'chance') by double-headed arrows emphasizes the fact that 'the "diamond" is a mutually reinforcing system. The effect of one determinant is contingent on the state of others' (Porter, 1990, p. 72). As Figure 3.3 shows, each of the determinants can be broken down into several subcomponents:

- *Factor conditions* in a particular country are a combination of 'given' and 'created' factors:

> The factors most important to competitive advantage in most industries, especially the industries most vital to productivity growth in advanced economies, are not inherited but are *created within a nation, through processes that differ widely across nations and among industries . . . Thus, nations will be competitive where they possess unusually high quality institutional mechanisms for specialized factor creation.*
>
> (Porter, 1990, pp. 74, 80, emphasis added)

The most important include the levels of skills and knowledge of the country's population and the provision of sophisticated physical infrastructure, including transport and communications.

- *Demand conditions*, especially those in the home market, are regarded as being particularly important. 'The home market usually has a disproportionate impact on a firm's ability to perceive and interpret buyer needs . . . proximity to the right type of buyers . . . [is] . . . of decisive importance in national competitive advantage' (Porter, 1990, pp. 86, 87). However, the extent to which a nation's firms are connected into international markets also increases national competitiveness.

- *Related and supporting industries* which are internationally competitive constitute a third major determinant of national competitive advantage. 'Perhaps the most important benefit of home-based suppliers . . . is in the *process of innovation and upgrading*. Competitive advantage emerges from close working relationships between world-class suppliers and the industry' (Porter, 1990, p. 103). However, Porter (p. 107) stresses that 'the benefits of both home-based suppliers and related industries . . . depend on the rest of the "diamond". Without access to advanced factors, home demand conditions that signal appropriate directions of product change, or active rivalry, for example, proximity to world-class domestic suppliers may provide few advantages'.
- *Firm strategy, structure and rivalry*, Porter's fourth competitive determinant, relates quite closely to Whitley's ideas on business systems:

> The way in which firms are managed and choose to compete is affected by national circumstances . . . Some of the most important are attitudes toward authority, norms of interpersonal interaction, attitudes of workers toward management and vice versa, social norms of individualistic or group behaviour, and professional standards. These, in turn, grow out of the education system, social and religious history, family structures, and many other often intangible but unique national conditions.
>
> (Porter, 1990, p. 109)

Porter lays great emphasis on the importance of intense rivalry between domestic firms, arguing that this creates strong pressures on firms to innovate in both products and processes, to become more efficient and to become high-quality suppliers of goods and services:

> Vigorous local competition not only sharpens advantages at home but pressures domestic firms to sell abroad in order to grow . . . Toughened by domestic rivalry, the stronger domestic firms are equipped to succeed abroad. It is rare that a company can meet tough foreign rivals when it has faced no significant competition at home.
>
> (Porter, 1990, p. 119)

In addition to the four primary competitive determinants which together form his 'diamond', Porter attributes secondary importance to two other components:

- *The role of chance* – for example, the occasionally random occurrence of innovations or the 'historical accidents' which may create new entrepreneurs – are seen to be important 'because they create discontinuities that allow shifts in competitive position' (Porter, 1990, p. 124).
- *The role of government.* Porter explicitly refuses to regard government as a competitive determinant of the same order as the four primary determinants of his 'diamond'. While describing the various policies which governments might implement, he sees government as merely an 'influence' on his four determinants, a contingent, rather than a central, factor; a contributor to the environment in which, as in the biological realm, the selective survival of species occurs.

One of Porter's distinctive contributions to the discussion of the sources of national competitiveness is the importance he attributes to *localized geographical clustering*. This is a topic we discussed at some length in both Chapter 1 and Chapter 2. But Porter was the first writer within the business management field to give it a central position as a competitive influence. His treatment is conventional. It expresses, in his own rather distinctive terminology, the kinds of interdependencies outlined in Chapter 1. In Porter's (1990, pp. 154, 155, 157, emphasis added) words

> Competitors in many internationally successful industries, and often entire clusters of industries, are often located in a single town or region within a nation . . . Geographic concentration of firms in internationally successful industries often occurs because the influence of the individual determinants in the 'diamond' and their mutual reinforcement are heightened by close geographical proximity within a nation . . . *Proximity, then, elevates the separate influences in the 'diamond' into a true system.*

Porter's analysis of the competitive advantage of nations has attracted much attention. On the one hand its 'recipes' have been adopted by many governments, both national and local in their attempts to improve their competitive position. On the other hand, it has attracted considerable criticism:

- It is highly reductionist in compressing immense complexity into a simple four-pointed 'diamond'.
- Its underplaying of the role of the state in pursuit of national competitiveness is a significant omission.[7] *All* states perform a key role in the ways in which their economies operate, although they differ substantially in the specific measures they employ and in the precise ways in which such measures are combined.
- It neglects the influence of the transnationalization of business activity on national 'diamonds':

> there is ample evidence to suggest that the technological and organizational assets of TNCs may be influenced by the configuration of the diamonds of the foreign countries in which they produce and that this, in turn, may impinge upon the competitiveness of the resources and capabilities in their home countries.
>
> (Dunning, 1992, p. 142)

Dunning therefore suggests that the activity of TNCs should be incorporated as an additional exogenous variable affecting the diamond.

Do states compete?

Quite apart from the specific criticisms directed at the nature and form of Porter's competitive model there is the more general question to be addressed. Is it correct to think of nations as being in competition with each other just as firms compete with other firms? The generally accepted view among both policy-makers and academics has been in the affirmative. Books, government reports, newspaper articles, television programmes – especially in the United States, but also elsewhere – resound with the language and imagery of the competitive struggle between states for a bigger slice of the global economic pie. Much of this concern focuses on the perceived loss of economic standing by the United States and European countries *vis-à-vis* Japan and the east and southeast Asian NIEs.

Expressed in the simplest terms, if the goals of business organizations are to achieve maximum (or at least satisfactory) profits, one of the goals of nation-states is to maximize the material welfare of their societies. In an increasingly integrated and interdependent global economy, it is argued, nations are forced to *compete* with one another in a struggle to attain such goals. States compete to enhance their international trading position and to capture as large a share as possible of the gains from trade. They compete to attract productive investment to build up their national production base which, in turn, enhances their international competitive position. States, then, can be regarded as *competition states* in which they take on some of the characteristics of firms as they strive to develop their strategies to create competitive advantage.[8]

In stark contrast to this conventional view, Paul Krugman (1994) has described the claim that states are competition states as 'a dangerous obsession'. In a highly polemical and wide-ranging critique of both the concept and some of its proponents, notably in the United States, Krugman attacks both the empirical evidence and the policies based upon the competitiveness concept.[9] The main points of his argument are as follows:

- *Nations are not like firms*:

  > The bottom line for a corporation is literally its bottom line: if a corporation cannot afford to pay its workers, suppliers, and bondholders, it will go out of business . . . Countries, on the other hand, do not go out of business. They may be happy or unhappy with their economic performance, but they have no well-defined bottom line. As a result, the concept of national competitiveness is elusive.
  >
  > (Krugman, 1994, p. 31)

- *International trade is not a zero-sum game*:

  > Coke and Pepsi are almost purely rivals . . . if Pepsi is successful, it tends to be at Coke's expense. But the major industrial countries, while they sell products that compete with each other, are also each other's main export markets and each other's main suppliers of useful imports. If the European economy does well, it need not be at US expense.
  >
  > (Krugman, 1994, p. 34)

- *Empirical evidence does not support the concept*:

  > As a practical, empirical matter the major nations of the world are not to any significant degree in economic competition with each other. Of course, there is always a rivalry for status and power – countries that grow faster will see their political rank rise. So it is always interesting to *compare* countries. But asserting that Japanese growth diminishes US status is very different from saying that it reduces the US standard of living.
  >
  > (Krugman, 1994, p. 35)

Krugman takes a highly cynical view of the motives of the perpetrators of 'the competitive obsession' (it sells books; it shifts responsibility for domestic failures on to external forces; it is a useful device for political leaders, he claims). But there is more to his argument than this:

> Thinking and speaking in terms of competitiveness poses three real dangers. First, it could result in the wasteful spending of government money supposedly to enhance . . . [national] . . . competitiveness. Second, it could lead to protectionism and trade wars. Finally, and most important, it could result in bad public policy on a spectrum of important issues . . . To make a harsh but not entirely unjustified analogy, a government wedded to the ideology of competitiveness is as unlikely to make good economic policy as a government committed to creationism is to make good science policy . . . competitiveness is a meaningless word when applied to national economies. And the obsession with competitiveness is both wrong and dangerous.
>
> (Krugman, 1994, pp. 41, 43, 44)

Not surprisingly, given the forcefulness of Krugman's argument, there is considerable controversy surrounding it.[10] Some argue that making the specific comparison between states and firms is too narrow a perspective. Rapkin and Strand (1995, p. 4) assert that 'there is more to the concept of national competitiveness than implied by the analogy to corporations' and criticize Krugman for collapsing the 'multiple rhetorics of competitiveness' into a 'single voice' and for 'attribut[ing] to that voice a claim not made by most who use the competitiveness concept' (p. 8). In fact, the nuances of the debate between contested interpretations of 'competitiveness' are beyond the scope of this book. But whether Krugman is right or wrong in his analysis, there seems little likelihood of policy-makers actually heeding his warnings and refraining from both the rhetoric and the reality of competitive policy measures. As long as the concept of national (and local) competitiveness remains in currency then no single state is likely to opt out. This is the classic 'Prisoners' Dilemma' of game theory, where a choice which seems rational to the *individual* in his or her search for maximum advantage produces an outcome which is *collectively* less beneficial. It involves 'second guessing' what others are likely to do. In the context of international economic policy-making, although all states might benefit from co-operation a powerful incentive exists to try to gain at the expense of other states.

An ideal-type framework

All states perform a key role in the ways in which their economies operate, although they differ substantially in the specific measures they employ and in the precise ways in which such measures are combined. Although a high level of contingency may well be involved (no two states behave in exactly the same way), certain regularities in basic policy stance can be identified. These will reflect the kinds of cultural, social and political structures, institutions and practices in which it is embedded. The precise policy mix adopted by a state will be influenced by:

- *The nation's political and cultural complexion and the strength of institutions and interest groups.* In general, conservative governments are less inclined to pursue interventionist policies than liberal or socialist governments. However, much depends on the power of institutions and interest groups within the national economy: for example, business and financial interests, labour unions, environmental groups. The particular form of political structure – whether centralized or federal – will also be important. An especially significant factor, therefore, is the degree of consensus or conflict between institutions and interest groups, a situation in which the nation's history and culture will be significant influences.
- *The size of the national economy, especially that of the domestic market.* This is especially relevant for trade policies: the larger the domestic market the less important relatively is external trade likely to be and *vice versa*. However, much will depend upon:
- *the nation's resource endowment, both physical and human.* A weak natural resource endowment will necessitate the import of essential materials which, in turn, must be paid for by exports of other, usually manufactured, products. A strong endowment of particular types of raw material may well influence the kind of economic policy pursued. Similarly, the size, composition and skill level of a nation's potential labour force will constrain the kinds of policy which may be pursued.
- *The nation's relative position in the world economy, including its level of economic development and degree of industrialization.* A nation's 'degrees of freedom' depend very much on its relative position in the world economy and, in particular, on the extent of its dependence on external trade and investment. The kinds of industry and trade policies pursued will also tend to reflect the nation's degree of industrialization. The policy emphasis of established industrial nations will differ from that of nations in the process of industrializing.

Figure 3.4 outlines one 'ideal-type' framework for classifying national economic-political systems.[11] As always, such simplified typologies need to carry a 'health

Market-ideological state	Plan-ideological state
Driven by the 'new right' economic and social policies of the 1980s. Based on a reversion to the state-civil society relations of the epoch of competitive capitalism. Policy choices based on ideological dogma.	The state owns and controls most or all economic units. Resource allocation/investment decisions are primarily a state function. State controls redistribution of wealth/income. Policy choices based on ideological dogma.
Market-rational/regulatory state	**Plan-rational/developmental state**
The state regulates the parameters within which private companies operate. The state regulates the economy in general but investment, production and distributive decisions are the preserve of private companies, whose actions are disciplined by the market. The state does not concern itself with what specific industries should exist and does not have an explicit industry policy.	The state regulation of economic activity is supplemented by state direction of the economy. The economy itself is largely in private ownership and firms are in competition, but the state intervenes in the context of an explicit set of national economic and social goals. High priority placed on industry policy and on promoting a structure that enhances the nation's economic competitiveness.

Figure 3.4 A typology of national economic-political systems
Source: Based on material in Johnson (1982); Henderson and Appelbaum (1992, Figure 1); Henderson (1993)

warning': 'These four constructs (market-rational, plan-rational, market-ideological, and plan-ideological) should be regarded as ideal types; actual existing political economies combine them in various historically contingent ways. Still, for any particular society, one will typically dominate, facilitating an overall characterization of its prevailing political economy' (Henderson and Appelbaum, 1992, p. 20). The four cells in Figure 3.4 should not be seen as representing a static situation. For example, the number of countries which can be placed in the *plan-ideological* category, based upon state owner-ship and control together with highly centralized central planning, is now very much smaller than it was before the collapse of the Soviet-led system at the end of the 1980s. On the other hand, there was certainly a resurgence of the *market-ideological* state during the 1980s, led by the 'new right' governments in the United Kingdom and the United States. Whether these two countries still remain in that rather extreme category or whether they have reverted to the market-rational mode is a matter for debate. Never-theless, the category should be retained as a possible ideal type.

However, it is the two bottom cells of Figure 3.4 which reflect the two dominant political economy models at the present time: the *market-rational (regulatory)* state and the *plan-rational (developmental)* state. Johnson (1982, pp. 19–20) describes the dif-ferences between these two models as follows:

A regulatory, or market-rational, state concerns itself with the forms and procedures – the rules, if you will – of economic competition, but it does not concern itself with substantive matters. For example, the United States government has many regulations concerning the antitrust implications of the size of firms, but it does not concern itself with what indus-tries ought to exist and what industries are no longer needed. The developmental, or plan rational, state, by contrast has as its dominant features precisely the setting of such substantive social and economic goals . . .

The government will give greatest precedence to industrial policy, that is, to a concern with the structure of domestic industry and with promoting the structure that enhances the nation's international competitiveness. The very existence of an industrial policy implies a strategic, or goal-oriented, approach to the economy. On the other hand, the market rational state usually will not even have an industrial policy (or, at any rate, will not recognize it as such). Instead, both its domestic and foreign economic policy, including its trade policy, will stress rules and reciprocal concessions (although perhaps influenced by some goals that are not industrially specific, goals such as price stability or full employment).

Market-rational perspective
- To maintain economic stability
- To provide physical infrastructure, especially those with high fixed costs (e.g. harbours, railways, airports)
- To supply 'public goods' (e.g. defence and national security, education, legal system)
- To contribute to the development of institutions for improving markets for labour, finance, technology
- To offset or eliminate price distortions in cases of demonstrable market failure
- To redistribute income to the poorest sufficient to meet basic needs

Plan-rational perspective
- To place top priority on economic development (defined in terms of growth, productivity, competitiveness, rather than in terms of welfare)
- To sustain private property and the market
- To guide the market with instruments formulated by an élite economic bureaucracy, led by a pilot agency or 'economic general staff'
- To create institutions to facilitate extensive consultation and co-ordination with the private sector. Such consultations constitute an essential part of the process of policy formulation and implementation
- To maintain a separation between a 'ruling' bureaucracy and 'reigning' politicians in order to sustain the autonomy of the state within a stable political environment. Associated with a 'soft authoritarianism', often linked with a virtual monopoly of political power in a single political party or institution over a long period of time

Source: Based on material in Wade (1990a, pp. 11, 25–26).

Box 3.2   Contrasting views of the 'proper' economic functions of government in market-rational and plan-rational systems

The essence of each of these two governance models is their differing conception of the 'proper' role of government in regulating the economy. Box 3.2 summarizes the major features of the market-rational view of the 'proper' economic functions of government (based upon neoclassical concepts) and of the developmental state view (based on the east Asian NIEs). Allowing for the 'health warning' about the use of simple ideal-type models it is not unreasonable to regard many of the current political-economic tensions in the world economy as a conflict between states occupying different positions within the market-rational/plan-rational space of Figure 3.4 and Box 3.2.

Trade, foreign investment and industry policies

> there is hardly a single policy issue without its international dimension.
>
> (Knight, in Carter, 1981, p. viii)

At one extreme, the macroeconomic policies pursued by governments to control domestic demand or to manage the money supply have extremely important implications for the distribution and redistribution of economic activity. At a more basic level, governments are generally the providers of the physical infrastructure of national economies – roads, railways, airports, seaports, telecommunications systems – without which private sector enterprises, whether domestic or transnational, could not operate. They are the providers, too, of the human infrastructure: in particular of an educated labour force as well as of sets of laws and regulations within which enterprises must operate. Between these two extremes of government involvement in the workings of the economy lie those policies whose explicit purpose is to influence the level, composition and distribution of production and trade. It is these policies – concerned with trade, foreign investment and industry – which form the substance of this section.

National governments have at their disposal an extensive kit of regulatory tools with which to control and to stimulate economic activity and investment within their own boundaries and to shape the composition and flow of trade and investment at the international scale. Although often viewed separately, trade policies, policies towards foreign investment and industry policies overlap to a very considerable degree. The boundaries between them may be blurred. Each may reinforce – or counteract – the other. They may be employed as part of a deliberate, cohesive, all-embracing national economic strategy as in the plan-rational/developmental state or, alternatively, indi-vidual policy measures may be implemented in an *ad hoc* fashion with little attempt at co-ordination as in the market-rational/regulatory state model.

Trade policies

Of all the measures used by nation-states to regulate their international economic position, policies towards trade have the longest history. The shape of the emerging world economy of the seventeenth and eighteenth centuries was greatly influenced by the mercantilist policies[12] of the leading European nations. The successful challenge to Britain's industrial supremacy in the late nineteenth century by the United States and Germany was based on a strongly protectionist trade policy by these nations. The deep world recession of the 1930s was also characterized by national retreat behind trade barriers. Since the late 1970s a new wave of protectionism – *neomercantilism* – has swept the world economy as governments have attempted to 'manage' trade in a variety of ways. Indeed, a whole new area of strategic trade policy has emerged in recent years, particularly in the United States, as we shall see later.

Major types of trade policy

Figure 3.5 summarizes the major types of trade policy pursued by national govern-ments. In general, policies towards imports are restrictive whereas policies towards exports, with one or two exceptions, are stimulatory. Policies on imports fall into two distinct categories: tariffs and non-tariff barriers (NTBs):

- *Tariffs* are, essentially, taxes levied on the value of imports which increase the price to the domestic consumer and make imported goods less competitive (in price terms) than otherwise they would be.

In general, the tariff level rises with the stage of processing. Tariffs tend to be lowest on basic raw materials and highest on finished goods, as Table 3.1 shows. The purpose of such 'tariff escalation' is to protect domestic manufacturing industry while allow-ing for the import of industrial raw materials. Thus, although tariffs may be regarded simply as one means of raising revenue, their major use has been to protect domestic

Policies towards imports
1. Tariffs
2. Non-tariff barriers
Import quotas (e.g. 'voluntary export restraint', 'orderly marketing agreements')
Import licences
Import deposit schemes
Import surcharges
Rules of origin
Anti-dumping measures
Special labelling and packaging regulations
Health and safety regulations
Customs procedures and documentation requirements
Subsidies to domestic producers of import-competing goods
Countervailing duties on subsidized imports
Local content requirements
Government contracts awarded only to domestic producers
Exchange rate manipulation

Policies towards exports
Financial and fiscal incentives to export producers
Export credits and guarantees
Setting of export targets
Operation of overseas export promotion agencies
Establishment of Export Processing Zones and/or Free Trade Zones
'Voluntary export restraint'
Embargo on strategic exports
Exchange rate manipulation

Figure 3.5   Major types of trade policy

industries: either 'infant' industries in their early delicate stages of development or 'geriatric' industries struggling to survive in the face of external competition. Largely through the successive rounds of international negotiation in GATT (see next section), the general level of tariffs in the world economy has declined very substantially. Figure 3.6 shows that in 1940 the average tariff on manufactured products was around 40 per cent; today the average tariff is around 4 per cent.

Table 3.1   The changes in import tariffs on developing country products

	Tariff (%)	
	Mid-1980s	Post-1995*
All industrial products (excluding oil)		
Raw materials	2.1	0.8
Semi-manufactured products	5.3	2.8
Finished products	9.1	6.2
Natural-resource-based products		
Raw materials	3.1	2.0
Semi-manufactured products	3.5	2.0
Finished products	7.9	5.9

Note:
*Based on the commitments given in the Uruguay Round.
Source: Based on material in GATT (1994, *The Results of the Uruguay Round: Market Access for Goods and Services*).

- *Non-tariff barriers (NTBs)*. While tariffs are based on the value of imported products, non-tariff barriers are more varied: some are quantitative, some are technical.

Although, in general, tariffs have continued to decline, the period since the mid-1970s has seen a marked increase in the use of non-tariff barriers. Indeed, today NTBs are

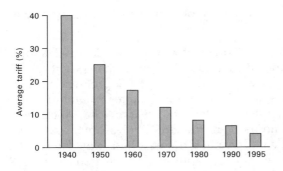

Figure 3.6 Reduction in tariffs, 1940–95

probably more important than tariffs in influencing the level and composition of trade between nation-states. Certainly much of what has been termed the 'new protectionism' consists of the increased use of NTBs. As Figure 3.5 shows, such trade barriers can take a variety of forms. The most important type of NTB is the import quota, a limit on the quantity of a particular product which may be imported. In some cases, quotas are established as part of so-called 'orderly marketing agreements' or 'voluntary export restraints'. Such devices have become extremely important in each of the industry cases described in Part III of this book. In addition to quotas, importers may be required to seek import licences, to pay deposits or be subject to import surcharges. 'Rules of origin' requirements may be imposed in order to restrict imports from specific countries and to encourage local production. Companies suspected of setting their prices in overseas markets at levels below those in their domestic markets may be subjected to anti-dumping penalties.

A second group of NTBs consists of various regulatory and bureaucratic devices such as special labelling or packaging regulations, health and safety requirements, customs procedures and documentation. A third category overlaps closely with some of the industry and foreign investment policies discussed later. For example, domestic producers may be subsidized to compete more effectively against imports; firms may have to comply with specific local content requirements; government contracts may be confined to domestic producers. There is no doubt that the greatly increased operation of NTBs has led to intensified tensions and stresses in the world trading system.

Figure 3.5 also lists the major export policies which may be pursued. Again, some of these may overlap with industry and foreign investment policies. In particular, financial and fiscal advantages may be granted to exporting firms. In addition, specific tax and tariff concessions may be applied: export earnings may be tax-free or taxed at a lower rate, tariffs may be waived or reduced on those imports which are essential for export activities. Governments may operate an export credit guarantee scheme, fix export targets and operate export promotion offices overseas. Particular geographical areas may be set aside as export-processing or free-trade zones. As we shall see, these have proliferated rapidly in many developing countries in recent years.

Most export policies are, of course, aimed at stimulating a nation's exports but there are some instances where policies are adopted which restrict exports. One example is the kind of 'orderly marketing arrangement' mentioned earlier, whereby the exporting nation 'voluntarily' restricts its exports of a good to the other nation (often the term 'voluntary' is hardly appropriate). Another example of export restriction occurs in the case of strategically or militarily sensitive items. Many countries operate such selective export embargoes. Finally, one measure common to both import and export policy is the manipulation of the nation's currency exchange rate. Devaluation of the currency makes exports cheaper and imports more expensive and *vice versa*. However, the ability of a government to manipulate its exchange rate in a controlled way in a world of floating and volatile currencies is very limited.

Bhagwati (1988) divides this proliferating array of non-tariff barriers into two broad types:

- *'High track' restraints*, which bypass the GATT rule of law. These are the 'visibly and politically negotiated' restraints negotiated by trading partners: the voluntary export restraints and orderly marketing arrangements which have proliferated in a variety of industries, including textiles and clothing (see Chapter 9), automobiles (Chapter 10), consumer electronics (Chapter 11), steel, footwear, machine tools, etc.
- *'Low track' restraints* which 'capture' and 'pervert' the GATT rules. Such measures include anti-dumping provisions and countervailing duties. These play 'legitimate roles in a free trade regime . . . but not if they are captured and misused as protectionist instruments' (Bhagwati, 1988, p. 48).

It has been estimated that NTBs affect more than a quarter of all industrialized country imports and are even more extensively used by developing countries. Kostecki (1987) identified 113 major voluntary export restraint agreements in operation in the late 1980s, excluding the most comprehensive case of the Multi-Fibre Arrangement in textiles and clothing. Of the 113 cases, almost 90 per cent protected the EC, US and Canadian markets and almost two-thirds of the arrangements involved exports from Asia, primarily Japan and South Korea. Currently there is a whole series of smouldering trade disputes in a whole range of industries. Some of these disputes will be discussed in some detail in Chapter 4 and in the case study chapters of Part III.

The international trade regulatory framework: from GATT to WTO

Trade policy is unique in that it is set within an *international* institutional framework. The General Agreement on Tariffs and Trade (GATT) was one of the three international institutions formed in the aftermath of the Second World War which constituted the framework of the postwar world economy. Initially there were 23 signatories; in mid-1996, 123 nations belonged fully to the World Trade Organization (WTO), the successor to the GATT established in January 1995 whilst another 27 states are seeking to join (including China). More than 90 per cent of all world trade is now covered by the WTO framework.[13]

The GATT was in fact introduced as a temporary framework in 1947 in anticipation of the establishment of a fully fledged international trade organization (ITO). However, because of disagreements between leading economic powers, this never happened. Between 1947 and 1994, when the Uruguay Round was finally concluded, there were eight rounds of multilateral trade negotiations.

Table 3.2 summarizes the major characteristics and outcomes of each of the rounds. In some, the outcomes were relatively modest, as was the number of participating states, especially in the first few rounds. However, between the Dillon Round of 1960–61 and the Kennedy Round of 1963–67 some major changes occurred. The number of participating countries increased from 39 to 74; the size and extent of tariff reductions agreed in the Kennedy Round was unprecedentedly large; it was the first to broaden the agenda beyond tariffs (introducing an anti-dumping code). It also led subsequently to the formal inclusion within the GATT of preferential treatment in favour of the developing countries. Prior to the mid-1960s, GATT was most concerned with trade between the developed nations. As a result, widespread dissatisfaction emerged among the developing countries with the state of world trade in respect of

Table 3.2 The evolution of the GATT/WTO international trade framework

Round	Countries	Major outcomes
Geneva Round (1947)	23 countries	Concessions on 43 tariff lines
Annecy Round (1949)	29 countries	Modest tariff reductions
Torquay Round (1950–51)	32 countries	8,700 tariff concessions
Geneva Round (1955–56)	33 countries	Modest tariff reductions
Dillon Round (1960–61)	39 countries	Tariff reductions following formation of EEC 4,400 tariff concessions exchanged
Kennedy Round (1963–67)	74 countries	Average tariff reduction of 35 per cent by developed countries Some 30,000 tariff lines bound Agreement on anti-dumping and customs valuation Moves to incorporate preferential treatment for developing countries
Tokyo Round (1973–79)	99 countries	Average tariff reduction of one-third by developed countries Codes of conduct established for interested GATT members on specific non-tariff measures
Uruguay Round (1986–94)	103 countries (1986) 117 by end 1993 124 by early 1995	Average tariff reduction of one-third by developed countries Agriculture, textiles and clothing brought into the GATT Creation of the WTO Agreements on services (GATS), intellectual property (TRIPs), trade-related investment (TRIMs) Most Tokyo Round codes enhanced and made part of GATT-1994, i.e. apply to all members of the WTO

Source: Based on Hoekman and Kostecki (1995, Table 1.2).

their own commodities and products. This led to the establishment, in 1964, of UNCTAD, whose major role was to promote the trading interests of the developing nations. A particularly sensitive issue was the access of developing country exports to developed country markets. Pressure led, in 1965, to the adoption within GATT of a generalized system of preferences (GSP) under which exports of manufactured and semi-manufactured goods from developing countries would be granted preferential access to developed country markets. In fact, there were a number of exclusions from the GSP, of which one of the most important was textiles. As we shall see in Chapter 9, trade in textiles is strongly affected by special trading agreements. Nevertheless, the GSP did mark a major shift in international trade policy, particularly in manufacturing industry.

The Tokyo Round of the 1970s involved almost 100 countries and achieved a reduction in tariffs comparable to the previous Kennedy Round. It also created a series of 'codes' which dealt with specific issues. However, it was the Uruguay Round, started in 1986 and eventually concluded in 1994 after a series of delays and near breakdowns, which constituted the most ambitious and wide ranging of all the GATT rounds. For the first time, the Uruguay Round addressed a number of additional trade issues. It brought agriculture, textiles and clothing into the GATT, while special agreements were concluded in services (GATS – the General Agreement on Trade in Services), intellectual property (TRIPs – Trade-Related Aspects of Intellectual Property Rights) and trade-related investment (TRIMs – Trade-Related Investment

Measures). There was a further large reduction in overall tariff levels. The major organizational change was the creation of a new World Trade Organization (WTO) which came into being in January 1995, almost fifty years after the original proposal to create an international trade organization foundered. The WTO incorporates the GATT itself together with the new areas of responsibilty.

The WTO represents a *rule-oriented* approach to multilateral trade co-operation rather than one which is based upon results (e.g. market share, volume of trade, etc.).[14] 'Rule-oriented approaches focus not on outcomes, but on the rules of the game, and involve agreements on the level of trade barriers that are permitted as well as attempts to establish the general conditions of competition facing foreign producers in export markets' (Hoekman and Kostecki, 1995, p. 24). The fundamental basis of the WTO (and the GATT) is that of *non-discrimination* This has two components. First, the 'most-favoured nation principle' (MFN) states that a trade concession negotiated between two countries must also apply to all other countries; that is all must be treated in the same way. The MFN is 'one of the pillars of the GATT' and 'applies unconditionally, the only major exception being if a subset of Members form a free-trade area or a customs union or grant preferential access to developing countries' (Hoekman and Kostecki, 1995, p. 26). Secondly, the 'national treatment rule' requires that imported foreign goods are treated in the same way as domestic goods. This is also unconditional in the GATT but not in the GATS (see Chapter 12).

Thus, national trade policies have to be seen within this international institutional context, although this does not necessarily mean that nations invariably abide by the rules of the WTO. What it does mean is that there are international pressures on national governments in their pursuit of trade policies which are not present in the other policy areas discussed below. On the other hand, the boundaries between trade policies and industry and foreign investment policies have become increasingly blurred, as we shall see, so that it becomes increasingly difficult to set trade policies apart from the other policy areas. There are inevitably wide divergences of opinion over the WTO and the extent to which it is likely to achieve the objectives set out in the Uruguary Round. Several of the areas of agreement will take a number of years to implement and there are major obstacles to such implementation. In general, the developing countries complain that the developed countries are dragging their feet in carrying out agreements and *vice versa*. There is fierce disagreement over whether the WTO should incorporate rules relating to labour standards and the environment (as desired by some developed countries). More broadly, there is continuing debate over the whole issue of so-called 'free trade'. We shall return to these topics in Part IV of the book.

Foreign investment policies

In a world of transnational corporations and of complex flows of investment at the international scale, national governments have a clear vested interest in the effects of foreign investment, whether positive or negative. From a national viewpoint, foreign investment is of two types: *outward* investment by domestic enterprises and *inward* investment by foreign enterprises. Few national governments operate a totally closed policy towards foreign investment although the degree of openness varies considerably. In general, the developed market economies tend to be less restrictive in their policies towards foreign investment than the developing market economies. One obvious reason is the fact that the developed economies are the major sources of, as

---

Policies relating to inward investment by foreign firms
- Government screening of investment proposals
- Exclusion of foreign firms from certain sectors or restriction on the extent of foreign involvement permitted
- Restriction on the degree of foreign ownership of domestic enterprises
- Insistence on involvement of local personnel in managerial positions
- Compliance with national codes of business conduct (including information disclosure)

- Insistence on a certain level of local content in the firm's activities
- Insistence on a minimum level of exports
- Requirements relating to the transfer of technology
- Locational restrictions on foreign investment

- Restrictions on the remittance of profits and/or capital abroad
- Level and methods of taxing profits of foreign firms

- Direct encouragement of foreign investment: competitive bidding via overseas promotional agencies and investment incentives

---

Policies relating to outward investment by domestic firms
- Necessity for government approval of overseas investment projects
- Restrictions on the export of capital (e.g. exchange control regulations)

Figure 3.7 Major types of foreign direct investment policy

well as the dominant destinations for, the world's foreign investment. There are, of course, some exceptions to this general pattern.

Figure 3.7 summarizes the major types of national policy towards foreign investment. Most national policies are concerned with inward investment although governments may well place restrictions on the export of capital for investment (for example, through the operation of exchange control regulations) or insist that proposed overseas investments be approved before they can take place. A far more extensive battery of policies exists in the case of inward investment. The inward investment policies listed in Figure 3.7 fall into four broad categories:

- The first category relates to the *entry* of foreign firms into a national economy. Governments may operate a 'screening' mechanism to attempt to filter out those investments which do not meet national objectives, either economic or political. Foreign firms may in fact be excluded entirely from particularly sensitive sectors of the economy or the degree of foreign penetration in a sector may be limited to a certain percentage share. More generally, there may be a restriction on the extent to which individual firms may be owned or controlled by a foreign enterprise. Government may insist that only joint ventures involving indigenous capital may be permitted, possibly on a 51:49 per cent basis in favour of domestic firms. Another possibility is for government to require foreign firms to employ local personnel in managerial positions. Generally speaking, of course, foreign firms must comply with prevailing national codes of business conduct – to be good corporate citizens – including those relating to the disclosure of information about the firm's activities. This latter point is frequently a major bone of contention between TNCs and national governments.

- The second category of policies shown in Figure 3.7 relates to the *operations* of foreign firms. A particularly common requirement is for such firms to meet a certain level of 'local content' in their manufacturing activities. Such a requirement is designed to increase the positive effects of a foreign investment on indigenous suppliers and to reduce the level of imported materials and components. Conversely, a government may insist on a foreign firm exporting a specified proportion of its output. One of the major elements in the foreign investment 'package' is that of technology. As we shall see in later chapters, there is much dispute about the extent to which TNCs do, in fact, transfer technology beyond their own corporate boundaries. Governments may wish to stimulate such technological diffusion through restrictive or stimulative measures.

- The third set of policies relates to government attitudes towards corporate *profits and the transfer of capital*. All governments are concerned to minimize the outflow of capital; on the other hand, TNCs invariably wish to remit at least part of their profits and capital abroad. Similarly, TNCs aim to minimize their liability to taxation; national governments wish to maximize their tax yield. Hence, variations in the restriction on the remittance of capital and profits, together with variations in the level and methods of taxing TNC profits, are extremely important.

- The final category of policies towards foreign investment shown in Figure 3.7 aims to *stimulate* inward investment. Indeed, an increasingly common feature of today's world economy – of developed as well as of developing economies – is the scramble to attract foreign investment. Competitive bidding via overseas promotional agencies and investment incentives has become endemic throughout the world (see Chapter 8). The important point about such international (and inter-regional) competition is that for certain types of investment it is truly global in extent. For the cost-oriented transnational investments, for example, countries and localities in Europe or North America may well be in direct competition with locations in Asia and Latin America.

Historically, there were very large differences in the policy positions adopted by countries towards inward direct investment. At the broadest level, it could be said that developed countries had a more liberal attitude towards inward investment than developing countries, although there were exceptions within each broad group. For example, among developed countries France had a much more restrictive stance than most other European countries. Among developing countries, Singapore had a particularly open policy, far more so than most other Asian countries. In the past two decades, however, national FDI policies have tended to converge in the direction of liberalization:

> In many cases, this liberalization of foreign investment policies has been part of broader, market-oriented reforms of economic policy and has proceeded in parallel with trade liberalization, deregulation and privatization.
>
> The recent trend to more open investment policies has been particularly evident in the removal or relaxation of regulatory barriers to the entry of FDI . . . There has also been a shift away from the imposition of performance requirements and a liberalization of regulations concerning the transfer of funds. In addition, there has been increasing acceptance of standards of non-discriminatory treatment of foreign investors and of international standards on matters such as compensation in case of expropriation . . . At the same time, however, there are several qualifications to this liberalization trend. First, the trend has not been homogeneous and significant differences between foreign investment regimes persist. Second, virtually all countries maintain some restrictions, often of a

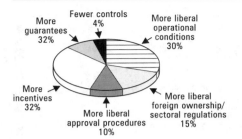

	1991	1992	1993	1994	1995
Number of countries that introduced changes in their investment regimes	35	43	57	49	64
Number of changes	82	79	102	110	112
Of which: In the direction of control	2	0	1	2	6
In the direction of liberalization or promotion	80	79	101	108	106

Figure 3.8 Changes in national regulation towards inward direct investment
Source: Based on UNCTAD (1996b, Table V.1, Figure V.1)

sectoral nature, on the entry of foreign investment. In this connection, an issue that has attracted attention is the existence of reciprocity requirements with regard to the entry and treatment of foreign investment.

(WTO, 1996, pp. 61–62)

Although national differences still exist, therefore, they are now rather less stark than in the past. Figure 3.8 summarizes the major regulatory changes towards FDI between 1991 and 1995.

In the case of foreign direct investment there is no international body comparable to the WTO which provides a set of rules for international trade. However, the Uruguay Round negotiations of the GATT included efforts to agree on a set of 'trade-related investment measures' (TRIMs). In effect some of the industrialized countries, led by the United States, wished to prohibit or restrict a number of the measures listed in Figure 3.7, notably local content rules, export performance requirements and the like. The advocates of TRIMs argued that such measures restrict or distort trade. The opponents, including many developing countries, saw such measures as essential elements of their economic development strategies. They, in turn, wished to see a tightening of the regulations against the restrictive business practices of transnational corporations. As a result, the TRIMs agreement is 'not surprisingly, a compromise. It explicitly affirms that GATT disciplines . . . apply to investment policies in so far as this directly affects trade flows . . . Thus, TRIMs that violate the GATT's national treatment rule or its prohibition on the use of QRs [quantitative restrictions] are banned' (Hoekman and Kostecki, 1995, p. 121). The TRIMs agreement is due to be reviewed within five years.

In fact, there is quite a long history of attempts to introduce an international framework relating to FDI and TNCs (apart from those agreed bilaterally or within the context of regional trade blocs). Examples include: the OECD Guidelines for Multinational Enterprises (first introduced in 1976); the International Labour Organization Tripartite Declaration of Principles Concerning Multinational Enterprises and Social Policy (1977); the United Nations Code of Conduct for Transnational Corporations initiated in 1982).[15] Despite all these initiatives, however, relatively little progress has been made in creating an international regulatory framework for direct investment.

Industry policies
National policies towards trade or foreign investment are explicitly concerned with international or crossborder issues and, therefore, are most obviously relevant to our

Investment incentives:
  Capital-related
  Tax-related
	State ownership of production assets

	Merger and competition policies

Labour policies:
  Subsidies
  Training
	Company legislation

	Taxation policies

State procurement policies	Labour regulation:
	  Labour union legislation
	  Immigration policies

Technology policies	
Small firm policies	National technical and product standards
Policies to encourage industrial restructuring	Environmental regulations
Policies to promote investment	Health and safety regulations

**Some or all of these policies may be applied
either generally or, more commonly, selectively.
Selectivity may be based on several criteria:**

1. Particular sectors of industry, e.g.
(a) to bolster declining industries
(b) to stimulate new industries
(c) to preserve key strategic industries

2. Particular types of firm, e.g.
(a) to encourage entrepreneurship and new firm formation
(b) to attract foreign firms
(c) to help domestic firms against foreign competition
(d) to encourage firms in import-substituting or export activities

3. Particular geographical areas, e.g.
(a) economically depressed areas
(b) areas of 'growth potential' or new settlement

Figure 3.9   Major types of industry policy

interest in the internationalization of economic activity. But there is a third policy area
– industry policy – which, although essentially concerned with internal issues, also
has broader international implications. Indeed, it is becoming increasingly apparent
that the boundaries between trade policy, foreign investment policy and industry
policy are often extremely blurred. Figure 3.9 lists the major types of regulatory
industry policies which may be used by national governments.

The most obvious stimulatory measures are the various financial and fiscal incentives governments may offer to private sector firms. As Figure 3.9 shows, the financial
measures most commonly used fall into two categories. On the one hand, governments may provide capital grants or loans to firms to supply part or all the investment
required for a particular productive venture. The other major financial, or rather fiscal,
incentive employed by governments is that of tax concessions. Under this banner a
whole variety of measures may be employed. For example, firms may be permitted to
depreciate or write down their capital investment against tax at an accelerated rate;
they may be granted tax reductions or even tax exemptions. Such concessions may be
for a specified period: the so-called 'tax holiday'.

Governments may also use various types of employment and labour policy to help
shape and encourage industrial activity within their boundaries. Most governments are

concerned to stimulate employment and, therefore, to reduce unemployment. In pursuit of such aims, firms may be encouraged to increase their labour force by direct subsidy. Training may be paid for – or even provided in – government establishments to provide a labour force with appropriate skills.

National governments tend to be the largest individual customers in any economy for an enormous variety of goods and services (including, of course, defence). Thus government policies of procurement are extremely important. The award of large government contracts or, conversely, their withdrawal or cancellation, may make or break a private sector enterprise and have enormous employment repercussions.

The rapid and far-reaching technological developments (to be discussed in Chapter 5) have led most governments to try to stimulate research and development in key sectors and to encourage technological collaboration between firms. The perceived need to stimulate entrepreneurial activity has produced a whole battery of policies to encourage small and medium-sized enterprises. Governments may also attempt to restructure firms – and even entire industries – to improve their international competitiveness.

Regulation of national industrial activity can also take a variety of other forms, as the right hand box of Figure 3.9 suggests. State ownership of productive assets is present in many countries although a current trend in many market economies is towards increased privatization. Entry into particular sectors may be discouraged through the operation of merger and competition policies although, again, there is a current trend towards the *deregulation* of certain industries such as telecommunications and financial services. Company legislation in general is designed to regulate the ways in which companies can be formed and operate and may be reinforced by specific taxation policies. Regulation of the labour market may be pursued through the encouragement or discouragement of labour union activity or through immigration policies. Technical and product standards are usually defined in specific national terms as are health and safety regulations and environmental regulations.

As Figure 3.9 suggests, the various stimulatory and regulatory policies may be applied *generally* across the whole of a nation's industries or they may be applied *selectively*. Such selectivity may take a number of forms: particular sectors of industry, particular types of firms (including, for example, the efforts to attract foreign firms), particular geographical locations.

International economic integration: regional economic blocs[16]

The general picture

Throughout this chapter we have focused on the nation-state. However, within the last few decades a significant additional element has been added to the politics of the world economic system: the emergence of regional economic blocs. The fundamental basis of such blocs is that of *trade*. Short of complete political union, we can identify four types of politically negotiated regional economic arrangements of increasing degrees of integration:

- *The free-trade area* in which trade restrictions between member states are removed by agreement but where member states retain their individual trade policies towards non-member states.

- *The customs union* in which member states operate both a free-trade arrangement with each other and also establish a common external trade policy (tariffs and non-tariff barriers) towards non-members.
- *The common market* in which not only are trade barriers between member states removed and a common external trade policy adopted but also the free movement of factors of production (capital, labour, etc.) between member states is permitted.
- *The economic union* is the highest form of regional economic integration short of full-scale political union. In an economic union, not only are internal trade barriers removed, a common external tariff operated and free factor movements permitted but also broader economic policies are harmonized and subject to supranational control.

	Free-Trade Area	Customs Union	Common Market	Economic Union
Removal of trade restrictions between member states	✓	✓	✓	✓
Common external trade policy towards non-members		✓	✓	✓
Free movement of factors of production between member states			✓	✓
Harmonization of economic policies under supranational control				✓

Figure 3.10   Types of regional economic integration

As Figure 3.10 shows, the progression is cumulative; each successive stage of integration incorporates elements of the previous stage, together with the additional element which defines that particular stage.

The number of regional trading arrangements has grown dramatically. Between 1948 and 1994, 109 regional trading arrangements (many of them bilateral in nature) were notified to the GATT. Although some of these were initially established (at least in embryonic form) between thirty and forty years ago there was a marked upsurge of such arrangements during the late 1980s and early 1990s. No fewer than 33 regional trading agreements were notified to GATT between 1990 and 1994 alone. Some of the major developments are shown in Box 3.3. Other, more tentative, regional trading arrangements are at least on the agenda for discussion. The United States has ambitions for a free-trade area encompassing the whole of the Americas (from Anchorage to Tierra del Fuego) while it has even been proposed that there might be a Transatlantic Free Trade Agreement – a 'TAFTA' – between NAFTA and the EU.

The vast majority of the regional economic groupings fall into the first two categories of the classification shown above (the free-trade area and and the customs union). There is a small number of common market arrangements but only one group – the European Union – which comes close to being a true economic union. In fact, not only is there enormous variation in the scale, nature and effectiveness of these regional trade groupings but also there is, in some cases, a considerable overlap of membership of different groups, especially in Latin America. Such diversity must be borne in mind when considering the likely geographical effects of regional integration both internally (on member states and communities) and externally (on the rest of the world economy). Table 3.3 shows the major regional agreements in force in the mid-1990s.

Regional trading blocs are essentially *discriminatory* in nature. As such, they go against the general principle of non-discrimination established by the GATT. However, provision for such integration was incorporated into Article XXIV of the GATT, which allowed for the creation of free-trade areas and customs unions, subject to certain provisos. Most regional blocs have a strongly defensive character; they represent

In Europe
- In the 1980s, membership of the European Community expanded further from nine to twelve
- In 1991, an agreement was reached with all but one of the members of the European Free Trade Association (EFTA) to form a European Economic Area (EEA)
- The Treaty on European Union was signed at Maastricht in December 1991
- The Single European Community Market legislation came into being in January 1993
- In 1995, the European Union expanded its membership from twelve to fifteen
- In 1995, free-trade agreements were implemented between the European Union (EU) and the three Baltic republics of Estonia, Latvia and Lithuania

In North America
- In 1989, the United States and Canada implemented the Canada–United States Free Trade Agreement
- In 1994, the North American Free Trade Agreement (NAFTA) came into being, whereby Mexico joined Canada and the United States in the first example of a developing country becoming fully integrated in a free-trade area with highly developed countries

In Latin America
- In 1995, the MERCOSUR customs union between Argentina, Brazil, Paraguay and Uruguay came into being
- Further development of the Andean Pact (involving Bolivia, Colombia, Ecuador, Peru and Venezuela) occurred
- A free-trade area involving Mexico, Colombia and Venezuela came into force

In the Pacific Basin, where, apart from AFTA, there is no formal regional trade bloc
- The eighteen member states of the Asia-Pacific Economic Forum agreed to remove all regional trade barriers by 2020
- Vietnam became the seventh member of AFTA

Source: Press reports.

Box 3.3   Acceleration of regional trading arrangements

an attempt to gain advantages of size in trade by creating large markets for their producers and protecting them, at least in part, from outside competition. Consequently, the most important of the regional blocs – particularly the European Union and NAFTA – have a very considerable influence on patterns of world trade. The classic analysis of the trade effects of regional blocs (specifically of customs unions) identifies two opposing outcomes:

- *Trade diversion* occurs where, as the result of regional bloc formation, trade with a former trading partner (now outside the bloc) is replaced by trade with a partner inside the bloc.
- *Trade creation* occurs where, as the result of regional bloc formation, trade replaces home production or where there is increased trade associated with economic growth in the bloc.

Table 3.3   Major regional economic blocs

Name	Membership	Date	Type
ANCOM (Andean Common Market)	Bolivia, Colombia, Ecuador, Peru, Venezuela	1969 (revived 1990)	*Customs union*
AFTA (ASEAN Free Trade Agreement)	Brunei Darussalam, Indonesia, Malaysia, Philippines, Singapore, Thailand, Vietnam	1967 (ASEAN) 1992 (AFTA)	*Free-trade area* – to be achieved by 2003
CARICOM (Caribbean Community)	Antigua & Barbuda, Bahamas, Barbados, Belize, San Cristobal, Dominica, Grenada, Guyana, Jamaica, Montserrat, St Kitts & Nevis, St Lucia, St Vincent & the Grenadines, Trinidad & Tobago	1973	*Common market*
EFTA (European Free Trade Association)	Iceland, Norway, Lichtenstein, Switzerland	1960	*Free-trade area*
EU (European Union)	Austria, Belgium, Denmark, France, Finland, Germany, Greece, Ireland, Italy, Luxembourg, The Netherlands, Portugal, Spain, Sweden, United Kingdom	1957 (European Common Market) 1992 (European Union)	*Common market* (aims to be *economic union* by late 1990s)
MERCOSUR (Southern Cone Common Market)	Argentina, Brazil, Paraguay, Uruguay (free-trade agreement with Bolivia, Chile)	1991	*Common market*
NAFTA (North American Free Trade Agreement)	Canada, Mexico, United States	1994	*Free-trade area*

In addition, regional trading blocs also have a major influence on flows of *investment* by transnational corporations, an issue we will explore more fully in later chapters. The effects of regional integration on direct investment, like that on trade, can also be conceptualized in terms of 'creation' and 'diversion'. In the latter case, the removal of internal trade (and other) barriers may lead firms to realign their organizational structures and value-adding activities to reflect a regional rather than a strictly national market. This, by definition, 'diverts' investment from some locations in favour of others.

Integration within the global triad: EU, NAFTA, APEC
In Chapter 2 we observed the strong tendency for a disproportionate share of global production, trade and investment to be concentrated in three 'mega-regions' – the so-called global triad of North America, Europe and east and southeast Asia (see Figures 2.24 and 2.25). Such concentration reflects, first and foremost, the basic economic-geographical processes we discussed in Chapter 1: the preference for proximity to markets and suppliers and a general tendency to 'followership' in location decision-making. The question we pose here is: how do *political* processes of economic integration map on to the three triad regions? Are all three regions going down the same path of closer formal integration as opposed to the 'natural' integration processes stimulated by geographical proximity? Given the widely held view that this is, indeed, the case it is instructive to find that the form of

regional economic integration is not only very different in each of the triad regions but also that there is no strong reason to believe that they are likely to go down the same path in the future. Each is the creature of a specific set of historico-geographical circumstances. To support this view let us look briefly at each of the three regions in turn.

The European Union

The EU (Figure 3.11) is by far the most highly developed and structurally complex of all the world's regional economic blocs. Although initially established as a six-member European *Economic* Community by the Treaty of Rome in 1957 it was always more than simply an 'economic' institution. The initial stimulus was the desire to bring together France and Germany in such a way that traditional emnities could no longer find their outlet in another round of European wars and also to strengthen western Europe in the face of the perceived Soviet threat. The EEC developed out of an earlier, more modest development: the European Coal and Steel Community (ECSC) which explicitly tied together the French and German coal and steel industries as part of the broader political agenda.

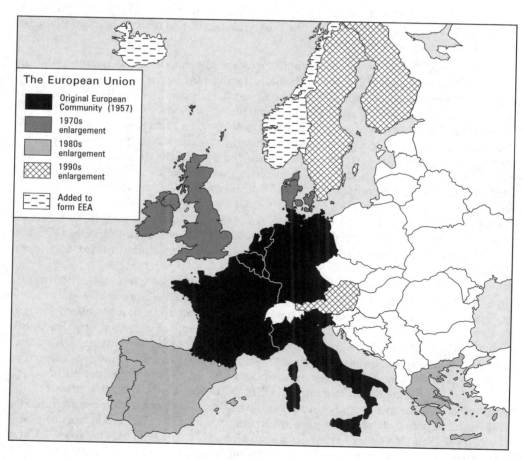

Figure 3.11 The European Union

From the initiation of the EEC in 1957, we can distinguish four major stages of development:

- *1958–68.* The elimination of customs duties between the six founder member states (Belgium, France, West Germany, Italy, Luxembourg, The Netherlands) and, in 1968, the introduction of a common external tariff.
- *1973–86* (a) The enlargement of the community with the accession of Denmark, Ireland and the United Kingdom in the 1970s and of Greece, Portugal, Spain in the 1980s. (b) The establishment of preferential trading agreements with EFTA countries; with countries around the Mediterranean rim; with countries in Africa, the Caribbean and the Pacific (the so-called ACP nations – all former European colonies) under the Lomé Convention.[17]
- *1986–92* (a) A renewed attempt to complete the Single European Market. By the mid-1980s it had become apparent that a single market did not actually exist. (b) A drive by the 'core' members of the EC to move towards full economic and monetary union. The signing of the Treaty on European Union at Maastricht in December 1991 created the European Union.
- *1992–present* (a) The further enlargement of the EU to fifteen with the accession of Austria, Finland and Sweden. (b) The attempts to implement fully both the Single European Market and the Maastricht Treaty in the context of both internal divisions and the pressures to enlarge the EU.

The stimulus to complete the Single European Market was the argument by the European Commission that the multiplicity of internal barriers to trade and factor movement which still remained was weakening the community's ability to operate in highly competitive global markets. Despite the provisions of the Treaty of Rome, individual EC countries were increasingly resorting to tactics which prevented or delayed the import of certain products from other member nations: for example, the insistence on particularly stringent health and safety checks on products, the heavy bureaucracy of customs and frontier procedures, even the awarding of government contracts to domestic producers and the contravention of EC regulations on the subsidizing of domestic industries. Each individual member of the EC was guilty of some such practices although some were more guilty than others. The commission argued that the 'costs of "non-Europe"' amounted to a significant loss of potential GDP and of jobs as well as a lessened ability to compete with the United States and Japan (Cecchini, 1988).

The Single European Act proposed the removal of three major sets of barriers – physical, technical and fiscal – together with the liberalization of financial services, the opening of public procurement and other measures. Such internal liberalization and deregulation, it was argued, would create a virtuous circle of growth for the European Community as a whole, its member states and for those business firms which successfully take advantage of the changes. The virtuous growth scenario is shown in Figure 3.12. Such a scenario was based upon a host of assumptions, not all of which were especially realistic. In fact, it is well-nigh impossible to measure precisely the actual effects of the single market process. The European Commission claimed that by 1996 some 90 per cent of the legislation to complete the single market was in place. The commission also claimed that European Union GDP in 1994 was between 1.1 and 1.5 per cent higher than it would have been without the single market (the Cecchini prediction was an increase of 4.5 per cent over five years) and that between 300,000

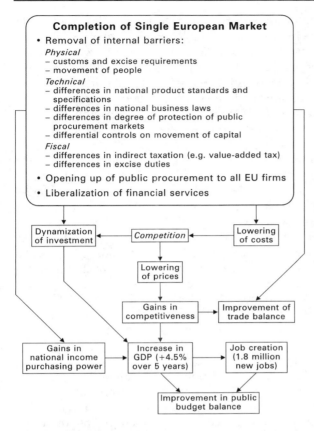

Completion of Single European Market

- Removal of internal barriers:

  Physical
  - customs and excise requirements
  - movement of people

  Technical
  - differences in national product standards and specifications
  - differences in national business laws
  - differences in degree of protection of public procurement markets
  - differential controls on movement of capital

  Fiscal
  - differences in indirect taxation (e.g. value-added tax)
  - differences in excise duties

- Opening up of public procurement to all EU firms

- Liberalization of financial services

Figure 3.12 The virtuous circle of growth: predicted outcomes of the completion of the Single European Market
Source: Based, in part, on Cecchini (1988, Chart 10.1)

and 900,000 people in work would have been unemployed without the single market (the Cecchini prediction was the creation of 1.8 million new jobs).[18]

However, even if such an optimistic outcome were to occur for the EC as a whole, there is bound to be great *geographical unevenness* in those outcomes. There would be winners and there would be losers at both national and subnational levels. Consequently, the Single European Act also included a social dimension – a reinforcement of existing EC social and regional programmes:

> The SEA required that the resources devoted to the EC's Structural Funds (designed to stimulate employment, reduce regional disparities in welfare and modernise agriculture) be increased substantially; consequently they were doubled over the period from 1988 to 1993 in parallel with the elimination of economic barriers . . . [a particular concern was the possibility of 'social dumping'] . . . This concept . . . encapsulates the fear that free movement of goods, services, capital and labour within a single market will lead to ruthless rivalry between companies, countries and regions . . . All this led to calls for a minimum (not common) level of social protection throughout the Community . . . in 1989, all member states but Britain adopted the 'Community Charter of the Fundamental Social Rights of Workers', the so-called 'Social Charter' . . . But the conversion of this ambitious statement of social intent into concrete action has lagged far behind the measures adopted to construct the Single Market.
>
> (Wise, 1994, p. 93)

The Treaty on European Union (TEU), signed at Maastricht in late 1991, marked a much more ambitious political agenda aimed at creating a fully fledged economic union. In particular it

- strengthened social provisions by a) incorporating the Social Charter; b) the enlargement of the EC Structural Funds; and c) the creation of a new Cohesion Fund to assist poorer areas of the union;
- set out the mechanisms for the creation of a single European currency and monetary system. This is supposed to be implemented by 1999 provided that member states are able to meet the strict convergence criteria stipulated in the Maastricht Treaty.

The issue of the EMU crystallizes some of the most difficult political problems within the EU, notably the sensitive issue of national sovereignty. The pros and cons of a single European currency are finely balanced. The major benefit would be the reduced costs and uncertainties associated with having to deal with fifteen separate currencies within a single market. Set against this is the fear that an individual state's ability to use financial and fiscal mechanisms to deal with periodic economic crises would be greatly reduced. In fact, it is unlikely that all member states will be able to meet the convergence criteria within the proposed timescale, leading to the probability of the creation of 'insiders' and 'outsiders'.

The other major issue is that of the possible enlargement of the EU. Several Mediterranean countries, such as Turkey, Malta and Cyprus, have long-standing applications to become members. They have been joined, more recently, by the 'new' market economies of central and eastern Europe. The community has undergone several phases of enlargement: from six, to nine, to twelve and to fifteen member states. Any further enlargement would be difficult under any circumstances; it will be doubly difficult given the scale and complexity of the current internal agenda. A particular concern is that the nature of the applicant states would place very severe strains on the EU's budget, quite apart from making the process of decision-making even more complex. Thus, one view is that a 'variable geometry' or 'variable speed' European Union is likely to emerge with a core of states, fully integrated economically and financially, surrounded by various groups of countries with different degrees of integration.

The North American Free Trade Agreement

The NAFTA is, without doubt, a highly significant political-economic development. By integrating two highly developed and one large developing country into a single free-trade area it changes the economic map of North America quite radically. The income gap between the United States and Mexico is very much greater than that between the richest and poorest states within the EU. The NAFTA came into force in 1994, but its origins can be traced back into the 1980s. One important building block, although this was not its intent, was the Canada–United States Free Trade Agreement (CUSFTA) signed in 1988 and implemented in 1989. Hitherto, the United States had not taken advantage of the GATT article allowing for exceptions to the principle of non-discrimination in the case of free-trade areas and customs unions.

In part, this change of stance reflected the United States' desire to have a 'lever' to achieve more open trade in the world economy by showing that it was prepared to make discriminatory agreements. In other words, it was part of the shift in US trade policy discussed in Chapter 4. Other United States motives included the desire to reduce trade barriers with Canada, to improve the treatment of US direct investment in Canada and to achieve specific deals on government procurement policies, trade in services and other areas. To the Canadian government a free-trade agreement offered the prospect of obtaining more secure access to the US market with less risk of being inhibited by various non-tariff barriers, of securing an effective dispute settlement procedure and of protecting its cultural industries by agreement. As the CUSFTA was being signed, two other developments were also occurring. President George Bush had made freer trade with Mexico a campaign issue in 1988. At the same time, President Carlos Salinas of Mexico made clear his determination to negotiate a free-trade area with the United States. Within a short time of bilateral talks starting, Canada had joined in an obvious defensive response.

The arguments in favour of creating the NAFTA varied between the three parties. For the United States, it formed part of its long-term objective of ensuring stable economic and political development in the Western Hemisphere and also gave access to Mexican raw materials (especially oil), markets and low-cost labour. It also promised further US leverage in a world of increasing regional integration. The Canadian government was anxious to consolidate the recent CUSFTA. The motives of the Mexican government were primarily to help to lock in the economic reforms of the previous few years, to create a magnet for inward investment, not only from the United States but also from Europe and Asia, and to secure access to the United States and Canadian markets.

It is interesting to note the different kinds of debate which surrounded the NAFTA compared with that surrounding the CUSFTA:

> Notably, although the Canada–US agreement sparked a heated debate in Canada, it was barely noticed in the United States. The United States has almost three times as much trade with Canada as with Mexico, yet the similarities between Canadian and US institutions made the prospect of free trade with Canada a fairly routine issue for the general US electorate. In Canada, however, there were more concerns about system differences, namely that the United States represented a threat to the Canadian welfare system.
>   . . . The NAFTA discussion in the United States was much more politically salient, and the debate was highly charged. Its opponents argued that the NAFTA meant a fundamental threat to US domestic institutions, especially ones affecting the environment and labour standards . . . NAFTA opponents in the United States also argued that NAFTA would have major consequences for US employment. Ross Perot made the memorable prediction that NAFTA would give rise to 'a giant sucking sound' as jobs left the United States for Mexico. In fact, much of the popular debate over NAFTA focused on its employment impact rather than its effects on welfare.
>
> (Lawrence, 1996, pp. 72–73)

A similar fear was expressed in Canada. One politician stated that 'this whole concept is a nightmare of US continentalists come true: Canada's resources, Mexico's labour, and US capital' (quoted in McConnell and MacPherson, 1994, p. 179). In Mexico, the fear was expressed that the country would become even more dominated by the United States.

The NAFTA took effect in 1994. The Clinton administration, which took office in 1993, used the 'fast-track' route to get the agreement through the US Congress. The main provisions of the NAFTA are summarized in Box 3.4. Note that, in addition to the various trade provisions, two 'side agreements' (on the environment and on labour standards) were incorporated to meet US and Canadian concerns. However, in contrast to the EU, there are no social provisions.

The aims of the NAFTA are gradually to eliminate most trade and investment restrictions between the three countries over a ten to fifteen-year period. The possibility of other countries joining the NAFTA is left open to negotiation but it is important to stress that the NAFTA is not a customs union. It does not incorporate a common external trade policy. Each of the three NAFTA members is free to make free-trade agreements with other states outside the NAFTA (as Mexico is in the process of doing). In sum, the NAFTA is light years removed from the far more advanced European Union and there is not the remotest possibility that the NAFTA will evolve into anything approaching the EU. Each is a reflection of its specific circumstances.

General provisions
- Tariffs reduced over a ten to fifteen-year period, depending on the sector
- Investment restrictions lifted (except for oil in Mexico; cultural industries in Canada; airline and radio communications in the United States)
- Immigration is not covered, with the exception that movement of some white-collar workers is to be eased
- Any member state can leave the agreement with six months' notice
- The agreement allows for the inclusions of any additional country
- Government procurement to be opened up over ten years
- Dispute resolution panels of independent arbitrators to resolve disagreements
- Some 'snap-back' tariffs allowed if surge in imports hurts a domestic industry

Sector-specific provisions
- *Agriculture*: Most tariffs between the United States and Mexico removed immediately. Tariffs on 6 per cent of products – corn, sugar, some fruits and vegetables – fully eliminated after fifteen years. For Canada, existing agreement with the United States applies
- *Automobiles*: Tariffs removed over ten years. Mexico's quotas on imports lifted over same period. Cars eventually to meet 62.5 per cent local content rule in order to be free of tariffs
- *Energy*: Mexican ban on private sector exploration continues, but procurement by state oil company opened up to the United States and Canada
- *Financial services*: Mexico gradually to open up financial sector to US and Canadian investment. Barriers to be eliminated by 2007
- *Textiles*: Agreement eliminates Mexican, US and Canadian tariffs over ten years. Clothes eligible for tariff breaks to be sewn with fabric woven in North America
- *Trucking*: North American trucks can be driven anywhere in the three countries by year 2000

Side agreements
- *Environment*: The three countries liable to fines, and Mexico and the United States sanctions, if a panel finds repeated pattern of not enforcing environmental laws
- *Labour*: Countries liable for penalties for non-enforcement of child, minimum wage and health and safety laws

Other arrangements
- The United States and Mexico to set up a North American Development Bank to help finance the clean-up of the United States–Mexico border
- The United States to spend roughly $90 million in the first eighteen months retraining workers losing their jobs because of the agreement

Source: Based on material in *The Financial Times* (17 November 1993).

Box 3.4 Major provisions of the North American Free Trade Agreement

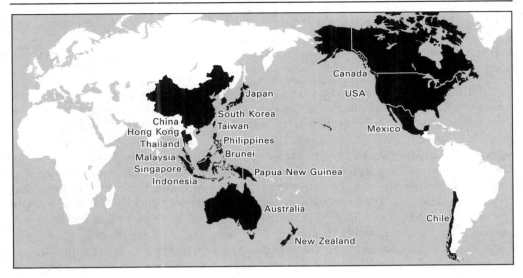

Figure 3.13   The Asia-Pacific Economic Co-operation Forum (APEC)

The Asia-Pacific Economic Co-operation Forum

Two of the three global triad regions now have politically based economic integration arrangements, though they differ greatly in the depth of that integration. No such regional arrangement exists in east and southeast Asia other than AFTA (established as ASEAN by six southeast Asian countries in 1967, and recently committed to free trade between its members by early in the twenty-first century). On the other hand, there have been various attempts over the past thirty years to develop inter-nation co-operation within what has become the world's fastest growing economic region.[19] Today, the major focus is on the Asia-Pacific Economic Co-operation Forum (APEC) which was established in 1989 on the initiative of the Australian government. Figure 3.13 shows the eighteen-member composition of APEC. The membership of APEC, as Figure 3.13 shows, is extremely diverse, involving not only the obvious east and southeast Asian states themselves (including China and Taiwan) but also Australia and New Zealand on the one hand and the United States, Canada and Mexico on the other. Not surprisingly, the individual agendas of particular members vary considerably:[20]

- The United States wishes both to signal its commitment to Asia and to strengthen its own economic and political interests there. It does not wish to concede Asia to Japan and to find itself confined to the Western Hemisphere.
- Japan is keen to develop further its relationships with ASEAN countries but does not wish to alienate the United States (still its most important ally despite economic tensions between the two countries). But Japan wishes to avoid the danger of a protectionist North American trade bloc and thus prefers to have the NAFTA group as part of APEC.
- The smaller Asian countries are fearful of dominance by any one of the three biggest economies in Pacific Asia: Japan, the United States, China. Yet they also wish to see all three involved in the region.
- However, Malaysia is concerned to preserve 'Asianness' and has proposed the establishment of an East Asia Economic Caucus (EAEC) which would exclude the North American countries, Australia and New Zealand.

- China is keen to become more widely accepted in the international economic system, particularly given its failure to be admitted to the WTO.
- Australia, which initiated the whole concept of APEC, sees it as legitimizing its role as a Pacific economy.

Since 1993, APEC has held annual summits of its national leaders to set the future direction of the organization. In 1994, at the Indonesian summit, it was agreed that the member states would commit themselves to the achievement of free trade and investment by the year 2020 at the latest. The advanced countries agreed to meet that objective by 2010. However, such liberalization is to be conducted in an open, multilateral manner and not in an inward-looking manner. 'APEC was promoted by several nations who are deeply committed to the multilateral trading system and concerned about the possibilities of a world divided into blocks that discriminate against outsiders. Thus, while APEC is itself a regional arrangement, it has the paradoxical mission of combating (preferential) regionalism' (Lawrence, 1996, pp. 87–88). Such 'open regionalism' creates the possibility of new members. At the same time as addressing trade liberalization measures, APEC has also drawn up an agenda to discuss wider issues, including competition policy, environmental protection, dispute procedures. Of course, agreeing to talk about such issues is a far cry from actually implementing measures. The 1996 summit in the Philippines did not produce any significant progress, even in the already-agreed area of trade liberalization. Unlike the formal integration agreements shown in Table 3.3, APEC is, at present, little more than a loose collection of states. Cynics have translated the APEC acronym as '*A Perfect Excuse to Chat*' but this is to denigrate what could be a significant initiative. It is more appropriate to see APEC, as Higgott (1993, p. 303) does, as being 'engaged in the early stages of institution building for cooperation'. Nevertheless, it would seem that the creation of a formal region-wide economic bloc in the Asia–Pacific region, along the lines of the EU or even of the NAFTA, is not a serious probability.

Conclusion

The aim of this chapter has been to assert the continuing significance of the nation-state as a major influence in the global economy and to describe some of the ways in which it operates. This does not mean, of course, that the role and functions of the state have not changed. On the contrary, it was recognized that the position of the state is being redefined in the context of a polycentric political-economic system in which national boundaries are more permeable than in the past. Nevertheless, the nation-state continues to contribute significantly towards the shaping and reshaping of the global economic map. We focused on three major aspects of the role of the state.

First, we explored the role of the state as a container of distinctive political, cultural, social and economic processes and institutions and as a regulator of economic activity. All states operate a broad variety of policies aimed at influencing the level and nature of economic activities within, and across, their borders. But the precise mix of policies and how they are implemented varies according to a number of variables. In particular, the policy stance varies according to the ideological complexion of the state. Within the matrix of ideal-type positions we focused especially on the market-rational/regulatory state on the one hand and the plan-rational/developmental state on the other.

Secondly, we set out the three major policy areas of trade, foreign direct investment and industry. The boundaries between these three policy areas are becoming increasingly blurred but there are some important distinctions between them. In particular, national trade policies operate within the long-established international regulatory framework of the GATT/WTO, whose objectives are to sustain a multi-lateral, non-discriminatory trade regime. The general trend since the late 1940s has been for tariffs to decline dramatically through the series of GATT rounds but, at the same time, for non-tariff barriers to proliferate. No such international framework exists in the other two policy areas discussed, although attempts continue to try to devise such a framework for international investment. In general, states have become increasingly liberal in their attitudes towards foreign direct investment and more cut-throat in their efforts to attract such investment. Although the policy emphasis and policy mix vary according to the the state's *general* position on the ideological spectrum the *specificities* of particular states create variety even within a given category. Thus, for example, although most NIEs can be categorized as developmental states their precise behaviour is highly differentiated; there is no standard 'east Asian model' as is so often suggested (see the discussion in Chapter 4).

Thirdly, we addressed the issue of regional integration.While there has certainly been an acceleration in the number of regional trading agreements in recent years, most are very limited in the depth and extent of their integration. Very few have proceeded beyond the first stage of a simple free-trade arrangement. Only the European Union has gone very far along the road of economic integration. Nevertheless, we should not underestimate the importance of the NAFTA or even the more limited developments in the Asia–Pacific region. Ultimately, however, regional blocs of whatever degree of economic integration originate, and are given legitimation by, nation-states which continue to be important building blocks in the global economy.

Notes for further reading

1. There is, currently, a considerable debate within international political economy focused around three alternative theoretical paradigms: realism, liberal-pluralism and neo-Marxism. *Realism* places the state at the centre of the analysis and sees it as continuing to be the dominant actor and pre-eminent power centre in the global political system. *Liberal-pluralism* sees the state as having a decreasing role and influence because of the rise of other significant global actors, including transnational corporations together with a variety of other international organizations. *Neo-Marxism* stresses the significance of global capitalism as creating the structures within which all actors operate, whether they are states, TNCs or other international organizations. The fundamental political processes are conceived as the expression of class conflicts. McGrew (1992) provides a concise discussion of these different approaches while the exchange between Krasner (1994) and Strange (1994), together with Strange (1996), illustrates the conflict between more traditional approaches to world politics and those based in international political economy. See also Underhill (1994).
2. The term 'governance' refers to a *general* process: 'the control of an activity by some means such that a range of desired outcomes is attained'. It is not, however, 'just the province of the state. Rather it is a function that can be performed by a wide variety of public and private, state and non-state, national and international institutions and practices' (Hirst and Thompson, 1996, p. 184).
3. See Dicken, Peck and Tickell (1997) for a discussion of this issue. The concept of the 'hollowing out' of the state is discussed by Jessop (1994).
4. Both Taylor (1994) and Agnew and Corbridge (1995) discuss the general notion of states as 'containers' and the nature and significance of territoriality and space in geopolitics.

5. The 'classic' sources on economic sociology are Zukin and DiMaggio (1990), Granovetter and Swedberg (1992), Smelser and Swedberg (1994).
6. Hofstede (1980) was the pioneer of comparative and systematic studies of cultural variations in business practices. More recently, Whitley (1992a; 1992b) and Hampden-Turner and Trompenaars (1994) have examined national differences in business cultures.
7. This is especially noticeable given the experience of, say, the east and southeast Asian NIEs. For a comment on this, see Stopford and Strange (1991), Henderson and Appelbaum (1992), Henderson (1993).
8. See Guisinger (1985).
9. 'Competitiveness: a dangerous obsession' is one of a series of polemical papers on aspects of the international economy written by Krugman and collected in Krugman (1996).
10. See, for example, the contributions in Rapkin and Avery (1995).
11. The original idea comes from Dahrendorf (1968) who distinguished between two ideal types of political economy: the *market-rational* and the *plan-rational*. He equated the latter with the state socialist command economies. In 1982, in his classic work on Japan, Johnson argued that these command economies were better described as *plan-ideological* systems, and he applied the term *plan-rational* to the economies of east Asia, notably Japan and the then leading NIEs, particularly South Korea, Taiwan and Singapore. Subsequently, Henderson and Appelbaum (1992) suggested that a fourth ideal type can be recognized – the *market-ideological* – which, they suggest, was epitomized by the Reagan and Thatcher 'new right' administrations of the USA and the UK respectively during the 1980s.
12. Mercantilist trade policies formed a cornerstone of international political (and military) relationships. They are based on the notion that a nation's wealth and influence depend upon its ability to regulate and control its external trade at the expense of its rivals.
13. Hoekman and Kostecki (1995) provide an extremely comprehensive account of the evolution of the international regulatory framework for trade, from the GATT through to the WTO. They provide details on each of the 'rounds' of trade negotiations together with the agreements arising from the Uruguay Round which was concluded in 1994.
14. See Hoekman and Kostecki (1995) for fuller details.
15. The UNCTAD *World Investment Report, 1996*, Part Three (UNCTAD, 1996b), and the WTO *Annual Report, 1996, Volume 1* contain a detailed discussion of international policy arrangements towards FDI.
16. Recent treatments of regional economic blocs are provided by Cable and Henderson (1994), Gibb and Michalak (1994), Gamble and Payne (1996), Lawrence (1996), OECD (1996b). Hoekman and Kostecki (1995, Chapter 9), discuss regional blocs within the context of the WTO.
17. The Lomé Convention was a broad-ranging scheme embracing not only trade but also development aid. It allowed for the free entry, at least in theory, of manufactured goods from the ACP countries. However, manufacturing industry is a small proportion of these countries' trade; the most important element of the Lomé Convention related to commodities and agricultural products.
18. Quoted in *The Financial Times* (30 October 1996).
19. Haggard (1995, Table 3.4), lists the various co-operation bodies within the Asian–Pacific region.
20. Lawrence (1996, pp. 86–87).

CHAPTER 4

National variations in policy stance

The specific policies adopted by individual states would each merit a book in themselves. Here, the intention is merely to sketch in the major empirical features of the kinds of policies adopted by some of the leading industrialized and newly industrializing economies. These specific cases should be seen within the kinds of conceptual framework discussed in the previous chapter.

The older industrialized economies of Europe and the United States

Substantial differences exist between the continental European countries, on the one hand, and the Anglo-Saxon countries of the United States and the United Kingdom on the other. They occupy different positions in the ideal-type framework of Figure 3.4. Historically, the major difference between most continental European countries and the United States has been the centrality of some kind of industrial policy in the former and the absence of such a policy in the latter. The United States and the United Kingdom are firmly located in the market-rational/regulatory category, if no longer in the market-ideological category. The positions of the continental European countries are more varied, although they tend to occupy a place at or near the boundary between the market-rational/regulatory and the plan-rational/developmental state. France, for example, has long displayed 'developmental state' characteristics whereas this is less so for most of the others. Whitley (1994a, p. 172), for example, distinguishes between those cases

> where the state coordinates economic development by its control over the flow of credit through the banking system and seeks to manage the growth and decline of sectors, from those where it plays a less active role and the banks operate as more autonomous agents. The former, state-coordinated, economies are represented by post-war Japan and France while the latter are represented by post-war West Germany and many other continental European societies.

Of course, those European countries which belong to the European Union are involved in at least some degree of policy convergence although, again, significant national differences continue to exist in the ways in which the state is involved in managing the economy.

One of the most significant trends to emerge during the 1980s in many industrialized countries was that of so-called *deregulation*:

> The concept of deregulation – as it has been used by most commentators, whether politicians, journalists or academics – is deceptively simple. It means the lifting or abolishing of government regulations on a range of economic activities in order to allow markets

115

to work more freely, as in classical capitalist economic theory. Proponents of this image of deregulation believe . . . that government regulations have come to distort the efficient working of markets. Deregulation of this sort has been intended, firstly, to reduce the costs of conforming with regulations . . . these costs are thought to more than counter-weigh the social, economic and political benefits of the particular regulations in question. In addition, even relatively 'costless' regulations are often seen as counterproductive in that they distort market actors' calculations of allocative efficiency at the microeconomic level.

(Cerny, 1991, p. 173)

Virtually all industrialized countries have followed the deregulation band-waggon although to varying degrees. In fact, as Cerny (1991, p. 174) rightly points out, the issues are far less simple than the 'deregulationists' claim. Because no activity can exist without some form of regulation (otherwise anarchy would prevail) 'deregula-tion cannot take place without the creation of new regulations to replace the old'. In effect, what is often termed *de*regulation is really *re*regulation. Outside the United States, deregulation has also been associated with a strong move towards *privatization* of a whole range of activities in which the state has been centrally involved. The selling off of publicly owned assets, whether in the private goods sectors or in the sphere of public goods (such as healthcare), has occurred widely.

France

Within Europe, France has had the most explicit state industrial policy, a reflection of its long tradition of strong state involvement that goes back to the seventeenth century. France's industrial policy in the postwar period has been an integral part of the series of national economic 'indicative' plans. A major component of French industrial policy has been the promotion of 'national champions' in key industrial sectors. In some cases, this was sought through state ownership of large-scale enter-prises. In its desire for technological independence, especially from the United States, France invested massively, but selectively, in certain sectors. In the late 1960s and early 1970s, for example, four *grands programmes* were launched in the nuclear, aerospace, space technology and electronics industries. Subsequently, the focus was narrowed to high-growth sectors in energy conservation equipment, office informa-tion systems, robotics, biotechnology and electronics as well as to the problem sector of textiles.

The French government has also exerted considerable influence on its industry through its purchasing policies and through its control of the major financial institu-tions. Both powers were used extensively. For example, government monopoly of purchasing in the heavy electrical engineering sector was used to enforce reorganiza-tion of the industry. State control of financial institutions enabled the French govern-ment to steer investment funds to targeted firms and sectors. In a number of ways, therefore, including the encouragement of industrial mergers, the French government intervened strongly to reshape the national industrial structure and to increase the country's international competitiveness.

Germany

The major exception among the continental European nations to a centralized ap-proach to industrial policy has been Germany. In part, at least, this reflects the fact that Germany is a federal political unit with power divided between the federal government and the provinces (*Länder*). Although often described as 'light', the federal government's role is far from insubstantial:

> [It] pursues active industrial intervention to achieve its objectives. Subsidies to industry (including tax concessions, grants, loans and interest remission), initially viewed as temporary exceptions to the rule of market competition in the determination of resource allocation, came to be viewed as legitimate industrial policy instruments . . . Within the West German framework of the social market economy, the state has a responsibility to regulate and intervene to improve the working of markets (as well as combat cartels and promote social justice).
>
> (Hesselman, 1983, p. 203)

The German economy is characterized both by a considerable degree of competition between domestic firms and also by a high level of consensus between various interest groups, including labour unions, the major banks and industry. The provincial governments have played an important role in the economy including, in recent times, the aiding of large manufacturing companies in distress. Like the French, the German government has also intervened to stimulate technological development in key sectors. During the 1970s, in particular, government financial support for research and development became focused on the development of advanced technologies – especially computer technologies – with wide application.

The major challenge facing Germany today, of course, is to cope with the fundamental transformation of the economy brought about by the reunification of the former West and East Germanies in 1990. Putting together the strongest economy in Europe with one which, for half a century, had existed in a completely different economic-ideological sphere, is an immense undertaking. As Gretschmann (1994, p. 471) points out,

> It came as a shock when Germans realised that unification was harder, more complex, more time-consuming, and more expensive than anyone might have figured in 1989. Today, it is clear that the costs of the largest takeover in history, unification, involve: (1) *permanent trade deficits*, caused by rising imports, combined with the diversion of exports to the new *Länder*; (2) *high structural unemployment* in the East, linked to the poor level of competitiveness of the former GDR economy, whose former markets have furthermore fallen away, combined with the high wage expectations of the population as it emerged from communism; (3) *high inflation* resulting from the huge public deficits required to pay, over the next decade, for the monetary transfers to the new *Länder*, which in turn are necessary to keep up social control and to build up infrastructure; (4) *high interest rates* set by the Bundesbank to stabilize the value of the DM in a situation of national fiscal laxity.

United Kingdom

Industry policy in the United Kingdom contrasts in a number of ways with that of the continental European countries. 'The UK industrial system seems to possess little of the "cement" provided by government links with industry in France and by the banks, interfirm cooperation and harmonious labour relations in Germany' (Shepherd, Duchene and Saunders, 1983, p. 18). Perhaps the most consistent feature of UK industry policy in the postwar period has been its *in*consistency. This only partly reflects changes associated with switches of government; even the same government has often adopted variable policies. In the early 1950s government involvement in the economy was primarily macroeconomic: the management of domestic demand and the creation of an amenable 'business climate'.

This policy stance altered markedly with the change to a Labour government in the mid-1960s when, for the one and only time, an attempt was made to operate a national plan. The attempt was short lived, largely because of external pressures on the country's balance of payments and a currency crisis. Successive changes of government in 1970 and 1974 brought, first, a return to government disengagement from economic intervention and then a renewal of government involvement in the form of an explicit industrial strategy. After 1979 the policy emphasis changed again as the Thatcher government's strongly anti-interventionist–market-ideological policy took hold. In concert with these large-scale policy fluctuations the degree of direct state ownership of productive enterprises also waxed and waned.

Even during the more active phases of government involvement in UK industry the precise focus also varied a good deal. In the late 1960s a key element was the encouragement of mergers in sectors such as motor vehicles, electrical engineering – through the Industrial Reorganization Corporation – and steel. In the late 1970s 'small is beautiful' became a dominant theme as government sought to stimulate indigenous investment. The Labour government's strategy of the 1970s was a sector-based approach to industry policy, with selective assistance becoming increasingly important. Subsequently, the focus shifted to one of stimulating advanced technology, including micro electronics, fibre-optics and robotics.

One policy thread, common to all the European states, including the United Kingdom, has been that of *regional policy*: of designating specific geographical areas for special assistance. All the major industrial nations of western Europe operate a regional industrial policy to varying degrees. In general, the aim is to try to solve problems of economic development – especially unemployment – by providing financial and other incentives for firms to locate in specially designated areas. In France the major thrust of regional policy has been to offset the economic dominance of Paris by developing countermagnets based on major provincial centres, and to stimulate the economies of depressed agricultural regions such as Brittany and of declining industrial regions such as Lorraine. In Italy – where regional differences in economic health are especially acute – policy has focused on the south (the Mezzogiorno). In the United Kingdom broad regional policies have subsequently been replaced by more selective policies, including a specific emphasis on the problems of the inner cities. For the member states of the European Union, all such regional policies are now set within that broader institutional context (see Chapter 3).

A major thrust of regional policies in Europe and, increasingly, of national industry policies has been the enticement of internationally mobile investment. The shortage of such investment in recent years has greatly intensified the degree of competitive bidding both between individual nations and also between parts of the same nation. TNCs, especially the very large ones, are assiduously courted, often at the highest government levels. National and regional agencies set out their stalls in lavish advertising and in overseas offices. It follows from such efforts by European countries to encourage foreign investment that their general attitude towards TNCs is one of openness. Even France, which has been most suspicious of American and Japanese investment in particular, has adopted a more liberal attitude, especially towards investments which bring new technology, increase exports or create employment in depressed areas. Several European governments, however, have imposed various kinds of *performance requirement* on some foreign investors. Japanese companies in particular have been required to satisfy local content requirements (the chapters on automobiles and electronics in Part III deal with this). Since the implementation of the Single European Market legislation in 1992, such performance requirements are now set at the EU level rather than the individual state level.

During the 1980s the policy emphasis of virtually all European national governments began to shift away from a direct interventionist position to one of greater disengagement. In both the United Kingdom and France, for example, previously state-owned enterprises – including those nationalized by the Mitterrand government of France in 1982 – were progressively privatized. The former emphasis on nurturing 'national champions' was dropped. The rescuing of 'lame duck' firms and industries was largely abandoned. Some key sectors – such as finance and telecommunications in the United Kingdom – were deregulated and opened up to competition. These policy shifts at the national scale reflect a wave of change which spread through Europe during the 1980s. Even more significant, however, from our particular perspective in this book, is the movement which gathered momentum during the second half of the 1980s towards a Single European Market and the development of the European Union.

United States of America

The policy stance of the United States reflects both the sheer scale and wealth of the domestic economy and also a basic philosophy of non-intervention by the federal government in the private economic sector. As far as industry is concerned the role of the federal government has generally been a *regulatory* one whose aim is to ensure the continuation of competition. Thus, 'the United States in the postwar period has not had anything that most Americans would think of as an "industrial policy"' (Diebold, 1982, p. 158). The United States remains the clearest example of a regulatory or market-rational state although, as we shall see, strong internal pressures currently exist for a more strategic policy approach.

Apart from the system of investment tax credits and accelerated depreciation schemes, action at the federal level has been based on macroeconomic policies of a fiscal and monetary kind. The aim has been to create an appropriate investment climate in which private sector institutions could flourish. This has not, however, prevented the federal government from rescuing specific firms – especially very large ones – from disaster. Prominent examples include Lockheed in 1971 and Chrysler in 1979. At the other end of the size spectrum, the Small Business Administration has provided aid to stimulate new and small firms. Federal procurement policies are generally non-discriminatory but the sheer size of federal government purchases, particularly in the defence and aerospace industries, has exerted an enormous influence on US industry. Entire communities and regions are heavily dependent on the work created by federal defence contracts.

Direct federal involvement in regional or area economic development is relatively limited, however. Most of the direct efforts to stimulate industrial development are carried out at the state and local level. All states employ a battery of investment incentives to entice new investment, and most spend vast sums on promotional activities. The result is frenzied and cut-throat competitive bidding between states. The most common form of state aid to industry is the industrial development bond, which is a device to raise funds for investment, although states also operate loan and loan guarantee schemes. States also differ considerably in the tax burden they impose on business. Quite apart from the fact that the level of tax varies from one state to another, states may also offer specific tax concessions to business. States also differ greatly in their attitude towards, and legislation of, labour unions. A substantial number of states, particularly in the south, operate 'right to work' legislation.

As we saw in Chapter 2, there has been a very rapid increase in foreign investment in the United States in the past few years. The federal attitude towards foreign investment is extremely liberal but it does not itself attempt to attract such investment to the United States. However, states and local governments certainly do; indeed, such efforts have greatly intensified in what Glickman and Woodward (1989) graphically term 'the mad scramble for the crumbs'. Almost all states have foreign promotional budgets, some have opened offices abroad and many advertise for potential investors in the international business press. However, the late 1980s and early 1990s saw an upsurge of concern within the United States over what some see as the 'buying up of America'.[1] So far, such concerns have not produced any new national policy although foreign-owned firms have come under increasing scrutiny by federal bodies over such issues as tax avoidance through transfer pricing.

One particularly interesting aspect of American trade policy which has had important implications for global shifts in manufacturing industry is the operation of *offshore assembly provisions* (OAPs). Under certain tariff provisions, an American firm may export domestic materials or components for processing overseas and then reimport the processed product on payment of duty only on the value of the foreign processing. OAPs are especially important in certain types of manufacturing, for example electronics and metal industries. Although applicable to any country, OAP imports to the United States have come to be dominated by a small number of developing countries, notably Mexico and the newly industrializing countries of east and southeast Asia. There is little doubt that the existence of OAPs has been one of the forces stimulating US firms to seek out cheap labour production locations in certain manufacturing industries, notably electronics (see Chapter 11).

United States policy towards international trade in the postwar period has been one of urging liberalization and the reduction of tariffs. As the strongest economy it has been, like Britain in the nineteenth century, the leading advocate of free trade. Even so, the federal government has intervened with the use of tariff and non-tariff barriers to protect particular interests. It has, for example, negotiated orderly marketing arrangements with various countries, notably Japan, in sectors such as textiles (see Chapter 9), automobiles (Chapter 10) and electronics (Chapter 11). The use of non-tariff barriers on a bilateral basis by the United States was explicitly specified in the Trade Act 1974. This Act in effect heralded the emergence of a 'new protectionism' in the United States and the beginnings of a movement towards a *strategic trade policy* (STP).[2] The term strategic 'trade' policy is rather misleading in that it implies too narrow a focus. STP involves far more than just trade policy; it encompasses issues of strategic industry policy and, by extension, FDI policy as well. In other words, it consists of some of the elements of a plan-rational/developmental system. The demand in the United States has been, increasingly, for a shift away from 'free' trade towards 'fair' trade – 'fairness' being defined by the United States itself. According to this view, other countries – notably Japan, other Asian NIEs and most European countries – are themselves engaged in unfair trade practices so it is only reasonable for the United States to take a similar stance.

During the 1970s and 1980s, the United States became increasingly embroiled in a whole series of trade disputes – with Japan, the east and southeast Asian NIEs and the

European Community. Such disputes focused upon US allegations of unfair trading practices by these countries in specific industrial sectors. In that respect, policies were no different from the similar measures being implemented throughout the later 1970s and the 1980s by individual European governments. A major new step in US trade policy came with the introduction of the Omnibus Trade and Competitiveness Act 1988 (OTCA), stimulated by the persistence of a massive trade deficit and the difficulties being encountered by a number of particular American industries. The OTCA incorporated a strongly *unilateralist* approach to trade negotiations, rather than the multilateralist approach enshrined in the GATT, and hitherto strongly supported by the United States. The key clause – what some have called the 'crowbar' – was the so-called 'Super 301' clause.[3] The aim is to achieve reciprocal access to what the United States defines as unfairly restricted markets.

The difference between the 1974 and 1988 Acts in this respect was that the Super 301 clause was directed to *entire countries*, not just individual industries. In its first application in May 1989 three countries – Japan, India and Brazil – were named as priority countries. Others were 'warned' that their trade practices were being watched. If, over a defined period of time, the named countries do not abandon what the United States defines as 'unjustifiable' or 'unreasonable' trade practices, the United States may retaliate unilaterally by restricting access to the US market for all goods from the countries in question. In a separate, but related, set of measures the United States negotiated with Japan a 'Structural Impediments Initiative' which aims to remove what are seen to be structural restrictions on access to the Japanese market.

The shifts towards a more strategic trade policy are particularly evident in high-technology sectors, seen to be at the centre of a country's competitive position. The basic rationale is that, in imperfectly competitive markets, governments must intervene in favour of their domestic firms. As Ostry (1990, p. 60) points out, the argument is based upon two issues: 'the "first mover advantage" that a country or firm captures by preempting foreign rivals . . . [which] . . . provides the opportunity for firms and countries to consolidate and extend their competitive advantage' and the issue of externalities or spillovers that enhance the competitiveness of other parts of the domestic economy. Thus, the United States shows some signs of moving towards a more developmental position in some respects. But there is no question that, in general, it remains very much a market-rational/regulatory state in its overall stance. In the eyes of the world, however, the United States is increasingly being seen as having a unilateralist tendency which is very much at odds with its traditional multilateral trading stance. There is concern that, from time to time, the United States tends to introduce *extraterritorial* trade legislation to achieve broader political objectives. The most recent such measure is the Helms–Burton law which penalizes foreign companies from doing business with Cuba. Of course, as in the case of European countries, the United States is now embedded within a regional trading group: the NAFTA.

Japan[4]

The basic difference in government's economic role between Japan on the one hand and the western – especially the Anglo-Saxon – nations on the other can be summarized most concisely in the two following quotations:

> Historically, the State assumed an active economic role in Western economies in order to correct what were considered to be the private sector's economic and social failures.

Japanese historical tradition, on the other hand, grants to government a legitimate role in shaping and helping to carry out industrial policy. Japanese businessmen share with government leaders and officials a sense of the importance of co-ordinated national development and are generally amenable to and, in fact, expect government intervention to advance this goal. Senior businessmen view government guidance of industry as a normal state of affairs. They may not always enjoy the process or approve of the specific actions but they see the process as legitimate and, in the main, useful.

(Magaziner and Hout, 1980, p. 29)

The Japanese government is extremely intrusive into the privately owned and managed economy, but it does this through market-conforming methods and in co-operation rather than confrontation with the private sector . . . Japan's first priority is, above all, developmental – meaning the effort by the government to secure Japan's economic livelihood through public policies based on such criteria as long-term dynamic comparative advantage and international competitive ability. The Japanese government's most important contributions to the economy are think-tank functions and supervision and co-ordination of the structural changes necessary to keep Japan competitive in world markets.

(Johnson, 1985, pp. 61, 62)

There is a high level of consensus between the major interest groups in Japan on the need to create a dynamic national economy. To some extent, this consensus is a cultural and institutional characteristic of Japanese society, with its deep roots in familism. But it also reflects the poor physical endowment of Japan and the limited number of options which faced the country when, in the 1860s, it suddenly emerged from its feudal isolation. In other words, consensus is also a pragmatic stance built up over more than a hundred years. Given virtually no natural resources and a poor agricultural base, Japan's only hope of economic growth lay in building a strong manufacturing base both domestically and internationally through trade. In this process, the state has played a central role not through direct state ownership but rather by *guiding* the operation of a highly competitive domestic market economy. Indeed, there is relatively little state-owned enterprise in Japan and a generally much smaller public sector than in most western economies.

The key government institution concerned with both industry policy and trade policy – the two are seen to be inextricably related in Japan – is the *Ministry of International Trade and Industry (MITI)*. After its establishment in 1949, MITI became the real 'guiding hand' in Japan's economic resurgence, although its role has often been misunderstood and exaggerated in the west. Box 4.1 identifies the major roles played by MITI in the Japanese economy. Until the 1960s Japan operated a strongly protected economy and it was not until 1980 that full internationalization of the Japanese economy was reached. During the 1950s and early 1960s MITI, together with the Ministry of Finance, exerted very stringent controls on all foreign exchange and over the import of technology.

In fact, imported technology played a most significant part in the rebuilding of the Japanese economy. Technology was imported largely through licensing from foreign suppliers and not via the direct investment of foreign firms in Japan itself. The technologies were chosen to meet the needs of particular industries – those regarded by MITI as being the ones necessary to achieve national objectives. The selected industries were further aided by preferential financing and tax concessions and were also protected from foreign competition. Within Japan, however, intense competition and rivalry were encouraged between rival Japanese firms with the result that domestic production costs were kept down and efficiency increased. Within the selected industries, MITI encouraged mergers to create large-scale enterprises, although such

1) *Constructs medium-term econometric forecasts of the development of, and needed changes in, the Japanese industrial structure*
- Establishes indicative plans or 'visions' of the desirable goals for the private sector
- Makes specific comparisons of cost structures of Japanese and foreign competitors

2) *Arranges for preferential allocation of capital to selected strategic industries*
- Involves governmental and semi-governmental banks
- Ministry of Finance guides commercial banks to co-ordinate their lending policies with MITI's industrial strategy
- Financial support of an industry implies guidance (though not control) by MITI

3) *Targets key industries for the future and puts together a package of policy measures to promote their development*
- Pre-early 1980s the major measure was protection against foreign competition in the Japanese domestic market
- Protectionism abandoned in early 1980s. Emphasis now on financial assistance, tax breaks, incentives given through administrative guidance, anti-trust relief (to facilitate 'research cartels')

4) *Formulates industrial policies for 'structurally recessed industries'*
- MITI designates a specific industry as 'structurally recessed'
- The ministry responsible for that sector formulates a stabilization plan specifying how the capacity to be scrapped should be shared between enterprises. The plans must be drawn up in consultation with the Industrial Structure Council of MITI
- Costs of scrapping production facilities are shared between the government and the private sector

Source: Based on Johnson (1985, pp. 66–67)

Box 4.1   The major roles of the Japanese Ministry of International Trade and Industry (MITI)

moves were not always successful. For example, MITI failed in its attempt radically to restructure the automobile industry.

Initially, MITI focused its energies on the basic industries of steel, electric power, shipbuilding and chemical fertilizers but then progressively encouraged the development of petrochemicals, synthetic textiles, plastics, automobiles and electronics. The results, in terms of growth in these sectors, were remarkably impressive. The Japanese economy was transformed from one based on low-value, low-skill products such as cheap clothing and textiles to a basis of high-value, capital-intensive products. The foundation of this transformation was the clearly targeted, selective nature of Japanese industry policy together with a strongly protected domestic economy. By the early 1970s Japan was the world leader in the production of steel, ships and consumer goods such as motor cycles and cameras. However, it was already becoming evident that the spectacular growth of Japanese industry had created substantial problems within the country itself.

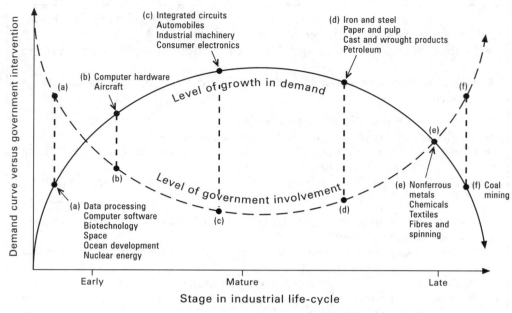

Figure 4.1   The relationship between Japanese government intervention and stage in the industry life-cycle
Source: Okimoto (1989, Figure 1.5)

In 1971 a new industrial policy began to emerge which attempted to meet the problems of environmental pollution, urban congestion and rural depopulation and so on by shifting the focus of Japanese industry towards high-technology, knowledge-intensive industries. Selective government assistance was moved away from the established capital-intensive industries, such as automobiles and steel, and towards the newly emerging high-technology sectors. In 1974 MITI published the first of its *long-term visions* of how the Japanese industrial structure ought to evolve to meet changed circumstances, both domestically and internationally. Appropriately in the Land of the Rising Sun, 'sunset' industries were to be scaled down and 'sunrise' industries encouraged.

In effect, MITI has used an industrial life-cycle model as the basis for deciding its strategic priorities (Figure 4.1). As Okimoto (1989, p. 50) has shown,

> although the degrees of intervention and selection of policy instruments vary by industry, MITI intervention tends to follow a curvilinear trajectory: that is, extensive involvement during the early stages of an industry's life cycle when market demand is still small, falling off significantly as the industry reaches full maturity and demand reaches its peak, and rising again as the industry loses its comparative advantage and faces the problems of senescence – saturated markets, the loss of market share and excess capacity.

Postwar Japanese economic policy has been strongly mercantilist. Growth in exports, particularly in the manufacturing sector, has been a major focus along with the building of a strong domestic economy. Among other things, manufactured exports provided the foreign exchange necessary for Japan to import the industrial raw materials which are in such short supply domestically (including, of course, oil). A key element in Japan's 'guiding hand' on the market economy has been the specific treatment of foreign direct investment. For much of the postwar period, both inward and outward investment were extremely closely regulated. The technological rebuilding of the Japanese economy was based on the purchase and licensing of foreign technology and *not* on the entry of foreign branches or subsidiaries. Although the inward investment laws have now been liberalized and foreign firms do indeed operate within Japan, their relative importance remains very small as we have seen.

As far as overseas investment by Japanese firms is concerned, the situation, as we saw in Chapter 2, changed dramatically during the 1970s and 1980s when Japanese overseas investment grew at a spectacular rate. This was consistent with MITI's policy of internationalizing the Japanese economy and reflected the economically strategic role which Japanese overseas investment came to play:

> For Japan . . . overseas production has suddenly emerged as a national requirement encompassing practically the entire spectrum of her industries and enterprises, small and large alike. The segments of industrial activities that are no longer suitable, environmentally or otherwise, for the Japanese economy need to be transplanted abroad, and overseas resources must now be developed more directly to ensure supplies . . . Furthermore, overseas investment is now viewed as an essential device by which to upgrade Japanese industry.
>
> (Ozawa, 1979, pp. 228–29)

Thus, overseas investment by Japanese firms came to be seen as an integral part of Japanese industrial policy. It is not something that has 'just happened': it has been positively encouraged.

Such encouragement has been enhanced by the perceived need for Japanese companies to respond to two major developments in their external environment. The first was the upsurge in protectionist measures in North America and Europe in such industries as automobiles and electronics from the mid-1970s (see Chapters 10 and 11). The 'laser-like' targeting of Japanese industrial policy resulted in a major backlash from various western economies. As a direct result of the erection of non-tariff barriers, Japanese firms began to invest heavily in production facilities overseas. The second external stimulus to increased overseas direct investment was 'endaka'; the major rise in the value of the Japanese yen which resulted from the 1985 Plaza Agreement among the Group of Five international finance ministers. This political decision stimulated an upsurge in overseas investment by Japanese firms to take advantage of lower production costs, particularly in east and southeast Asia.

During the 1990s, therefore, Japanese policy has been especially exercised by the problem of a high-value currency, with contentious trading relationships with the United States and Europe and also with the deep domestic recession which accompanied the collapse of the so-called 'bubble economy' at the end of the 1980s. The Japanese government role has changed over time: strong and broad-based intervention in the immediate postwar period has now become increasingly selective. Thus, although the government agencies such as MITI are far from the monolithic institutions they are often alleged to be, there can be no doubt of their importance in stimulating Japanese industry and trade. However, the role of MITI is now less prominent than it was and some degree of deregulation in the economy has been occurring, though not as extensively as in the United States and the United Kingdom. But Japan undoubtedly remains a *developmental state,* though not a centrally planned economy. It is a guided market economy in which intense domestic competition between firms is encouraged.[5]

Newly industrializing economies in east and southeast Asia and Latin America[6]

Although they are frequently grouped together, the world's newly industrializing economies are a highly heterogeneous collection of countries. They vary enormously in size (both geographically and in terms of population); in their natural resource endowments; in their cultural, social and political complexions. But they all tend to have one feature in common: the central role of the national state in their economic development. Despite many popular misconceptions, none of today's NIEs is a free-wheeling market economy in which market forces have been allowed to run their unfettered course. They are, virtually without exception, *developmental* states; market economies in which the state performs a highly interventionist role. Having said that, the precise role of the state – the degree and nature of its involvement – varies greatly from one NIE to another. In some cases, state ownership of production is very substantial; in others it is insignificant. In some cases, the major policy emphasis is upon attracting foreign direct investment; in others such investment is tightly regulated and the policy emphasis is upon nurturing domestic firms. Thus, although the recurring central theme which runs through the current economic behaviour of all NIEs is the role of the state, each individual NIE performs a specific variation on that general theme; a reflection of its specific historical, cultural, social, political and economic complexion.[7]

Most observers go along with the notion of the central role of the state in the industrialization of the Latin American economies but the critics would point to their relative lack of success compared with the east and southeast Asian economies. According to this view, the Latin American path has tended to be one of state-led inward-orientation while that of the east and southeast Asian economies has been that of 'hands-off' outward-orientation. Allegedly, the state's role in east and southeast Asia has been restricted to that of 'getting the fundamentals right' and in facilitating the free operation of market forces. This, certainly, was the conclusion of the 1993 World Bank study *The East Asian Miracle*. However, there is a very powerful counter-view[8] which argues that to play down the role of the state in the economic development of these 'miracle' economies is to ignore the evidence. As we shall see, the role of the state has indeed been central to the industrialization of the east and southeast Asian economies but in ways which are far from uniform from one country to another.

Types of industrialization strategy

Broadly speaking, a developing country may pursue one or more of three basic types of strategy:

- Exports of indigenous commodities.
- Import-substituting industrialization (ISI) – the manufacture of products which would otherwise be imported, based upon protection against such imports.
- Export-oriented industrialization (EOI).

Which of these strategies is, or can be, pursued depends upon a number of factors: the economy's resource endowment (both physical and human); its size (particularly of its domestic market); its international context (especially the rate of growth of world trade and the policies of TNCs); and the attitude of the national government. For example, not all developing countries possess a natural resource endowment which could form the basis of a local processing industry. Even those which have such an asset may experience difficulty in setting up a local industry. Both developed country tariffs and also international freight rates tend to be higher on processed than on unprocessed materials. In addition, where TNCs are involved it may be corporate policy to locate processing operations elsewhere. Of course, neither Singapore nor Hong Kong, for example, has the material base to support such a strategy anyway.

The general pattern of industrialization beyond the commodity export phase has, with few exceptions, been one of an initial emphasis on import substitution followed eventually by a shift to export-oriented policies. During both the 1920s and 1930s and the period after 1945 the developing countries faced huge economic problems. In particular, their embryonic manufacturing industries faced the threat of being still-born by the competition of imports from the more efficient developed country producers. As a result, a number of countries – especially the larger Latin American countries such as Brazil and Mexico – pursued an explicit policy of *import substitution*. In the postwar period large Asian countries such as India, as well as smaller ones like South Korea, Taiwan and the Philippines, began to follow a similar path. (As we have seen, Japan also pursued a highly protective policy stance.) The aim of import substitution was to protect a nation's infant industries so that the overall industrial structure could be developed and diversified and dependence on foreign technology and capital reduced. To this end, many of the policies listed in Figure 3.5, Figure 3.7 and Figure 3.9 were employed by national governments. In particular, very high tariffs were imposed on those sectors chosen for protection. Import quotas and other devices, including licences, deposits and multiple exchange rates, were also used, as were incentives to encourage domestic production.

The import-substitution strategy, in theory, is a long-term *sequential* process involving the progressive domestic development of industrial sectors through a combination of protection and incentives:

Stage 1: Domestic production of consumer goods.
Stage 2: Domestic production of intermediate goods.
Stage 3: Domestic production of capital goods.

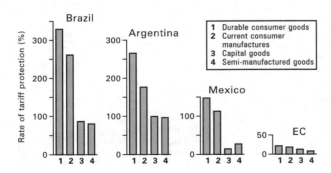

Figure 4.2   Import-substitution policies: tariffs on consumer goods, intermediate goods and capital goods in the early 1960s
Source: United Nations (1964, *Economic Bulletin for Latin America*, 9, 1, Table 5)

Invariably, the process began with the heavy protection of domestic consumer goods industries to stimulate local production. As Figure 4.2 shows, tariffs on consumer goods in the major Latin American countries in the early 1960s were many times greater than the tariffs on intermediate and capital goods (and very much higher than EC tariffs). As a result, domestic consumer goods' production in at least some of the countries pursuing an import-substitution policy grew considerably, although much depended on the size of the domestic market. Although dependence on imported consumer goods certainly declined, however, dependence on the import of intermediate and capital goods – and, therefore, on foreign technology – increased. In most cases progression beyond the production of consumer goods did not occur to the extent anticipated. Hence, various critics described import-substituting industrialization as 'half-way' industrialization or as 'getting stuck' at the consumer goods stage. The hoped-for domestic multiplier effects and the stimulus of a broader industrial structure did not necessarily occur. Where the domestic market was small, local production of consumer goods could not achieve appropriate economies of scale so that domestic prices remained high. The necessarily high level of imports of intermediate and capital goods imposed balance of payments constraints on many developing economies. Yet there were strong pressures from domestic vested interests – especially the protected consumer goods manufacturers – against reducing the protection afforded to that sector in favour of other manufacturing sectors as originally envisaged.

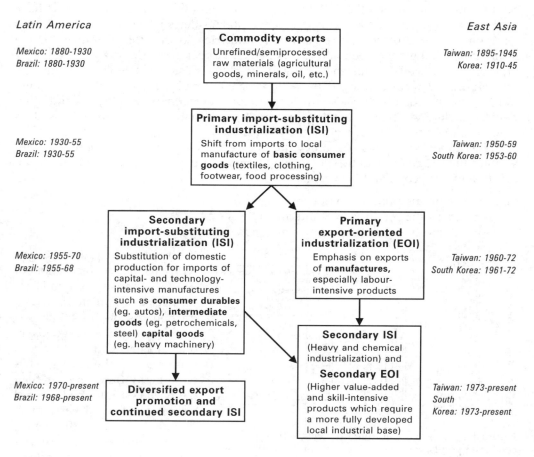

Figure 4.3   Paths of industrialization in Latin America and east Asia: common and divergent features
Source: Based on material in Gereffi (1990, Figure 1.1 and p. 17)

The realization that an import-substituting strategy could not, on its own, lead to the desired level of industrialization began to dawn in a growing number of countries; some during the 1950s, rather more during the 1960s. Generally it was the smaller industrializing countries which first began to shift towards a greater emphasis on *export orientation* because of the constraints imposed upon ISI policy by a small domestic market. Increasingly, an export-oriented, outward-looking industrialization strategy became the conventional wisdom among such international agencies as the Asian Development Bank and the World Bank.

A shift towards export-based industrialization was made possible by such developments as: the rapid liberalization and growth of world trade during the 1960s; the 'shrinkage' of geographical distance through the enabling technologies of transport and communications; and the global spread of the transnational corporation and its increasing interest in seeking out low-cost production locations for its export platform activities. Such export orientation was based upon a high level of government involvement in the economy. The usual starting point was a major devaluation of the country's currency to make its exports more competitive in world markets. The whole battery of export trade policy measures shown in Figure 3.5 was invariably employed by the newly industrializing economies. In effect, these amounted to a subsidy on exports which greatly increased their price competitiveness. Of course, the major domestic resource on which this export-oriented industrialization rests has been that of the labour supply – not only its relative cheapness but also its adaptability and, very often, its relative docility. Indeed, in many cases, the activities of labour unions have been very closely regulated.

In fact, the 'paths of industrialization' followed by individual NIEs have been rather more complex than this simple sequence suggests. Gereffi (1990) proposes a five-phase sequence of industrialization based upon the experiences of the Latin American and east Asian NIEs (Figure 4.3). As Figure 4.3 shows, 'each of the two regional pairs of NICs has followed a sequence of development strategies that closely approximates the ISI and EOI ideal types . . . plus a "mixed" strategy in the most recent period' (Gereffi, 1990, p. 18). Gereffi draws a series of conclusions from his analysis:

- The distinction commonly drawn between inward-oriented Latin American industrialization strategies and outward-oriented east Asian industrialization strategies is misleading. 'While this distinction is appropriate for some periods, a historical perspective shows that each of these NICs has pursued *both* inward- and outward-oriented approaches . . . rather than being mutually exclusive alternatives, the ISI and EOI development paths in fact have been complementary and interactive' (Gereffi, 1990, p. 18).
- The initial stages of industrialization were common to NIEs in both regions; 'the subsequent divergence in the regional sequences stems from the ways in which each country responded to the basic problems associated with the continuation of primary ISI' (Gereffi, 1990, p. 21).
- 'The duration and timing of these development patterns vary by region. Primary ISI began earlier, lasted longer, and was more populist in Latin America than in East Asia . . . The East Asian NICs began their accelerated export of manufactured products during a period of extraordinary dynamism in the world economy . . . [After 1973] . . . the developing countries began to encounter stiffer protectionist

measures in the industrialized markets. These new trends were among the factors that led the East Asian NIEs to modify their EOI approach in the 1970s' (Gereffi, 1990, p. 21).

- Some degree of convergence in the strategies of the Latin American and east Asian NIEs began to occur in the 1970s and 1980s. Each 'coupled their previous strategies from the 1960s (secondary ISI and primary EOI respectively) with elements of the alternate strategy in order to enhance the synergistic benefits of simultaneously pursuing inward- and outward-oriented approaches' (Gereffi, 1990, p. 22).

The attraction of *foreign direct investment* has been an integral part of both import-substituting and export-oriented industrialization in many developing countries, although the extent to which particular countries have pursued this strategy varies considerably. In general, the Latin American NIEs have been more restrictive in their attitudes to foreign direct investment than the Asian NIEs, although there have been recent shifts towards more liberal investment policies in Latin America. In general, ownership requirements in Latin America have tended to be stricter than in most east and southeast Asian countries and the number of sectors in which foreign involvement is prohibited rather greater. Despite such differences in national attitude, most developing countries engaged in what has been called 'the battle for rapid industrialization' are involved in fiercely competitive rivalry to attract foreign firms. The competition is with both developed and developing countries and few punches are pulled in the comparisons drawn in the national promotional literature. The resource which most developing countries stress at the primary export-oriented industrialization phase is undoubtedly that of their *labour force*. Virtually every glossy brochure – and some are very glossy indeed – put out by national economic development agencies eulogizes its quantity, quality and, with some recent exceptions, its relative cheapness. Those industrializing countries located close to major world markets such as North America or the EU, or which have special trading relationships with them, also emphasize this key locational attribute of proximity.

Export-processing zones

Amongst all the measures used by developing countries to stimulate their export industries and to attract foreign investment one device in particular – the export-processing zone (EPZ) – has received particular attention.[9] An EPZ can be defined as:

> a relatively small, geographically separated area within a country, the purpose of which is to attract export-oriented industries, by offering them especially favourable investment and trade conditions as compared with the remainder of the host country. In particular, the EPZs provide for the importation of goods to be used in the production of exports on a bonded duty free basis.
>
> (UNIDO, 1980, p. 6)

EPZs are, in effect, *export enclaves* within which special concessions apply:

- Special investment incentives and trade concessions.
- Exemption from certain kinds of legislation.
- Provision of all physical infrastructure and services necessary for manufacturing activity: roads, power supplies, transport facilities, low-cost/rent buildings.
- Waiver of the restrictions on foreign ownership, allowing 100 per cent ownership of export-processing ventures.

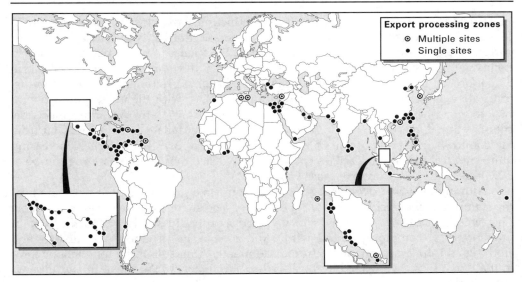

Figure 4.4   Export Processing Zones in developing countries
Source: ILO (1988, Table 20); UNCTAD (1994, Table IV.2); press reports

Within developing countries EPZs have been located in a variety of environments. Some have been incorporated into airports, seaports or commercial free zones or located next to large cities. Others have been set up in relatively undeveloped areas as part of a regional development strategy.

It is often difficult to distinguish between EPZs and the many other similar kinds of zone such as freeports or free-trade zones. Strictly speaking, freeports and free-trade zones are commercial zones only; their functions involve warehousing and transhipment of goods with no change in the nature of the goods themselves. EPZs, on the other hand, are set up for actual manufacturing: the processing and/or assembly of export products from primarily imported materials and components. There are very many more freeports and free-trade zones throughout the world than there are export-processing zones. Figure 4.4 shows the geographical distribution of EPZs. Almost all of the 200 or so EPZs in operation were established after 1971. Before the mid-1960s there were only two EPZs in the developing countries – in India and in Puerto Rico. Numerically, some 90 per cent of all EPZs in the developing countries are located in Latin America and the Caribbean and Asia. The EPZs themselves vary enormously in size, ranging from geographically extensive developments to a few small factories; from employment of more than 30,000 to little more than 100 workers. Total employment in developing country EPZs is roughly 4 million.

Hong Kong and Singapore are, in effect, entire free zones but with export-processing activities concentrated in a number of industrial estates. The other major concentrations are in Taiwan (four EPZs), Malaysia (eleven EPZs), South Korea (three EPZs) and the Philippines (four EPZs). The most interesting recent development in Asia, however, is the establishment of EPZs in the People's Republic of China as the focus of its plan to industrialize with the participation – strictly controlled – of foreign capital (see below). While the People's Republic engaged in its first experiment in attracting foreign investment, Taiwan was busy moving 'up-market' with its construction of the Hsinchu Science-Based Industrial Park, close to Taipei and adjacent to two universities. Unlike the earlier generation of EPZs in Taiwan and most current ones in the developing world, Taiwan's science park is being developed exclusively for high-technology industries. Similar developments are occurring in Singapore. So far, at least, EPZs have not played such a prominent part in the industrialization programmes of South American countries, with the exception of one or two countries such as Colombia. But the most important Latin American industrial country, Brazil, has only one EPZ and this is located at Manaus in Amazonia, far from the country's economic centre of gravity. The number of EPZs has been increasing quite significantly in the Caribbean but it is in Mexico that the largest programme of EPZ-type activity has occurred (see Figure 2.31).

Although EPZs in developing countries come in a great variety of sizes and bear the stamp of their specific national context, they also share many features in common. The overall pattern of incentives to investors is broadly similar, as is the type of industry most commonly found within the zones. The production of textiles and clothing and the assembly of electronics dominate. Almost half the total labour force in the Asian EPZs is engaged in the electronics industry. In the Mexican *maquiladoras* 60 per cent of the workforce is employed in electrical assembly and a further 30 per cent in textiles and clothing. The characteristics of the labour force itself are similarly uniform with a dominance of young female workers. Some of the implications of these characteristics will be explored in later chapters.

Having outlined some of the major features of NIE policies in general terms we need to acknowledge the substantial diversity that exists between individual countries. In order to do this, the following sections present a brief sketch of five NIEs: South Korea, Taiwan, Singapore, China and Mexico.

South Korea[10]

South Korea (officially the Republic of Korea) came into being in 1948, following the partition of Korea into two parts. From 1910 to 1945, Korea had been a Japanese colony with whose economy it was very tightly integrated. Between 1948 and 1988, when political liberalization occurred, South Korea was governed by a succession of authoritarian, military-backed and strongly nationalistic governments. These governments, particularly the one led by Park Chung Hee (1961–79), operated a strong state-directed economic policy articulated through a series of five-year plans. As Figure 4.3 showed, the emphasis changed over time from primary ISI, through primary EOI, secondary ISI and secondary EOI. Two important developments during the 1950s helped to provide the basis for these strategies: the land reform of 1948–50, which removed the old landlord class and created a more equitable class structure, and the redistribution of Japanese-owned and state properties to well-connected individuals which helped to create a new Korean capitalist class (Koo and Kim, 1992).

A powerful economic bureaucracy was created, with a key role played by the new Economic Planning Board (EPB). At the same time, the financial system was placed firmly in the hands of the state (until the early 1980s); the banks were nationalized and the Bank of Korea brought under the control of the Ministry of Finance. This highly centralized 'state-corporatist' bureaucracy, in effect, 'aggressively orchestrated the activities of "private" firms' (Wade, 1990a, p. 320). As Amsden (1989, p. 14) observes, 'where Korea differs from most other late industrializing countries is in the discipline its state exercises over private firms'. In particular, the state made possible – and actively encouraged – the development of a small number of extremely large and highly diversified firms – the *chaebol* – which continue to dominate the Korean economy. The model for such firms was the Japanese *zaibatsu*, the giant family-owned firms which had been so important in the pre-Second World War development of the Japanese economy. The *chaebol*

> are highly centralized, most being owned and controlled by the founding patriarch and his heirs through a central holding company. A single person in a single position at the top exercises authority through all the firms in the group. Different groups tend to specialize in a vertically integrated set of economic activities.
>
> (Wade, 1990a, p. 324)

As a result, the Korean economy is very highly concentrated and oligopolistic while the small and medium-sized firm sector is relatively underdeveloped. Not only this, but many of these smaller firms are tightly tied into the production networks of the *chaebol*.

By controlling the financial system, particularly the availability of credit, the Korean government was able to operate a strongly interventionist economic policy. The *chaebol* were consistently favoured through their access to preferential financing (including the preferential allocation of subsidized loans) and very strong, long-term relationships were developed between the state and the *chaebol*. From the 1960s Korean policy had a strong sectoral emphasis as the state decided which particular industries should be supported through a battery of measures, including financial subsidy and protection against external competition. As Wade (1990a, p. 334) points out, however, this was less a case of 'picking winners' than 'making winners':

> The governments of Taiwan, Korea and Japan have not so much *picked* winners as *made* them. They have made them by creating a larger environment conducive to the viability of new industries – especially by shaping the social structure of investment so as to encourage productive investment and discourage unproductive investment, and by controlling key parameters on investment decisions so as to make for greater predictability. The instruments included protection to modulate international competition, restrictions on capital outflow so as to intensify reinvestment within the national territory and drive the exports of goods rather than capital, and controls on domestic financial institutions. In this environment lumpy and long-term investment projects were undertaken which would probably not have been undertaken in an economy with free trade and capital movements, because they would not have been consistent with short-term profit maximization.

The precise sectoral focus changed over time as Table 4.1 indicates. From an emphasis on consumer products during the primary import-substituting industrialization phase, the emphasis shifted, first, to chemicals, petroleum and steel and, subsequently, to automobiles, shipbuilding and electronics.[11]

A major need of the developing Korean economy was, of course, access to modern technologies which, for the most part, had to be acquired from abroad. Like Japan at a similar stage Korea, for the most part, eschewed the channel of inward foreign investment for such technology transfer. Indeed, South Korea adopted the most restrictive policy towards inward foreign investment of all the four leading Asian NIEs. Until 1983 it operated strict rules on foreign direct investment which restricted the permitted level of foreign ownership, specified a minimum export performance and local content level. Korean government policy has been to build a very strong domestic sector.

Table 4.1   Changing sectoral focus in Korea's developmental strategy

Developmental phase	Major industries
Primary import-substituting industrialization	Food; beverages; tobacco; textiles; clothing; footwear; cement; light manufacturing (e.g. wood, leather, rubber, paper products)
Primary export-oriented industrialization	Textiles and apparel; electronics; plywood; wigs Intermediate goods (chemicals, petroleum, paper, steel products)
Secondary import-substituting industrialization and secondary export-oriented industrialization	Automobiles Shipbuilding Steel and metal products Petrochemicals Textiles and apparel Electronics Videocassette recorders Machinery

Source: Based on material in Gereffi (1990, Table 1.6).

Starting in the early 1980s, however, the emphasis of Korean economic policy shifted towards a greater degree of (restricted) liberalization. State control of the financial system was eased in 1983. The domestic market was opened up to a greater degree of imports. Inward foreign direct investment began to be encouraged following the 1984 Foreign Capital Inducement Law which greatly increased the number of manufacturing industries open to foreign investment. Some relaxation of the country's extremely stringent labour laws occurred. South Korea has had the most restrictive labour laws and practices of all the east and southeast Asian NIEs:

> The need for deep, repressive labour controls in South Korea followed from a development strategy that encouraged large-scale industry and proletarian employment systems. Lacking either Singapore's corporatist union structure or Taiwan's enterprise paternalism, proletarianized Korean workers were better positioned to form locally independent organizations and unions, despite higher level political controls at the federation level. As union pressure mounted during the 1970s, the government's only available response was one of repression. The vicious cycle of repression and protest that issued from this early pattern was eventually to contribute to the democratic opening of the mid-1980s. In the shorter term, however, repression sufficed to scuttle demands for greater provision for social welfare and social insurance, to hold wages down, and to confine social policy primarily to education and training.
>
> (Deyo, 1992, p. 299)

In 1988, the military regime was replaced by a democratically elected government, although one which seems not entirely to have lost the authoritarian habit. Some attempts have been made to persuade the *chaebol* to change some of their practices, but with only limited success. Most significantly, in the mid-1990s Korea applied to join the OECD (Organization for Economic Co-operation and Development), membership of which will put increased pressure on Korea to make major changes to its financial system. Overall, Korea has followed a consistent developmental state strategy based upon, in particular, an alliance between the state and the *chaebol*. However, as the Korean economy has become increasingly internationalized, and as the internal pressures for such things as labour market reform have intensified, this cosy alliance is having to change. In the latter case, the Korean government's hastily implemented legislation in early 1997 to reduce employment security generated serious labour unrest.

Table 4.2 Changing sectoral focus in Taiwan's developmental strategy

Developmental phase	Major industries
Primary import-substituting industrialization	Food; beverages; tobacco; textiles; clothing; footwear; cement; light manufacturing (e.g. wood, leather, rubber, paper products)
Primary export-oriented industrialization	Textiles and apparel; electronics; plywood; plastics Intermediate goods (chemicals, petroleum, paper, steel products)
Secondary import-substituting industrialization and secondary export-oriented industrialization	Steel Petrochemicals Computers Telecommunications Textiles and apparel

Source: Based on material in Gereffi (1990, Table 1.6).

Taiwan[12]

Taiwan shares a number of common features with Korea. First, like Korea, Taiwan was a Japanese colony (from 1895 to 1945) and was tightly integrated into the Japanese economic system. A substantial industrial base and physical infrastructure were established by the Japanese to utilize local labour and materials; land reform was instituted. Secondly, Taiwan also has a difficult external political situation to face: the claim by the Peoples' Republic of China over Taiwan as an integral part of the mainland. Thirdly, Taiwan has followed a broadly similar developmental path to that of Korea (compare Tables 4.1 and 4.2, for example). Both can be described as 'authoritarian corporatist' states. But there are also some significant differences between the Taiwanese and the Korean experiences. In fact, Taiwan shares some common features with Hong Kong, notably the massive influx of Chinese population (including many actual or potential entrepreneurs) from mainland China at the time of the communist revolution in 1949 and the greater importance of small entrepreneurial firms in the domestic economy. At the same time, Taiwan – like Korea – has operated strict labour laws (though in a less repressive manner). 'Labour unions are tightly circumscribed . . . The right to strike is prohibited by martial law' (Wade, 1990b, p. 267).

Wade divides the state-led Taiwanese development experience into three phases: 1945–60; 1960 to the early 1970s; 1970s to present. The period 1945–60 was the 'Initial Nationalist Period' during which the incoming government, retreating from the communist forces on the mainland, acquired all the former Japanese-owned assets:

> Given the country's lack of raw materials and a population then growing at over 3 per cent a year, raising living standards required labor-intensive manufacturing. Recapturing the mainland – which remained a central preoccupation of the government during the 1950s – required the development of some upstream industries. Over the 1950s the basis was laid for production of plastics, artificial fibers, cement, glass, fertilizer, plywood and many other industries, but above all, textiles.
>
> (Wade, 1990a, p. 77)

The characteristic approach was for the state itself to set up such new upstream industries and then either continue to operate them under state ownership or to transfer them to private entrepreneurs. A distinctive feature of Taiwan's development, then, was a heavy direct involvement in production through state ownership. Textiles (building on the large number of immigrants with expertise in the industry), plastics and synthetic fibres formed the dominant focus of Taiwan's industrialization strategy in this initial period. In each case, the state played the initiating role. An explicit statement in the Second Four-Year Plan (1958–61) showed the government's determination to steer the direction of investment: 'the Government should positively undertake to guide and help private investments so that they do not flow into enterprises which have a surplus production and a stagnant market' (quoted in Wade, 1990a, pp. 81–82). There was, in this phase, a very strong emphasis on import substitution.

From 1960 until the early 1970s, Taiwan concentrated on a dual developmental strategy of establishing new export sectors and continuing import substitution. Much greater emphasis was placed on the development of heavy, capital goods industries. The Fourth Plan contained the following statement:

> For further development, stress must be laid on basic heavy industries (such as chemical wood pulp, petrochemical intermediates, and large-scale integrated steel production) instead of end product manufacturing or processing. Industrial development in the long run must be centered on export products that have high income elasticity and low transportation cost. And around these products there should be development of both forward and backward industries, so that both specialization and complementarity may be achieved in the interest of Taiwan's economy.
>
> (Quoted in Wade, 1990a, p. 87)

During this period, the level of state ownership of productive activities declined: from 56 per cent of total manufacturing output in 1952 to 21 per cent in 1970. Even so, the level of public ownership was considerably higher than in other east and southeast Asian NIEs.

The third period of Taiwanese policy identified by Wade was marked by a number of changes of emphasis. The world recession of the 1970s, together with the thawing of the relationship between the United States and China, necessitated an intensified emphasis on export orientation which, it was believed, required a high degree of state involvement. Sectoral priorities included, in particular, petrochemicals, electrical machinery, electronics, precision machine tools, computer terminals and peripherals. The government also addressed how such sectors should best be organized. 'Some subsectors were identified as suitable for development by local firms, others as requiring joint ventures with foreign companies and public enterprises (especially petrochemicals), and still others as suitable for a mix of foreign and local private firms (electronics)' (Wade, 1990a, p. 96). Following the oil price shock of the mid-1970s, the Taiwanese government initiated major investments in the heavy and chemical industries in order to reduce the country's exposure to external supply shocks.

Although an emphasis on new technology had existed since the 1950s, there was now an intensification of the drive to upgrade the educational and technological levels in the economy and to move the balance of the economy towards 'non-energy intensive, nonpolluting, and technology-intensive activities like machine tools, semiconductors, computers, telecommunications, robotics, and biotechnology' (Wade, 1990a, p. 98). In 1973, the government set up the Industrial Technology Research Institute (ITRI); in 1980 it established the Hsinchu Science-Based Industry Park 'where foreign and domestic high-technology firms operate in close proximity to ITRI laboratories and where the government is willing to take up to 49 per cent equity in each venture' (p. 98).

Like Korea, Taiwan has controlled the inflow of foreign direct investment into the domestic economy, although rather less strictly. The foreign sector is important to the

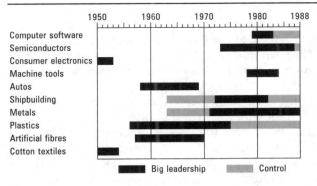

Figure 4.5   State control and leadership episodes in Taiwan's industrial development
Source: Based on Wade (1990b, Figure 9.1)

Taiwan economy (around 6.6 per cent of GDP, as Table 2.10 showed) but far less so than in the case of the southeast Asian NIEs. Initially, Taiwan seems to have been less restrictive than Korea but it has became rather more selective since the 1970s. In particular, foreign firms have been discouraged from an involvement in labour-intensive industries but encouraged to invest in higher-level activities. Taiwan does, indeed, attempt to lure foreign investment into selected sectors but also operates several of the performance requirements shown in Figure 3.7.

Figure 4.5 shows the changing pattern of state leadership of different industries in Taiwan. Lall summarizes the Taiwanese developmental experience as follows:

> Like Hong Kong, it had a large influx of capital, skills and entrepreneurship from mainland China after the revolution. Its development strategy had elements of the 'Korean style' attempts to select and promote local industries, by protection, credit allocation and selectivity in letting in FDI . . . However, its selectivity was far less detailed than that of the Republic of Korea and it did not attempt to create giant conglomerates or to push so heavily into advanced technologies or capital-intensive activities. The less intense relationship of Taiwan Province of China with private industry meant that many of its more ambitious forays into heavy industry had to be led by the public sector, and it has the largest public sector of the NIEs.
>
> The strength of Taiwan Province of China has lain in its small and medium-sized enterprises that tapped its large base of human capital and the infant industry promotion offered by the government to grow and diversify in skill-intensive activities. The resulting industrial structure is 'lighter' than that of the Republic of Korea.
>
> (Lall, 1994, p. 77)

Singapore[13]

Singapore is by far the smallest of all the east and southeast Asian NIEs. Like both Korea and Taiwan, Singapore had a very long history as a colony (within the British system). Unlike the two larger Asian countries, it was not as tightly integrated into its imperial system although it played a highly significant role as a commercial entrepôt in southeast Asia, a reflection of its strategic geographical position. Singapore became fully independent in 1965 when it separated from Malaysia. Since then, although Singapore is a parliamentary democracy, it has been governed by one political party (the Peoples' Action Party) and, until quite recently was led by one powerful individual, Lee Kuan Yew.

After separation from Malaysia the option of pursuing an import-substituting industrialization strategy was a non-starter. Consequently, from the outset in 1965, the Singapore government pursued a very clear policy of export-oriented, labour-intensive manufacturing development. Export orientation reflected both the lack of a sizeable domestic market (even today it has a population of only around 3 million) and of indigenous natural resources. Concentration on manufacturing – especially labour-intensive manufacturing – was adopted because of the need to reduce a very high unemployment rate in a society which, at the time, had one of the fastest population growth rates in the world. The twin pillars of the policy were those of complementary economic and social planning, the latter being much more overt than in other Asian NIEs.

The particular ways in which Singapore has operated its export-oriented policy have been substantially different from those of Korea and Taiwan. Most significantly, the central pillar was based on a strategy of attracting foreign direct investment. As a result, the Singaporean economy is overwhelmingly dominated by foreign firms, notably from Japan and the United States. As Table 2.10 showed, FDI accounts for 73 per cent of the country's GDP. 'Unlike Hong Kong, there was a weak tradition of local entrepreneurship, and there was no sudden influx of technical and entrepreneurial know-how from China' (Lall, 1994, p. 76). The most explicit industrialization measures, therefore, were those of incentives to inward investors. The basic incentive was that of corporate tax exemption for a period of five to ten years for firms granted 'pioneer' status. This was a sectorally selective process, with particular attention being devoted initially to electronics, petroleum and shipbuilding industries. The government agency responsible was the Economic Development Board (EDB) which plays an extremely influential role in the Singapore economy. In addition, profits earned on export activities were taxed at only one-tenth of the standard rate. With a few exceptions, Singapore operated a free port policy with little use of trade protectionist measures. The second set of direct measures used to promote industrial development was the establishment of a high-quality physical infrastructure.

At the same time, a series of social policy measures was introduced aimed at creating an amenable environment for foreign investment. Major housing programmes, partly funded through the state's compulsory savings scheme (the Central Provident Fund) were undertaken. More specifically, the government effectively incorporated the labour unions into the governance system by establishing a National Trades Union Council. 'Strikes and other industrial action were declared illegal unless approved through secret ballot by a majority of a union's members. In essential services, strikes were banned altogether . . . These labour market regulations resulted in the creation of a highly disciplined and depoliticised labour force in Singapore' (Yeung, 1998).

Thus, through a whole battery of interlocking policies, the Singapore government created a very high growth, increasingly affluent industrial society in which foreign firms played the dominant economic role in production but within a highly regulated political and social system. However, during the second half of the 1970s, the government became increasingly convinced of the need to modify its industrialization strategy. Rising internal expectations, together with an increasingly difficult international economic environment, led the government to declare Singapore's 'Second Industrial Revolution'. Although this continued to be strongly oriented towards inward investment, there was a greater emphasis than hitherto on the encouragement of domestic firms. The explicit aim was to upgrade the country's technological and skill level; to move it away from low-skill assembly sectors. Four broad types of policy were implemented as part of the Second Industrial Revolution: fiscal incentives were targeted towards higher-technology activities; local manufacturers who could not upgrade were to be encouraged to relocate outside Singapore; wage levels were allowed to rise to encourage a more efficient use of labour; and a massive expansion was initiated in education and training as well as in research and development.

Mainly in response to the unexpected (but, in fact, short-lived) economic shock of 1985 when, for the first time since independence, Singapore experienced a decline in GDP growth, the government initiated a new policy thrust. A major feature of the 'New Direction' for the Singapore economy was to be a reduced dependence on the manufacturing sector and a shift towards the business services sectors. A Services Promotion Division was established within the EDB with the objective of identifying new investment opportunities in the services sector and promoting them through the EDB's international network of offices. Incentive packages were introduced to attract foreign firms prepared to set up service operations in Singapore. A specific part of that policy was the Operational Headquarters Scheme. As the 1986 'New Directions' report stated:

> We must move beyond being a mere production base. Manufacturing companies should not come to Singapore solely to make or assemble products designed elsewhere. We should also encourage operational headquarters to be set up here, to do research and development work, provide administrative, technical and management services to subsidiaries in the region, and manage their treasury activities. It would then become worthwhile to establish a plant in Singapore to produce for export.
>
> (Quoted in Dicken and Kirkpatrick, 1991, p. 176)

Singapore sets out to market itself as a global business centre on the basis of the very high quality of its physical and human infrastructure and its strategic geographical location. Government policies are geared towards this goal which also includes an explicit strategy to 'regionalize' the Singaporean economy by encouraging domestic firms to set up operations in southeast Asia while Singapore develops as the 'control centre' of a regional division of labour. One element in this is the promotion of the Singapore–Johor Bahru (Malaysia)–Batam (Indonesia) Growth Triangle. More ambitiously, the government is driving a series of initiatives using government-linked corporations to develop major infrastructural projects in Asia.[14] At the same time, the emphasis on research and development and technological upgrading continues through the creation of a number of specialist institutes, the National Technology Board and a major Science and Technology Park close to the National University.

China[15]

In discussing the development of South Korea, Taiwan and Singapore (as well as the other emerging NIEs in Asia) it is necessary to assert that the state has played a key role. In the case of China, of course, such a claim is unnecessary. As a centrally controlled command economy there is no doubt about the state's centrality. The point about China is that, after several decades of self-imposed separation from the world economy, it has become an immensely significant regional – and global – player. What makes China so significant in the long run is its sheer size – some 1.2 billion people – and its massive economic potential. Whether that potential will be realized, and how far China will come to dominate the Asian economy, is a matter for speculation. What is important, here, is to outline the dramatic changes in Chinese policy towards the rest of the world since 1979.

The Peoples' Republic of China (PRC) came into being in 1949 with the replacement of the nationalist government by a communist government led by Mao Zedong. For the next thirty years, China followed a policy of economic self-reliance. 'A country should manufacture by itself all the products it needs whenever and wherever possible . . . [self-reliance] also means that a country should carry on its general economic construction on the basis of its own human, material and financial resources' (1960s' Policy Statement, quoted in Cannon and Jenkins, 1990, p. 10).

This policy of self-reliance was pursued through a series of major – often extreme – initiatives. Initially, the new government followed the example of the USSR in establishing a Five Year Plan (1953–57). This relatively successful policy was jettisoned in 1958 when Mao announced the 'Great Leap Forward' – a total transformation of economic planning with the emphasis on small-scale and rural development. Although this initiated the notion of rural industrialization, the GLF was disastrous in its consequences with mass famine one of the results. In 1966, policy changed again with the introduction of the 'Cultural Revolution' – a phase which lasted for some ten years with, again, disastrous human and social implications. The period after Mao's death in 1976 was one of political hiatus which was eventually resolved by the emergence of Deng Xiaoping as leader. It was under Deng's leadership that China began to jettison the self-reliance policy of the previous thirty years and to make links with the world market economies. This has been done, however, without substantial political change. In the words of the new Party Constitution of 1997, it is 'Socialism with Chinese characteristics'. The draconian response to the Tiananmen Square demonstrations in May–June 1989 was one reflection of what that means.

The pivotal year was 1979 when China began its 'open policy' based upon a carefully controlled trade and inward investment strategy. This was set within the so-called 'Four Modernizations' (concerned with agriculture, industry, education, and science and defence). The open policy 'has not proceeded in a linear fashion. The pace, extent, forms and direction of opening have vacillated over the past fifteen years with advances and retreats which have their origins in both economic and political factors' (Thoburn and Howell, 1995, p. 170). A central element of the new policy has been the opening up of the Chinese economy to foreign direct investors. As we saw in Chapter 2, FDI has grown very rapidly indeed in China since the early 1980s and now accounts for 18 per cent of GDP (Table 2.10). The organizational form of these investments varies from wholly owned foreign subsidiaries to equity joint ventures with Chinese partners and other partnership arrangements.

The most distinctive feature of the open policy, however, is the explicit use of *geography* in its implementation. Partly in order to control the spread of capitalist market ideas and methods within Chinese society, and partly to make the policy more effective through external visibility and agglomeration economies, FDI has been steered to specific locations. Initially, the foci were the four Special Economic Zones[16] established in 1979 at Shenzhen, Zhuhai, Shantou and Xiamen (Figure 4.6). Apart from the desire to attract United States, Japanese and European investment,

> The four original zones were located with a view to maximizing their attraction to investment from ethnic Chinese living outside China. Shenzhen, the largest, was located in Guangdong province immediately adjacent to Hong Kong. Zhuhai was set up beside the Portuguese enclave of Macau, also in Guangdong. Shantou, in the north-east of Guangdong province was established in an area with many links with South-east Asian Chinese communities and Xiamen SEZ in Fujian province was intended to attract Taiwanese investors.
>
> (Thoburn and Howell, 1995, p. 173)

The Chinese Special Economic Zones share some features in common with Export Processing Zones (see above), although they are generally much larger. They offer a package of incentives, including tax concessions, duty-free import arrangements

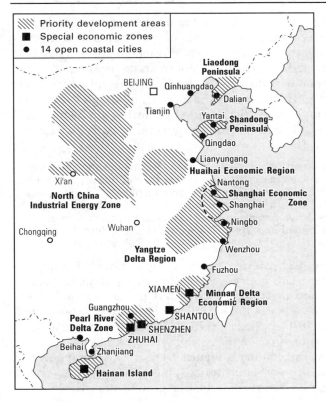

Figure 4.6   The spatial expression of China's 'open policy': Special Economic Zones, open cities and priority development areas
Source: Based on Phillips and Yeh (1990, Figure 9.4)

and serviced infrastructure. But the Chinese SEZs play additional roles: 'The SEZs had the initial objective of manufacturing to produce goods for export to earn foreign exchange but, very importantly, they have been regarded as social and economic laboratories, in which foreign technological and managerial skills might be observed and adopted, albeit selectively' (Phillips and Yeh, 1990, p. 236). Certainly, the focus on southern Guangdong province is also related to the incorporation of Hong Kong and Macau into the PRC in the late 1990s. Well before the political control of Hong Kong reverted to China, Hong Kong businesses had set up thousands of factories in the municipalities just across the border. The original SEZs were also located in areas well away from the major urban and industrial areas in order to control the extent of their influence. However, as Figure 4.6 shows, there has been considerable development of other kinds of externally oriented locations in the form of 'open coastal cities' and 'priority development areas'.

Despite these inflows of foreign capital and technologies, China remains a centrally controlled command economy in which state-owned enterprises (SOEs) predominate, at least in employment terms though not in terms of their share of industrial output. The SOE share of China's industrial output has fallen from 65 per cent in 1985 to around 30 per cent in 1995 (*The Economist*, 14 December 1996). It is clear that the organizational structure of the Chinese economy is in a state of flux with much increased variety of forms. But the central issue is the likely direction of economic policy after Deng Xiaoping's death in 1997. As in the past, the key lies in the internal political power struggles between the 'modernizers' who wish to sustain and develop the open policies of the recent past and those who wish to retain a degree of isolation.

Mexico[17]

Our final case example of national industrialization policies is drawn from the Americas. Following the financial collapse of 1982 – when Mexico's economic turmoil precipitated the international debt crisis – Mexico has undergone dramatic political, social and economic change as the state has attempted to integrate the national economy more strongly into the global economy. But the path of transition from a strongly inward-oriented industrialized policy position to an export-oriented position has been far from smooth. Two basic characteristics are important to an understanding of the Mexican case. The first is its location next door to the world's dominant political and economic power, the United States. The second is the fact that although Mexico has been governed, since 1929, by a single party (the Partido Revolucionario Institucional – PRI) reformist pressures both inside and outside the governing party have intensified, particularly since the 1980s. Thus,

> what was once seen as a predictable and stable political system became the arena for new tensions and conflicts. The onset of economic crisis coincided with, and in part, led to the emergence of a new governing elite of young technocrats, the independent mobilization of a new business class . . . , the growth of grassroots popular movements and the unification of the left . . . All these developments implied a serious erosion of corporatist ties between state and society which had underpinned Mexican stability since the 1930s.
>
> (Harvey, 1993, p. 4)

For almost sixty years from 1929, Mexico pursued a predominantly import-substituting industrialization policy. This policy was, for the most, part quite successful. The large domestic market and a strong physical and human resource base, together with proximity to the United States market, allowed Mexico to grow at high rates, especially during the 1950s–1970s period. The high returns from the country's oil reserves, in particular, had a major effect on the economy. It was the collapse of oil prices in 1981 which precipitated the country's financial crisis. One component of Mexico's export policy during the import-substitution phases was the *maquiladora* programme which created a very specific form of industrial growth along the Mexico–United States border (see Figure 2.31). Between 1982 and 1985, government policy emphasized stabilization rather than structural adjustment but external pressures from the IMF resulted in a pronounced shift in policy emphasis:

> This round of crisis and reconciliation led to important shifts within the Mexican administration. Economic difficulties had already strengthened the hand of the technocrats within the administration . . . the stabilization efforts of 1985–86 were accompanied by the initiation of trade reform and negotiations that led to accession to GATT. In the Solidarity Pact of 1988 . . . the government used the much-needed stabilization programme to counter strong resistance from the private sector to further trade policy reform. The government cut the maximum tariff level from 100 per cent to 20 per cent, the average tariff level dropped to just over 10 per cent.
>
> (Haggard, 1995, p. 81)

During the second half of the 1980s, the Mexican government pursued a wide-ranging programme of liberalization involving both the deregulation of some areas of the economy and the privatization of state-owned enterprises. By 1993, some 90 per cent of Mexico's state-owned enterprises had been privatized either wholly or partially although the government retained control of some key companies, including PEMEX, the state oil company. The single most important act of the Salinas administration, which came to power in 1988, was to take Mexico into the NAFTA (see Chapter 3). This symbolized the government's headlong attempt to tie the domestic economy firmly into the global system and, in particular, to gain greater benefits from a formal association with the United States. One result of the major domestic reforms and the opening up of the economy was a very rapid buildup of speculative foreign capital ('hot money') which drove up the value of the peso to unsustainable levels. In

December 1994, the new Zedillo administration was forced into a massive currency devaluation which heralded a new economic crisis for the country.

Lall neatly summarizes the nature and impact of Mexican industrial policy over the longer term as follows:

> A long history of inward-oriented industrialization, combined with a sizeable internal market, a respectable base of natural and human resources, and a location next to the world's most powerful industrial nation, enabled Mexico to build up a large and diversified industrial sector. For several decades Mexican manufacturing value added ranked third in the developing world . . . The 1980s witnessed a considerable slowdown accompanied by severe fluctuation, as macroeconomic disturbances and accompanying stabilization policies took their toll on the economy.
>
> In the mid-1980s, Mexico launched a sweeping and thorough process of economic liberalization to restructure and upgrade its industrial sector, changing its trade orientation, privatizing many State enterprises and generally reducing the role of the State in economic activity.
>
> [However] . . . in general there was a tradition of excessive technological dependence in Mexican industry in comparison to countries like the Republic of Korea or Taiwan Province of China. There was a high level of reliance on imported capital goods for a country of its industrial size and sophistication. This was accompanied by a similar and widespread reliance on inflows of foreign know-how, licences and expertise through much of industry. Despite the nationalistic stance of the government, there was a relatively large presence of foreign multinational enterprises in the advanced sectors of Mexican industry . . . It . . . restricted the ability of Mexican industry to move into technologically more dynamic or sophisticated industries. While the Republic of Korea used import substitution to foster industrialization, it also pursued a strategy of independent industrial technological development. Mexico was never as export-oriented, nationalistic, State-led and technologically ambitious as the Republic of Korea.
>
> (Lall, 1994, p. 81)

Conclusion

The aim of this chapter has been to outline the diversity of ways in which states occupying different positions within the global economy, and having different political-ideological stances, have attempted to influence their national economies. The chapter has shown the broad relevance of the distinction made in Chapter 3 between developmental and regulatory states although, as the individual cases demonstrate, the boundary between these two ideal types is often rather fuzzy. It is also apparent that states often shift within these two categories and also move towards the boundary between them at different times. Although there are broad similarities between groups of countries – for example, between developed economies on the one hand and developing economies on the other – there is enormous internal diversity within those two broad groupings. This is an especially important point to make in respect of the east and southeast Asian economies which are all too readily lumped together as if they have followed a single developmental policy path. As this chapter has shown, this is far from being the case, even though it is certainly true that these economies share some common features. As in other aspects of the global economy, it is extremely important to beware simple stereotypes.

Notes for further reading

1. See, for example, Tolchin and Tolchin (1988), Glickman and Woodward (1989), Graham and Krugman (1989). Particular concern has been directed at the perceived 'Japanese invasion'

(see, for example, Choate, 1990). There was even a best-selling novel – Crichton (1992) – whose plot was based on this same fear.

2. See Stegemann (1989), Ostry (1990), Richardson (1990), Tyson (1993), Yoffie (1993), Krugman (1986; 1990; 1996).

3. The Trade Act 1974 contained a clause 301 which met the GATT criteria of dealing with unfair trade practices. The 1988 Act adapted this clause to a strongly unilateral measure. For a very critical view of the Omnibus Trade and Competitiveness Act 1988, see Bhagwati (1989). He calls it the 'ominous' trade and competitiveness Act.

4. Reviews of Japanese economic policy are provided by Johnson (1982; 1985), Dore (1986), Inoguchi and Okimoto (1988), Okimoto (1989), Odagiri (1992).

5. As Porter (1990, p. 414) observes, 'overall, what has most separated Japanese policy from French "Indicative Planning" and other past efforts at national economic planning is a much greater stress on competition'.

6. Gereffi and Wyman (1990) have edited an excellent collection of essays which, together, analyse the diverse 'paths of industrialization' in Latin America and east Asia. See also Stallings (1995, Part II), Brohman (1996).

7. Douglass (1994, p. 543) argues with some force that it is highly misleading to group the Asian NIEs into an 'undifferentiated model of the "developmental state". A closer examination reveals significant differences among them in terms of the state's relations to capital, labour, and the external economy'.

8. See, for example, Lall (1994), Wade and Evans (1994), Singh (1994; 1995).

9. See ILO (1988), Sklair (1989), Chant and McIlwaine (1995).

10. Among the most useful accounts of South Korean industrialization policy are Amsden (1989), Wade (1990a; 1990b), Koo and Kim (1992).

11. Korean government policies in the automobile and electronics industries will be discussed in Chapters 10 and 11.

12. This account of Taiwan's developmental strategies is based primarily on Wade (1990a; 1990b). See also Lall (1994).

13. For discussions of Singapore's developmental policies see Lim and Pang (1986), Dicken (1987), Lim (1988), Dicken and Kirkpatrick (1991), Rodan (1991), Lall (1994), Ramesh (1995), Yeung (1998).

14. The 'regionalization' strategy of Singapore is discussed in detail by Yeung (1998).

15. Useful accounts of Chinese economic development policy are provided in Wong, Lau and Li (1988), Cannon and Jenkins (1990), Crane (1990), Dwyer (1994), Benewick and Wingrove (1995).

16. The Special Economic Zones are analysed by Crane (1990), Phillips and Yeh (1990), Thoburn and Howell (1995), Wong and Chu (1985).

17. Mexican industrialization policies are discussed in Sklair (1989), Villareal (1990), Harvey (1993), Lall (1994), Haggard (1995).

CHAPTER 5

Technology: the 'great growling engine of change'

Technology and economic transformation

Technological change is at the heart of the process of economic growth and economic development. As Joseph Schumpeter (1943, p. 83) pointed out many years ago, 'the fundamental impulse that sets and keeps the capitalist engine in motion comes from the new consumers' goods, the new methods of production or transportation, the new markets, the new forces of industrial organization that capitalist enterprise creates'. Technological change is the 'prime motor of capitalism'; the 'great growling engine of change' (Toffler, 1971); the 'fundamental force in shaping the patterns of trans-formation of the economy' (Freeman, 1988); the 'chronic disturber of comparative advantage' (Chesnais, 1986). Although technologies, in the form of inventions and innovations, originate in specific places, they are no longer confined to such places. Innovations spread or diffuse with great rapidity under current conditions. Indeed, one of the most significant sets of innovations is in the sphere of communications, which itself facilitates such technological diffusion. As we shall see, however, this does *not* signal the 'death of distance' or 'the end of geography'. Indeed, there continues to be a pronounced geography of knowledge creation and a strong geographical localization of innovative activity.

Technology is, without doubt, one of the most important contributory factors underlying the internationalization and globalization of economic activity:

> technological change, through its impact on the economics of production and on the flow of information, is a principal factor determining the structure of industry on a national scale. This has now become true on a global scale. Long-term technological trends and recent advances are reconfiguring the location, ownership, and management of various types of productive activity among countries and regions. The increasing ease with which technical and market knowledge, capital, physical artefacts, and managerial control can be extended around the globe has made possible the integration of economic activity in many widely separated locations. In doing so, technological advance has facilitated the rapid growth of the multinational corporation with subsidiaries in many countries but business strategies determined by headquarters in a single nation.
>
> (Brooks and Guile, 1987, p. 2)

However, in looking specifically at technology in this chapter, we need to beware of adopting a position of technological determinism. It is all too easy to be seduced by the notion that technology 'causes' a specific set of changes, makes particular structures and arrangements 'inevitable' or that the path of technological change is linear and sequential. Technology in, and of, itself does not cause particular kinds of change. In one sense, then, technology is an *enabling* or *facilitating* agent: it makes possible new structures, new organizational and geographical arrangements of economic activities, new products and new processes, while not making particular outcomes inevitable.

But in certain circumstances technology may, indeed, be more of an *imperative*. In a highly competitive environment, once a particular technology is in use by one firm, then its adoption by others may become virtually essential to ensure competitive survival. More generally, as Freeman (1982, p. 169) points out, for business firms 'not to innovate is to die'.

In this chapter we focus only on certain aspects of technology and technological change: those which specifically influence the processes of internationalization and globalization of economic activity. The chapter is divided into four major parts:

- First, some of the broad characteristics of technological change are discussed in order to identify the key technologies and their evolution over time.
- Second, we focus on the 'space shrinking' technologies of transport and communication which are obviously central to the processes of internationalization and globalization.
- Third, we look at technological changes in both products and processes and explore the extent to which totally new forms of production technology and organization are occurring.
- Fourth, we focus explicitly on the geography of innovation, in particular on the tendency of innovative activity and knowledge creation to be geographically localized in 'technology districts'.

The process of technological change: an evolutionary perspective[1]

Technological change is a form of *learning*: of how to solve specific problems in a highly diverse, and often volatile, environment. It is, however, more than a narrowly 'technical' process. Technology is also not independent or autonomous; it does not have a life of its own. *Technology is a social process* which is socially and institutionally embedded. It is created and adopted (or not) by human agency: individuals, organizations, societies. The ways in which technologies are used – even their very creation – are conditioned by their social and their economic context. In effect, from the viewpoint taken here, this means the values and motivations of capitalist business enterprises operating within an intensely competitive system. Choices and uses of technologies, therefore, are influenced by the drive for profit, capital accumulation and investment, increased market share and so on.

A typology of technological change
Freeman and Perez (1988) identify four broad types of technological change, each of which is progressively more significant:

- *Incremental innovations*: the small-scale, progressive modification of existing products and processes:

   > They may often occur, not so much as the result of any deliberate research and development activity, but as the outcome of inventions and improvements suggested by engineers and others directly engaged in the production process, or as a result of initiatives and proposals by users ('learning by doing' and 'learning by using') . . . Although their combined effect is extremely important in the growth of productivity, no single incremental innovation has dramatic effects, and they may sometimes pass unnoticed and unrecorded.
   >
   > (Freeman and Perez, 1988, p. 46)

- *Radical innovations*: discontinuous events which may drastically change existing products or processes. A single radical innovation will not, however, have a widespread effect on the economic system; 'its economic impact remains relatively small and localized unless a whole cluster of radical innovations are linked together in the rise of new industries and services, such as the synthetic materials industry or the semiconductor industry' (Freeman, 1987, p. 129).
- *Changes of 'technology system'*: 'these are far-reaching changes in technology, affecting several branches of the economy, as well as giving rise to entirely new sectors. They are based on a combination of radical and incremental innovations, together with *organizational* and *managerial* innovations affecting more than one or a few firms' (Freeman and Perez, 1988, p. 46). Freeman (1987) suggests that the following five 'generic' technologies have created such new technology systems:

  1) information technology
  2) biotechnology
  3) materials technology
  4) energy technology
  5) space technology.

- *Changes in the techno-economic paradigm*: the truly large-scale revolutionary changes which are

  > the 'creative gales of destruction' that are at the heart of Schumpeter's long wave theory. They represent those new technology systems which have such pervasive effects on the economy as a whole that they change the 'style' of production and management throughout the system. The introduction of electric power or steam power or the electronic computer are examples of such deep-going transformations. A change of this kind carries with it many clusters of radical and incremental innovations, and may eventually embody several new technology systems. Not only does this fourth type of technological change lead to the emergence of a new range of products, services, systems and industries in its own right – it also affects directly or indirectly almost every other branch of the economy . . . the changes involved go beyond specific product or process technologies and affect the input cost structure and conditions of production and distribution throughout the system.
  >
  > (Freeman, 1987, p. 130)

Long waves[2]

The notion that global economic growth occurs in a series of long waves of more or less fifty-years' duration is generally associated with the work of the Russian economist N.D. Kondratiev in the 1920s, although he did not invent the idea. Figure 5.1 outlines the kind of long-wave sequence commonly envisaged. Four complete K-waves are identified with the implication that we are now in the early stages of a fifth. Each wave lasts for approximately fifty years and appears to be divided into four phases: prosperity, recession, depression, recovery. Each wave tends to be associated with particularly significant technological changes around which other innovations – in production, distribution and organization – cluster and ultimately spread through the economy. Such diffusion of technology stimulates economic growth and employment, although technology alone is not a sufficient cause of economic growth: demographic, social, industrial, financial and other demand conditions have 'also to be right'. At some point, however, growth slackens: demand may become saturated or firms' profits squeezed through intensified competition. As a result, the level of new investment falls, firms strive to rationalize and restructure their operations and

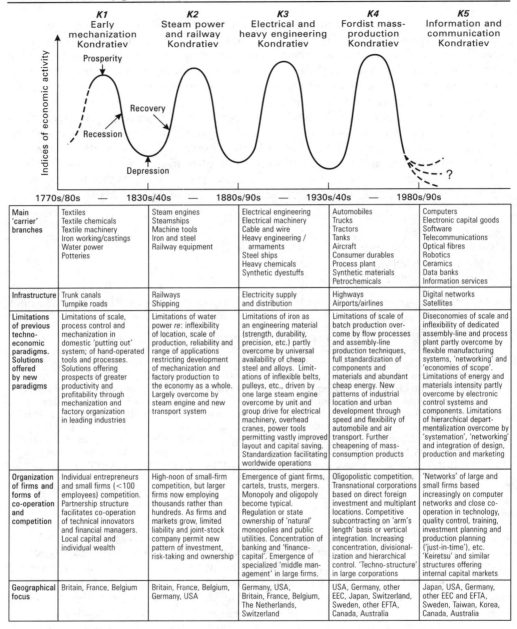

	K1 Early mechanization Kondratiev	**K2** Steam power and railway Kondratiev	**K3** Electrical and heavy engineering Kondratiev	**K4** Fordist mass-production Kondratiev	**K5** Information and communication Kondratiev
Main 'carrier' branches	Textiles Textile chemicals Textile machinery Iron working/castings Water power Potteries	Steam engines Steamships Machine tools Iron and steel Railway equipment	Electrical engineering Electrical machinery Cable and wire Heavy engineering / armaments Steel ships Heavy chemicals Synthetic dyestuffs	Automobiles Trucks Tractors Tanks Aircraft Consumer durables Process plant Synthetic materials Petrochemicals	Computers Electronic capital goods Software Telecommunications Optical fibres Robotics Ceramics Data banks Information services
Infrastructure	Trunk canals Turnpike roads	Railways Shipping	Electricity supply and distribution	Highways Airports/airlines	Digital networks Satellites
Limitations of previous techno-economic paradigms. Solutions offered by new paradigms	Limitations of scale, process control and mechanization in domestic 'putting out' system; of hand-operated tools and processes. Solutions offering prospects of greater productivity and profitability through mechanization and factory organization in leading industries	Limitations of water power *re*: inflexibility of location, scale of production, reliability and range of applications restricting development of mechanization and factory production to the economy as a whole. Largely overcome by steam engine and new transport system	Limitations of iron as an engineering material (strength, durability, precision, etc.) partly overcome by universal availability of cheap steel and alloys. Limitations of inflexible belts, pulleys, etc., driven by one large steam engine overcome by unit and group drive for electrical machinery, overhead cranes, power tools permitting vastly improved layout and capital saving. Standardization facilitating worldwide operations	Limitations of scale of batch production overcome by flow processes and assembly-line production techniques, full standardization of components and materials and abundant cheap energy. New patterns of industrial location and urban development overcome through speed and flexibility of automobile and air transport. Further cheapening of mass-consumption products	Diseconomies of scale and inflexibility of dedicated assembly-line and process plant partly overcome by flexible manufacturing systems, 'networking' and 'economies of scope'. Limitations of energy and materials intensity partly overcome by electronic control systems and components. Limitations of hierarchical depart-mentalization overcome by 'systemation', 'networking' and integration of design, production and marketing
Organization of firms and forms of co-operation and competition	Individual entrepreneurs and small firms (<100 employees) competition. Partnership structure facilitates co-operation of technical innovators and financial managers. Local capital and individual wealth	High-noon of small-firm competition, but larger firms now employing thousands rather than hundreds. As firms and markets grow, limited liability and joint-stock company permit new pattern of investment, risk-taking and ownership	Emergence of giant firms, cartels, trusts, mergers. Monopoly and oligopoly become typical. Regulation or state ownership of 'natural' monopolies and public utilities. Concentration of banking and 'finance-capital'. Emergence of specialized 'middle management' in large firms.	Oligopolistic competition. Transnational corporations based on direct foreign investment and multiplant locations. Competitive subcontracting on 'arm's length' basis or vertical integration. Increasing concentration, divisional-ization and hierarchical control. 'Techno-structure' in large corporations	'Networks' of large and small firms based increasingly on computer networks and close co-operation in technology, quality control, training, investment planning and production planning ('just-in-time'), etc. 'Keiretsu' and similar structures offering internal capital markets
Geographical focus	Britain, France, Belgium	Britain, France, Belgium, Germany, USA	Germany, USA, Britain, France, Belgium, The Netherlands, Switzerland	USA, Germany, other EEC, Japan, Switzerland, Sweden, other EFTA, Canada, Australia	Japan, USA, Germany, other EEC and EFTA, Sweden, Taiwan, Korea, Canada, Australia

Figure 5.1   Kondratiev long waves and their basic characteristics
Source: Based in part on material in Freeman and Perez (1988, Table 3.1)

unemployment rises. Of course, a central assumption of the long-wave idea is that eventually the trough of the wave will be reached and economic activity will turn up again. A new sequence will be initiated on the basis of key technologies – some of which may be based on innovations which emerged during recession itself – and new investment opportunities.

Although there is some disagreement over the precise mechanisms and timing involved, it is generally agreed that major technological changes are associated with long waves. In fact, each of the waves is associated with changes in the techno-economic paradigm; with the culmination of a technology revolution, as one set of techno-economic practices is displaced by a new set. This is not a sudden process but one which occurs gradually and involves the ultimate 'crystallization' of a new paradigm. Essentially,

> a new techno-economic paradigm develops initially within the old, showing its decisive advantages during the 'downswing' phase of the previous Kondratiev cycle. However, it becomes established as a dominant technological regime only after a crisis of structural adjustment, involving deep social and institutional changes, as well as the replacement of the motive branches of the economy.
>
> (Freeman and Perez, 1988, p. 47)

As Figure 5.1 shows, however, the process of change involves more than technical change alone. Each phase is also associated with characteristic forms of economic organization, co-operation and competition. The trajectory of organizational change has followed a path from an early focus on individual entrepreneurs in K1, through small firms, but of larger average size in K2, to the monopolistic, oligopolistic and cartel structures of K3, the centralized, hierarchical TNCs of K4 and, it is argued, the 'network' and alliance organizational forms of K5. (These are issues we will explore in Chapter 7.) Each successive K-wave also has a specific geography; technological leadership in one wave is not necessarily maintained in succeeding waves:

> One of the most compelling facts of history is that there have been enormous differences in the capacity of different societies to generate technical innovations that are suitable to their economic needs . . . individual societies have themselves changed markedly over the course of their own separate histories in the extent and intensity of their technological dynamism.
>
> (Rosenberg, 1982, p. 8)

The technological leaders of K1 were Britain, France and Belgium. In K2 these were joined by Germany and the United States. K3 saw leadership firmly established in Germany and the United States although the other earlier leaders were still prominent and had been joined by Switzerland and The Netherlands. By K4 Japan, Sweden and the other industrialized countries were in the leadership group. K5 has seen a more prominent role in technological leadership by Japan and, more unexpectedly, by the emergence of two of the east Asian NIEs – Taiwan and South Korea – to prominent technological positions in specific areas.

There is a further dimension to the geography of technological leadership:

> just as on the international stage, so within each of the leading national economies: the locus of the leading-edge innovative industries has switched from region to region, from city to city. From the birth of New IT in the second Kondratiev and on through the third, Berlin and the Boston–New York Corridor were the main global centres of innovation; during the fourth, they were supplemented or supplanted by new urban centres such as Southern California, Silicon Valley, the Western Crescent around London, the Stuttgart–Munich Corridor, and the Tokaido Megalopolis.
>
> (Hall and Preston, 1988, p. 6)

Information technology: a key generic technology[3]

The fifth Kondratiev cycle, which appears to have begun in the 1980s and 1990s, is associated primarily with the first of the five 'generic' technologies referred to above: *information technology* (IT). 'The contemporary change of paradigm may be seen as a shift from a technology based primarily on cheap inputs of energy to one predominantly based on cheap inputs of information derived from advances in microelectronic and telecommunications technology' (Freeman, 1988, p. 10). Information technology, therefore, is the new techno-economic paradigm around which the next wave of technological and economic changes will cluster. But, as Hall and Preston (1988, p. 30) point out, information technology in itself is nothing new: 'for thousands of years, since the first cave paintings and the invention of writing, humans have used tools and techniques to collect, generate and record data'. Consequently, they identify three main phases of information technology:

- Simple pictorial representation and written language, evolving eventually into printing: its basic elements were paper, writing instruments, ink and printing presses.
- Mechanical, electromechanical and early electronic technologies which developed during the late nineteenth and early twentieth centuries: the basic elements were the telephone, typewriter, gramophone/phonograph, camera, tabulating machine, radio and television.
- Microelectronic technologies, which emerged only in the second half of the twentieth century: the basic elements are computers, robots and other information-handling production equipment, and office equipment (including facsimile machines).

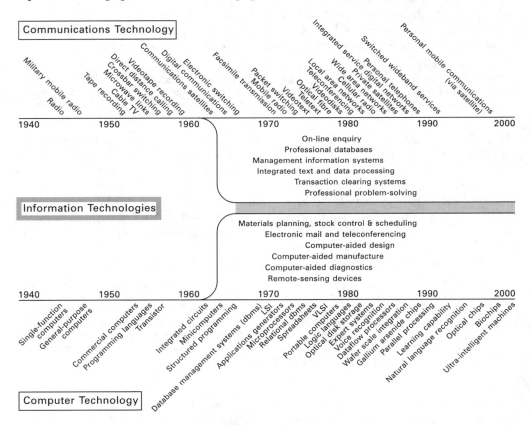

Figure 5.2   Information technology: the convergence of communication technologies and computer technologies
Source: Based on Freeman (1987, Figure 2)

Hall and Preston regard the first of these as 'old IT'; the second and third together as 'new IT'. They then employ a further term 'convergent IT' to refer to the newest advances of the 1970s and 1980s, whereby computers and telecommunications became integrated into a single system of information processing and exchange.

It is this quality of the *convergence* of two initially distinct technologies which is of the greatest importance for developments in today's (and tomorrow's) global economy. It is this kind of information technology which is most significant for the processes of internationalization and globalization of economic activities. When we use the term 'information technology' or 'IT' in this and the following chapters it is the 'convergent IT' which is involved. Figure 5.2 shows the nature of this convergence between communications technology, which is concerned with the transmission of information, and computer technology, which is concerned with the processing of information. As the diagram indicates, it is not until the early 1960s that we can clearly identify convergent information technology.

The 'space-shrinking' technologies[4]

A fundamental prerequisite of the evolution of international production and of the transnational corporation is the development of technologies which overcome the frictions of space and time. The most important of such enabling technologies – and the most obvious – are the technologies of transport and communications. Neither of these technologies can be regarded as the cause of international production or of the TNC; rather, they make such phenomena feasible. But without them, today's complex global economic system simply could not exist. Indeed, both the geographical and organizational scale at which any human activity can occur is directly related to the available media of transport and communication. Similarly, the degree of geographical specialization – the spatial division of labour – is constrained by these media.

Transport and communication technologies perform two distinct, though closely related and complementary roles:

- *Transport systems* are the means by which materials, products and other tangible entities (including people) are transferred from place to place.
- *Communication systems* are the means by which information is transmitted from place to place in the form of ideas, instructions, images and so on.

For most of human history, transport and communications were effectively one and the same. Prior to the invention of electric technology in the nineteenth century, information could move only at the same speed, and over the same distance, as the prevailing transport system would allow. Electric technology broke that link, making it increasingly necessary to treat transport and communication as separate, though intimately related, technologies. Developments in both have transformed our world, permitting unprecedented mobility of materials and products and a globalization of markets.

Major developments in transport technology

In terms of the time it takes to get from one part of the world to another there is no doubt that the world has 'shrunk' dramatically (Figure 5.3). For most of human history, the speed and efficiency of transport were staggeringly low and the costs of overcoming the friction of distance prohibitively high. Movement over land was

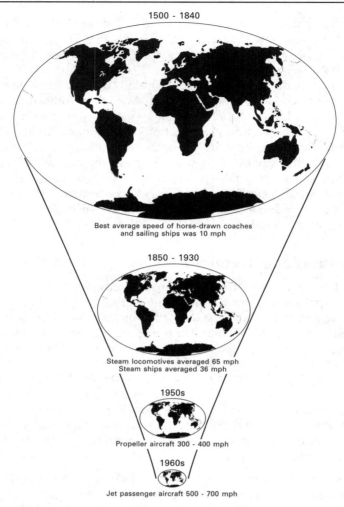

Figure 5.3   Global shrinkage: the effect of changing transport technologies on 'real' distance
Source: Based on McHale (1969, Figure 1)

especially slow and difficult before the development of the railways. Indeed, even as late as the early nineteenth century, the means of transport were not really very different from those prevailing in biblical times. The major breakthrough came with two closely associated innovations: the application of steam power as a means of propulsion and the use of iron and steel for trains and railway tracks and for ocean-going vessels. These, coupled with the linking together of overland and ocean transport and the cutting of the canals at Suez and Panama, greatly telescoped geographical distance at a global scale. The railway and the steamship introduced a new, and much enlarged, scale of human activity. The flow of materials and products was enormously enhanced and the possibilities of geographical specialization were greatly stimulated. Such innovations were a major factor in the massive expansion in the global economic system during the nineteenth century.

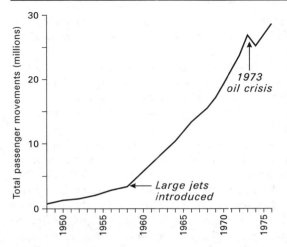

Figure 5.4 The 'take-off' of air traffic after the introduction of the jet aircraft
Source: Based on Cherry (1978, Figure 3.16)

The twentieth century, and especially the past few decades, has seen an acceleration of this process of global shrinkage. In economic terms, the most important developments have been the introduction of commercial jet aircraft, the development of much larger ocean-going vessels (superfreighters) and the introduction of containerization, which greatly simplifies transhipment from one mode of transport to another and increases the security of shipments. Of these, it is the jet aircraft which has had the most pervasive influence, particularly in the development of the TNC. It is no coincidence that the take-off of TNC growth and the (more

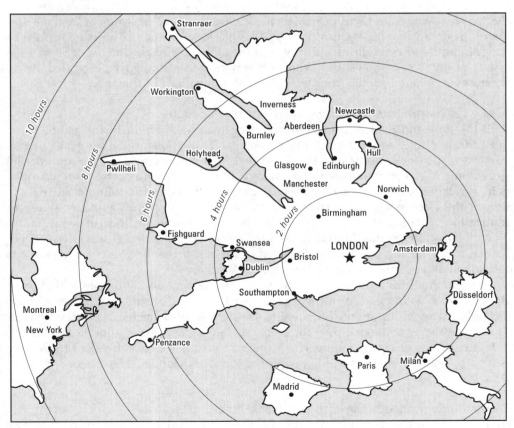

Figure 5.5 The unevenness of time-space convergence
Source: Based on Dicken and Lloyd (1981, Figure 2.7)

literal) take-off of commercial jets both occurred during the 1950s. As a consequence, in terms of time, New York is now closer to Tokyo than it was to Philadelphia in the days of the thirteen colonies. Figure 5.4 gives some indication of the effect of the introduction of large commercial jets on international air traffic from and to London. From the late 1950s the growth curve was virtually a straight line with a very steep upward slope.

However, although the world has shrunk in relative terms, we need to be aware that, contrary to the impression given by Figure 5.3, such shrinkage is highly uneven. The technological developments in transport (and in communications) tend to be geographically concentrated. What the geographer Donald Janelle called *time-space convergence* affects some places more than others. While the world's leading national economies and the world's major cities are being pulled closer together others – less industrialized countries or smaller towns and rural areas – are, in effect, being left behind. Figure 5.5 shows this process of differential geographical shrinkage for just one country, Britain, in the 1960s. It shows the fastest travel times from London to other parts of the country. The effects of the major investments in transport tech-nologies of that time were to pull the bigger cities, like Manchester and Birmingham, Glasgow and Edinburgh, closer to London. But the effect of differential investment in transport technologies was very different for other places. For example, the former cotton textiles town of Burnley, though only 25 miles north of Manchester, became twice as far away from London as Manchester. The small Welsh town of Pwllheli was almost as far away from London as were New York or Montreal. The same process of geographically uneven shrinkage applies at the global scale. The time-space surface is highly plastic; some parts shrink whilst other parts become, in effect, extended. By no means everywhere benefits from technological innovations in transport.

Major developments in communications technology

Both the time and relative cost of transporting materials, products and people have fallen dramatically as the result of technological innovations in the transport media. However, such developments have depended, to a considerable degree, on parallel developments in communications technology. In the nineteenth century, for example, neither rail nor ocean transport could have developed as they did without the innova-tion of the electric telegraph and, later, the oceanic cable. Only with the ability to transmit information at great speed – for example to co-ordinate flows of commodities on a global scale – could the potential of the transport technologies be fully realized. Similarly, the far more complex global transport system of the present day depends fundamentally on telecommunications technology.

The communications media are, however, fundamentally significant in their own right. Indeed, as implied in our earlier discussion of the central role of information technology, communications technologies should now be regarded as the key technol-ogy transforming relationships at the global scale. 'The new telecommunications tech-nologies are the electronic highways of the informational age, equivalent to the role played by railway systems in the process of industrialization' (Henderson and Cas-tells, 1987, p. 6).

Transmission channels: satellites and optical fibres

Global communications systems have been transformed radically during the past twenty or thirty years through a whole cluster of significant innovations in informa-tion technology (see Figure 5.2). Probably the most important catalyst to enhanced global communications was the development of satellite technology. The use of satellites for commercial telecommunications dates only from 1965, when the Early Bird or Intelsat I satellite was launched. It was the first 'geostationary' satellite, located above the Atlantic Ocean and capable of carrying 240 telephone conversations or two television channels simultaneously. Since then the carrying capacity of the communications satellites has grown exponentially. Intelsat IV carried 6,000 simultaneous telephone conversations; Intelsat V carries 12,000 as well as television channels; Intelsat VIII (launched in 1995) carries 22,500 two-way telephone circuits and three television channels. Its capacity can be increased to 112,500 two-way telephone circuits with the use of digital multiplication equipment.[5] The Intelsat system is a multi-nation consortium of 122 countries whose satellites are positioned to provide complete global coverage. In addition to Intelsat, there are regional systems, such as Eutelsat, which serves Europe, and others in Asia, Latin America and the Middle East and also private satellite systems, such as the PanAm Global Satellite System. The PanAm system was the first private system to provide global satellite services.

Satellite technology, together with a whole host of other communications technologies, is making possible quite remarkable levels of global communication of conventional messages and also the transmission of data. In this respect, the key element is the linking together of computer technologies with information-transmission technologies over vast distances. It has become possible for a message to be transmitted in one location and received in another on the other side of the world virtually simultaneously. Consequently,

> communications costs are becoming increasingly insensitive to distance. The crucial fact is the economics of satellite communication. Within the beam of a satellite it makes no difference to costs whether you are transmitting for five hundred miles or five thousand miles. The message goes from the earth station up twenty-two thousand three hundred miles to the satellite and down again twenty-two thousand three hundred miles. It makes no difference whether the two points on earth are close together or far apart . . . The important point about satellites is that their existence sets a limit on the extent to which costs are a function of distance. Many other technologies may compete with satellites, but in the end, satellite communication will ultimately be cheaper. Whatever that distance, it makes no difference how much further one communicates, the costs will be the same. Under those circumstances, the cost of access to any particular data base or information service becomes largely independent of its location. That does not make it free nor even necessarily cheap. There are costs for compiling the data and costs for manipulating it.
>
> (de Sola Pool, 1981, pp. 162–63)

Satellite communications are now being challenged by a new technology: optical fibre cables. Optical fibre systems have a very large carrying capacity, and transmit information at very high speed and with a high signal strength. 'Each hair-like strand can now accommodate up to 60,000 simultaneous telephone calls (as opposed to 6–7,000 for a much wider coaxial cable)' (Graham and Marvin, 1996, p. 18).

Figure 5.6 shows the rapid increase in satellite and cable capacity across the Atlantic Ocean; the basic skeleton of the optical fibre communications system in the Pacific; and the 'global digital highway' scheme designed to link the world's three major markets of North America, western Europe and Japan, using a network of optical fibre cables capable of carrying 100,000 simultaneous messages. Within these very large-scale developments other smaller, but highly significant, changes are occurring. In the early postwar development of telecommunications the major medium was the telex. Within the last few years this has been substantially displaced by the fax

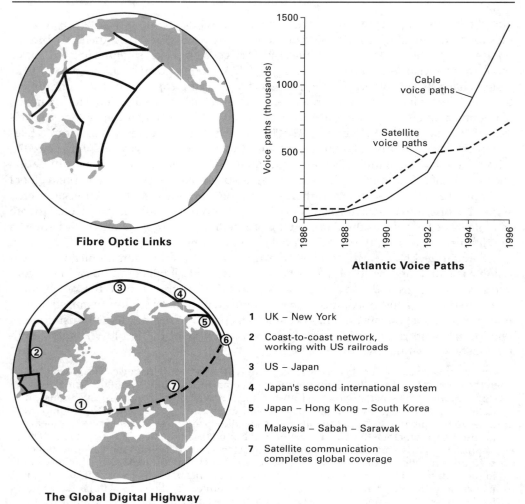

Fibre Optic Links

Atlantic Voice Paths

1   UK – New York

2   Coast-to-coast network,
    working with US railroads

3   US – Japan

4   Japan's second international system

5   Japan – Hong Kong – South Korea

6   Malaysia – Sabah – Sarawak

7   Satellite communication
    completes global coverage

The Global Digital Highway

Figure 5.6   Developments in global telecommunications: satellites and optical fibre networks
Source: Based on material in *The Sunday Times* (6 December 1987, p. 79); Warf (1989, Figure 3);
Graham and Marvin (1996, Figure 1.6)

(facsimile machine), which has experienced quite phenomenal growth and, even more
recently, by electronic mail using the Internet.

   One result of all these developments has been a sharp decline in the cost of
telecommunications services. Not only are transmission costs by satellite insensitive
to distance but also the user costs have fallen dramatically. In the 1960s the annual
cost of an Intelsat telephone circuit, giving a connection from one point on the earth's
surface to any other, was more than $60,000. In the late 1980s the same facility cost
only $9,000. In 1973, the price of a three-minute international telephone call from
London to New York was £13.73 (at 1994 prices). In 1993, the same call cost £1.78 (*The
Financial Times*, 17 September 1994). Even so, the cost of international telecommunications remains far higher than should be the case in purely technological terms. This is

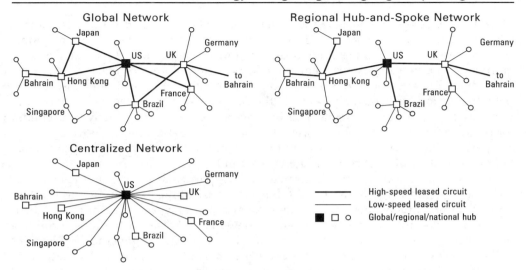

Figure 5.7 Major types of international leased telecommunications networks for a United States TNC
Source: Langdale (1989, Figure 1)

because the telecommunications industry has been very highly regulated at the national scale, despite the moves towards deregulation in some parts of the world. Early in 1997, however, the WTO engineered an agreement to liberalize global telecommunications markets which will greatly intensify competition as formerly protected national markets are opened up to the entry of foreign companies.

Nevertheless, only the very large organization, whether business or government, yet has the resources to utilize fully the new communications technologies. For the TNC, however, they have become essential to its operations.[6] All TNCs operate immense international telecommunications networks. They are the major users of international leased telecommunications networks, which permit them to transmit their internal communications at great speed to other parts of their international corporate network. 'One reason for the importance of international leased networks for large TNCs is that the unit cost of a leased circuit falls as usage increases; major TNCs thus have a substantial advantage over smaller companies in being able to channel large volumes of information in their leased network' (Langdale, 1989, p. 506). The spatial form of such networks may be categorized into three broad types (Figure 5.7), although a TNC may well use a mixture of types. The larger and more extensive a TNC's operations, the more likely it is to use either regional hub-and-spoke or global networks. The nodes in these networks tend to be the major world cities and it is significant that NIEs such as Singapore are investing huge sums in their communications infrastructures to position themselves strategically in the global communications network. It is the possession of such instantaneous global communications systems that enables the TNC to operate globally, whether it is engaged in manufacturing, resource exploitation or business services.

The mass media

Developments in the communications media have revolutionized the potential for large organizations to operate over vast geographical distances and, as such, have played a key role in facilitating the development of the TNC. But there is another sphere – that of the *mass media* – in which innovations have transformed the global economy and are facilitating the globalization of markets. Large business firms require large markets to sustain them; global firms aspire to global markets. The existence of such markets obviously depends on income levels, but it depends, too, on potential customers becoming aware of the firm's offerings and being persuaded to purchase them. Even where consumer incomes are low the ground may be prepared for possible future ability to purchase by creating a desirable image. The mass media are particularly powerful means both of spreading information and of persuasion; hence their vital importance to the advertising industry.

On a global scale, it is the electronic media – particularly radio and television – which are the most significant. In part, this is because of their vividness and sense of immediacy and involvement. But an important characteristic of the electronic media is that they make no demands on literacy, a demand which even the most primitive news sheet makes. Access to the electronic media is largely governed by level of affluence. Not surprisingly, therefore, there are enormous geographical variations in the availability of radio and television. At one extreme, the United States had 2,118 radio receivers and 815 television receivers per thousand population in the mid-1990s. At the other extreme, India had a mere 80 radio receivers and 37 TV receivers per thousand.

Perhaps more than any other innovation in the mass media it was the development of the transistor radio receiver which initially had the most revolutionary effects, especially in developing countries. Not only is it portable but it is also relatively cheap. Largely because of this, the sequence of development of the mass media has been rather different in the developing countries:

> In the Western world, the pattern has been newspapers, radio, television. In Africa and much of Asia, the first contact the ordinary man [sic] has with any means of mass communication is the radio. It is the transistor which is bringing the people of remote villages and lonely settlements into contact with the flow of modern life.
>
> (Hachten, 1974, p. 99)

Today, it is television which has the most dramatic impact on people's awareness and perception of worlds beyond their own direct experience. Although the ownership of TV receivers remains very low in many developing countries there is a great deal of collective listening and viewing in public places. Hence, the actual reach of the electronic media may be greater than the figures on sets per thousand population suggest.

The electronic media transmit messages of all kinds. From our viewpoint, the most important point is that a very large proportion of these messages are commercial messages aimed at the consumer. Commercial advertising is a feature of most radio and television networks outside the centrally planned economies. Even in systems which are state controlled some advertising is often included. Indeed, 'television has developed primarily as a commercial medium . . . commercial advertising is carried by all but a handful of the world's . . . television systems' (Dizard, 1966, pp. 12, 13). Thus, there is a very close relationship between the global spread of advertising agencies and the diffusion of the electronic media.[7] The communications media, in effect, open the doors of national markets to the heavily advertised products of the transnational producers.

These trends have been underway for several decades. The 1980s, however, saw a major 'phase shift' in the mass media:

During the 1980s new technologies transformed the world of media. Newspapers were written, edited and printed at a distance, allowing for simultaneous editions of the same newspaper tailored to several major areas . . . But the decisive move was the multiplication of television channels, leading to their increasing diversification. Developments of cable television technologies, to be fostered in the 1990s by fiber optics and digitization, and of direct satellite broadcasting dramatically expanded the spectrum of transmission and put pressure on the authorities to deregulate communications in general and television in particular. It followed an explosion of cable television programming in the United States and of satellite television in Europe, Asia, and Latin America . . . In the US the number of independent TV stations grew during the 1980s from 62 to 330. Cable systems in major metropolitan areas feature up to 60 channels, mixing network TV, independent stations, cable networks, most of them specialized, and pay TV. In the countries of the European Union, the number of TV networks increased from 40 in 1980 to 150 by the mid-1990s, one-third of them being satellite broadcasted.

(Castells, 1996, pp. 337, 338, 339)

Prior to the diversification wave of the 1980s, there was a high level of standardization in the kinds of TV programme available. It was this kind of 'mutual experience' which led Marshall McLuhan to coin the metaphor of the *global village* in which certain images are shared and in which events take on the immediacy of participation. Although in one sense the world may not have shrunk for the rural peasant or the urban slum dweller with no adequate means of personal transport, it had undoubtedly shrunk in an indirect sense. It was now possible to be aware of distant places, of lifestyles, of consumer goods through the vicarious experience of the electronic media. But the increasing segmentation of TV messages made possible by the communications revolution of the 1980s leads Castells (1996, p. 341) to question the continuing validity of the global village idea:

the fact that not everybody watches the same thing at the same time, and that each culture and social group has a specific relationship to the media system, does make a fundamental difference vis-à-vis the old system of standardized mass media . . . While the media have become indeed globally interconnected, and programs and messages circulate in the global network, *we are not living in a global village, but in customized cottages globally produced and locally distributed.*

The Internet: the 'skeleton of cyberspace'[8]

Within the space of just a few years, these developments in the mass communications media have been joined by an even more revolutionary technology which provides the potential for *interactive* communications on a global basis. As a result, 'there is a new geography in the making. It is composed of a vast web of telecommunications whose fixed points are computers of many sorts. Its traffic is information, and the vehicles that navigate its highways are intelligent software animals, sometimes called knowbots. This is cyberspace . . . ' (Batty and Barr, 1994, p. 699).

The 'skeleton' of this emerging cyberspace is the Internet, whose origins go back some twenty-five years and are to be found within the US Department of Defense. It spread initially through the linking of more specialized computer networks and now consists of a complex multilevel geographical structure; a series of 'backbone' networks into which regional and local networks are linked. As Figure 5.8 shows, growth of the Internet – measured in terms of the number of registered computers ('hosts') – has been exponential.

Although much of the discussion of the Internet is very heavy on hype, there is no doubt that the incredibly rapid development of a mass-user computer-based

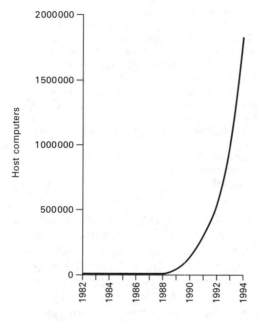

Host computers

2000000

1500000

1000000

500000

0

1982 1984 1986 1988 1990 1992 1994

Figure 5.8 Exponential growth of the Internet
Source: Based on Ogden (1994, Figure 1)

communications system is indeed revolutionary. Almost overnight, it seems, anybody who has a PC and telephone line can link into a global communications network which provides both interpersonal or interorganizational communication through electronic mail (email) and also access to a cornucopia of information through the World Wide Web. (The large TNCs have been operating an *intra*net system for some time through the kinds of leased telecommunications networks shown in Figure 5.7). However, for the world at large, the geographical pattern of growth of the Internet so far has been very uneven indeed. Usage is dominated by the United States (65 per cent of the hosts in 1994) and is then very much related to the structure of the global economy, as might be expected, with developing countries (including virtually the whole of Africa) being highly under-represented. The lower than expected level of Internet connection in east Asia is explained by Batty and Barr as the outcome of relatively strict regulations on telecommunications. It seems unlikely that this regulatory constraint can persist indefinitely.

It is too early to assess the eventual impact of the Internet and it is already clear that the system has grown so fast that there are major problems of effective usage. The instantaneous character of the Internet is often contradicted by long connection delays because of the sheer number of people trying to use it:

> The upshot is that the Net still resembles a congested city street: because users pay only for their car and its fuel, rather than for the inconvenience their presence on the road imposes on others, they have no incentive to limit the use of their car in order to avoid traffic jams.
>
> (*The Economist*, 19 October 1996)

Summary

Technological developments in communications media have transformed space-time relationships between all parts of the world. Of course, not all places are equally affected. Consistent with the nature of the time-space convergence process, as defined by Janelle, is its inherent geographical unevenness. In general, the places which benefit most from innovations in the communications media are the 'important' places. New investments in communications technology are market related; they go to where the returns are likely to be high. The cumulative effect is to reinforce both certain communications routes at the global scale and to enhance the significance of the nodes (cities/countries) on those routes. For example, although developing countries contain around 75 per cent of the world's population they have only around 12 per cent of the world's telephone lines. 'A new geography of rich and poor is emerging with the poor now those deprived of access to . . . communications technology' (Batty and Barr, 1994, p. 711). There is an additional factor which limits the universal spread of new communications technologies. In virtually all countries of the world, governments regulate the communications industries within their borders. Today, however, there is a strong trend towards the deregulation of telecommunications in several countries. Within this geographically uneven communications surface there is also a social dimension. Not everybody – whether they are business firms or private individuals – has equal access. Despite the general decline in communications costs driven by technological change, the costs of usage are far from trivial.

Nevertheless, although we need to beware of the hype which surrounds the 'information revolution', there is no doubt that epochal changes are occurring through the development of digital technologies:

> The rapid advance of digital systems, based on the 'ones and zeros' of binary computer language, is sweeping away the remaining differences between data processing and telephony, and leading to the dawn of a new information age, epitomized by the explosive growth of the Internet and internal corporate intranets . . . Digital technology has made it possible to convert text, sound, graphics and moving images into coded digital messages which can be combined, stored, manipulated and transmitted quickly, efficiently, and in large volumes over wired and wireless networks without loss of quality. As a result, electronic commerce and the multimedia revolution are driving the computing and telecommunications worlds into ever-closer competition and cooperation. 'The coming era of digital personal communications is an era of converging technologies, converging products, converging media and converging industries.'
>
> (*The Financial Times*, 5 March 1997)

Technological changes in products and processes

When thinking of 'products' most of us tend to think of consumer products. But, for industry as a whole, most products are themselves intermediate in nature; they form the inputs to subsequent stages in the production chain (see Chapter 1). Bearing in mind this intricate relationship between product and process, however, it is useful to look at them separately in the first instance.

Product innovation and the product life-cycle[9]

In an intensely competitive environment, the introduction of a continuous stream of new products is essential to a firm's profitability and, indeed, survival. 'Long-run growth requires either a steady geographical expansion of the market area or the continuous innovation of new products. In the long run only product innovation can avoid the constraint imposed by the size of the world market for a given product' (Casson, 1983, p. 24). The idea that the demand for a product will decline over time is captured in the concept of the product life-cycle (PLC).

The essence of the product life-cycle is that the growth of sales of a product follows a systematic path from initial innovation through a series of stages: early development, growth, maturity and obsolescence (Figure 5.9). When a new product is first introduced on the market the total volume of sales tends to be low because customers' knowledge is limited and also they tend to be uncertain about the product's quality and reliability. Assuming that the new product gains a foothold in the market (and very many do not get beyond this initial stage) it then enters a phase of rapid growth as overall demand increases. Such growth is likely to have a ceiling,

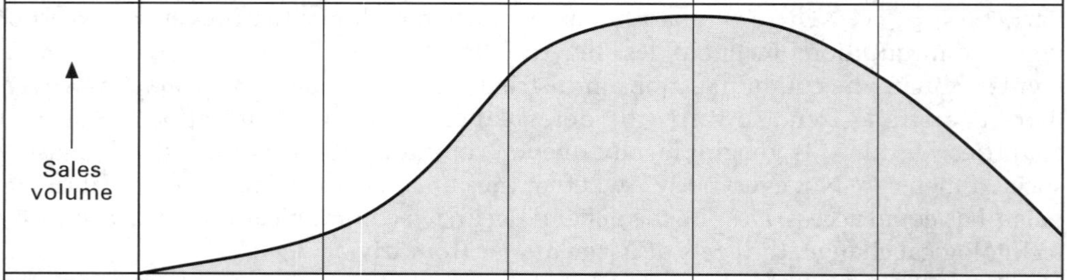

	Initial development	Growth	Maturity	Decline	Obsolescence
Sales volume (↑)					
Demand conditions	Very few buyers	Growing number of buyers	Peak demand	Declining demand	Steep fall-off in demand
Technology	Short production runs Rapidly changing techniques	Introduction of mass production methods Some variation in techniques but less rapid change	Long production runs and stable technology Few innovations of importance		
Capital intensity	Low	High because of high rate of obsolescence		High because of large quantity of specialized equipment	
Industry structure	Entry is 'know-how' determined Numerous firms supply specialist services Few competitors	Growing number of competing firms Increasing vertical integration	Financial resources critical for entry Number of firms starts to decline	General stability at first, followed by exit of some firms	
Critical production factors	Scientific and engineering skills External economies (access to specialist firms)	Management, capital	Semi-skilled and unskilled labour, capital		

Figure 5.9   The product life-cycle and its changing characteristics
Source: Based, in part, on Hirsch (1967, Table II (1))

however; the product attains maturity in which demand levels out. Eventually, demand for the product will slacken as the product becomes obsolescent.

The kind of development path suggested by the product life-cycle concept clearly has very important implications for the growth of firms and for their profit levels. The product life-cycle implies that all products have a limited life; that obsolescence is inevitable. Of course, the rate at which the cycle proceeds will vary from one product to another. In some highly ephemeral products the cycle may run its course within a single year or even less, in others the cycle may be very lengthy. However, there is growing evidence that the general length of product cycles is tending to become shorter. Thus, in order to continue to grow and to make profits firms need to innovate on a regular basis (or to acquire innovations from other firms). There are three major ways in which a product's sales may be maintained or increased. One way is to introduce a new product as the existing one becomes obsolete so that 'overlapping' cycles occur. An alternative is to find ways of extending the cycle for the existing product either by making minor modifications in the product itself to 'update' it or by finding new uses for it. Thirdly, changes may be made to the production technology itself to make the product more competitive. Whichever strategy is pursued, however, innovation and technological change are fundamental. In so far as product cycles are shortening in many industries, this implies increasing pressure on firms to develop new products.

The production process and technology

In today's intensely competitive global environment product innovation alone is inadequate as a basis for a firm's survival and profitability. Firms must endeavour to operate the production process as efficiently as possible. Recent developments in technology – and, especially, in information technology – are having profound effects upon production processes in all economic sectors. Three major, and closely interrelated, decisions are involved in the production process (Smith, 1981):

- The *technique* to be adopted. This decision concerns both the particular technology used and also the way in which the various inputs or factors of production are combined. It is almost always possible to vary the precise combination of, say, labour and capital according to their relative availability and cost. However, there are limits to such substitution of factors. Some production processes are intrinsically more capital intensive than others and *vice versa*. Closely related to the question of technique is that of
- the *scale* of production. In general, the average cost of production tends to decline as the volume of production increases. The extent of such economies of scale varies considerably from one industry to another. They are much greater, for example, in automobile production than in the manufacture of fashion garments. Technique and scale are, themselves, closely related. Table 5.1 identifies seven types of production process ranging from the manufacture of individual units to customers' requirements through to mechanized and automated mass production and continuous-flow processing. In each case, both the technique and the scale of production differ and so, too, does the type of labour required.
- *Location*:

> The choice of location cannot be considered in isolation from scale and technique. Different scales of operation may require different locations to give access to markets of different sizes . . . Different techniques will favour different locations, as firms tend to gravitate toward cheap sources of the inputs required in the largest quantities, and location itself can influence the combination of inputs and hence the technique adopted.
> (Smith, 1981, pp. 23–24)

Clearly, therefore, a firm which is seeking to reduce its production costs or to increase its efficiency and productivity can seek such economies at different points in the production process. It can attempt to purchase lower-cost inputs. In the case of material inputs this increasingly involves a shift to supplies in developing countries. In the case of labour, a relatively immobile factor of production, the search for lower costs may involve the physical relocation of production to a cheap labour location.

Table 5.1 A classification of production processes

Type of process	Industry examples
1) Craft-type unit production to individual customer needs	various
2) Craft-type batch production	aircraft; construction
3) Manual assembly	electronics assembly; garments
4) Mechanized assembly	automobiles; home appliances
5) Mechanized processing	textiles
6) Automated processing	pulp and paper; standardized metal working (e.g. ball bearings)
7) Continuous processing	petrochemicals, refining

Source: Based in part on Storper and Walker (1984, pp. 34–36).

Production processes and the product life-cycle

We looked earlier at the product life-cycle as a way of understanding the drive for the continuous innovation and introduction of new products. The nature of the production process itself also tends to vary systematically according to stages in the product life-cycle (Figure 5.9). Each stage will tend to have different production process attributes: of technology, of capital intensity, of labour force characteristics and of industry structure. During the *early stage* of the cycle, production technology tends to be volatile with frequent changes in product specification. Production tends to be in short runs or batches. There is a tendency to rely on specialist suppliers and subcontractors. Capital intensity is relatively low. The most important type of labour is scientific and engineering. Entry into the new industry is determined largely by 'know-how' rather than by financial resources.

By the time the product has progressed to the *growth stage* (assuming that it does) some important changes have occurred in the production process. The rapidly growing demand for the product (see Figure 5.9) permits the introduction of mass production and assembly-line techniques. Even though the technology may be evolving it is less volatile than in the earlier stage of the cycle. Capital intensity at the growth stage is considerably higher and the key type of labour is now managerial. The need is for administrative and marketing, rather than scientific, skills. In the growth stage also the number of firms engaged in the industry tends to be increasing but with a high casualty rate. There is a tendency towards increased vertical integration as firms seek to ensure the stability of component supplies and to exert greater influence over distribution of their product, often through acquisition and merger.

In the *mature stage* of the cycle demand has reached its peak (Figure 5.9) and the market is becoming saturated. The major emphasis is on reducing the costs of production through long production runs. By this stage the technology is stable with few important changes. Long runs require the installation of high-volume specialist equipment which, in turn, increases the capital intensity of the industry. Stable technology and the application of a finer division of labour alter the labour force focus. In the mature stage the emphasis is on semi-skilled and unskilled labour performing routine, repetitive tasks in a mechanized manner. Labour costs become an increasingly significant element in production costs. The major barrier to entry into the industry is finance. Acquisition and merger are important mechanisms of entry and exit.

Hence, each stage in the product life-cycle has a significant influence on the nature of the production process. Each stage has particular production characteristics. One of the most important of these is the way in which the relative importance of the major production factors changes. In general, as the cycle proceeds, the emphasis shifts from product-related technologies to process technologies and, in particular, to ways of minimizing production costs. In this respect, the relative importance of labour costs, especially of semi-skilled and unskilled labour, increases. More generally, different types of geographical location are relevant to different stages of the product cycle.

This view of systematic changes in the production process as a product matures is appealing and has some validity. There undoubtedly are important differences in the nature of the production process between a product in its very early stages of development and the same product in its maturity. But this linear, sequential notion of change in the production process is overly simplistic. At any stage, the production process may be 'rejuvenated' by technological innovation. There may not necessarily be a linear sequence leading from small-scale production to standardized mass production. Taylor (1986, p. 753) is especially critical of the *technological determinism* which is implied in the product life-cycle: although 'there is an appealing logic . . . that derives from its simplicity. As the same time, however, simplicity is also its greatest weakness'.

Flexibility: the world 'after Fordism'

This point leads us to consider the major recent developments that have been occurring in the technology of production processes and, particularly, those associated with the new techno-economic paradigm of information technology. Most technological developments in production processes are, as we observed earlier, gradual and incremental: the result of 'learning by doing' and of 'learning by using'. But periods of radical transformation of the production process have occurred throughout history. We are now in the midst of such a radical transformation.

Over the long timescale of the development of industrialization, the production process has developed through a series of stages each of which represents increasing efforts to mechanize and to control more closely the nature and speed of work. The stages generally identified are:

- *Manufacture*: the collecting together of labour into workshops and the division of the labour process into specific tasks.
- *Machinofacture*: the application of mechanical processes and power through machinery in factories. Further division of labour.
- *Scientific management* ('Taylorism'): the subjection of the work process to scientific study in the late nineteenth century. This enhanced the fineness of the division of labour into specific tasks together with increased control and supervision.
- *'Fordism'*: the development of assembly-line processes which controlled the pace of production and permitted the production of large volumes of standardized products.
- *'After-Fordism'*: the development of new flexible production systems based upon the deep application of information technologies.

These stages in the production process map fairly closely on to the long-wave sequence shown earlier in Figure 5.1. The first Kondratiev wave was associated with the transition from manufacture to 'machinofacture'. The application of scientific management principles to the production process emerged in the late phase of K2 and developed more fully in K3. The bases of Fordist production were established during K3 but reached their fullest development during K4. The fifth Kondratiev is seen as marking the transition from Fordism to a new regime, the crossing of what Piore and Sabel (1984) termed the 'Second Industrial Divide'. However, there is considerable disagreement over the precise form of this new regime; hence the use here of the neutral term 'after-Fordism'.

The Fordist system was characterized by very large-scale production units using assembly-line manufacturing techniques and producing large volumes of standardized products for mass market consumption. It was a type of production especially characteristic of particular industrial sectors, notably automobiles. Not all sectors, nor all production processes, lent themselves to such a system of mass production but it was seen to be the main characteristic of the K4 phase. Many now argue that this Fordist system of production (and its associated organizational structures) entered a period of 'crisis' from about the mid-1970s and that it has been replaced by new modes of production. The most important characteristic of this new system is claimed to be *flexibility*: of the production process itself, of its organization within the factory and of the organization of relationships between customer and supplier firms (see Chapter 7).

The key to production flexibility lies in the use of *information technologies* in machines and operations. These permit more sophisticated control over the production process. With the increasing sophistication of automated processes and, especially, the new flexibility of electronically controlled technology, far-reaching changes in the process of production need not necessarily be associated with increased scale of production. Indeed, one of the major results of the new electronic and computer-aided production technology is that it permits rapid switching from one part of a process to another and allows the tailoring of production to the requirements of individual customers. 'Traditional' automation is geared to high-volume standardized production; the newer 'flexible manufacturing systems' are quite different:

> Flexible automation's greatest potential for radical change lies in its capacity to manufacture goods cheaply in small volumes . . . In the past batch manufacturing required machines dedicated to a single task. These machines had to be either rebuilt or replaced at the time of product change. Flexible manufacturing brings a degree of diversity to manufacturing never before available. Different products can be made on the same line at will . . . The strategic implications for the manufacturer are truly staggering. Under hard automation the greatest economies were realised only at the most massive scales. But flexible automation makes similar economies available at a wide range of scales. A flexible automation system can turn out a small batch or even a single copy of a product as efficiently as a production line designed to turn out a million identical items.
>
> (Bylinsky, 1983, pp. 53–54)

Clearly, the potential of such flexible technologies is immense and their implications enormous for the nature and organization of economic activity at all geographical scales, from the local to the global. Perez (1985) identifies three major tendencies:

- A trend towards information intensity rather than energy or materials intensity in production.
- Much enhanced flexibility of production which challenges the old best-practice concept of mass production in three central respects:
  1) a high volume of output is no longer necessary for high productivity. This can be achieved through a diversified set of low-volume products;
  2) because rapid technological change becomes less costly and risky the 'minimum change' strategy in product development is less necessary for cost effectiveness;
  3) the new technologies allow a profitable focus on segmented, rather than mass, markets. Products can be tailored to specific local conditions and needs.
- A major change in labour requirements in terms of both volume and type of labour.

The current situation, therefore, would seem to be that of a diversity of production processes and technologies but one in which the relative importance of specific processes is changing. Thus, we can find a trend towards:

- An *increasingly finer degree of specialization* in many production processes, enabling their fragmentation into a number of individual operations
- An *increasing standardization and routinization* of these individual operations, enabling the use of semi-skilled and unskilled labour. This is especially apparent during the mature stage of a product's life-cycle.

- An *increasing flexibility in the production process*, which is altering the relationship between the scale and the cost of production, permitting smaller production runs, increasing product variety and changing the way production and the labour process are organized.

The precise mix will vary from one economic sector to another, as we shall see in the industry case studies of Part III.

What does this all mean in terms of the broader question of what comes 'after-Fordism'? There are strongly opposed interpretations of the nature of Fordism itself (for example, the extent to which it really constituted an all-embracing system of production, even in its heyday) and of what it is being replaced by.[10] Is it a variant on Fordism, 'neo-Fordism', in which automated control systems are applied within a Fordist structure? Or is it a totally new 'post-Fordism', in which the new technologies create quite different forms of production organization? It is a debate which stretches way beyond the bounds of technology and technological change into the realms of the social organization of production, of the ways in which the state regulates economic activity, and the nature of consumption and markets. This is part of the broader regulationist debate referred to in Chapter 2 (see Note 3). More specifically, it reminds us that the processes involved are more than narrowly technological. As Figure 5.1 shows, there is also an *organizational* dimension to be considered. To repeat an earlier statement, technology is always socially embedded.

Flexible specialization: the rebirth of craft-based production?

Some assert unequivocally that flexible specialization is becoming the norm; the dominant style of production displacing Fordism. This is the 'post-Fordist' view which sees the hegemony of Fordism as being replaced by a new regime of flexible production and smaller organizational units. It is the viewpoint most closely associated with the work of Piore and Sabel whose 1984 book, *The Second Industrial Divide*, triggered off much of the 'after-Fordist' debate. As we have seen, Fordism was associated overwhelmingly with very large, vertically integrated firms producing standardized goods at very large volumes to benefit from economies of scale in production and selling to mass consumer markets. Piore and Sabel's interpretation of the development of flexible production technologies is that it leads to the resurgence of small, independent entrepreneurial firms emancipated from the tyrannies of mass production by the new flexibility which permits small-scale operations to serve small (perhaps local) markets.

This craft-based, 'flex-spec' interpretation of the changes in the production system also sees it as heralding a process of reskilling of labour as opposed to the relatively low skills characteristic of Fordism. The deintegration of the production system also goes hand-in-hand with a deintegrated organizational structure which then develops as horizontal networks of inter-related, specialist firms. The tendency, it is argued, will be for such networks to display a very strong *geographical localization* in a series of 'new industrial districts'.[11] Piore and Sabel drew their initial inspiration from empirical work on the so-called 'Third Italy' – the northeast central region of the country centred on Emilia Romagna – with its dense network of small, artisanal firms in the knitwear and other craft industries. They claimed to find similar flexible specialization regions in such areas as southern Germany, particularly in Baden Wurttemburg.

However, not all agree with this diagnosis. For example, Gertler (1988), Sayer and Walker (1992) and Harrison (1994) are all critical of the simplistic view which replaces

Fordism with a system based on flexibility. There are many different types of 'flexibility': flexibility in volume of output, product flexibility, employment flexibility, flexible working practices, flexible machinery, flexible forms of organization. As Sayer and Walker point out, these have tended to get confused in much of the literature. The argument is not that flexible technologies do not exist – clearly they do – but that they are not as straightforward, as universal or as pervasive as is often claimed. Equally, the Fordist system was never as all-pervasive as is sometimes claimed; less rigid and smaller-scale production has always coexisted with mass-production methods.

Japanese-inspired versions of 'after-Fordism': Toyotism or Fujitsuism?[12]

Rather different organizational successors to Fordism have been suggested by writers impressed by the demonstrable economic success of the Japanese economy. 'While some features of Japanese industry resemble those of the flexible specialization model, there are others which either contradict it or are simply absent from it' (Sayer and Walker, 1992, p. 212). Although the Japanese production system has undoubtedly built heavily upon the new IT-based technologies, including the intensive use of robotic equipment, it can be argued that it is the 'organizational' technologies developed within the Japanese system which have been far more important to its

Table 5.2   The major characteristics of craft production, Fordist mass production and Japanese flexible production

Characteristic	Craft production	Fordist mass production	Japanese flexible production
Technology	Simple, but flexible tools and equipment using unstandardized components	Complex, but rigid single-purpose machinery using standardized components. Heavy time and cost penalties involved in switching to new products	Highly flexible methods of production using modular component systems. Relatively easy to switch to new products
Labour force	Highly skilled workers in most aspects of production	Very narrowly skilled professional workers design products but production itself performed by unskilled/semi-skilled 'interchangeable' workers. Each performs a very simple task repetitively and in a predefined time and sequence	Multi-skilled, polyvalent workers workers operate in teams. Responsibilities include several manufacturing operations plus responsibility for simple maintenance and repair
Supplier relationships	Very close contact between customer and supplier. Most suppliers located within a single city	Distant relationships with suppliers, both functionally and geographically. Large inventories held at assembly plant 'just in case' of disruption of supply	Very close relationships with a functionally tiered system of suppliers. Use of 'just-in-time' delivery systems encourages geographical proximity between customers and suppliers
Production volume	Relatively low	Extremely high	Extremely high
Product variety	Extremely wide – each product customized to specific requirements	A narrow range of standardized designs with only minor product modifications	Increasingly wide range of differentiated products

Source: Based in part on material in Womack, Jones and Roos (1990).

postwar economic success. Many of these were in place before the onset of the breakup of Fordism in the west and reflect, in particular, the conditions within which the Japanese economy was rebuilt after 1945.

There are several aspects of the contemporary Japanese production system which, taken together, may be regarded as an alternative both to classic Fordism and the craft-based/flexible specialization model described above. Table 5.2 summarizes some of the contrasts between the three systems.

In particular,

- The organization of production is *flexible*, both in the use of facilities and, especially, in the way in which the *labour force* is organized:

> In Japan, work teams, job rotation, learning-by-doing, and flexibility have been used to replace the functional specialization, task fragmentation and rigid assembly-line production of US Fordism . . . there are few job classifications, work rules overlap, and production is organized on the basis of teams. Since tasks are allocated by team, workers can cover for each other and experiment with new allocations and machine configurations . . . Shopfloor learning is a basic characteristic of post-Fordist production in Japan . . . Management in post-Fordist Japan can be characterized as comprised of many 'little brains' sharing information, as opposed to the one 'big brain' directing many 'appendages' of Fordism . . . Learning by doing at many levels makes the Japanese firm an information-laden enterprise with problem-solving capabilities which far exceed its Fordist counterparts.
>
> (Kenney and Florida, 1989, pp. 144, 145)

Table 5.3 The characteristics of 'just-in-case' and 'just-in-time' systems

'Just-in-case' system	'Just-in-time' system
Characteristics	
Components delivered in large, but infrequent batches	Components delivered in small, very frequent, batches
Very large 'buffer' stocks held to protect against disruption in supply or discovery of faulty batches	Minimal stocks held – only sufficient to meet the immediate need
Quality control based on sample check after supplies received	Quality control 'built in' at all stages
Large warehousing spaces and staff required to hold and administer the stocks	Minimal warehousing space and staff required
Use of large numbers of suppliers selected primarily on the basis of price	Use of small number of preferred suppliers within a tiered supply system
Remote relationships between customer and suppliers	Very close relationships between customer and suppliers
No incentive for suppliers to locate close to customers	Strong incentive for suppliers to locate close to customers
Disadvantages	
Lack of flexibility – difficult to balance flows and usage of different components	Must be applied throughout the entire supply chain
Very high cost of holding large stocks	Reliance on small number of preferred suppliers increases risk of interruption in supply
Remote relationships with suppliers prevents sharing of developmental tasks	
Requires a deep vertical hierarchy of control to co-ordinate different tasks	

Source: Based on material in Sayer (1986).

- There is an obsessive preoccupation with *quality control* as an intrinsic element at all stages of the production process. The concept of total quality management (TQM) involves building in quality from the beginning rather than checking for faults at the end. There is a continuous drive towards 'zero tolerance' of faults. This requires the development of a particular set of attitudes within the workforce at all levels.
- Customer–supplier relationships take on a particularly close form as firms attempt to reduce inventory to a minimum through the use of '*just-in-time*' (JIT) systems. Schonberger (1982) contrasted this with the 'just-in-case' systems characteristic of Fordism. Table 5.3 summarizes the differences between the two systems. The JIT system

> emerged in the post-war period as Japanese car manufacturers – particularly Toyota – attempted to adapt US practices to Japanese conditions . . . Just-in-time is first and foremost a novel form of integrating the parts of a manufacturing system, involving an approach to time economy different than that of JIC. It is literally a system in which tasks are done just when needed, in just the amount required to meet desired output levels . . . Ideally, all machines, even the most expensive, should be run no faster than the speed necessary to produce the required output . . . Instead of pushing production through at maximum speed in long runs, each operation is done just-in-time to meet projected orders. Buffer stocks are very small and are only replenished to replace parts moved downstream. Workers at the end of the line are given output instructions on the basis of short-term order forecasts, and they instruct the workers immediately upstream to produce the parts they will need just-in-time, and those workers in turn instruct workers upstream to produce just-in-time, and so on . . . In short, it is a pull rather than a push system.
>
> (Sayer and Walker, 1992, pp. 170, 171)

A particular interpretation of the Japanese 'Toyotist' system sees the entire process as being *lean*:

> lean production . . . is 'lean' because it uses less of everything compared with mass production – half the human effort in the factory, half the manufacturing space, half the investment in tools, half the engineering hours to develop a new product in half the time. Also it requires keeping far less than half of the needed inventory on site, results in many fewer defects, and produces a greater and ever growing variety of products.
>
> (Womack, Jones and Roos, 1990, p. 13)

The use of the term 'Toyotism' to describe the Japanese alternative to Fordism reflects the leading role played by the automobile industry in general, and of Toyota in particular (see Chapter 10).

A more recent interpretation of the nature and development of the Japanese post-Fordist system is provided by Kenney and Florida (1989) in their notion of 'Fujitsuism', the label being derived from one of the leading Japanese information technology companies, Fujitsu:

> Fujitsuism is the term we use to refer to the way Japan is moving beyond post-Fordism to a new model of industrial organization which is particularly well suited to the information age. The rise of Fujitsuism revolves around three basic dimensions:
>
> - the use of ITs to transform traditional manufacturing
> - the linkage of innovation to production; and
> - new ways of organizing demand and channelling consumption.
>
> (Kenney and Florida, 1989, p. 145)

These various technological/organizational innovations developed within the postwar Japanese economy are generally accepted as contributing towards the country's spectacular economic success. Where interpretations differ is, first, on how far the Japanese system is transferable to other national and local contexts and, secondly, on the labour implications of the system. On the first point, it is worth noting that the source of many of the practices of 'just-in-time/total quality control' production was, in fact, western engineers, such as the American William Deming. Some of the ideas were imported into Japan during the 1950s and 1960s (having been largely ignored in the United States and Europe) as part of the process of technological learning and adapted to local conditions. In that sense, therefore, they form part of a broader – but specific – set of institutional practices: the distinctive Japanese 'business system' which is unlikely to be totally transferable to other contexts.

The diffusion of these practices outside Japan is occurring in two ways. One is through the overseas expansion of Japanese firms themselves and their attempts to transfer their domestic practices to different contexts. The evidence suggests great variability in the extent to which such firms are able to implement such transfers without a considerable degree of adaptation to local conditions, producing what Abo (1994) calls 'hybrid factories'. The second way is through the 'demonstration effect' – often provided by Japanese firms operating overseas but also through the acceptance and promotion of these ideas as 'best practice' by influential writers in management. The evidence suggests that adoption has been uneven, both geographically and between different industries.

The second controversial issue concerns the implications of the Japanese system – whether it is labelled 'Toyotism' or 'Fujitsuism' – for labour. Here views are strongly polarized as Peck and Miyamachi[13] have shown. The positive interpretation claims that the Japanese flexible production system is more humane than the Fordist mass-production system because

- it involves the reskilling of workers, notably through the requirements of multitask operations within work teams, job rotation and 'learning by doing'
- it provides workers with a substantial degree of control and involvement
- team working and self-management reduce alienation
- individualized payments systems create significant incentives for workers
- it is based upon long-term (lifetime) employment contracts.

On the other hand, critics of the Japanese system argue that, far from being more humane, it is actually highly exploitative of the labour force:

- The multiskilled, team-based workforce is subject to strict managerial control. Work teams are used as the means to extend such control.
- The emphasis on continuous improvement places great pressure on workers.
- The individual payments system is used as part of a managerial strategy to 'divide and rule' the workforce.
- Long-term contracts apply only to 'core' workers in large companies; the remaining workforce is peripheralized with no job security.

The conclusion to be drawn from this discussion of the various 'after-Fordism' alternatives is that it is unwise to seek a single hegemonic alternative at all. 'It is premature to diagnose a global trend toward a unique and well-defined successor to Fordism as a paradigm for production organization' (Coombs and Jones, 1989, p. 115). It follows from this reservation that the prediction of much of the post-Fordist literature (especially in its flexible specialization form) that the new flexible technologies herald the rebirth of the small firm at the expense of the large is an exaggeration. The flexible technologies change the possibilities for firms of all sizes but in different ways. It is more satisfactory, therefore, to see current trends as consisting of a mix of the several systems described in this section. In any case, precisely how production systems are actually devised and implemented will vary greatly between specific national and, indeed, local contexts. It is to this issue that we turn now.

The geography of innovation: innovative milieux and technology districts[14]

> conditions of knowledge accumulation are highly localized.
>
> (Metcalfe and Diliso, 1996, p. 58)

In the earlier discussion of Kondratiev long waves it was observed that each K-wave has a distinctive geography. The point being made there was that technological leadership has tended to shift over time, both nationally and regionally. The purpose of this final section is rather different. In the light of our discussion of the epochal change from a predominantly Fordist system to some kind of after-Fordist system grounded in information-technology-based flexibility we explore the processes underlying the *localization* of technological development. In so doing we use, and develop in a specific way, some of the concepts of spatial agglomeration introduced in Chapter 1.

Innovation – the heart of technological change- is fundamentally a *learning* process. Such learning – by 'doing', by 'using', by observing from, and sharing with, others – depends upon the accumulation and development of relevant knowledge of very wide variety. Of course, the development of highly sophisticated communications systems facilitates the diffusion of knowledge at unprecedented speed and over unprecedented distances. The fact remains, however, that knowledge is produced in specific places and often used and enhanced most intensively in those same places. A key concept, therefore, is that of the *innovative milieu* which forms the specific sociotechnological context within which innovative activity is embedded. It consists of a mixture of both tangible and intangible elements: both the economic, social and political institutions themselves, the knowledge and know-how which evolves over time in a specific context (the 'something in the air' notion identified many decades ago by Alfred Marshall), and the '*conventions*, which are taken-for-granted rules and routines between the partners in different kinds of relations defined by uncertainty' (Storper 1995, p. 208).

The scale of such milieux may vary from the national down to the local. Distinctive national innovation systems exist, as Nelson (1993) has demonstrated. But at the heart of such national systems we invariably find geographically localized innovative milieux. Consequently

> Geography plays a fundamental role in the process of innovation and learning, since innovations are in most cases less the product of individual firms than of the assembled resources, knowledge and other inputs and capabilities that are localized in specific places. The clustering of inputs such as industrial and university R&D, agglomerations of manufacturing firms in related industries, and networks of business-service providers may create scale economies, facilitate knowledge-sharing and cross-fertilization of ideas and promote face-to-face interactions of the sort that enhance effective technology transfer . . . Two main features help explain the advantage of spatial agglomeration in this context: the involvement of inputs of knowledge and information which are essentially

'person'embodied', and a high degree of uncertainty surrounding outputs. Both require intense and frequent personal communications and rapid decision-making, which are arguably enhanced by geographic proximity between the parties taking part in the exchange. Indeed, in the present era of rapid global dissemination of codified knowledge, we may even argue that tacit, and spatially more 'sticky' forms of knowledge are becoming more important as a basis for sustaining competitive advantage.

(Malmberg and Maskell, 1997, pp. 28–29)

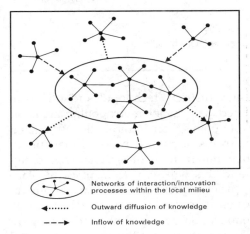

Networks of interaction/innovation processes within the local milieu

Outward diffusion of knowledge

Inflow of knowledge

Figure 5.10 Processes involved in the local accumulation of knowledge
Source: Based on Malmberg, Solvell and Zander (1996, Figure 3)

Figure 5.10 is a diagrammatic representation of the processes involved in the local accumulation of knowledge and the innovative process (Malmberg, Solvell and Zander, 1996). The diagram shows three elements:

- The tendency for a *high intensity of interaction within the local milieu* based upon the trial-and-error nature of problem-solving and the need to exchange information on a face-to-face basis.
- The *barriers which exist to the outward diffusion of locally embedded knowledge.* 'In essence, the ability to gain access to informal and formal networks for knowledge exchange and to accumulate social capital is by and large reserved for insiders in a milieu, and this is the ultimate barrier to diffusion of knowledge' (Malmberg, Solvell and Zanders, 1996, p. 93).
- The *inflow of knowledge* from outside the local milieu derived from 'both resources brought in by outsiders and initiatives taken by incumbents to tap resources from the outside' (p. 93).

Although such local innovative milieux are based upon strongly place-specific networks, the development of transnational corporations means that some of the actors may be both 'insiders' and 'outsiders' at the same time. Indeed, TNCs have a strong incentive to attempt to become insiders in order to tap into local innovative systems. This linking together of territorially based local networks with non-territorially based TNC networks is a distinctive and important feature of the structure of the global economy[15] and is one to which we will return again.

Local innovative milieux, therefore, consist essentially of a *nexus of untraded interdependencies* set within a temporal context of *path-dependent* processes of technological change (Storper, 1995). We outlined the major elements of these processes in general terms in Chapter 1. The point of emphasizing the 'untraded' nature of the interdependencies within such milieux is to distinguish the 'cement' which binds this kind of localized agglomeration from that which may be associated with the minimization of transaction costs (e.g. of materials and components transfers) through geographical proximity. Storper (1992) uses the term *technology district* to differentiate geographical clusters based upon 'product based technological learning' from those based on other types of industrial district. A similar idea, though expressed in a rather different way, is captured in the concept of the *technopole* (Castells and Hall, 1994).

Table 5.4 Some leading technology districts/technopoles

United States	Europe	Asia
Southern California (including Silicon Valley)	M4 Corridor, London	Tokyo
Boston, MA	Munich	Seoul–Inchon
Austin, TX	Stuttgart	Taipei–Hsinchu
Seattle, WA	Paris-Sud	Singapore
Boulder, CO	Grenoble	
Raleigh–Durham, NC	Montpellier	
	Nice/Sophia Antipolis	
	Milan	

Table 5.4 lists some of the major technology districts/technopoles identified in empirical research. Many of them are associated with major metropolitan areas although some have developed outside the metropolitan sphere in rather less urbanized areas. Most are the outcome of the historical process of cumulative, path-dependent growth processes although a few are the deliberate creations of national technology policy. But whatever their specific origin – and this will vary from place to place because of historical and geographical contingencies – these technological agglomerations form one of the most significant features of the contemporary global economy. Storper, in fact, concludes that the global economy can be depicted as consisting, in part, of a *mosaic of technology districts* which constitute the leading edge of national economic growth. This mosaic of localized technology districts exists both within and across the boundaries of national innovation systems.

Conclusion

The aim of this chapter has been to identify some of those features of technological change which are most important in the internationalization and globalization of economic activity. Technological change is at the dynamic heart of economic growth and development; it is fundamental to the evolution of a global economic system. We focused on four specific aspects of technological change.

First, we explored the process of technological change as an evolutionary process in which much change is gradual and incremental, often unnoticed but none the less extremely significant. But there are periodic radical transformations of existing technologies – revolutionary developments in clusters of technologies – which dramatically alter not only products and processes in one industry but which also pervade the entire socioeconomic system. These are the shifts in the technoeconomic paradigm which seem to be associated with the long waves of economic change.

Second, we concentrated on what is undoubtedly the major technological driving force today: the convergence of two initially distinct technologies – computer technology and communications technology – into a single, though complex, strand: information technology. IT is transforming both the technologies of transport and communication and also the technologies of products and processes. IT is capable of spreading into all sectors of the economy and to all types and sizes of organization but it is still the very large business organization, particularly the TNC, which is reaping the greatest benefits.

Third, we argued that the claim that we are shifting from one hegemonic (Fordist) system to another hegemonic (post-Fordist) system is far too sweeping and simplistic to capture the complex reality of a world based upon increased flexibility of production and organization. There are a number of alternatives to Fordism which, although they are all based upon the new flexibilities, take on rather different forms. We need to recognize the existence of such diversity. Part of that diversity is related to the fourth focus of the chapter: the strongly localized nature of innovation and technological change. The path-dependent nature of technological change and the social conditions within which such change occurs give major importance to the *geography* of the process.

In concluding this chapter, then, we should again remind ourselves that technological change, in itself, is not deterministic. We must not assume that a particular technology will lead inevitably and irrevocably to a particular outcome. More realistically

> A frontier of new possibilities has been defined: a frontier which identifies the types of new products and services that can be made available. That frontier is itself a product of past choices . . . Specific choices within the frontier of technological possibilities are not the product of technological change; they are, rather, the product of those who make the choices within the frontier of possibilities. *Technology does not drive choice; choice drives technology.*
>
> (Borrus, quoted in Cohen and Zysman, 1987, p. 183, emphasis added)

Notes for further reading

1. This kind of perspective on technological change is based upon the work of Perez (1985), Freeman (1982; 1987), Dosi *et al.* (1988). See also Metcalfe and Diliso (1996).
2. Major contributions to the literature on long waves have been made by Schumpeter (1939), Mensch (1979), Mandel (1980). More recent important work includes Freeman, Clark and Soete (1982), van Duijn (1983), Freeman and Perez (1988), Hall and Preston (1988).
3. Useful general introductions to information technology can be found in Forester (1985; 1987). More advanced treatments are provided by Freeman (1987), Hall and Preston (1988), Hepworth (1989), Castells (1996).
4. For broad-ranging discussions of these technologies, see Hall and Preston (1988), Brunn and Leinbach (1991), Castells (1996), Graham and Marvin (1996).
5. These data are from Baylin (1996).
6. See Craig (1981), Schiller (1982), Bakis (1987), Hepworth (1989). A detailed study of the use of international leased networks by TNCs is provided by Langdale (1989).
7. A detailed study of US commercial involvement in Latin American television is provided by Wells (1972). More recent studies of the internationalization of television are provided by Negrine and Papathanassopoulos (1990), Sklair (1995).
8. Batty and Barr (1994) provide an excellent discussion of the development of the Internet. See also Ogden (1994), Warf (1995), Castells (1996).
9. The concept of the product life-cycle has been applied in a number of different ways. For its treatment within the marketing literature, see Majaro (1982), Paliwoda (1986). Van Duijn (1983) traces the historical evolution of the related concept of an industry life-cycle. The product life-cycle has also been employed by Hirsch (1967; 1972), Wells (1972), Vernon (1966; 1979) to explain international trade and international production. Its use to explain variations in regional growth is criticized by Storper (1985), Taylor (1986).
10. There is a huge literature on this subject. Important contributions are provided by Piore and Sabel (1984), Blackburn, Coombs and Green (1985), Gertler (1988), Coombs and Jones (1989), Kenney and Florida (1989), Schoenberger (1988a, 1989), Sayer and Walker (1992), Amin (1994), Ruigrok and van Tulder (1995).
11. The major features of this debate are captured in Piore and Sabel (1984), Scott (1988a), Sabel (1989), Amin and Robins (1990), Lovering (1990), Henry (1992), Amin (1994), Harrison (1994), Malmberg (1996), Storper (1995; 1997).

12. Florida and Kenney (1989) make the most explicit argument about the distinctive post-Fordist nature of the Japanese system in their concept of 'Fujitsuism'. More general accounts of the Japanese production system are provided by Schonberger (1982), Aoki (1984), Dore (1986), Sayer (1986b), Womack, Jones and Roos (1990), Fruin (1992), Sayer and Walker (1992), Peck and Miyamachi (1995). For a powerful critique of *lean production*, see Williams et. al. (1992).

13. This section is based mainly on Table 1 in Peck and Miyamachi (1995). See also Tomaney (1994).

14. This section draws primarily on Storper (1995), Malmberg, Solvell and Zander (1996), Malmberg and Maskell (1997). Other very useful contributions are provided by Nelson (1993), Castells (1996), de la Mothe and Paquet (1996), Metcalfe and Diliso (1996), Storper (1992; 1997).

15. See Amin and Thrift (1992).

CHAPTER 6

Transnational corporations: the primary 'movers and shapers' of the global economy

Introduction

More than any other single institution it is the transnational corporation which has come to be regarded as the primary shaper of the contemporary global economy. It has been the rise of the TNC – especially of the massive global corporation – which is seen to pose the major threat to the autonomy of the nation-state. As shown in some detail in Chapter 2, there has not only been a massive growth of foreign direct investment but also the sources and destinations of that investment have become increasingly diverse. But FDI is only one measure of TNC activity. Because the FDI data are based on ownership of assets they do not capture the increasingly intricate ways in which firms engage in international operations through various kinds of collaborative ventures and through the different ways in which they co-ordinate and control production chain transactions. It is for this reason that we adopt a much broader definition of the TNC than that normally used in the conventional literature:

> A transnational corporation is a firm which has the power to co-ordinate and control operations in more than one country, even if it does not own them.

The aim of this and the following chapter is to explore the implications of this broader definition of the TNC, particularly in the context of the increasingly complex networks of inter-relationships within which all TNCs are embedded.

The significance of the TNC, especially the very large global corporation, lies in three basic characteristics:

- its co-ordination and control of various stages of individual production chains within and between different countries;
- its potential ability to take advantage of geographical differences in the distribution of factors of production (e.g. natural resources, capital, labour) and in state policies (e.g. taxes, trade barriers, subsidies, etc.);
- its potential geographical flexibility – an ability to switch and to reswitch its re-sources and operations between locations at an international, or even a global, scale.

Hence, the articulation (co-ordination and configuration) of production chains and, therefore, much of the changing geography of the global economy is sculptured by the TNC through decisions to invest or not to invest in particular geographical locations. It is moulded, too, by the resulting flows of materials, components and finished products, as well as of technological and organizational expertise, between geographically dispersed operations. Although the relative importance of TNCs varies considerably – from industry to industry, from country to country and between different parts of the same country – there are now few parts of the world in which TNC influence, whether direct or indirect, is not important. In some cases, indeed, TNC influence on an area's economic fortunes can be overwhelming.

However, in making these claims about the significance of TNCs it should not be assumed that TNCs are all of a kind; identical economic beings which stamp an identical footprint on the landscape. On the contrary, TNCs are highly differentiated not only in size and geographical extent but also in the ways in which they operate. Far from being the 'placeless' organizations often claimed, TNCs continue to reflect many of the basic characteristics of the home country environments in which they remain strongly embedded, despite the growing extent of their international operations.

Technological changes, particularly in the space-shrinking technologies of transport and communications, help to make possible the internationalization of economic activity and the development and geographical spread of transnational corporations. But they do not explain why business firms internationalize their activities by engaging in operations outside their countries of origin. For this we need to focus upon the TNC itself as an organizational form. It is clear that some of the explanations which have been offered to explain the growth of international production do not hold up when we look closely at the nature of transnational activity. For example, in the kind of world assumed by international trade theory transnational corporations simply could not exist. Traditional trade theory assumes that the factors of production – including capital – are geographically immobile and that there are no economies of scale in production. Both these assumptions are contravened by the very existence of transnational corporations. Similarly, the simple notion that capital will flow from areas of surplus to areas of deficit is not borne out by the fact that most transnational investment in the world economy is between capital-rich areas and only about a quarter of the total is located in the developing countries. We have to look elsewhere for an explanation.

The chapter is organized into two major parts:

- First, we outline some of the broad theoretical explanations of why firms should attempt to internationalize their activities beyond that of exporting their products to foreign markets.
- Secondly, we confront the commonly held view that TNCs are becoming 'placeless' and reject the claim that such firms are converging towards some common model.

Why *not* internationalize? Some general explanations[1]

TNCs are essentially capitalist enterprises (a small number of TNCs are state-owned enterprises but they are in the minority). As such they must behave according to the basic 'rules' of capitalism. The most fundamental of these is the drive for *profit* which is at the heart of all capitalist activity. Of course, business firms may well have a variety of motives other than profit, such as increasing their share of a market, becoming the industry leader or simply making the firm bigger. But, in the long run, none of these is more important than the pursuit of profit itself. A firm's profitability is the key barometer to its business 'health'; any firm which fails to make a profit at all over a period of time is likely to go out of business (unless 'rescued' by government or acquired by another firm). At best, therefore, firms must attempt to increase their profits; at worst, they must defend them.

Of course, a capitalist market economy is an intensely competitive economy. One firm's profit may be another firm's loss unless the whole system is growing sufficiently strongly to permit all firms to make a profit. Even so, some will make a larger profit than others. A key feature of today's world, of course, is that *competition is increasingly global* in its extent. Firms are no longer competing largely with national rivals but with firms from across the world. The pursuit is for *global profits*. Expressed in the simplest terms, profit (P) is the difference between the revenue (R) which a firm receives from selling its products and the cost (C) of producing and distributing the firm's goods or services: $P = R - C$. Obviously, therefore, profit can be increased either by raising R or reducing C, or by a combination of the two. The internationalization of a firm's operations may be motivated by each of these. The process is a totally logical extension of the firm's 'normal' mode of expansion: from local, to regional, to national and then to international scales of operation. This does not mean, of course, that all firms will inevitably internationalize. Some will find themselves unable to do so because of internal or external constraints; others may choose not to do so and to remain within a limited geographical niche.

In this section, we outline two broad theoretical approaches. The first is a macro-level approach based upon the nature of the capitalist system itself. The second approach is set at the microlevel of the firm.

A macrolevel approach: internationalization of the circuits of capital[2]

One of the most useful attempts to explain the internationalization of economic activity at the general level is based upon the concept of the circuits of capital. This, in turn, is embedded within Marx's conceptualization of the capitalist system as a whole. From this perspective, the internationalization of economic activity and its major vehicle, the TNC, can be regarded simply as being part of

> a general trend of internationalisation inherent in the expansive nature of capitalism which tends to create a specifically capitalist world economy. Increasingly, this capitalist world economy is subject as a whole to the laws of motion of capitalism . . . international firms must be understood in terms of the internationalisation of capital and the accumulation of capital.
>
> (Radice, 1975, pp. 17, 18)

The 'laws of motion' of capitalism are derived primarily from the drive to enhance profits and to accumulate capital through increasing appropriation of surplus value from the process of production. To Marxist writers the basis of the extraction of surplus value, or profit, is the exploitation of human labour power by capitalist firms which own the means of production. The internationalization of production, from this perspective, is the extension of the system of labour exploitation and class struggle to a global scale.

The capitalist economic process can be envisaged as a continuous circuit, basically a very simple idea, as Figure 6.1 shows. Its essence is that 'the money at the end of the process is greater than that at the beginning and the value of the commodity produced is greater than the value of the commodities used as inputs' (Harvey, 1982, p. 69). Thus, money (M) is used to purchase 'commodities' (C) in the form of raw materials and labour. These inputs are transformed in the process of production (P) and acquire increased value (C'). When exchanged for money (M') this

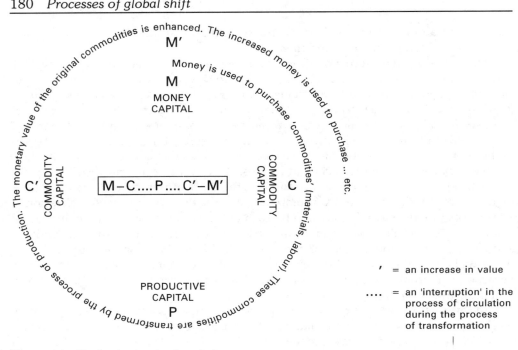

Figure 6.1   The basic circuit of capital

increased value can be used to purchase a further round of inputs for the production process, and so the circuit proceeds .

The basic circuit of capital shown in Figure 6.1 can be expanded into three distinct circuits:

1) *The circuit of money capital:* as in Figure 6.1:

$$M - C \ldots P \ldots C' - M'$$

2) *The circuit of productive capital:*

$$P \ldots C' - M' - C' \ldots P'$$

3) *The circuit of commodity capital:*

$$C' - M' - C' \ldots P' \ldots C''$$

In fact, the three circuits are part of a completely interconnected whole:

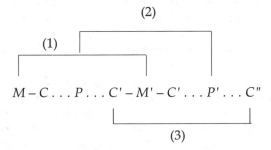

From the viewpoint of our discussion, the important point is that each of these three circuits of capital has become *internationalized*. This is the argument particularly associated with the work of Christian Palloix. He suggests that a clear historical sequence can be identified:

- The *circuit of commodity capital* was the first of the three circuits to become internationalized in the form of world trade.
- The *circuit of money capital* was the second to become internationalized in the form of the flow of portfolio investment capital into overseas ventures.
- The *circuit of productive capital* was the most recent to become internationalized in the form of the massive growth of transnational corporations and of international production.

A major merit of this circuits of capital approach to the internationalization of economic activity is that it emphasizes the totally *interconnected* nature of finance, production and commodity trade. It is too easy to regard these as completely separate phenomena, yet – as Chapter 2 showed – they are, indeed, components of a single system. The TNC itself is the clearest manifestation of this fact. The very large TNCs, in particular, display at the microscale the interlocking circuits of capital. They not only engage in international production but also generate a significant proportion of international trade within their own organizational boundaries and engage in sophisticated international financial transactions (including foreign currency dealings). The circuits of capital approach, by emphasizing the interconnection and interpenetration of the processes, also helps to capture something of the complexity of foreign direct investment flows, including the intricate cross-investments which we have observed to be occurring between the industrialized economies.

An explanatory approach based at this macrolevel, therefore, helps to illuminate several basic features of the internationalization process. But it is insufficiently specific to deal with such questions as the precise form (geographical, organizational, sectoral) of transnational corporate activity. This is certainly not to deny the value of the circuits of capital approach even to non-Marxists, who can readily accept the broad framework it provides even if they do not necessarily subscribe to the entire explanatory package, particularly the overwhelming emphasis on class struggle and the exploitation of labour by capital. It serves to remind us that, in trying to explain the internationalization of economic activity, we are dealing with the workings of a dynamic capitalist market system.

Microlevel approaches: the search for an integrative framework

An alternative approach to understanding the internationalization of economic activity through the TNC is to take a *firm-specific*, rather than a general system, view. Instead of focusing upon the capitalist system as a whole, therefore, we now adopt a perspective based upon the capitalist firm. Before the early 1960s, there was no adequate theory of the TNC at the firm-specific level. Since then numerous writers have contributed to this mode of explanation.[3] However, there was a strong tendency in the past to base such explanations on a stereotypical TNC – the very large, oligopolistic, mostly American enterprise – and theoretical conclusions have been extrapolated to apply to all TNCs. But TNCs from different source nations and of different sizes tend to differ considerably in their characteristics and behaviour. It is unwise to build an explanation of all TNCs on the study of just one type, no matter how important it may be. Any satisfactory explanation, therefore, needs to be sufficiently general to encompass the diversity of the world's TNC population.

However, there is no single accepted microlevel theoretical explanation. Here, we outline three of the more important contributions: by Stephen Hymer, Raymond Vernon and John Dunning.

Stephen Hymer: the undisputed pioneer[4]

Before Hymer's pioneering study in 1960 there was no genuine and specific theory of why firms engaged in international production. As noted in the Introduction to Part II, foreign direct investment was treated as just another variant of international capital theory. It was Hymer who led the way, drawing his inspiration from a completely different source: industrial organization theory and, especially, that part which dealt specifically with barriers to entry. Hymer began from the assumption that, in serving a particular market, domestic firms would have an intrinsic advantage over foreign firms. Domestic firms, he argued, would have a better understanding of the local business environment: the nature of the market, business customs and legislation, and the like. Given such domestic advantage, Hymer argued that a foreign firm wishing to produce in that market would have to possess some kind of firm-specific advantage which would offset the advantages held by indigenous firms. Such advantages were essentially those of firm size and economies of scale, market power and marketing skills (for example, brand names, advertising strength), technological expertise (either product, process or both), or access to cheaper sources of finance. On these bases, then, a foreign firm would be able to out-compete domestic firms in their own backyard.

Hymer's contribution was truly seminal. It was the first time that the firm had been taken as the specific focus of explanation and that international production (rather than international trade) had been the explicit object of analysis. He emphasized the importance of market imperfections in stimulating the internationalization of production. He showed how, once established, the control of overseas productive assets itself became a source of competitive advantage:

> the pioneering conceptual insight of Hymer was to break out of the arid mould of international trade and investment theory and focus attention upon the [TNC] per se. This permits us to treat FDI as a modality by which firms extend their territorial horizons abroad. The unique feature of FDI is a mechanism by which the [TNC] maintains control over productive activities outside its national boundaries, that is, FDI means international production, rather than international exchange. Until Hymer articulated the process of FDI as an international extension of industrial organization theory, it was not possible to understand why the MNE transfers intermediate products such as knowledge or technology across different nations while still retaining property rights over such assets. Today, it is widely recognized that the theory of FDI (i.e. international production) is primarily about the transfer of nonfinancial and ownership-specific assets by the [TNC], which needs to appropriate and control the rate of use of its internalized advantages.
>
> (Dunning and Rugman, 1985, p. 228)

Of course, Hymer's theory had its limitations. It was much better at explaining why and how firms might begin to become internationalized as producers and less good at explaining their subsequent development from an established international position. But, like all theories, it needs to be evaluated in terms of the level of understanding prevailing at the time he was writing. By that criterion, his contribution was immense. His particular focus on the internal ownership-specific characteristics of firms has become an accepted part of the theoretical literature.

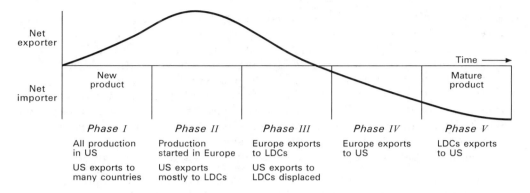

Figure 6.2 The product life-cycle and its effects on the location of United States production and trade
Source: Wells (1972, Figure 1, p. 15)

Raymond Vernon and the product life-cycle[5]

The concept of the product life-cycle, which we have already encountered in Chapter 5 (see Figure 5.9), was specifically adapted as an explanation of the evolution of *international production* by Raymond Vernon in 1966. Vernon's major contribution was to introduce an explicitly *locational* dimension into the product cycle concept which, in its original form, had no spatial connotation at all. Figure 6.2 shows Vernon's PLC model based upon the United States experience (it should be read in conjunction with Figure 5.9).

Vernon's starting point was the assumption that producers are more likely to be aware of the possibility of introducing new products in their home market than producers located elsewhere (the parallel with Hymer is close in this respect). The kinds of new products introduced, therefore, would reflect the specific characteristics of the domestic market. In the United States case, the high average-income level and high labour costs tended to encourage the development of new products which catered to high-income consumers and were labour-saving (both for consumer and producer goods). Examples often quoted include vacuum cleaners, washing machines, non-iron shirts and the like.

In this first phase of the product cycle, as Figure 6.2 shows, all production is located in the United States and overseas demand is served by exports. But this situation is unlikely to last indefinitely. US firms would eventually set up production facilities in the overseas market either because they saw an opportunity to reduce production and distribution costs or because of a threat to their market position. Such a threat might come from local competitors or from government attempts to reduce imports through tariff and other trade barriers.

It follows from the nature of the product cycle model that the first overseas production of the product will occur in other high-income markets. In the specific case of US investment this tended to be western Europe and Canada. The newly estab-lished foreign plants would come to serve these former export markets and thus displace exports from the United States, which would be redirected to other areas in which production had not begun (phase II in Figure 6.2). Eventually, the production cost advantages of the newer overseas plants would lead the firm to export to other, third-country markets (phase III) and even to the United States itself (phase IV) from these plants. Finally, as the product becomes completely standardized, production will be shifted to low-cost locations in developing countries (phase V). It is interesting to note that when Vernon first suggested this possibility he regarded it as a 'bold projection'. At that time (the mid-1960s) there was relatively little evidence of developing country export platforms serving European and US markets.

This locational interpretation of the product life-cycle became common currency, although even by the early 1970s Vernon himself was beginning to voice some doubts. He shifted the emphasis of his product cycle model to that of the *oligopolistic* behaviour of TNCs and suggested that the development sequence consisted of three phases: innovation-based oligopolies; mature oligopolies; and senescent oligopolies:

- The *innovation phase* will still tend to be located in the country of origin but it was recognized that this would no longer be just the United States. European innovations and Japanese innovations would tend to be rather different from US innovations, perhaps emphasizing materials-saving considerations rather than labour-saving ones.
- In the *mature oligopoly* phase, firms will tend to react to match the actions of their major competitors. Locationally, they will tend to pursue a *follow-the-leader* strategy[6] whereby a move by one major firm to establish production facilities in a particular country is likely to be followed by the other major firms in the same industry. This leads to a clustering of investment decisions in both time and space. In the mature oligopoly phase, firms exert their production and market power to prevent the entry of competitors.
- In the *senescent oligopoly* phase, production locations are determined largely by geographical differences in costs. In this phase, the tendency to seek out low-cost locations at a global scale becomes especially strong.

How valid is the product life-cycle as an explanation of the locational evolution of TNCs? There is no doubt that a good deal of the *initial* overseas investment by US firms did fit the product cycle sequence quite well. But it can no longer explain the majority of international investment by TNCs. As these firms have become more complex globally it is unrealistic to assume a simple evolutionary sequence from the home country outwards. Even within strongly innovative TNCs the initial source of the innovation and of its production may be from any point in the firm's global network as we shall see later. In addition, as we saw in Chapter 2, much of the world's international investment is reciprocal or cross-investment between advanced industrial countries. Such investment cannot easily be explained in product cycle terms. However, although its current relevance is limited, Vernon's adaptation of the PLC model represented an extremely important step in the development of a body of theory. Like Hymer's contribution it should be evaluated in those terms.

John Dunning's 'eclectic' paradigm

Most of the microlevel theoretical explanations of FDI and of international production are partial. In contrast, John Dunning has proposed a framework which attempts to integrate various strands of explanation of international production. Dunning proposed a set of three general and inter-related principles which, he suggests, are fundamental to an understanding of international production. Because the three principles themselves are derived from a variety of theoretical approaches – the theory of the firm, organization theory, trade theory and location theory – Dunning labels his approach *eclectic*.

Table 6.1   The eclectic paradigm and types of international production

Types of international production	(O) Ownership advantages (the 'why' of MNE activity)	(L) Location advantages (the 'where' of production	(I) Internalization advantages (the 'how' of involvement)	Strategic(s) goals of MNEs	Illustration of types of activity that favour MNEs
Natural resource seeking	Capital, technology, access to markets; complementary assets; size and negotiating strengths	Possession of natural resources and related transport and communications infrastructure; tax and other incentives	To ensure stability of supplies at right price; control markets	To gain privileged access to resources *vis-à-vis* competitors	(a) Oil, copper, bauxite, bananas, pineapples, cocoa, hotels (b) Export processing, labour-intensive products or processes
Market seeking	Capital, technology, information, management and organizational skills; surplus R&D and other capacity; economies of scale; ability to generate brand loyalty	Material and labour costs; market size and characteristics; government policy (e.g. with respect to regulations and to import controls, investment incentives, etc.)	Wish to reduce transaction or information costs, buyer ignorance or uncertainty, etc., to protect property rights	To protect existing markets, counteract behaviour of competitors; to preclude rivals or potential rivals from gaining new markets	Computers, pharmaceuticals, motor vehicles, cigarettes, processed foods, airline services
Efficiency seeking (a) of product (b) of processes	As above, but also access to markets; economies of scope, geographical diversification and international sourcing of inputs	(a) Economies of product specialization and concentration (b) Low labour costs: incentives to local production by host governments	(a) As for second category plus gains from economies of common governance (b) The economies of vertical integration and horizontal diversification	As part of regional or global product rationalization and/or to gain advantages of process specialization	(a) Motor vehicles electrical appliances, business services, some R&D (b) Consumer electronics, textiles and clothing, cameras, pharmaceuticals
Strategic asset seeking	Any of first three that offer opportunities for synergy with existing assets	Any of first three that offer technology, markets and other assets in which firm is deficient	Economies of common governance; improved competitive or strategic advantage; to reduce or spread risks	To strengthen global innovatory or production competitiveness; to gain new product lines or markets	Industries that record a high ratio of fixed to overhead costs and which offer substantial economies of scale or synergy
Trade and distribution (import and export merchanting)	Market access; products to distribute	Source of inputs and local markets; need to be near customers; after-sales servicing, etc.	Need to protect quality of inputs; need to ensure sales outlets and to avoid underperformance or misrepresentation by foreign agents	Either as entry to new markets or as part of regional or global marketing strategy	A variety of goods, particularly those requiring contact with subcontractors and final consumers
Support services	Experience of clients in home countries	Availability of markets, particularly those of 'lead' clients	Various (see above categories)	As part of regional or global product or geographical diversification	(a) Accounting, advertising, banking, producer goods (b) Where spatial linkages are essential (e.g. airlines and shipping)

Source: Dunning (1993, Table 4.2, pp. 82–83).

According to Dunning, a firm will engage in international production when all of the following three conditions are present:

- A firm possesses certain *ownership-specific advantages* not possessed by competing firms of other nationalities.
- Such advantages are most suitably exploited by the firm itself rather than by selling or leasing them to other firms. In other words, the firm *internalizes* the use of its ownership-specific advantages.
- There must be *location-specific factors* which make it more profitable for the firm to exploit its assets in overseas, rather than in domestic, locations.

Table 6.1 summarizes the major characteristics of each of these three elements and shows how they may vary according to particular types of international production.

Ownership-specific advantages are assets which are internal to a firm. They are those

> which an enterprise may create for itself (e.g. certain types of knowledge, organization and human skills) or can purchase from other institutions, but over which, in so doing, it acquires some proprietary right of use. Such ownership-specific assets may take the form of a legally protected right, or of a commercial monopoly, or they may arise from the size, diversity or technical characteristics of firms.
>
> (Dunning, 1980, p. 9)

As we have seen, Stephen Hymer was the first to suggest that foreign direct investment could occur only if the investing firm possessed a particular advantage over indigenous firms. The most obvious ones relate to size and market power. Large firms, in general, are in a better position to obtain their production inputs at favourable rates than smaller firms. They generally have better access to finance either from their own retained earnings or because of a better credit rating on the financial markets.

Technology – of production, of marketing and of organization in general – is a particularly important source of advantage. Technology, in the broadest sense of 'know-how', is an intangible asset easily transferable from one location to another. Caves (1971), for example, emphasizes the advantage which a particular brand image may give over lesser known competitors. As he points out, a characteristic of many large firms, especially in consumer goods industries, is that they endeavour to differentiate their products from those of their competitors. This is done by making minor modifications to design or presentation and by reinforcing such differences through mass advertising. The practice of product differentiation is readily transferable from one market to another.

Why should a firm *internalize* the use of its ownership-specific advantages by investing in overseas production? It may simply export its products at 'arm's length' through the usual trade channels. Alternatively, a technology may be licensed to indigenous firms in foreign countries in return for the payment of fees or royalties. Both these alternatives are used extensively. The main reason such alternatives may not be followed lies in the nature of the markets for materials, for intermediate goods and for finished products. In the miraculous world of neoclassical economics, markets are assumed to operate perfectly. If this were to be so then there would be no advantage to a firm in attempting to bypass the market. The global economic system would consist of a whole series of discrete transactions between independent buyers and sellers.

But the world is not like this. *Markets are imperfect.* The greater the imperfection the greater will be the incentive for a firm to perform the function of the market itself by *internalizing* market transactions.[7] The most obvious example of such internalization is vertical integration in which a firm decides to control either its own sources of supply or the destination of its outputs. In both cases, the functions of independent material suppliers or of wholesale and retail merchants are absorbed – internalized – within the firm.

The major incentive for a firm to internalize markets is *uncertainty.* The greater the degree of uncertainty – whether over the availability, price or quality of supplies or of the price obtainable for the firm's product – the greater the advantage for the firm to control these transactions. Internalization is especially likely to occur in the case of knowledge. We have seen already that innovation and technological change are vital elements in a firm's ability to remain competitive and profitable. Many firms, especially large firms but also all those .in high-technology industries, spend huge sums of money on research and development. To ensure a satisfactory return on such investment and to protect against predators, firms have a strong incentive to retain the technology and its use within their own boundaries. Rather than sell or lease the technology to another firm overseas the firm sets up its own production facilities and exploits its technological advantage directly. Thus,

> there is a special reason for believing that internalization of the knowledge market will generate a high degree of multinationality among firms. Because knowledge is a public good which is easily transmitted across national boundaries, its exploitation is logically an international operation; thus unless comparative advantage or other factors restrict production to a single country, internalization of knowledge will require each firm to operate a network of plants on a world-wide basis.
>
> (Buckley and Casson, 1976, p. 45)

The third element in the international production question is that of *location-specific factors,* 'those which are available, on the same terms, to all firms whatever their size and nationality, but which are specific in origin to particular locations and *have to be used in those locations*' (Dunning, 1980, p. 9, emphasis added). Thus, in the absence of more favourable locational conditions overseas, a firm would serve foreign markets by exports from a domestic base. Several major types of location-specific factors are especially important in the context of international production, although their precise significance will vary according to the type of activity involved:

- markets
- resources
- production costs
- political conditions
- cultural/linguistic affinities.

In Dunning's 'eclectic model', the propensity to engage in international production is influenced by these three inter-related variables: ownership-specific advantages, internalization and location-specific factors. For international production to occur, *all three* conditions must be satisfied. Quite how they are satisfied, however, will vary according to the type of investment involved, as Table 6.1 reveals. It is also important to point out that ownership-specific advantages and location-specific factors are not necessarily independent of each other. This is a point we shall return to later.

Dunning's self-styled eclectic theory has been criticized as being merely 'a list of factors likely to be important in the explanation of the modern . . . [TNC] . . . rather than the explanation itself. Theoretical relations between the different factors too often remain untheorized' (Taylor and Thrift, 1986, p. 11). But this is to devalue Dunning's contribution rather too heavily. Seen as a pragmatic framework which attempts to integrate significant elements of other bodies of explanation, Dunning's approach is extremely useful as a conceptual structure within which specific cases can be examined.

Types of international production

Following our earlier breakdown of profit into revenue and cost components, we can classify overseas investments by business firms into two broad categories:

- market-oriented production
- supply or cost-oriented production.

In fact, the boundary between them is by no means as clear as this dichotomy suggests.

Market-oriented production

Most foreign direct investment in both manufacturing and service industries is designed to serve a specific geographical market by locating inside that market. The good or service produced overseas may be virtually identical to that being produced in the firm's home country although there may well be modifications to suit the specific tastes or requirements of the local market. In effect, such specifically market-oriented investment is a form of horizontal expansion across national boundaries. Two attributes of markets are especially important:

- The most obvious attraction of a specific market is its *size*, measured, for example, in terms of per capita income. Figure 6.3 shows the enormous variation in income levels (represented by gross national product (GNP) per capita) on a global scale. Per capita GNP in the developed economies as a whole averaged $23,420 in 1994; in the lowest income group of developing countries, the average per capita income was a mere $360. In many cases, of course, income levels were very much lower than this. The largest geographical markets in terms of incomes, although not in terms of population, are obviously the United States and western Europe. Such variations in per capita GNP provide a crude indication of how the *level* of demand will vary from place to place across the world.
- Countries with different income levels will tend to have a different *structure* of demand. It has long been recognized that the type and mix of goods demanded tend to vary systematically with income. As incomes rise, so does the aggregate demand for goods and services. But such increased demand does not affect all products equally. In the economist's terminology, different products have different income elasticities of demand. (Income elasticity of demand is simply the way in which the quantity of a good demanded responds to changes in the incomes of consumers.) One would expect populations in countries with low income levels to spend a larger proportion of their income on primary products (basic necessities) and, conversely, countries with high income levels to spend a higher proportion of their income on 'higher-order' manufactured goods and services. Thus, countries with different per capita income levels will tend to differ greatly in both the magnitude and the nature of their consumption patterns.

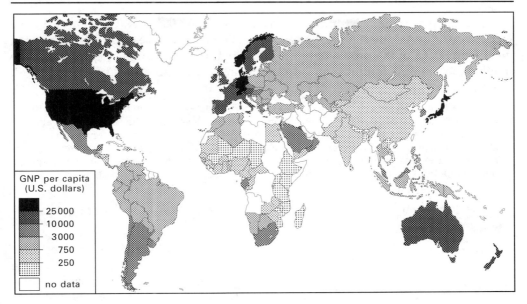

Figure 6.3   World variations in per capita income
Source: World Bank (1996, Table 1)

Supply or cost-oriented production

Supply considerations are obviously the dominant motivation for firms in the natural resource industries. Such firms, of necessity, must locate at the sources of supply which, themselves, tend to be highly localized geographically. Often, such investments form the first element in a sequence of vertically integrated operations whose later stages (processing) may be located quite separately from the source of supply itself. In many cases – in line with the predictions of Weberian location theory – final processing occurs close to the market. Supply-oriented overseas investments have a very long history; many of the early transnational investments by American, British and continental European firms in the late nineteenth and early twentieth centuries were motivated by the desire to obtain and ensure raw material supplies. However, cost-oriented foreign investments by manufacturing firms are a relatively recent phenomenon and are very closely related to the kinds of technological developments discussed in Chapter 5.

In discussing the production process in Chapter 5 we noted, for example, that the relative importance of the various production factors tends to vary according to the stage in the product's life-cycle and, especially, with the maturity of the technology. More generally, the precise mix of production factors varies from one industry to another. In one sense, therefore, the key consideration is the relative importance of the individual factors in the firm's cost structure. But there is more to it than this. A particular factor of production may well be the most important element in a firm's total costs yet it may exert a negligible locational influence if its cost does not vary over space. If a factor costs the same everywhere it has a zero locational cost.[8]

Technological changes in production processes and in transport have evened out the significance of location for some of the traditionally important factors of production (for example, natural resources). Many now hold the view that, at least at the global scale, the single most important location-specific factor is *labour*. As we have seen, labour – especially semi-skilled and unskilled labour – is especially important in the mature stage of the product cycle and has become increasingly significant as the standardization and routinization of certain production processes have developed. Hence,

> with the trend towards greater locational capability, labour moves to the forefront because of its degree of spatial differentiation. As capital develops its capability of locating more freely with respect to most commodity sources and markets, it can afford to be more attuned to labour force differences. Under the pressure of competition this becomes a necessity. The reasons for labour's persistent geographic distinctiveness lie in the unique nature of labour as a 'factor of production' – its embodiment in human beings.
>
> (Storper and Walker, 1983, pp. 3–4)

The locational significance of labour as a production factor[9] is reflected in a number of ways:

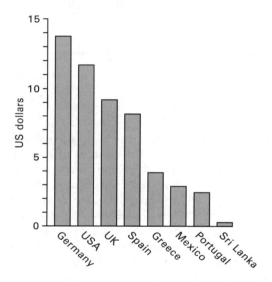

Figure 6.4   Hourly earnings in selected countries
Source: International Monetary Fund, *Yearbook of Financial Statistics, 1995*; United Nations, *Statistical Yearbook, 1995*

- Geographical variations in *wage costs*. International differences in wage levels can be staggeringly wide, as Figure 6.4 shows. These figures should be treated with some caution; they are averages across the whole of manufacturing industry and are therefore affected by the specific industry mix. Some industries have much higher wage levels than others. Even so, the contrasts are striking. Compared with high-cost countries like Germany ($13.70 per hour) and the United States ($11.70), manufacturing earnings in Greece, Mexico and Portugal were all below $4 per hour, whilst in Sri Lanka hourly earnings were measured in cents rather than dollars.

- Geographical differences in *labour productivity*. Spatial variations in wage costs are only a partial indication of the locational importance of labour as a production factor. The 'performance capacity' of labour varies enormously from place to place, a reflection of a number of influences: including education, training, skill, motivation, as well as the kind of machinery and equipment in use.

- Geographical variations in the extent of *labour controllability*. Largely because of historical circumstances, there are considerable geographical differences in the degree of labour 'militancy' and in the extent to which labour is organized through labour unions. The fact that most firms are very wary of 'highly organized' labour regions is demonstrated by their tendency to relocate from such regions or to make new investments in places where labour is regarded as being more malleable.[10]

● A further significant dimension of labour as a factor of production is that it tends to be far *less mobile geographically*, particularly over great distances, than other factors such as capital or technology. In general, labour is strongly 'place-bound' although the strength of the bond varies a great deal between different types of labour. On average, male workers are more mobile than female workers, skilled workers are more mobile than unskilled workers; professional white-collar workers are more mobile than blue-collar workers. But there are exceptions to such generalizations as shown by the substantial waves of labour migration at different periods of history and towards particular kinds of geographical location. Such flows do not, however, contradict the basic point that labour is a factor which is strongly differentiated spatially and deeply embedded in local communities in quite distinctive ways.[11]

Global variations in production costs are a significant element in the international investment-location decision. This is obviously the case for supply-oriented investments but it is also a critical consideration for market-oriented investments as well. In that case there is always a trade-off to be calculated between the benefits of market proximity on the one hand and locational variations in production costs on the other. But the problem is not merely one of variations in production costs at a single moment in time or even the obvious point that such costs change over time. A particularly important consideration is the *uncertainty* of the level of future production costs in different locations. One way of dealing with such uncertainty is for the TNC to locate similar plants in a variety of different locations and then to adopt a flexible system of production allocation between plants. However, this strategy may well be complicated further by the volatility of currency exchange rates between different countries. What appears to be a least-cost source under one set of exchange rates may look very different if there is a major change in these rates.

TNC development as a sequential process

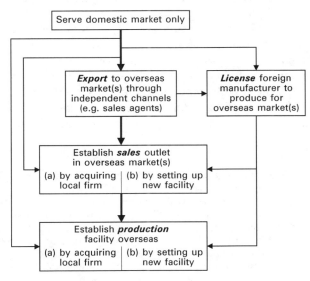

Figure 6.5   Sequential development of a transnational corporation

At the level of the individual firm, we can explain why international production occurs and why TNCs exist at all in terms of specific combinations of Dunning's three sets of conditions. But is there an identifiable *evolutionary sequence* of TNC development? Does the transition from a firm producing entirely for its domestic market to one engaged in overseas production follow a systematic development path? The answer to these questions is both 'yes' and 'no'. Yes, in the sense that certain common patterns of development are evident. No, in the sense that we should not expect all firms to follow the same sequence or in the sense that all firms will inevitably become TNCs. It is useful to consider the broad path of TNC development, however, because it helps to give some sense of the dynamics of the processes involved.

Figure 6.5 illustrates the kind of sequence most commonly identified. It begins with the assumption that, initially, the firm is purely a domestic firm in terms both of production and of markets. In all national economies the majority of firms are of this type. However, the limits of the firm's domestic market may be reached and overseas markets may need to be penetrated to maintain growth and profitability. It is generally assumed that this is done initially through exports using the services of overseas sales agents. Such agents are, of course, independent of the exporting firm. However, the benefits of internalization which we discussed earlier may eventually stimulate the firm to exert closer control over its foreign sales by setting up sales outlets of its own in overseas locations. This may be achieved in one of two ways: by setting up an entirely new facility or by acquiring a local firm (possibly the previously used sales agency itself). Acquisition is, in fact, one of the most common methods of entry both to new product markets and to new geographical markets. It offers the attraction of an already-functioning business compared with the more difficult, and possibly more risky, method of starting from scratch in an unfamiliar environment. Eventually the time may come – though not inevitably – when the need is felt for an actual production facility overseas. Again, this may be achieved through either acquisition or 'greenfield' investment.

There is a good deal of anecdotal evidence to support this sequence of development among firms which actually became TNCs. For example, in the case of Japanese firms investing in Europe it is clear that actual manufacturing operations came rather late following a long period of development of Japanese service investments in the form of the general trading companies, banks and other financial institutions and the sales and distribution functions of the manufacturing firms themselves.[12] However, there is nothing inevitable about the progression through each of the stages. Figure 6.5 shows a number of possible variations on the main theme and also suggests that the general sequence may be 'short-circuited' at various points. A firm may bypass the intermediate stages and set up overseas production facilities as a first step. One way in which this may occur is through the acquisition of another domestic firm which already has foreign operations. Thus, a firm may become transnational almost incidentally. For example, a common means of involvement in overseas markets is through licensing agreements with local firms which are permitted to use a particular technology or to manufacture a specific product for a defined market on payment of fees and royalties. In some cases, such agreements may form the basis of other forms of direct foreign involvement.

Thus, the sequential process

> can be speeded up either through a faster progression from one step to the next or through leapfrogging directly to FDI. There is some evidence that this is happening. For example, a study of 228 outward direct investments from Australia (65 per cent of which originated in 1970–1979) showed that, in 39 per cent of the cases, there was no pre-existing host country presence . . . Support for the idea of leapfrogging is also provided by an empirical study of Swedish firms which moved into the Japanese market directly through FDI . . . Leapfrogging over steps in the traditional process is more likely to occur in high-technology firms: a survey of 807 British companies found that 'low technology manufacturing companies in general followed the process of exports before investment more frequently than high-technology companies'.
>
> (UNCTAD, 1996b, p. 98)

The sequential model of Figure 6.5 assumes that the process is market driven. It fits the conventional view of the TNC as developing progressively, from serving markets through exports to direct investment in overseas production facilities. The broader definition of the TNC introduced earlier (see p. 177) allows for other possibilities. Recall that the essence of this broader definition is the *co-ordination* of international production from one centre of strategic decision-making. It is quite possible for a firm to begin its transnational activities by co-ordinating production in overseas locations in order to serve the firm's domestic market rather than to serve overseas markets. This might subsequently evolve into a more elaborate transnational network of operation. We shall return to some of these alternative forms of transnational organization in the next chapter.

Transnational corporations are not 'placeless': the myth of the 'global corporation'[13]

A major ingredient of the 'globalization' scenario outlined in Chapter 1 is the idea that many TNCs are 'global corporations' whose ways of doing things have converged towards a single globally integrated model. The pressures of operating in a globally competitive environment, it is argued, are creating a uniformity of strategy and structure among TNCs. The implication is that all TNCs are moving inexorably along the same path. In so doing, it is argued, TNCs lose all identification with, or allegiance to, particular countries or communities. They become, in effect, *placeless*. For example, both Robert Reich and Kenichi Ohmae claim that TNCs have become – or are becoming – 'denationalized'. In Ohmae's (1990, p. 94) exhortation to managers

> Before national identity, before local affiliation, before German ego or Italian ego or Japanese ego – before any of this comes the commitment to a single, unified global mission . . . Country of origin does not matter. Location of headquarters does not matter. The products for which you are responsible and the company you serve have become denationalized.

In this section we challenge this scenario of placeless global corporations. Not only are there very few, if any, truly 'global' corporations but also TNCs remain strongly affected by specific national and local environments. In particular, the TNC's home environment remains fundamentally important to how it operates, notwithstanding the geographical extensiveness of its activities. *All* TNCs have an identifiable home base, which ensures that every TNC is essentially embedded within its domestic environment. Of course, the more geographically extensive its operations the more likely it will be to take on some characteristics of its host environments. But even where there is substantial local adaptation and local embeddedness, the influence of the firm's geographical origins remains very strong.

If there are 'global' corporations then we would expect to find them well represented among the world's largest TNCs. In fact, the evidence is very weak indeed. Table 6.2 lists the 100 largest TNCs in the world in 1994 in terms of total foreign assets. The companies are ranked by their 'index of transnationality' which is based upon three indicators: foreign sales, foreign assets and foreign employment. If the 'global corporation' hypothesis is valid then we would expect to find that at least the majority of these largest TNCs have the overwhelming majority of their assets and employment outside their home country. (Foreign sales alone are a poor indicator because they reflect only exports and not foreign production; their inclusion in the index will

Table 6.2 How 'global' are the top 100 transnational corporations?

Ra	Ri	Index	Company	Country	Industry	Foreign share of Assets	Employ-ment
60	1	92.3	Thomson Corp.	Canada	Publishing and printing	95.7	88.7
71	2	92.2	Solvay	Belgium	Chemicals	92.8	89.5
50	3	91.4	RTZ	UK	Mining	–	96.9
17	4	90.5	Roche Holdings	Switzerland	Pharmaceuticals	90.4	82.9
42	5	88.8	Sandoz	Switzerland	Pharmaceuticals	–	85.1
15	6	88.4	ABB	Switzerland	Electrical equipment	85.2	93.7
52	7	87.3	Electrolux	Sweden	Electronics	–	82.8
13	8	86.5	Nestlé	Switzerland	Food	65.6	96.9
24	9	85.0	Philips	The Netherlands	Electronics	87.7	83.0
23	10	84.5	Unilever	UK/The Netherlands	Food	77.5	89.9
58	11	80.2	Glaxo-Wellcome	UK	Pharmaceuticals	75.2	75.0
93	12	79.3	Akzo	The Netherlands	Chemicals	71.0	73.4
61	13	78.6	Seagram	Canada	Beverages	76.9	–
21	14	72.5	Bayer	Germany	Chemicals	81.7	38.2
87	15	72.3	Alcan	Canada	Metal products	58.8	71.8
67	16	72.0	Michelin	France	Rubber and plastics	61.1	–
73	17	68.0	Total	France	Petroleum	–	56.6
27	18	67.2	British Petroleum	UK	Petroleum	67.7	73.1
62	19	66.8	News Corporation	Australia	Publishing and printing	46.4	–
33	20	66.7	BAT Industries	UK	Tobacco	32.6	91.2
39	21	66.6	Volvo	Sweden	Automobiles	76.3	40.6
80	22	65.9	Cable & Wireless	UK	Telecommunications	–	75.3
54	23	65.2	Saint Gobain	France	Building materials	–	72.1
34	24	64.6	Hoechst	Germany	Chemicals	59.9	55.7
36	25	64.6	Ciba-Geigy	Switzerland	Chemicals	48.7	75.1
3	26	63.8	Exxon	USA	Petroleum	63.9	64.0
1	27	63.6	Royal Dutch Shell	UK/The Netherlands	Petroleum	62.5	74.5
29	28	63.3	Hanson	UK	Building materials	52.9	78.4
35	29	61.8	Rhone-Poulenc	France	Chemicals	68.1	56.9
6	30	60.4	Volkswagen	Germany	Automobiles	–	39.8
18	31	58.9	Alcatel Alsthom	France	Electronics	45.1	59.4
11	32	58.7	Mobil	USA	Petroleum	63.1	46.8
19	33	58.5	Sony	Japan	Electronics	–	57.7
94	34	57.5	Pechiney	France	Metals	49.5	58.0
91	35	57.4	LVMH Moet-Hennessey	France	Beverages	–	62.5
81	36	57.2	Digital Equipment	USA	Computers	56.7	52.7
10	37	56.7	Elf Aquitaine	France	Petroleum	–	49.1
5	38	56.4	IBM	USA	Computers	54.1	52.6
85	39	55.3	Thomson	France	Electronics	35.6	57.9
66	40	54.4	BMW	Germany	Automobiles	48.0	46.2
45	41	51.5	BASF	Germany	Chemicals	44.0	37.9
74	42	50.5	McDonalds	USA	Restaurants	–	–
55	43	50.0	Procter & Gamble	USA	Soaps/cosmetics	37.7	59.6
84	44	49.0	Carrefour	France	Trade	48.7	49.0
79	45	48.9	Minnesota Mining	USA	Mining	48.9	46.7
86	46	48.3	Sara Lee	USA	Food	49.6	58.2
78	47	48.1	Johnson & Johnson	USA	Pharmaceuticals	42.0	52.0
46	48	48.0	VIAG	Germany	Diversified	48.1	48.0
25	49	47.3	Siemens	Germany	Electronics	–	83.0
65	50	47.2	Robert Bosch	Germany	Auto parts	–	40.5

20	51	47.0	Fiat	Italy	Automobiles	38.1	38.2
48	52	45.3	Dow Chemical	USA	Chemicals	39.3	45.0
92	53	44.9	Alcoa	USA	Metals	–	50.9
44	54	44.2	Texaco	USA	Petroleum	45.9	35.8
26	55	43.7	Renault	France	Automobiles	–	28.9
89	56	43.6	Motorola	USA	Electronics	29.7	44.5
99	57	43.5	Norsk Hydro	Norway	Chemicals	34.1	50.0
96	58	43.0	Eastman Kodak	USA	Scientific and photo equipment	–	43.6
9	59	42.8	Daimler Benz	Germany	Transport and communications	42.0	24.0
100	60	42.7	Bridgestone	Japan	Rubber and plastics	–	58.0
98	61	42.5	United Technologies	USA	Aerospace	30.8	55.7
31	62	42.0	Du Pont	USA	Chemicals	–	32.7
76	63	42.0	Grand Metropolitan	UK	Food	–	42.0
69	64	41.6	Sharp	Japan	Electronics	–	67.7
59	65	41.4	Hewlett Packard	USA	Computers	45.9	40.1
51	66	41.0	Honda	Japan	Automobiles	–	21.2
28	67	41.0	Philip Morris	USA	Food	34.2	51.5
16	68	39.8	Matsushita	Japan	Electronics	–	42.3
82	69	37.3	Mannesmann	Germany	Industrial equipment	–	32.4
49	70	36.7	Xerox	USA	Scientific and photo equipment	26.4	36.7
68	71	33.5	Canon	Japan	Computers	33.5	48.6
14	72	32.2	Nissan	Japan	Automobiles	–	24.1
95	73	31.9	RJR Nabisco	USA	Food and tobacco	15.6	48.1
12	74	31.0	Mitsubishi	Japan	Diversified	–	31.0
77	75	30.3	BHP	Australia	Metals	32.2	25.0
40	76	30.3	Chevron	USA	Petroleum	37.8	23.2
72	77	29.8	Pepsico	USA	Food	30.7	29.8
30	78	29.5	Mitsui	Japan	Diversified	–	29.5
32	79	29.0	Nissho Iwai	Japan	Trading	–	29.0
2	80	28.6	Ford	USA	Automobiles	27.6	28.6
37	81	28.1	ENI	Italy	Petroleum	–	21.3
8	82	28.1	Toyota	Japan	Automobiles	–	16.0
22	83	27.7	Hitachi	Japan	Electronics	–	24.1
90	84	26.6	International Paper	USA	Paper	28.7	29.3
4	85	25.7	General Motors	USA	Automobiles	–	25.7
63	86	24.4	Nippon Steel Corp	Japan	Metal	–	29.7
38	87	24.2	Sumitomo	Japan	Trading	–	–
64	88	23.3	Amoco	USA	Petroleum	29.0	17.5
43	89	22.7	Itochu	Japan	Trading	–	26.7
70	90	22.4	Veba	Germany	Trading	20.0	18.8
53	91	21.2	ITT	USA	Diversified services	–	21.2
41	92	20.0	Toshiba	Japan	Electronics	–	20.0
88	93	20.0	Atlantic Richfield	USA	Petroleum	22.8	20.0
57	94	19.3	NEC Corporation	Japan	Electronics	19.5	11.6
47	95	19.1	Marubeni	Japan	Trading	–	19.1
97	96	17.0	Kobe Steel	Japan	Metals	–	17.0
7	97	16.7	General Electric	USA	Electronics	13.5	16.7
75	98	15.4	Chrysler	USA	Automobiles	–	19.8
83	99	13.3	GTE	USA	Telecommunications	13.7	13.3
56	100	10.8	AT&T	USA	Electronics	11.9	10.8

Notes:
Ra – rank by total foreign assets.
Ri – index of transnationality. Represents the average of foreign assets to total assets, foreign sales to total sales and foreign employment to total employment.
Source: Calculated from UNCTAD (1996b, Table I.12).

tend to inflate its value.) Even so, careful scrutiny of Table 6.2 reveals no clear evidence to support the view that even the 100 largest TNCs are 'global' in terms of these indicators.

Only 42 of the 100 companies have an index of greater than 50; a mere 13 have an index greater than 75. Significantly, the 13 most transnational firms originate from small countries (Switzerland, the UK, The Netherlands, Belgium, Canada). Conversely, the biggest TNCs in terms of total foreign assets all have relatively low transnational index scores. For example, the second largest TNC (in terms of foreign assets) Ford ranks 80th on the index of transnationality; GM (4th in terms of foreign assets) ranks 85th; IBM ranks 38th; VW ranks 30th; Toyota, the 8th-ranking TNC in foreign assets ranks 82nd on the broader index. On this measure, therefore, there is little evidence of TNCs having the share of their activities outside their home countries which might be expected if they are global firms.[14]

A rather broader approach to this question is provided by Hu (1992). He suggests the following criteria for evaluating the extent to which TNCs are 'stateless':

- In which nation or nations is the bulk of the corporation's assets and people located?
- By whom are the local subsidiaries owned and controlled, and in which nation is the parent company owned and controlled?
- What is the nationality of the senior positions (executive and board posts) at the parent company, and what is the nationality of the most important decision-makers at the subsidiaries in host nations?
- What is the legal nationality of the parent company? To whom would the group as a whole turn for diplomatic protection and political support in case of need?
- Which is the nation where the tax authorities can, if they choose to do so, tax the group on its worldwide earnings rather than merely its local earnings?

On the basis of an empirical analysis of a sample of TNCs, Hu (1992, p. 121) concludes that 'these criteria usually produce an unambiguous answer: that it . . . [the TNC] . . . is a national corporation with international operations (i.e. foreign subsidiaries)'. Thus, despite many decades of international operations, TNCs remain distinctively connected with their home base. Ford is still essentially an American company, ICI a British company, Siemens a German company:

> However great the global reach of their operations, the national firm does, psychologically and sociologically, 'belong' to its home base. In the last resort, its directors will always heed the wishes and commands of the government which has issued their passports and those of their families. A recent study of the boards of directors of the top 1000 US firms, for example, shows that only 12 per cent included a non-American – rather fewer, in fact, than in 1982 when there were 17 per cent . . . The Japanese firm with even one token foreign director would be hard to find. Even in Europe, with the exception of bi-national firms like Unilever, you do not find the top management reflecting by their nationality the geographical distribution of its operations.
>
> (Stopford and Strange, 1991, p. 233)

This is not to argue that TNCs necessarily retain a 'loyalty' to the states in which they originate. The nature of the embeddedness process is far more complex than that. The basic point is that TNCs are 'produced' through an intricate process of embedding in which the cognitive, cultural, social, political and economic characteristics of the national home base play a dominant part. TNCs, therefore, are 'bearers' of such characteristics, which then interact with the place-specific characteristics of the countries in which they operate to produce a set of distinctive outcomes. But the point is that the home-base characteristics invariably remain dominant. This is not to claim that TNCs from a particular national origin are identical. This is self-evidently not the case; within any national situation there will be distinctive corporate cultures, arising from the firm's own specific corporate history, which predispose it to behave strategically in particular ways. Take the case of the automobile industry which we will examine in detail in Chapter 10. As US companies, Ford and GM are quite distinctive from Toyota, Volkswagen, Fiat or Renault. But they are also different from each other. Similarly, Toyota and Nissan are distinctive, but not identical, Japanese automobile firms; the same point can be made about the French auto producers and so on. However, there are generally greater similarities than differences between firms from the same national base. As we saw in Chapter 3, nation-states act as 'containers' of distinctive assemblages of institutions and practices – 'business systems' in Whitley's terminology.

Such containers help to produce particular kinds of firms. In Dunning's framework discussed in the previous section, the ownership-specific advantages of firms are related to the location-specific endowments of nations. As Dunning (1979, pp. 283–84) points out,

> today's ownership advantages of enterprises may be the inheritance of yesterday's country specific endowments. This is especially true of those to do with national resources and government policy. Further, such endowments may not give enterprises the same advantages at all points of the product and/or investment development cycle. It is possible that, as a new product becomes more standardized, ownership advantages will diminish or change . . . [also] . . . As an enterprise increases its degree of multinationality, the country specific characteristics of the home country become less, and that of other countries more, important in influencing its ownership advantages.

Table 6.3 outlines some of the links which exist between the ownership-specific advantages of firms and the location-specific characteristics of the firm's *home* country. It is this link which helps to explain the different characteristics of TNCs from

Table 6.3 Links between selected ownership-specific advantages and country-specific characteristics

Ownership-specific advantages	Country characteristics favouring such advantages
Size of firm	Large, standardized markets Liberal regime towards mergers and concentration
Managerial expertise	Pool of managerial talent Educational and training facilities
Technology-based advantages	Good R&D facilities Government support of innovation Pool of scientific and technical labour
Labour and/or mature, small-scale intensive technologies	Large pool of labour (including technical labour) Appropriate consultancy services
Production differentiation	High-income national markets
Marketing economies	High-income elasticity of demand Highly developed marketing/advertising system Consumer-oriented society
Access to (domestic) markets	Large national market No restrictions on imports
Capital availability and financial expertise	Well developed, reliable capital markets Appropriate professional advice

Source: Based on Dunning (1979, Table 6).

Table 6.4   Differences between United States, German and Japanese TNCs

	United States TNCs	German TNCs	Japanese TNCs
Corporate governance and corporate financing	Constrained by volatile capital markets; short-termist perspectives. Finance-centred strategies	Relatively high degree of operational autonomy except during crises. Long-term perspectives. Conservative strategies	Bound by complex but reliable networks of domestic relationships. Long-term perspectives. Market-share centred strategies
	High risk of takeover	Low risk of takeover	Very low risk of takeover – mainly confined to within network
	90% of firm shares held mainly by individuals, pension funds, mutual funds. Less than 1% held by banks	Firm shares held mainly by non-financial institutions (40%). Significant role of regional bodies	High degree of cross-shareholdings within group. Lead bank performs a steering function
	Banks provide mainly secondary financing, cash management, selective advisory role	Banks play a lead role. Supervisory boards of companies are strongly bank influenced	Ratio of bank loan/ corporate liabilities = 60–70%
	Ratio of bank loan/ corporate financial liabilites = 25–35%	Ratio of bank loan/ corporate liabilities = 60–70%	
Research and development	Corporate R&D expenditure peaked in 1985 at 2.1% of GDP. Declining	Corporate R&D expenditure declined steeply in late 1980s/early 1990s. At 1.7% of GDP lower than USA and Japan	Corporate R&D grew very rapidly in 1980s. Overtook USA in 1989. Peaked at 2.2% of GDP. Real cuts made only as last resort
	Diversified pattern; innovation oriented	Narrow focus	High-tech and process orientation
	Some propensity to perform R&D abroad	Some propensity to perform R&D abroad	Very limited propensity to perform R&D abroad
Direct investment and intrafirm trade	Extensive outward investment. Substantial competition from inward investment	Selective outward investment. Moderate competition from inward investment	Extensive outward investment. Very limited competition from inward investment.
	Moderate intrafirm trade; high propensity to outsource	High level of intrafirm trade	Very high level of intrafirm and intragroup trade

Source: Based primarily on material in Pauly and Reich (1997).

different source nations. For example, the large domestic market and high level of technological sophistication of the US domestic economy have helped to produce the distinctive characteristics of United States TNCs. The lack of natural resources and the strong involvement of government and other institutions in technological and industrial affairs helps to explain the particular attributes of Japanese TNCs. Similarly, the specific characteristics of certain developing countries are reflected in the nature of their newly emerging TNCs.

Recent detailed empirical research by Pauly and Reich (1997) provides strong evidence to support the view that nationally-based differences between TNCs tend to persist contrary to the notion of a convergence towards a standard global model of firm structure and behaviour. They examine United States, Japanese and German TNCs in terms of several criteria: corporate governance and financing, research and development, foreign investment and intrafirm trade practices. Their results are summarized in Table 6.4.

The experiences of United States, German and Japanese firms analysed by Pauly and Reich provide strong arguments to counter the convergence hypothesis. Their evidence

> shows little blurring or convergence at the cores of firms based in Germany, Japan, or the United States . . . Durable national institutions and distinctive ideological traditions still seem to shape and channel crucial corporate decisions . . . there remain systematic and important national differences in the operations of [TNCs] – in their internal governance and long-term financing, in their R&D activities, and in their intertwined investment and trading strategies . . . the domestic structures within which a firm initially develops leave a permanent imprint on its strategic behavior . . . At a time when many observers emphasize the importance of cross-border strategic alliances, regional business networks, and stock offerings on foreign exchanges – all suggestive of a blurring of corporate nationalities – our findings underline, for example, the durability of German financial control systems, the historical drive behind Japanese technology development through tight corporate networks, and the very different time horizons that lie behind American, German, and Japanese corporate planning.
>
> (Pauly and Reich, 1997, pp. 1, 4, 5, 24)

So, one aspect of the 'TNCs are not placeless' argument is that the conditions in which firms develop in their home countries continue to exert a very strong influence on their subsequent behaviour when operating outside their home country. The other aspect of the argument that 'geography matters' concerns the extent to which firms operating in different countries take on some of the characteristics of those host environments. Although the influence of the home base is highly significant, this does not mean that it is totally deterministic of how firms operate abroad. For a whole variety of reasons – political, cultural, social – foreign firms invariably have to adapt some of their domestic practices to local conditions. It is virtually impossible to transfer the whole package of firm advantages and practices to a different national environment. We referred in Chapter 5 to the 'hybrid' nature of Japanese overseas manufacturing plants, for example. The same argument applies to United States firms operating abroad. Even in the United Kingdom, where the apparent 'cultural distance' between the US and the UK is less than in many other cases, there is a very long history of American firms having to adapt some of their business practices to local conditions. What results, therefore, is a varying mix of home-country and host-country influences. But although local adaptation almost invariably occurs, Pauly and Reich (1997, p. 25) are probably correct in observing that although TNCs originate from different home bases they 'appear to adapt themselves at the margins but not much at the core'.

Conclusion

The aim of this chapter has been to explore two specific aspects of transnational corporations. First, we explored some of the major theoretical explanations of why firms become 'transnational' by looking at both the macrostructural theories based upon the internationalization of the circuits of capital and the microscale theories cast at the level of the individual firm. In fact, both levels of explanation provide valuable insights into the internationalization process; it is important to develop explanatory frameworks which combine both general elements of the way in which the system as a whole operates with specific elements of how firms behave. One of the striking features is the fact that, at the firm-specific level, there was no real attempt at theorization until Hymer's pioneering study in the early 1960s.

Secondly, we challenged the popular conception of the 'placeless' transnational corporation. All TNCs have an identifiable home base and the characteristics of that base continue to exert an influence on how firms behave as they develop international networks of operations. We showed clear differences between TNCs of different nationalities; there is little evidence of TNCs converging towards a single model. Of course, as TNCs move into new environments they have to adapt, to a greater or lesser degree, to local circumstances. On both counts, however, it is very apparent that 'geography matters' to how TNCs co-ordinate and configure their production chains and, by extension, to the kind of impact they have on the states and communities within which they operate. This is an issue to which we turn specifically in the next chapter.

Notes for further reading

1. In the Second Edition of this book, I posed the question as 'why internationalize?' Bob Rowthorn rightly pointed out to me that a more accurate question is 'why *not* international-ize?' on the grounds that capitalist firms have an intrinsic tendency to expand the geograph-ical scope of their activities. He is, of course, correct and I have tried to make this more explicit.
2. For a detailed treatment of international production as the process of the internationaliza-tion of the various circuits of capital, see Palloix (1975; 1977), Jenkins (1984). Barr (1981) adapts and expands the concept at the level of the individual enterprise.
3. General reviews of the theoretical literature are provided by Pitelis and Sugden (1991), Dunning (1993). Key contributions are those by Buckley and Casson (1976), Hymer (1976), Vernon (1966; 1971; 1974; 1979), Dunning (1977; 1993).
4. Hymer's pioneering contribution was contained in his 1960 PhD thesis which was not published until after his tragic death in the 1970s (Hymer, 1976). Dunning and Rugman (1985) assess the significance of Hymer's work on the theory of foreign direct investment, as do some of the contributors in Pitelis and Sugden (1991).
5. Vernon's thinking on the product life-cycle evolved through a series of contributions: 1966, 1971, 1974, 1979. Hirsch (1967) and Wells (1972) both made significant contributions to the product cycle interpretation of international production and trade. Critical appraisals of the product cycle explanation of international production are to be found in Buckley and Casson (1976), Giddy (1978).
6. This particular hypothesis was investigated empirically by Knickerbocker (1973).
7. The initial concept of 'internalization' (though not the term itself) is generally attributed to Coase (1937). Its specific application to TNCs dates from the 1970s through the work of Buckley and Casson (1976) and Rugman (1981) amongst others.
8. See Smith (1981) for a discussion of the distinction between basic costs and locational costs.
9. But as Herod (1997, p. 2) argues, working people are much more than 'crude abstractions in which labour is reduced to the categories of wages, skill levels, location, gender, union membership and the like, the relative importance of which is weighed by firms in their locational decision-making'.
10. Peet (1983) explores this in terms of the 'geography of class struggle' and its effects on the location of manufacturing industry in the United States. At the global scale, part of the 'new international division of labour' thesis is the same theme of firms fleeing sites of class struggle in the industrialized countries and relocating their operations in not just low-cost but also 'benign' labour conditions. See Frobel, Heinrichs, Kreye (1980).
11. Peck (1996) provides an excellent discussion of the 'place' of labour within the capitalist market system. See also Herod (1996).
12. Mason (1994), Dicken, Tickell and Yeung (1997).
13. There is a growing literature on this topic. See, for example, Whitley (1994b), Hu (1992; 1995), Ruigrok and van Tulder (1995), Reich (1996), Pauly and Reich (1997).
14. Using slightly different data, Ruigrok and van Tulder (1995, Chapter 7) reach similar conclusions.

CHAPTER 7

'Webs of enterprise': transnational corporations within networks of relationships

Introduction

In Chapter 6 we outlined some of the theoretical explanations of why TNCs exist. We also strongly emphasized the continuing *diversity* of TNCs which arises, partly, from their specific geographical origins. TNCs are not placeless. In this chapter we take a broader perspective and focus on the *networks of relationships* which exist both within and between firms as they attempt to co-ordinate and configure their production chains. Three sets of relationships are explored:

- First, we examine TNCs as networks of *internalized* relationships and show how the structure of these networks – both organizational and geographical – changes as the firm's basic strategy changes.
- Secondly, we explore the extremely complex networks of *externalized* relationships which exist between independent and quasi-independent firms. This allows us to explore the increasingly diversified forms of interfirm relations, ranging from international strategic alliances, through sourcing and subcontracting links to more disaggregated network forms.
- Thirdly, we examine some of the ways in which these internalized and externalized networks map on to geographical space.

Networks of internalized relationships: inside the transnational corporation

It follows from our discussion in Chapter 6 that to talk of *the* transnational corporation runs the danger of universalizing what is, in fact, a highly differentiated kind of institution. We need constantly to bear this in mind as we focus upon the ways in which TNCs organizationally co-ordinate, and geographically configure, their production chains within, and across, their own organizational boundaries. (It will be useful to refer back to the general discussion of production chains in Chapter 1 and, especially, to look again at Figures 1.1 and 1.2.)

Coping with complexity: a diversity of organizational structures
One of the basic 'laws' of growth of any organism or organization is that as growth occurs its internal structure has to change. In particular, the *functional role* of its component parts tends to become more *specialized* and the links between the parts become more *complex*. As the size, organizational complexity and geographical spread of TNCs have increased so the internal inter-relationships between their

geographically separated parts have become a highly significant element in the global economy. However, the precise manner in which TNCs organize and configure their production chains arises from a number of inter-related influences:[1]

- The *firm's specific history*, including
  - characteristics derived from its *home-country embeddedness* (see the previous chapter),
  - its *culture and administrative heritage* in the form of accepted practices built up over a period of time which produce what Heenan and Perlmutter call its 'strategic predisposition'.
- The nature and complexity of the *industry environment(s)* in which the firm operates, including the nature of competition, technology, regulatory structures, etc.

Bartlett (1986, pp. 392–93) uses a physiological analogy to capture the complexity of this process:

> This analogy helps emphasize that while formal structure is critical in defining the basic anatomy, its role is by no means dominant. To be effective, anatomical changes must be accompanied and complemented by appropriate adaptations to the physiology (the organization's systems and decision processes), and to the psychology (the organization's culture and management values).

The traditional approach to changing organizational structures – based primarily on the hierarchical western (i.e. US) model – has depicted it as a sequential process whereby firms transformed their organizational structures from a *functional* form, in which the firm is subdivided into major functional units (production, marketing, finance, etc.), into a *divisional* form (either product based or area based). Each product division is responsible separately for its own functions, particularly of production and marketing, although some functions (especially finance) tend to be performed centrally for the entire corporation. Each product division usually acts as a separate profit centre. The main advantage of the divisional structure is usually seen to be one of a greater ability to cope with product diversity. Thus, as large US firms became increasingly diversified during the 1950s and 1960s they also tended to adopt a divisional structure.

Adoption of a divisional structure gave firms greater control over their increasingly diverse product environment. However, operating across national boundaries, rather than within a single nation, poses additional problems of co-ordination and control. Largely through trial and error, TNCs groped their way towards more appropriate organizational structures. Figure 7.1 shows four commonly used structures. Which one is actually adopted depends upon a number of factors including the age and experience of the enterprise, the nature of its operations and its degree of product and geographical diversity. The form most commonly adopted in the early stages of TNC development – at least when there are several overseas subsidiaries – is simply to add on an international division to the existing divisional structure (Figure 7.1a). This has tended to be a short-lived solution to the organizational problem if the firm continues to expand its international operations.

In such a hybrid structure problems of co-ordination inevitably arise. Tensions develop between the parts of the organization operated on product lines (the firm's domestic activities) and those organized on an area basis. The need arises for an organizational form which can integrate both the domestic and international operations of the firm. There are two obvious possibilities. One is to organize the firm on a

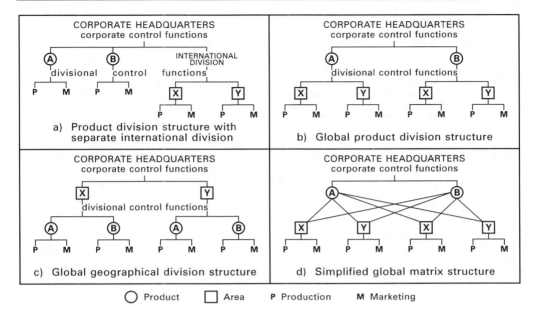

Figure 7.1 Types of organizational structure in transnational corporations

global product basis; in other words, to apply the product-division form throughout the world and to remove the international division (Figure 7.1b). The other possibility is to organize the firm's activities on a *worldwide geographical* basis (Figure 7.1c). But neither of these structures resolves the basic tension between product- and area-based systems. For such reasons some of the largest TNCs adopted sophisticated *global grid* or *global matrix* structures (Figure 7.1d) which contain elements of both product and area structures and involve dual reporting links.

There is plenty of evidence to support such a sequence, especially amongst US and some other western firms. Equally, however, there is also plenty of evidence to demonstrate far greater organizational diversity. Bartlett and Ghoshal (1989) have suggested a useful typology of three major ideal-type TNCs together with a fourth type which, they argue, is in the process of emerging. Table 7.1 summarizes the major features of each type:

- The *'multinational' organization model*: emerged particularly during the interwar period. Firms were stimulated by a combination of economic, political and social forces to decentralize their operations in response to national market differences. This ideal-type model is characterized by a decentralized federation of overseas units and simple financial control systems overlain on informal personal co-ordination. The company's worldwide operations are organized as a portfolio of national businesses (Bartlett and Ghoshal, 1989, p. 49). This was the kind of transnational organizational form used extensively by expanding European companies. Each of the firms' national units has a very considerable degree of autonomy and a predominantly 'local' orientation. It is able, therefore, to respond to local needs but its fragmented structure lessens scale efficiencies and reduces the internal flow of knowledge.

Table 7.1   Some ideal types of TNC organization: basic characteristics

Characteristics	'Multinational'	'International'	'Classic global'	'Complex global'
Structural configuration	Decentralized federation. Many key assets, responsibilities, decisions decentralized	Co-ordinated federation. Many assets, responsibilities, resources, decisions decentralized but controlled by HQ	Centralized hub. Most strategic assets, resources, responsibilities and decisions centralized	Distributed network of specialized resources and capabilities
Administrative control	Informal HQ–subsidiary relationship; simple financial control	Formal management planning and control systems allow tighter HQ–subsidiary linkage	Tight central control of decisions, resources and information	Complex process of co-ordination and co-operation in an environment of shared decision-making
Management attitude towards overseas operations	Overseas operations seen as portfolio of independent businesses	Overseas operations seen as appendages to a central domestic corporation	Overseas operations treated as 'delivery pipelines' to a unified global market	Overseas operations seen as integral part of complex network of flows of components, products, resources, people, information among inter-dependent units
Role of overseas operations	Sensing and exploiting local opportunities	Adapting and leveraging parent company competencies	Implementing parent company strategies	Differentiated contributions by national units to integrated worldwide operations
Development and diffusion of knowledge	Knowledge developed and retained within each unit	Knowledge developed at the centre and transferred to overseas units	Knowledge developed and retained at the centre	Knowledge developed jointly and shared worldwide

Source: Based upon material in Bartlett and Ghoshal (1989).

- The *'international' organization model*: came to prominence in the 1950s and 1960s through the large US corporations which expanded overseas to capitalize on their firm-specific assets of technological leadership or marketing power. This ideal type involves far more formal co-ordination and control by the corporate headquarters over the overseas subsidiaries. Whereas 'multinational' organizations are, in effect, portfolios of quasi-independent businesses, 'international' organizations see their overseas operations as appendages to the controlling domestic corporation. Thus, the international subsidiaries are more dependent on the centre for the transfer of knowledge and the parent company makes greater use of formal systems of control (Bartlett and Ghoshal, 1989, pp. 50–51). The 'international' TNC is well equipped to exploit the knowledge and capabilities of its parent company but its particular configuration and operating systems tend to make it less responsive than the 'multinational' model. It is also rather less efficient than the third ideal type –
- The *'classic global organization model*: was one of the earliest forms of international business (used, for example, by Ford and by Rockefeller in the early 1900s as well as

by Japanese firms in their much later internationalization drive of the 1970s and 1980s). It is based upon a tight centralization of assets and responsibilities in which the role of the local units is to assemble and sell products and to implement plans and policies developed at the centre. In this ideal-type model, overseas subsidiaries have far less freedom to create new products or strategies or to modify existing ones (Bartlett and Ghoshal, 1989, p. 51). Thus, the 'classic global' TNC capitalizes on scale economies and on centralized knowledge and expertise. But this implies that local market conditions tend to be ignored and the possibility of local learning is precluded.

Although each of these three ideal-type models developed initially during specific historical periods there is no suggestion that one was sequentially replaced by another. Each form has tended to persist, in either a pure or hybrid form, helping to produce a diverse population of TNCs. There is some correlation between organizational type and nationality of parent company but it is by no means perfect; it is better to regard firms of different national origins as having a predisposition to one or other form of organization (Heenan and Perlmutter, 1979). These ideal types and variations upon them are still apparent. However, new forms of transnational organization are also emerging that may – although not inevitably – replace some of the existing forms. As outlined above, each of the three ideal types possesses specific strengths but each also has specific weaknesses.

In Bartlett and Ghoshal's view, the dilemma facing TNCs in today's turbulent competitive environment is that they need the best features of each one of the three ideal types: to be globally efficient, multinationally flexible and capable of capturing the benefits of worldwide learning all at the same time. Hence, they argue, we are now seeing the emergence of a fourth ideal-type TNC:

- The *'complex global' organization model*:[2] characterized by an integrated network configuration and a capacity to develop flexible co-ordinating processes. Such capabilities apply both inside the firm which, it is argued, is displacing hierarchical governance relationships and also outside the firm through a complex network of interfirm relationships. In other words, it implies a blurring of traditional organizational boundaries. We will explore this issue in a later section in the broader context of networks of externalized relationships. Again, however, the identification of this organizational type does not imply an inevitable sequential development but merely that some firms are beginning to develop such a complex networked structure.

Strategic tensions: global integration – local responsiveness

As Alfred Chandler demonstrated many years ago, organizational structures are intimately related to organizational strategies. Firm strategies are especially influenced by the kind of economic activity in which it is engaged: in particular, the nature of competition in the industry, the prevailing technologies, the nature of the regulatory environment. Michael Porter (1986, p. 17), in fact, sees the *industry* as being the key dimension: 'Industries may vary along a spectrum from multidomestic to global in their competitive scope.' As Figure 7.2 shows, these two ends of the spectrum have different competitive characteristics which need to be addressed through appropriate strategies of national/local responsiveness or global integration respectively. The important point about Figure 7.2 is that it depicts two polar extremes of a *spectrum* of

MULTIDOMESTIC INDUSTRIES	GLOBAL INDUSTRIES
Competitive characteristics: • Competition in each country is essentially independent of competition in other countries. • The international industry becomes a collection of essentially domestic industries. **Strategy:** • The firm should manage its international activities as a portfolio. • National strategies should enjoy a high degree of autonomy. • The firm's strategy in a country should be determined largely by competitive conditions in that country, i.e. it should be a country-centred strategy. • International strategy collapses to a series of domestic strategies.	**Competitive characteristics:** • An industry in which a firm's competitive position in one country is significantly affected by its position in other countries and *vice versa*. • The industry is not merely a collection of domestic industries but a series of *linked* industries in which rivals compete against each other worldwide. **Strategy:** • The firm must *integrate* its activities on a worldwide basis to capture linkages between countries. • The global competitor must view its international activities as an overall system but must still maintain some country perspective.

Figure 7.2   The competitive and strategic characteristics of multidomestic and global industries
Source: Based on material in Porter (1986, Chapter 1)

industry configurations. In fact, not only may there be relatively few 'pure' examples at either end of the spectrum – most will fall at various points along it – but also there is unlikely to be just one strategic model in any industry. What we have in reality is a very *diverse* mix of TNC structures and strategies reflecting firm- and country-specific influences.

Yves Doz (1986b) summarizes both the advantages and the disdvantages of a globally integrated strategy from the TNC's perspective. The basic *advantages* are as follows:

• It increases the firm's oligopoly power through the exploitation of scale and experience effects beyond the size of individual national markets.
• It places the TNC in a better position to exploit the growing discrepancy between a relatively efficient market for goods (created by freer trade) and very inefficient markets for production factors.
• It increases the possibility of exploiting differences in tax rates and structures between countries and, therefore, the possibility of engaging in transfer pricing.
• The specialized and integrated function of individual country operations makes hostile government action less rewarding and less likely.

On the other hand, there are potential *costs and risks* of pursuing a globally integrated strategy:

• The TNC may be vulnerable to disruption of its entire operations (or part of them) because of labour unrest or government policy changes affecting a particular unit.
• Fluctuations in currency exchange rates may disrupt integration strategies, drastically altering the economies of intrafirm transactions of intermediate or final goods.
• Governments may impose performance requirements or other restrictions which impede the optimal operation of the firm's integrated production chain.
• The task of managing a globally integrated operation is more complex and demanding than that of managing separate national subsidiaries.

For TNCs in general, in fact, there is a basic tension between *globalizing* pressures on the one hand and *localizing* pressures on the other, which makes any notion of a 'pure' global integration or 'pure' multidomestic/nationally responsive strategy completely inappropriate. Each must contain elements of the other. Firms operating in so-called multidomestic industries must take account of global forces; conversely, firms operating in 'global' industries must be responsive to national and local differences. Significantly, many TNCs make much of this 'global–local tension' in their corporate literature.

- The Swedish transnational corporation, ABB, claims to have perfected 'the art of being local worldwide'.
- The American financial services TNC, J.P. Morgan, asserts that 'the key to global performance is understanding local markets'.
- The Anglo-Dutch firm, Unilever, describes itself as a 'multi-local multinational'.
- The Hong Kong Bank (HSBC) boasts of its 'local insight, global outlook'.
- The Japanese electronics firm, Sony, was one of the first to use the term 'glocalization' to describe its international corporate strategy.

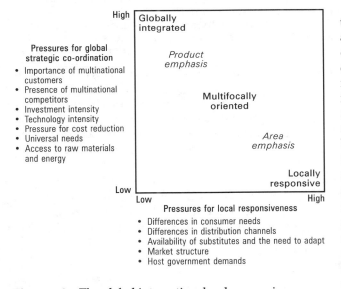

Figure 7.3 The global integration–local responsiveness framework
Source: Based on material in Pralahad and Doz (1987, Figure 2.2 and pp. 18–21)

Hence, the intensification of global competition in a world which retains a high degree of local differentiation creates, for all TNCs, an internal tension between globalizing forces on the one hand and localizing forces on the other. Figure 7.3 captures this basic tension within a 'global integration–local responsiveness' framework. The vertical axis shows the major pressures on firms to strive for global strategic co-ordination of their activities; the horizontal axis shows the countervailing pressures on firms to develop locally responsive strategies. Hence, TNCs 'must balance pressures for integrating globally with those for responding idiosyncratically to national environments . . . [however] . . . there may not be a single optimal point on the fragmentation–unification continuum but a range of tenable positions' (Kobrin, 1988, pp. 104, 107).

The geography of the transnational corporation: configuring the firm's production chain

> every business is a package of functions and within limits these functions can be separated out and located at different places.
>
> (Haig, 1926, p. 426)

This observation, made more than seventy years ago, is even more valid today. Over time, the 'limits' have become less and less restrictive mainly because of developments in the enabling technologies of transport and communications. But there is more to it than this. The enabling technologies set the outer limits of what is possible but they do not determine what actually occurs. The extent to which a firm separates out its component parts and how it locates them at different places depends upon a number of factors. One of these is the actual strategy being pursued. In discussing both the structural and strategic dimensions of TNC activity we have identified various possible alternatives. By definition, different strategies and different structures have different geographies which reflect the particular locational configuration of the firm's production chain functions. Indeed, the specific locational configuration tends to differ for *each element* within the chain.

Some elements may be geographically dispersed; others geographically concentrated. Some elements of the chain may be located in close geographical proximity to one another, others may be separately located:

> A firm may standardize (concentrate) some activities and tailor (disperse) others. It may also be able to standardize and tailor at the same time through the co-ordination of dispersed activities, or use local tailoring of some activities (e.g. different product positioning in each country) to allow standardization of others (e.g. production).
>
> (Porter, 1986, p. 35)

In fact, combining standardization and local tailoring is becoming increasingly possible with the emergence of flexible production technology (see Chapter 5).

The particular *spatial* form of this internal division of labour – precisely where the separate parts are located – is the result of the interaction between two sets of factors: organizational and technological factors on the one hand and the relevant location-specific factors on the other. We discussed location-specific factors in general terms in Chapter 6. The importance of the various factors differs, however, according to the particular locational requirements of the individual parts of a TNC's operations. *Different parts of the enterprise have different locational needs*; these needs can be satisfied in various types of geographical location. Each tends, therefore, to develop rather distinctive spatial patterns.

We can illustrate this by looking at three of the most important functions of the TNC:

- corporate and regional headquarters offices
- research and development facilities
- production units.

Each of these, as we shall see, displays certain geographical regularities: notably a highly uneven pattern both globally and locally. Other functions in the production chain, such as marketing and sales, distribution and service, tend to be distributed far more widely in accordance with the firm's geographical markets. Indeed, as competition intensifies, firms are increasingly placing an emphasis on the service component of their business. This is as true in industrial products as in consumer products. Such a service orientation (which involves investment in marketing, local distribution channels and the like) is spreading rapidly through many industries. A local market presence is becoming essential.

Corporate and regional headquarters

The *corporate headquarters* is the locus of overall control of the entire TNC, responsible for all the major strategic investment and disinvestment decisions that shape and direct the whole enterprise: which products and markets to enter or to leave, whether to expand or contract particular parts of the enterprise, whether to acquire other firms or to sell off existing parts. One of its key roles is financial; it is the corporate head-quarters which holds the purse strings and which decides on the level and allocation of the corporate budget between its component units. Headquarters offices are, above all, handlers, processors and transmitters of information to and from other parts of the enterprise and also between similarly high-level organizations outside. The most important of these are the major business services on which the corporation depends (financial, legal, advertising) and also, very often, major departments of government, both foreign and domestic.

Regional headquarters offices constitute an intermediate level in the corporate hier-archy with a geographical sphere of influence encompassing several countries. Re-gional headquarters perform a distinctive role: to co-ordinate and control the activities of the firm's affiliates (manufacturing units, sales offices, etc.) and to act as the inter-mediary between the corporate headquarters and its affiliates within its particular region. Branch affiliates generally report to these regional headquarters, which, in turn, report to the corporate head office. Thus, the regional headquarters act as a channel of communication, transmitting instructions from the corporate centre to its affiliates and information from the affiliates back to the centre. Certain decisions are delegated to the regional headquarters, which may also perform a regional marketing function. Regional headquarters are both co-ordinating mechanisms within the TNC and also an important part of the TNC's 'intelligence-gathering' system.

These characteristic functions of corporate and regional headquarters define their particular *locational* requirements:

- Both require a *strategic location* on the global transport and communications net-work in order to keep in close contact with other, geographically dispersed, parts of the organization.
- Both require access to *high-quality external services* and a particular range of *labour market skills*, especially people skilled in information processing.
- Since much corporate headquarters activity involves interaction with the head of-fices of other organizations, there are *strong agglomerative forces* involved. Face-to-face contacts with the top executives of other high-level organizations are facilitated by close geographical proximity. Such high-powered executives invariably prefer a location which is rich in social and cultural amenities.

At the global scale only a relatively small number of cities contains a large proportion of both corporate and regional headquarters offices of TNCs. Such *global cities* are sometimes described as the geographical 'control points' of the global econ-omy (we shall have more to say about them in the final section of this chapter). Figure 7.4 presents a simplified map of these centres (the links shown are diagrammatic only; they are intended solely to give an impression of a connected network of cities). Three global cities – New York, Tokyo and London – stand head and shoulders above all the others. Below them is a tier of other key cities in each of the three major economic regions of the world, western Europe, North America and Asia, with other represen-tation in Australia and Latin America.

One of the striking features of the geography of corporate headquarters is that very few, if any, major TNCs have moved their ultimate decision-making operations outside their home country. This is one further indicator of the continuing significance

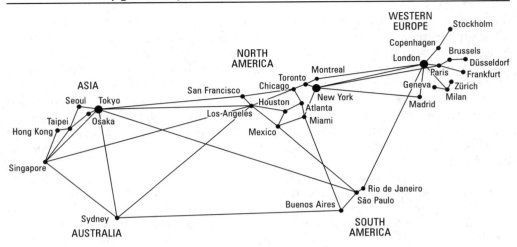

Figure 7.4   Major corporate and regional headquarters of TNCs
Source: Based, in part, on Cohen (1981, pp. 307–308); Friedmann (1986, Figure 1)

of the home base. On the other hand, there is quite a lot of evidence of TNCs relocating major divisional headquarters overseas. For example, in the early 1990s, several major US companies, including IBM, Hewlett Packard and Monsanto did exactly that as part of major corporate reorganization. In each case, however, there was no question of the overall headquarters function being relocated outside the United States.

Both the development of the Single European Market and the rapid growth of the east and southeast Asian economies have stimulated the need for regional headquarters in those areas. United States TNCs, which have had a presence in Europe for a very long time, have been establishing European headquarters to co-ordinate their regional operations. A number of Japanese TNCs have set up regional headquarters in Europe as the scale and extent of Japanese operations within Europe have increased (Aoki and Tachiki, 1992). In Asia, the Singapore government introduced an Operational Headquarters Scheme (see Chapter 4) which has attracted a considerable number of foreign TNCs to establish regional headquarters in Singapore (Dicken and Kirkpatrick, 1991).

Within individual countries, on the other hand, the locational pattern of both corporate and – especially – regional headquarters is far from static. Geographical decentralization of corporate headquarters out of the city centres of New York and London has certainly been occurring. In the case of London, most of these shifts are short distance to the less congested outer reaches of the metropolitan area. In the United States, on the other hand, there appears to be a much higher degree of locational change in headquarters functions. Detailed empirical research by Lyons and Salmon (1995, pp. 103–104) for the period 1974–89 shows a considerable degree of change:

> Although the highest concentration of corporate headquarters continues to be found among the four diversified national metropolitan regions, New York, Chicago, Los Angeles and San Francisco (48.7 per cent of the top 250 in 1989), only Los Angeles recorded an increase in concentration. The majority of the declines were concentrated in New York (–11.5 per cent), although . . . New York continues to be the most important centre of corporate headquarter locations. The major beneficiaries of New York's demise were a

select group of smaller regional diversified cities . . . in particular Atlanta, Dallas/Fort Worth, Philapelphia, and St Louis.

Nevertheless, apart from the United States, the fact that the locational needs of corporate and regional headquarters of TNCs are satisfied most readily in the very large city means that they tend not to be spread very widely within any particular country. In the United Kingdom, for example, there are very few corporate headquarters of major firms or regional headquarters of foreign TNCs outside London and the south east; in France few locate outside Paris. In Italy the most important centre is Milan, in the highly industrialized north, which is more important than Rome as a location for foreign TNCs.

Research and development facilities
In general, TNCs spend more on R&D than other firms, as part of their drive to remain competitive and profitable on a world scale. Innovation – of new products or new processes – is critically important for such firms in an increasingly competitive global economy. The R&D function is, therefore, highly significant for the TNC. Indeed, it has become even more important with the intensified pace and changing nature of technology.

The process of R&D is a complex sequence of operations in which three major phases can be identified (Figure 7.5). Each phase tends to have rather different

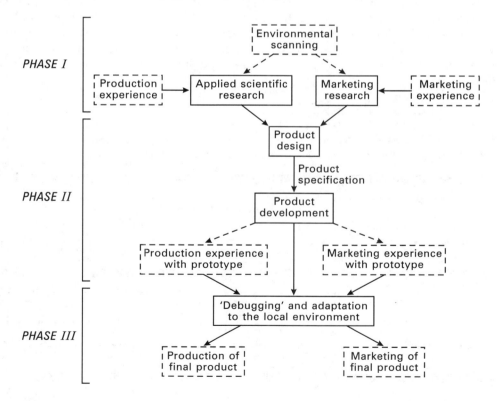

Figure 7.5 Major phases in the research and development process
Source: Based, in part, on Buckley and Casson (1976, Figure 2.7)

locational requirements although, in each case, the TNC has to reconcile several factors. One of these is the advantage of scale economies derived from concentrating R&D against the need to locate R&D closer to other corporate functions or to markets:

- In *phase I* the emphasis is on applied scientific and marketing research. The primary need is for access to the basic sources of science and marketing information – universities, research institutes, trade associations and the like.
- *Phase II* is concerned with product design and development. It tends to require large-scale teamwork, that is, access to a sufficiently large supply of highly qualified scientists, engineers and technicians.
- *Phase III* is the 'debugging' phase and the adaptation of the new product to local circumstances. Its locational requirements are for quick two-way contact with the users of the innovation: the production or marketing units themselves.

The type of R&D undertaken by TNCs in their overseas locations can be classified into three major categories (Figure 7.6). The lowest level of R&D activity is the *support laboratory* whose primary purpose is to adapt parent company technology to the local market and to provide technical backup. It is the equivalent of phase III in Figure 7.5 and is by far the most common form of overseas R&D facility. The *locally integrated R&D laboratory* is a much more substantial unit, in which product innovation and development are carried out for the market in which it is located. It is the equivalent of phase II in Figure 7.5. The *international interdependent R&D laboratory* is of a quite different order. Its orientation is to the integrated global enterprise as a whole rather than to any individual national or regional market. Indeed, there may be few, if any, direct links with the firm's other affiliates in the same country. Only a small number of technologically intensive global corporations operate international interdependent laboratories.

Support laboratory	
Function:	Technical service centre; translator of foreign manufacturing technology.
Reason for establishment:	Response to market growth; differing market conditions; expectation of continuing stream of technical service projects.

Locally integrated R&D laboratory	
Function:	Local product innovation and development; transfer of technology.
Reason for establishment:	Improved status of subsidiaries; concept of overseas operations as fully developed business entities; identification of new business opportunities outside home country. Frequently develop out of support laboratories.

International interdependent R&D laboratory	
Function:	Basic research centre; close links with international research programme; may or may not interact with the firm's foreign manufacturing affiliates.
Reason for establishment:	Operation of co-ordinated world R&D programmes as part of global product strategies involving the manufacture of a single product line for world markets. Units tend to be created by direct placement.

Figure 7.6   A typology of corporate R&D activities
Source: Based on material in Hood and Young (1982)

The kind of R&D activity, and its locational pattern, varies according to the specific market orientation of the TNC (Behrman and Fischer, 1980):

- TNCs with a strong *home market* orientation tend to carry out little overseas R&D other than of the support laboratory type. Such firms tend to regard their foreign sales as not requiring any further R&D beyond that carried out for their domestic market.
- *Host-market* TNCs – those oriented towards the national (or regional) market in which their overseas operations are located – operate both support laboratories and also higher-level locally integrated laboratories. The most important locational criteria are proximity to the firm's overseas markets and the fact that the firm's overseas operations are sufficiently substantial to justify separate R&D activities. Such activities tend to be located in the firm's biggest and most important overseas markets.
- *World-market* firms are the globally integrated corporations whose orientation is to world, rather than to national, markets. Their R&D activities include both support and locally integrated laboratories but, in addition, their adoption of a globally integrated production strategy leads them to establish specially designed international interdependent research laboratories. The major locational criteria for these world-market R&D activities are the availability of highly skilled scientists and engineers, access to sources of basic scientific and technical developments and an appropriate infrastructure:

> Universities were frequently mentioned as an important means of gaining access to the foreign scientific and technical communities that are of such great interest to the foreign exploratory laboratories of world-market companies. Every one of the world-market firms stressed the need for a strong local university system as a prerequisite for choosing an overseas location for R&D.
>
> (Behrman and Fischer, 1980, p. 21)

As in the case of their corporate headquarters, TNCs show a very strong preference for keeping their high-level R&D in their home countries. The Office of Technology Assessment (1994) calculated that only around 13 per cent of the total R&D performed by United States manufacturing TNCs is located abroad. In fact, most of the overseas R&D carried out by TNCs of all national origins is the relatively low-level adaptation of existing products and processes to local conditions. Although there has been some geographical dispersal of R&D, its actual extent is the subject of some dispute.[3] On the one hand, writers like Howells (1990, p. 504) identify a substantial degree of R&D dispersal globally:

> as more companies move from being 'host market' to 'world market' firms, the role of R&D has moved from a direct but secondary role of helping to serve the market via product modification towards a much more integrated mechanism in gaining new markets. Increasingly the sources of new ideas for new products and innovations are coming from the user firms and industries and if firms are to remain competitive and be able to move into new markets they must be able to maintain close relationships with their existing and potential customers.

Howells also points to another important influence stimulating the geographical spread of R&D by TNCs: the growing demand for skilled scientists. This particular labour market is intensely competitive and is forcing firms to extend their R&D networks in order to capture geographically dispersed scientific workers.

Others, however, contest the extent to which there has been a really significant *general* shift in the international location of R&D. Detailed empirical analyses of patent data for almost 600 firms by Patel (1995) produced the following conclusions:

- Only 43 firms in the sample (7.6%) located more than half of their technological activities outside their home country.
- More than 40 per cent of the sample performed less than 1 per cent of their technological activity abroad.
- More than 70 per cent of the sample performed less than 10 per cent of this activity abroad.
- Very little of the overseas R&D activity of firms from the United States, Japan, Germany, France and Italy is located outside the 'global triad'.
- Most of the apparent increases in overseas R&D came about through merger and acquisition rather than through internal growth and geographical expansion.

Why should such home-country bias in R&D persist? Why do TNCs show such a strong preference for keeping their major R&D activities close to their home base? Patel's (1995, p. 152) explanation is consistent with our discussion in Chapters 1 and 5 of the importance of 'untraded interdependencies':

> Two key features related to the launching of major innovations may help explain the advantages of geographic concentration: the involvement of inputs of knowledge and information that are essentially 'person-embodied', and a high degree of uncertainty surrounding outputs. Both of these are best handled through geographic concentration. Thus it may be most efficient for firms to concentrate the core of their technological activities in the home base with international 'listening posts' and small foreign laboratories for adaptive R&D.

Not only are corporate R&D activities strongly concentrated in TNCs' home countries but also the spatial pattern *within* nations is very uneven. The support laboratories are the most widely spread in so far as they generally locate close to the production units, although not every production unit has an associated support laboratory. But the larger-scale R&D activities tend to be confined to particular kinds of location. Their need for a large supply of highly trained scientists, engineers and technicians together with proximity to universities and other research institutions confines them to large urban complexes. These are often the ones which are also the location of the firm's corporate headquarters. A secondary locational influence is that of 'quality of living' for the highly educated and highly paid research staff: an amenity-rich setting including a good climate and potential for leisure activities as well as a stimulating intellectual environment.

Spatial patterns of corporate R&D in both the United States and the United Kingdom illustrate both these locational influences. In the United States, corporate R&D is still predominantly a big-city activity despite recent growth in smaller urban areas. The pull of the amenity-rich environment is illustrated by the considerable concentration of R&D activities in locations such as Los Angeles, San Francisco and San Diego in California, Denver–Boulder in Colorado and the 'Research Triangle' in North Carolina. In the United Kingdom corporate R&D, like corporate headquarters and regional offices, is disproportionately concentrated in southeast England. Within this region firms can be both close to London with all its intellectual, social and cultural facilities and also locate in some of the supposedly green and pleasant lands of the south east.

Production units

There are clearly some identifiable geographical regularities in the patterns of both corporate headquarters and R&D functions. This is because the locational needs of

a. Globally concentrated
   production

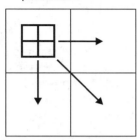

b. Host-market
   production

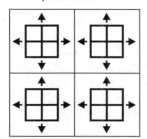

c. Product specialization for a
   global or regional market

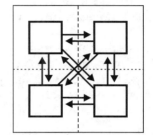

All production occurs at a single
location. Products are exported
to world markets.

Each production unit produces a
range of products and serves the
national market in which it is located.
No sales across national boundaries.
Individual plant size limited by the
size of the national market.

Each production unit produces only
one product for sale throughout a
regional market of several countries.
Individual plant size very large
because of scale economies offered
by the large regional market.

d. Transnational vertical integration

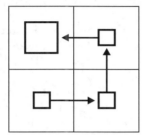

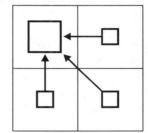

Each production unit performs a
separate part of a production
sequence. Units are linked across
national boundaries in a 'chain-like'
sequence – the output of one plant
is the input of the next plant.

Each production unit performs a
separate operation in a production
process and ships its output to a
final assembly plant in another
country.

Figure 7.7 Some major ways of organizing the geography of TNC production units

corporate offices and R&D laboratories are broadly similar for all firms, regardless of the particular industries in which they are involved. This is not so for production units. Consequently it is far more difficult to generalize about their locational tendencies. Their locational requirements vary considerably depending upon the specific organizational and technological role they perform within the enterprise and the geographical distribution of the relevant location-specific factors. It is certainly true that, compared with corporate headquarters and R&D facilities, production units of TNCs have become more and more dispersed geographically. But there is no single and simple trend or pattern of dispersal common to all activities, whether at the global scale or within individual nations. The pattern varies greatly from one industry to another. Figure 7.7 illustrates diagrammatically four types of geographical orientation which a TNC might adopt for its production units.

Globally concentrated production   Figure 7.7a presents the simplest case. All production is concentrated at a single geographical location (or, at least, within a single country) and exported to world markets through the TNC's marketing and sales networks.

This is a procedure consistent with the classic global strategy shown in Table 7.1. This is the kind of strategy followed by many Japanese companies until their relatively recent move towards more dispersed global production.

Host-market production (Figure 7.7b)   Here production is located in, and oriented directly to, a specific host market. Where that market is similar to the firm's home market the product is likely to be identical to that produced at home. Such production units have been variously termed 'miniature replicas', 'relay affiliates' or 'stand-alone' plants. The specific locational criteria for the setting up of host-market plants are:

● the size and sophistication of the market as reflected in income levels
● the structure of demand and consumer tastes
● the cost-related advantages of locating directly in the market
● government-imposed barriers to market entry.

In effect, this kind of production is import substituting. Most of the manufacturing plants established by United States' TNCs in Europe in the post-1945 period were of this kind, in many cases following a product life-cycle sequence (see Chapter 6). The more recent surge of European manufacturing investment in the United States is also directly host-market related. Similarly, the large markets of some developing countries, such as Brazil, have attracted considerable numbers of TNC manufacturing affiliates whose role is to serve that market directly.

   With changes in the 'space-shrinking' technologies, the establishment of a production unit in a specific geographical market becomes less necessary in purely cost terms. There are, however, two reasons for the continued development of host-market production:

● The need to be close to the market in order to be sensitive to variations in customer demands, tastes and preferences or to be able to provide a rapid after-sales service. As we have seen, sensitivity to local geographical differences continues to be an important issue even where TNCs pursue broadly global strategies.
● The existence of tariff and, particularly, non-tariff barriers to trade. Tariff barriers have been a significant locational factor from the very early days of transnational investment. Today it is the various kinds of non-tariff barrier which have become most prevalent (see Chapter 3). Both types of barrier have stimulated TNCs to jump over them and to establish direct production units to serve the local market. The recent growth of Japanese and other Asian manufacturing investment in Europe and in North America is substantially a response to the actual, or threatened, existence of non-tariff trade restrictions.

Product specialization for a global or regional market   During the last three decades or so a radically different form of production organization has become increasingly prominent. Figure 7.7c shows production being organized geographically as part of a rationalized product or process strategy to serve a global or a large regional market (such as the European Union or NAFTA). The existence of a huge internal market together with differences in factor endowments between member nations facilitates the establishment of very large, specialized units of TNCs which serve the entire regional market rather than single national markets. The key locational consideration, therefore, involves the 'trade-off' between

- the economies of large-scale production at one or a small number of large plants and
- the additional movement costs involved in assembling the necessary inputs and in shipping the final product to a geographically extensive regional market.

Transnational vertical integration of production   A rather different kind of rationalized production strategy involves geographical specialization by process or by semi-finished product. As we saw in Chapter 5, technological innovations in the production process permit a number of processes to be fragmented into separate parts and have led to a greater degree of standardization in some manufacturing operations. Parallel developments in the technologies of transport and communications have introduced a much enhanced flexibility into the geographical location of the production process. Hence, TNCs can locate production units to take advantage of geographical variations in production costs at a global scale. In other words, transnational vertical integration becomes feasible, in which different parts of the firm's production system are located in different parts of the world. Materials, semi-finished products, components and finished products are transported between geographically dispersed production units in a highly complex web of flows.

In such circumstances, the traditional market connection is broken. There is no direct link at all between the location of production itself and the national market in which the production unit is located. The output of a manufacturing plant in one country may become the input for a plant belonging to the same firm located in another country. Alternatively, the finished product may be exported to a third-country market or to the home market of the parent firm. The host country performs the role of an 'export platform'. The plants themselves are sometimes termed 'work-shop affiliates'; their role is to act as international sourcing points for the TNC as a whole. Figure 7.7d shows two simplified ways in which such international process specialization might be organized as part of a vertically integrated set of operations across national boundaries.

Such offshore sourcing and the development of vertically integrated production networks at a global scale were virtually unknown before the early 1960s. The pioneers were US electronics firms which set up offshore assembly operations in east and southeast Asia as well as in Mexico. Since then, the growth of such international production networks has been extremely rapid, although it is far more important in some types of activity than in others, as we shall see in the case study chapters of Part III. It was this kind of production relocation which formed the focus of the *new international division of labour* theory in the early 1980s.[4] There are several variants of the NIDL thesis but all assign the major role to the search by TNCs for cheap, controllable labour at a global scale. However, there are several factors influencing the location of offshore production in developing countries:

- The *labour intensity of the product or process in developed countries*. As we have seen, the wages gap between developed and developing countries can be immense (Figure 6.4) whereas differences in labour productivity may be far less. At the same time, many TNCs have experienced labour problems in their home countries and in some of their operations in other developed countries (e.g. resistance to particular kinds of task, especially the more monotonous or unpleasant operations on assembly lines). Labour militancy and union pressure – on wage levels, working hours, working conditions, fringe benefits – have also contributed towards creating an important 'push' factor.

- The *degree of standardization of the production process*. Industries differ considerably in the extent to which production processes can be standardized. This has important implications for the ability of the firm to utilize unskilled labour and to train it quickly. It is processes possessing a high degree of repetitiveness that can be most easily taught to an unskilled, and often uneducated, labour force.

- The *extent to which the production process can be fragmented into individual, self-contained operations and the importance of additional 'distance' costs*. Production processes differ in the extent to which they can be divided into discrete operations but also it is worth while doing so only if the additional costs of transporting the materials or components to the production site and back again are low enough. A critical consideration is the weight and bulk involved. Additional 'distance costs' may also be imposed by bureaucratic delays at national borders or at trans-shipment points.

- *Government policies towards offshore processing and export production*. Notably: 1) the adoption of export-oriented industrial policies by a number of developing countries, including the operation of export-processing zones. 2) Such host-country export-oriented policies have been reinforced by the operation of offshore assembly provisions by major developed countries which permit the export of domestic materials, their processing or assembly overseas, and their reimportation into the home country on payment of import duties only on the value-added overseas.

These practices of international intrafirm sourcing have become an increasingly important mechanism of global integration of production processes in which

> the more mobile factors, such as technology, management and equipment, are moved to the site of the least mobile. Through this method the multinational corporation is able to utilize the labour of the less developed countries in production processes formerly asssociated only with the more industrialized. It brings together both low-cost labour and advanced techniques.
>
> (Leontiades, 1971, p. 27)

However, the choice of location for a production unit at the global scale is by no means as simple as it is often made out to be. It is not just a matter of looking at differences in labour costs between one country and another or at the incentives offered as part of an export-oriented policy. Despite the enormous shrinkage of geographical distance that has occurred, the relative geographical location of parent company and overseas production unit may still be significant. The sheer organizational convenience of geographical proximity may encourage TNCs to locate offshore production in locations close to their home country even when labour costs there are higher than elsewhere. A clear example of this is Mexico, in the case of United States TNCs, and parts of southern and eastern Europe in the case of European TNCs.

Of course, just as geographical proximity may over-ride differentials in labour costs so, too, other locational influences may dominate in any particular case. For the largest TNCs the world is indeed their oyster. Their production units are spread globally, often as part of a strategy of *dual or multiple sourcing* of components or products. This is one way of avoiding the risk of over-reliance on a single source whose operations may be disrupted for a whole variety of reasons. In a vertically integrated production sequence in which individual production units are tightly interconnected, an interruption in supply can seriously affect the other units, perhaps those located at the other side of the world. In an extreme case, a whole segment of the TNC's operations may be halted.

The location of TNC production units at an *intranational* scale has been the subject of far less attention than their location at the international scale. Yet their distribution within a country is extremely important from the viewpoint of their economic and social impact, especially in terms of employment. Within developing countries TNCs prefer to locate their production units in the *core areas:* those having a relatively high intensity of economic activity and the necessary infrastructure. This means either the major urban centres or the export-processing zones.

Within developed economies which, we should remind ourselves, contain the bulk of TNC activities the distribution of production units generally follows that of industrial activity in general. Many such units are, of course, the branch plants of domestic TNCs. Both these and the branch plants of foreign TNCs have also shown some tendency to locate in areas of relatively high unemployment in order to tap large pools of labour. In some cases, too, TNCs have responded to government regional development policies in locating their production units. However, it may simply be that regional incentives merely reinforce prevailing corporate locational trends rather than actually initiating new ones.

The spatial division of labour within the TNC is extremely complex not only because of the intricacy of its organizational structures but also because of the varying extent to which different parts of the enterprise can be separated both functionally and spatially. Such separation varies according to the nature of the TNC's organizational strategy, the technological characteristics of the industry in which it is involved and the locational requirements of the component parts. Different parts of the TNC have different locational requirements and these can be satisfied in various types of geographical location.

Restructuring and reorganization within the TNC

Transnational corporate networks and their resulting spatial patterns are, by their very nature, in a continuous state of flux. Change is endemic, although the precise form change takes may well vary from one part of the TNC to another. At any one time, some parts may be growing rapidly, others may be stagnating, still others may be in steep decline. The functions performed by the component parts and the relationships between them may alter. Change itself may be the result of a planned strategy of adjustment to changing internal and external circumstances or the 'knee jerk' response to a sudden crisis. Whatever its origin, however, corporate change will have a specific spatial expression. The changes that occur within the TNC itself will be projected into particular kinds of impact on the localities in which the component parts are located, relocated, expanded or contracted.

Forces underlying reorganization and restructuring

The forces which may lead to corporate reorganization and restructuring and, hence, to spatial change, can be divided broadly into two categories – external and internal:

- *External* conditions. These may be negative pressures such as declining demand, increased competition in domestic or foreign markets, changes in the cost or availability of production inputs, militancy and resistance of labour forces in particular places, the pressure of national governments to modify their activities or even to cede control. Conversely, changes in external conditions may be positive rather than negative, for example the growth of new geographical markets or the availability of new production opportunities. A good illustration is the formation of regional economic groupings, such as the European Union or NAFTA which dramatically alter the pattern of investment opportunities. The creation of a large regional market made up of separate nation-states, each with its own resource endowment and production cost attributes but with free movement of materials and products across national boundaries, provides an unprecedented opportunity for TNCs to restructure their production activities to serve the regional market. Investments which had made sense in the context of an individual nation are no longer necessarily rational in the wider context (see Figure 7.7c).

• *Internal pressures* may also stimulate reorganization and rationalization. Such forces may relate to the enterprise as a whole or to one or other individual parts: sales may be too low in relation to the firm's target, production costs may be too high. In a global corporation the performance of individual plants in widely separate locations can be continuously monitored and compared with one another to assess their efficiency. Studies of corporate change often reveal the key influence of the 'new-broom factor': a change in top management – a new chief executive who undertakes a sweeping evaluation of the enterprise's activities and investments and makes changes which stamp his or her authority. More generally there may be a perceived need to free capital and managerial resources for more profitable activities or to introduce new technologies, whether in products or processes.

In reality, external and internal pressures may be so closely inter-related that it may be difficult to disentangle one from the other. More than this, as Schoenberger (1997) has shown, precisely how firms both identify and respond to changes in their circumstances is very much conditioned by the firm's *culture*. Drawing on detailed case studies of two large US corporations, Lockheed and Xerox, she observes

> . . . corporate cultures are not the source of resistance *tout court* . . . Corporate cultures do, however, shape the process of change. They are intimately involved in determining what kinds of change will be accepted and which refused, whatever their 'objective' desirability. Corporate strategies, for this reason, cannot be understood apart from the cultural processes underlying their production. These changes, moreover, play out through a highly conflictual process in which the power to define who and what the firm is in the world, how markets should be understood, how competition works, how different products and practices should be valued, and so on are all at stake. The strategic trajectory of the firm depends on how these conflicts are resolved.
>
> (Schoenberger, 1997, p. 204)

All the evidence suggests that complex corporate restructuring is occurring at all geographical scales, from the global to the local, as strategic decisions have to be made regarding the organizational co-ordination and geographical configuration of the TNC's production chain functions within a volatile environment. Decisions to centralize or to decentralize decision-making powers or to cluster or disperse some or all of the firm's functions in particular ways are, however, contested decisions. They are the outcome of power struggles within firms, both within their headquarters and between headquarters and affiliates. How they are resolved depends very much on the nature and the location of the dominant coalition. Such processes also have to be seen within the context of

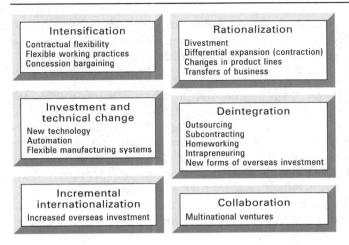

the fundamental tension facing TNCs: whether to globalize fully or whether to respond to local differentiation (see Figure 7.3).

Specifically, corporate restructuring may occur in a variety of ways (Figure 7.8) involving in some cases technological change, changes in work practices, rationalization of corporate activities, changes in the extent to which the value-added chain is internalized, and increased international investment.

Figure 7.8 Major forms of corporate restructuring
Source: Based on Enderwick (1989, Table 1)

The geography of reorganization and restructuring

Whether corporate reorganization is the result of a consciously planned strategy for 'rational' change or simply a reaction to a crisis (internal or external), its spatial outcome may take several different forms. In Figure 7.9 two broad categories of spatial-organizational change are identified:

- *In situ adjustment* to the existing network of production units is by far the most common form of adjustment. A major advantage of the multiplant firm is that it can make substantial adjustments *in situ* without necessarily engaging in locational shift. The capacity of an existing plant can be increased to achieve economies of scale or reduced (partial disinvestment) to shed surplus capacity; an existing plant's capital stock can be replaced with new technology. In such ways, the importance and even the actual function of production units can be altered as the TNC reallocates tasks among its existing geographically dispersed operations. Change at an existing plant may be either a gradual process of incremental adjustment or a more sudden change to its scale or function.
- *Locational shifts* explicitly involve abrupt change because they consist of either an increase or decrease in the number and location of plants operated by the enterprise

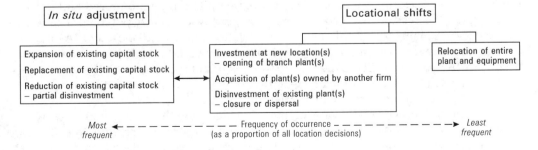

Figure 7.9 Reorganization, restructuring and spatial change: major types of investment-location decision

or even, in rare cases, the physical relocation of an entire plant. The most common locational shifts within a TNC's production network are: disinvestment at an existing plant; greenfield investment at a new location; acquisition of plants belonging to another firm.

Acquisition and merger are particularly important mechanisms of corporate adjustment and growth. Until recently most of this activity occurred within national boundaries but within the last two decades or so crossborder acquisitions and mergers have become one of the most important ways for firms to expand their activities internationally. In fact, this mode of entry is now far more important than new greenfield investments. United Nations data show that crossborder merger and acquisitions in 1995 amounted to $229 billion, twice as high as in 1988 (UNCTAD, 1996b, p. 10). The completion of the Single European Market led to a particularly high level of acquisition and merger activity as firms attempt to gain entry or to rationalize market competition. However, the importance of acquisition as an entry mechanism varies substantially between TNCs from different geographical origins. For example, Japanese firms have traditionally used acquisition to a far smaller extent than western firms, although recently Japanese firms have increased their acquisition levels.

Locational shifts arising from reorganization and restructuring have important repercussions for other plants within the firm. For example, a decision to establish a new branch plant in one country may be related to a reduction in scale or even closure of one or more plants in another country. Such locational adjustments may well be associated with the introduction of new technology at different locations from those at which existing technology is being replaced or with the shift of production to lower-cost locations. Similarly, the integration of acquired plants may alter the functions or the scale of existing plants. For the larger and more extensive TNCs each of these processes may be occurring simultaneously in different parts of the organization and in different geographical locations. Some existing plants may be expanding or contracting; in some cases this may be associated with functional changes – new plants may be opening and existing ones closed. The whole adjustment process is kaleidoscopic on a global scale.

Thus, reorganization, restructuring and the resulting spatial change are an inevitable aspect of the evolution of transnational corporations. The actual form such change takes depends upon forces both internal and external to the firm. The very large global corporations are developing into what Vernon calls *global scanners*. They use their immense resources to evaluate potential production locations in all parts of the world. The performance of existing corporate units can be monitored and evaluated against the rest of the corporate network and also against potential locations. Those existing plants which fall short of expectations may be disposed of. As plants become obsolete in one location they are closed down; whether or not new investment occurs in the same locality depends upon its suitability for the TNC's prevailing strategy. The chances are, in many cases, that the new investment will be made at a different location, quite possibly in a different country altogether.

We should, however, beware of over-exaggerating the speed and ease with which TNCs can and do restructure their operations. There are 'barriers to exit' – in many cases production units represent huge capital investments which cannot be written off lightly. But there are other kinds of costs – sunk costs – which impose additional exit barriers.[5] 'These are costs that cannot be recovered (for example by selling off surplus assets) or closed out in the short run (as with the case of pension liabilities) even when an operation is terminated' (Schoenberger, 1997, p. 88). Political pressures may also inhibit firms from closing plants, especially in areas of economic and social stress. On the other hand, TNCs do have a highly tuned capacity to *switch* and *reswitch* operations within their existing corporate network. They also have the resources to alter the shape of their spatial network through locational shifts.

Within these broad restructuring processes by TNCs, four general tendencies are especially apparent:

- A redefinition of the firm's core activities through stripping away activities that no longer 'fit' the firm's strategy.
- A repositioning of the firm's focus along the production chain to place a greater emphasis on downstream, service functions.
- A geographical reconfiguration of the firm's production chain activities internationally – and in some cases globally – to redefine the roles and functions of individual corporate units.
- An organizational reconfiguration of the firm's production chain activities which involves redefining the boundary between internalized and externalized transactions.

It is this latter point which forms the focus of the next section of this chapter.

Networks of externalized relationships

So far we have concentrated upon the organizational and geographical dimensions of the *internal* relationships between the constituent parts of TNCs. But TNCs are also locked into *external* networks of relationships with a myriad of other firms: transnational and domestic, large and small, public and private (see Figure 1.3). It is through such interconnections, for example, that a very small firm in one country may be directly linked into a global production network, whereas most small firms serve only a very restricted geographical area. Such inter-relationships between firms of different sizes and types increasingly span national boundaries to create a set of *geographically nested relationships from local to global scales*. These interfirm relationships are the threads from which the fabric of the global economy is woven. It is through such links that changes are transmitted between organizations and, therefore, between different parts of the global economy.

There is, in fact, a bewildering variety of interorganizational *collaborative* relationships. These are frequently multilateral rather than bilateral, polygamous rather than monogamous. Collaborative ventures are a long-established form of international business organization[6] that, in the past, involved specific relationships between conventionally organized, hierarchically structured firms as just one element in their competitive strategies. The position today is rather different. Not only have collaborative ventures in the more traditional sense moved to the centre of firms' strategies but also new forms of collaboration are emerging which are embedded within much flatter and looser network structures or webs of enterprise (Reich, 1991).

Business networks in east Asia
The idea that interfirm networks based upon long-term collaboration are something new is very much a western perspective. East Asian business organizations have long been embedded in complex structures of interorganizational networks.[7] Here we

examine briefly just two examples: the Japanese *keiretsu* and the Overseas Chinese family business network.

The Japanese keiretsu[8]

> Intercorporate alliances in the contemporary Japanese economy are marked by an elaborate structure of institutional arrangements that enmesh its primary decision-making units in complex networks of cooperation and competition . . . There is a strong predilection for firms in Japan to cluster themselves into coherent groupings of affiliated companies extending across a broad spectrum of markets.
>
> (Gerlach, 1992, p. xiii)

The precise composition and structure of such business groupings are immensely varied. Here we focus on the industrial groupings generally known as *keiretsu*. Gerlach (1992, p. 4) identifies five diagnostic characteristics of these groups:

- Transactions are conducted through alliances of *affiliated* companies. This creates a form of organization intermediate between vertically integrated firms and arm's length markets.
- Interfirm relationships tend to be *long term* and stable, based upon mutual obligations.
- These interfirm relationships are *multiplex* in form, expressed through cross-shareholdings and personal relationships as well as through financial and commercial transactions.
- Bilateral relationships between firms are embedded within a broader 'family' of *related companies*.
- Intercorporate relationships are imbued with *symbolic significance* which helps to sustain links even where there are no formal contracts.

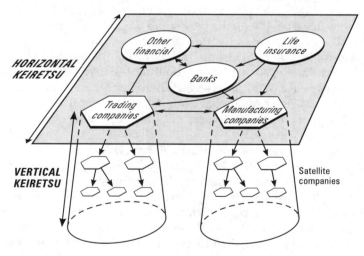

Figure 7.10   The basic elements of the Japanese *keiretsu*
Source: Based, in part, on Gerlach (1992, Figure 1.1)

Figure 7.10 shows the simplified structure of the two basic types of *keiretsu*:

- *Horizontal keiretsu* are highly diversified industrial groups organized around two key institutions: a core bank and a general trading company (*sogo shosha*). Three of the horizontal *keiretsu* groups (Mitsubishi, Mitsui, Sumitomo) are the successors of the prewar family-led *zaibatsu* groups which were abolished after 1945. The others are primarily bank centred.
- *Vertical keiretsu* are organized around a large parent company in a specific industry (for example, Toshiba, Toyota, Sony).

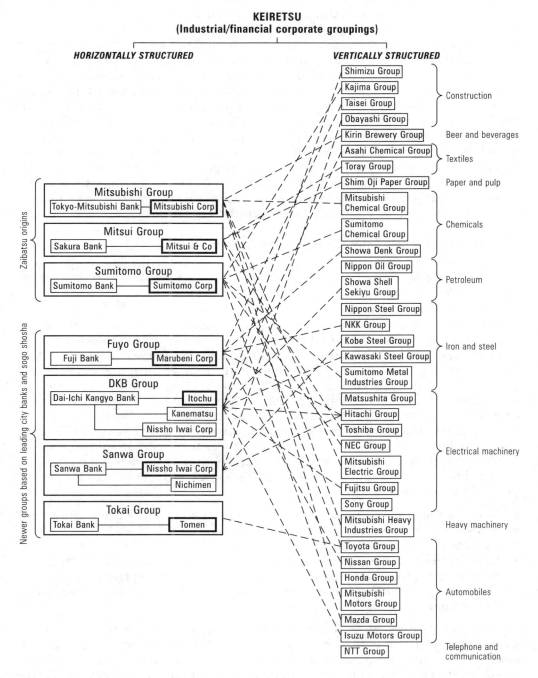

Figure 7.11 Major relationships between horizontal and vertical *keiretsu*
Source: Based on material in Dodwell Marketing Consultants (1992)

Although distinctive, the two types of group are not mutually exclusive. 'A *keiretsu* formed through horizontal ties accommodates enterprises that are being formed through vertical integration. Separately, enterprises which are of the latter type but comparatively independent, are often inclined to establish certain links with one horizontally held industrial group or the other' (Helou, 1991, p. 103). For example, the electronics manufacturer, Toshiba, is itself a parent company controlling hierarchically substantial numbers of satellite companies including parts suppliers, in a vertically integrated Toshiba group. At the same time, Toshiba is also a member of the horizontally integrated Mitsui industrial group. In fact, the webs of inter-relationships are extremely complex, as Figure 7.11 shows. The organizational scale of the leading *keiretsu* is immense. For example, the seven leading horizontal *keiretsu* shown in Figure 7.11 consist of around 900 separate companies but, in effect, they control, in total, some 12,000 companies. In the early 1990s, the 163 leading companies of the six major *keiretsu* effectively controlled more than 40 per cent of all Japanese non-financial enterprises and some 32 per cent of total assets (Helou, 1991).

The Overseas Chinese family business network[9]

> In many Western economies, the main efficiences in coordination derive from large-scale organization. In the case of the Overseas Chinese, the equivalent efficiences derive from networking.
>
> (Redding, 1991, p. 45)

A very different kind of business network is to be found within the Overseas Chinese entrepreneurial system which underpins much of the dynamic economic development not only of Hong Kong and Taiwan but also throughout much of southeast Asia. Redding (1991, p. 30) captures their essence in the phrase 'weak organizations and strong linkages'. The basis of the Overseas Chinese business system is to be found in the specifics of cultural and historical experience, in the set of norms and values derived from a common historical experience and implemented in particular contexts. Redding argues that three influences have been especially strong in influencing how Overseas Chinese businesses are managed and operated: Confucian value systems of familism and respect for authority; experience as refugees; experience of oppression.

On this basis, Redding (1991, p. 36) identifies five basic characteristics of Overseas Chinese business behaviour:

- Control of the firm must be retained in the long-term interest of family prosperity.
- Family assets must be protected by hedging of risks.
- Key decision-making should be confined to an inner circle.
- Dependence on outsiders ('non-belongers') for key resources must be limited.
- Interfirm transactions should be based on 'networks of interpersonal obligation. Personalizing trust-bonds substitutes for a system of legal contracts so normal in Western contexts'.

By its very nature the Overseas Chinese business is less visible than other, more formally structured businesses. Because of its extended network form, any approach based solely on the legal definition of the firm will fail to capture its full extent. It will also underestimate its significance as a form of transnational business activity. Yet the significance throughout Asia and, increasingly, in other parts of the world, of Overseas Chinese family business networks is immense and growing. Kao (1993) uses the neat phrase 'the worldwide web' to describe it. However, despite its common

Table 7.2 Contrasts between the large Japanese enterprise and the Chinese family business

Characteristic	Large Japanese enterprise	Chinese family business
Size	Large	Small/medium
Capital intensity	Varied	Low
Managerial discretion from owners	High	Low
Business specialization and managerial homogeneity	High	High in firms, medium in families
Strategic change	Incremental	Opportunistic
Growth focus	Sector share	Volume expansion and opportunistic diversification
Integration of different activities	Minority shareholding	Personal ties and ownership
Risk management	Extensive mutual dependence	Limiting commitment and maximizing flexibility

Source: Based on Whitley (1992a, Table 3.1).

features, the Overseas Chinese business firm is by no means homogeneous. The TNCs from the major countries of origin also bear the imprint of those home countries. Redding's study of firms in Hong Kong and Taiwan illustrates this point. In both cases, there is a tendency for firms to focus on one type of business (the diversified firm is rare) but the specifics vary. Taiwanese firms showed a much stronger bias towards manufacturing; Hong Kong firms were far more heavily involved in financial investment and property. As Yeung (1997, p. 7) notes, 'Chinese industrialists in Hong Kong are known for their entrepreneurship and high propensity to engage in risky business and overseas ventures'.

Table 7.2 summarizes the major differences between the large Japanese enterprise and the Chinese family business. Both are very different from the kinds of business system characteristic of the western economies. However, it is very clear that the notion of interfirm networking has become increasingly common among western firms in recent years. We look first at the apparent 'explosion' of strategic alliances.

International strategic alliances[10]

One of the most significant developments in the global economy in recent years has been the growth and spread of strategic alliances – various forms of collaboration between firms at an international scale. They arise from the kinds of changes which have been occurring in the global economy, notably: intensification of competition, acceleration in technological change, increased costs of developing, producing and marketing new products. The major objective of a strategic alliance is to enable a firm to achieve a specific goal which it believes that it cannot achieve on its own. In particular, an alliance involves the sharing of risks as well as rewards through joint decision-making responsibility for a specific venture. Strategic alliances are not the same as mergers, in which the identities of the merging companies are completely subsumed. In a strategic alliance only *some* of the participants' business activities are involved; in every other respect the firms remain not only separate but also often competitors.

Collaborative ventures between firms across national boundaries are nothing new. What is new is their current scale, proliferation and the fact that they have

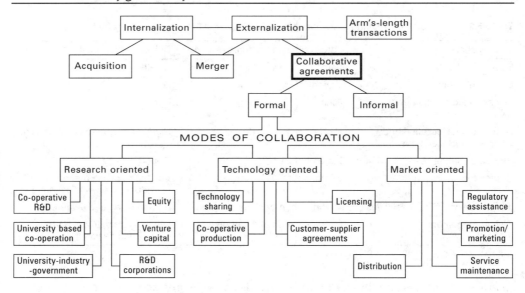

Figure 7.12   Types of interfirm collaboration
Source: Anderson (1995, Figure 1)

become *central* to the global strategies of many firms rather than peripheral to them. Most strikingly, the overwhelming majority of strategic alliances are between competitors. Of 839 agreements identified by Morris and Hergert (1987) between 1975 and 1986, 81 per cent were between two firms and no less than 71 per cent were between two companies in the same market. However, it appears that many companies are forming not just single alliances but *networks of alliances*. Relationships are increasingly polygamous rather than monogamous:

> few companies have only a single alliance. Instead, they form a series of alliances, each with partners that have their own web of collaborative arrangements. Companies like Toshiba, Philips, AT&T and Olivetti are at the hub of what are often overlapping alliance networks which frequently include a number of fierce competitors.
>
> (Business International, 1987, pp. 113–14)

As a result it becomes more and more difficult to establish the precise boundaries between firms. Figure 7.12 sets such collaborative agreements in their broader organizational context and also divides them into three broad types: research-oriented, technology-oriented, and market-oriented.

Although strategic alliances are not confined to particular sizes or types of firm, they are undoubtedly more common between large TNCs with extensive international operations. They also display a distinctive geography. The vast majority of strategic alliances are generated within and across the global triad of North America, Europe and Japan as major firms attempt to gain a presence in each of these three dominant markets. In addition, alliances are more common in some industries than others. As Table 7.3 shows the majority of strategic alliances formed during the 1980s were in sectors 'typified by high entry costs, globalization, scale economies, rapidly changing technologies, and/or substantial operating risks' (Morris and Hergert, 1987, p. 18).

Very often, the motivations for strategic alliances are very specific. In the case of R&D ventures, for example, co-operation is limited to research into new products and

Table 7.3   Strategic alliances by sector

Sector	Alliances		Major reasons for alliance
	Number	Per cent	
Information technology based	1660	39.7	Market access/structure
Microelectronics	383		Market access/structure
Telecommunications	366		Market access/structure
Software	344		Technology complementarity
Industrial automation	278		Technology complementarity
Computers	198		Market access/structure
Other	91		Market access/structure
Biotechnology	847	20.3	Technology complementarity
New materials technology	430	10.3	Technology complementarity
Chemicals	410	9.8	Market access/structure
Aviation/defence	228	5.5	Technology complementarity
Automotive	205	4.9	Market access/structure
Heavy electric/power	141	3.4	High cost risks
Instruments/medical technology	95	2.3	Reduction of innovation timespan
Consumer electronics	58	1.4	Market access/structure
Food and beverages	42	1.0	Market access/structure
Other	66	1.6	High cost risks
Total	**4182**	**100.0**	**Market access/structure**

Source: Based on material in Hagedoorn and Schakenraad (1990).

technologies while manufacturing and marketing usually remain the responsibility of the individual partners. Cross-distribution agreements offer firms ways of widening their product range by marketing another firm's products in a specific market area. Cross-licensing agreements are rather similar but they also offer the possibility of establishing a global standard for a particular technology, as happened, for example, in the case of compact disc players (Business International, 1987). Joint manufacturing agreements are used both to attain economies of scale and also to cope with excess or deficient production capacity. Joint bidding consortia are especially important in very large-scale projects in industries such as aerospace or telecommunications, where the sheer scale of the venture or, perhaps, the specific regulatory requirements of national governments put the projects out of reach of individual companies.

Advocates of strategic alliances claim that 'by co-operating, companies can combine complementary technologies, R&D capabilities, skills, products, market presence, production capacity etc., in a way that will strengthen each partner' (Business International, 1987, p. 3). But not everybody shares this rosy view. Robert Reich, for example, has been strongly critical, particularly of alliances between US and Japanese companies, on the grounds that they will severely damage the long-term competitiveness of US firms (Reich and Mankin, 1986). The fear is that entering into such alliances will result in the loss of key technologies by the US partners. More broadly, strategic alliances are clearly more difficult to manage and co-ordinate than single ventures; the potential for misunderstanding and disagreement, particularly between partners from different cultures, is great. Certainly many such alliances have relatively short lives. Nevertheless, the obvious attractions of international strategic alliances in today's volatile and competitive global economy are likely to guarantee their continued growth as a major organizational form.

Connecting the production chain: relationships between customers and suppliers

As we noted in Chapter 1, production chain transactions may be either *internalized* within a single, vertically integrated firm or be *externalized*, with various operations being performed by independent firms. In some industries, one mode or the other has tended to predominate although there have always been substantial differences between individual firms in the same industry. But not even the most highly integrated firm is totally self-sufficient. All firms acquire at least some of their inputs from outside suppliers. Possibly between 50 and 70 per cent of manufacturing costs are spent on purchased inputs (Schroeder, 1989). Today, all the evidence suggests that the boundary between internalization and externalization – between what a firm does for itself and what it has done for it by suppliers of intermediate goods and services – is shifting.

The general trend is for a greater proportion of inputs to be sourced from supplier firms. Some of these purchases will be, as it were, 'off-the-shelf' or 'catalogue' sourcing from independent suppliers at the arm's length market price. However, an increasingly significant proportion of such purchases is made on the basis of longer-term relationships whereby a customer firm *subcontracts* certain tasks to independent firms.

Subcontracting is a kind of half-way house between complete internalization of procurement on the one hand and arm's length transactions on the open market. It is

> a situation where the firm offering the subcontract requires another independent enterprise to undertake the production or carry out the processing of a material, component, part or subassembly for it according to the specifications or plans provided by the firm offering the contract. Thus, subcontracting differs from the mere purchase of ready-made parts and components from suppliers in that there is an actual contract between the two participating firms setting out the specifications for the order.
>
> (Holmes, 1986, p. 84)

Types and characteristics of subcontracting relationships

Box 7.1 sets out the basic elements of the subcontracting relationship. It shows that subcontracting occurs in both industrial and commercial spheres; that it can cover not only processes and components but also complete finished products. Generally, the firm placing the order or contract is known as the 'principal firm'; the firm carrying out the order is known as the 'subcontractor'. *Commercial subcontracting* involves the manufacture of a finished product by a subcontractor to the principal's specifications. The subcontractor plays no part in marketing the product, which is generally sold under the principal's brand name and through its distribution channels. The principal firm itself in this case may be either a producer firm, that is, one which is also involved in manufacturing, or a retailing or wholesaling firm whose sole business is distribution. This distinction is closely related to Gereffi's two primary 'drivers' of commodity chains: producer driven and buyer driven (see Figure 1.4). Whereas a producer firm may engage in both industrial and commercial subcontracting, retailers/wholesalers are confined to commercial subcontracting.

Industrial subcontracting can be subdivided into three types according to the motivation of the principal firm. *Speciality* subcontracting involves the carrying out, often on a long-term or even a permanent basis, of specialized functions which the principal chooses not to perform itself but for which the subcontractor has special skills and equipment. *Cost-saving* subcontracting is self-explanatory: it is based upon

Technical aspects of production
- Subcontracting *processes*   } *Industrial*
- Subcontracting *components*   } *subcontracting*
- Subcontracting *entire products* } *Commercial subcontracting*

Nature of the principal firm
- Producer firm (both industrial and commercial subcontracting)
- Retailing/wholesaling firm (commercial subcontracting)

Type of subcontracting (motivation of principal firm)
- Speciality subcontracting
- Cost-saving subcontracting
- Complementary or intermittent subcontracting

Types of relationship between principal firm and subcontractor
- Time period may be long term, short term, single batch
- Principal may provide some or all materials or components
- Principal may provide detailed design or specification
- Principal may provide finance, e.g. loan capital
- Principal may provide machinery and equipment
- Principal may provide technical and/or general assistance and advice
- Principal is invariably responsible for all marketing arrangements

Geographical scale
- Within-border, i.e. *domestic subcontracting*
- Crossborder, i.e. *international subcontracting*

Source: Based on material in Sharpston (1975), Germidis (1980).

Box 7.1   Major elements of the subcontracting relationship

differentials in production costs between principal and subcontractor for certain processes or products. *Complementary or intermittent* subcontracting is a means adopted by principal firms to cope with occasional surges in demand without expanding their own production capacity. In effect, the subcontractor is used as extra capacity, often for a limited period or for a single operation. The actual relationship between principal and subcontractor can also take a variety of forms, as Box 7.1 indicates. The length of time involved may be long or short. The principal's involvement may vary in terms of finance, technology, design, the provision of materials and equipment. Invariably, however, the principal is solely responsible for marketing the finished products or for arranging further assembly or processing. This kind of subcontracting is often termed *original equipment manufacture* (OEM).

Costs and benefits of subcontracting to the participants

The precise advantage of subcontracting to the *principal firm* depends very much on the type of subcontracting involved. In general, however, it offers a number of benefits:

- The firm may avoid having to invest in new or expanded plant.
- It offers a degree of flexibility: it is easier to change subcontractors than to close down or reduce the firm's own fixed capacity.

- By entering into a contractual agreement the principal firm gains a certain amount of control over the operation.
- It is one way of externalizing some of the risks and costs of certain operations.

From the viewpoint of the *subcontractor* there are both costs and benefits. The major *costs* include:

- bearing some of the risks involved. Small subcontracting firms may, in effect, perform a 'shock-absorbing' role for large firms. Subcontractors tend to be both expandable and expendable, particularly where they are small firms in an unequal power relationship with large firms
- devoting an excessive proportion of its total production to subcontract work for a particular customer. In effect, the subcontractor becomes part of a vertically integrated operation, but without the full benefits of such involvement. As such, its freedom to move into new products or new markets may be limited. The problem is greatest for the small subcontractor where the principal firm specifies the product in detail and where the subcontractor depends upon the principal for product and process development.

On the other hand, small firms may well *benefit* substantially from their subcontracting role:

- Access is gained to particular markets via brand names which would otherwise be unattainable.
- Continuity of orders (in some cases over a long period of time) is assured.
- Injection of capital in the form of equipment and access to technology.

In many respects, therefore, the subcontracting relationship is symbiotic – a division of labour between independent firms – in which each partner contributes to the support of the other. Subcontracting operations are especially important for small firms. Many firms actually start their lives as subcontractors to larger firms, and it is certainly an important channel through which small entrepreneurial firms can operate. Such observations have led to the view that large and small firms are related in a particular kind of unequal power relationship in which large firms dominate. There is often a great deal of truth in this but it is not the whole story. Subcontracting relationships are not confined solely to those in which large firms dominate the small. In some industries, such as aerospace and automobiles, very large firms act as subcontractors to other large firms.

The geographical dimension of subcontracting

As a process, subcontracting is as old as industrialization itself, if not older. The 'putting-out' system was a key element of most industries from their earliest stages. It depended, essentially, on close geographical proximity between firms and their subcontractors. The very fine and intricate network of subcontracting relationships, based on the externalization of transactions in the production chain, often led to the development of highly localized industrial districts. Such tight, functionally and transactionally based, geographical agglomerations of linked economic activities declined in most western industrial countries with increasing speed between the 1960s and early 1980s, although they persisted in Japan.

Indeed, Japan still has one of the most highly developed domestic subcontracting networks. Each large Japanese firm is surrounded by a constellation of small and medium-sized subcontracting firms which act as suppliers of components or perform specialist processes to the specification, and the timetable, laid down by the controlling large firm. Indeed, the Japanese subcontracting system, with its sharp distinction between the two major segments, has contributed a great deal to the international competitiveness of the Japanese economy, as discussed in previous chapters. However, conditions in the myriad of small subcontracting firms are very different from those in the major firms. The subcontracting segment has none of the much-lauded qualities of lifetime employment and corporate paternalism which exist in the major corporations. Competitiveness within the subcontracting segment is fierce; the small firms are very heavily subservient to the stringent demands of the principal companies.

One of the most significant developments of the last thirty years has been the extension of subcontracting across national boundaries: the emergence of *international subcontracting* as an important global activity. The revolution in transport and communications technology, together with developments in the production process itself, have created the potential for firms to establish subcontracting networks over vast geographical distances in the same way as TNCs have established offshore production units of their own. Relatively low transportation costs, plus the ability to control and co-ordinate the operation of a long-distance subcontracting system, have allowed firms to take advantage of very low labour costs in developing countries. Various kinds of international subcontracting relationship are possible. Figure 7.13 distinguishes between direct international subcontracting, that is, between independent firms located in separate countries (Figure 7.13a), or indirect, where the principal is an overseas affiliate of a transnational corporation and the subcontractor is either a local firm or perhaps an affiliate of another TNC (Figure 7.13b).

From its origins in the 1960s, the geographical extent of the international subcontracting networks, developed by western companies in particular, increased progressively to become truly global in scope. Of course, within that global spread there

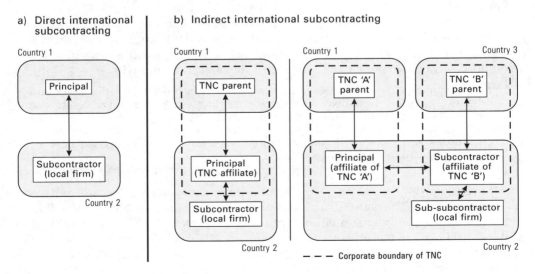

Figure 7.13 Types of international subcontracting relationship
Source: Based on Michalet (1980, pp. 51–52)

were also pockets of international subcontracting in closer geographical proximity to the principal firms' home bases. The obvious examples, referred to earlier in our discussion of intrafirm networks, were in such places as Mexico and the Mediterranean rim. The basic driving force was cost minimization: by locking into areas of very low labour costs firms could offset additional transport costs. The spread of *just-in-time* production systems, described in Chapter 5, is, arguably, changing this simple cost equation. Certainly the adoption of such systems, with their need for very frequent delivery of components, would suggest a possible geographical reconcentration of supplier firms and customers. The evidence is conflicting in this regard with a mixture of arrangements involving both long- and short-distance linkages, depending upon the particular circumstances in specific industries and firms.

One undoubted development is the tendency for many firms to move towards *closer functional relationships* with their suppliers. Rather than merely seeking out the lowest cost supplier and little else, there is a strong move towards the nomination of 'preferred suppliers' with whom very close relationships are developed. Such suppliers are increasingly being given greater responsibility for the quality of their outputs and, indeed, are playing a more direct role in the design of products. More generally, the procuring firm has a variety of options in relation to its suppliers. It can opt for *single sourcing* to gain economies of scale (and lower costs) but with the risk of putting all its procurement eggs in a single supply basket. Alternatively, it can opt for *dual or multiple sourcing* and spread its procurement network more widely.

Dynamic and flexible business networks

Conventionally, industrial subcontracting has involved a firm putting out certain aspects of its production chain operations to specialist suppliers while continuing to engage in production itself. Now, however, a new organizational form seems to be emerging: the *vertically disaggregated network organization* in which almost all functions in the production chain, other than those of central co-ordination and control, are contracted to independent firms but in which the final product is marketed under the lead company's brand name. These are dynamic and flexible business networks which involve complex relationships between independent firms, each of which performs a specialist role within a co-ordinated network. Figure 7.14 shows in a highly simplified form the major elements of such a dynamic network, whose

> major components can be assembled and reassembled in order to meet complex and changing competitive conditions . . . Business functions such as product design and development, manufacturing, marketing and distribution, typically conducted within a single organization, are performed by independent firms within a network . . . Because each function is not necessarily part of a single organization, business groups are assembled or located through brokers.
>
> (Miles and Snow, 1986, p. 64)

Such a network goes a long way beyond conventional subcontracting and interfirm relationships. The entire structure is relatively 'flat' and non-hierarchical. It is, to a certain extent, the interfirm equivalent of the 'complex global TNC' identified in Table 7.1. However, these interfirm networks are *relational* structures between independent and quasi-independent firms which are based upon a high degree of trust, something which takes time to develop. However, this does not mean that there are not *power* differentials within the network. There certainly are, although more so in some networks than in others. Miles and Snow (1986, p. 69) suggest that the network

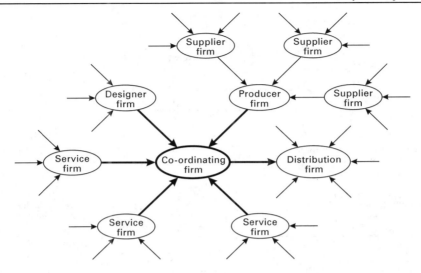

Figure 7.14 An outline of a dynamic and flexible business network
Source: Based, in part, on Miles and Snow (1986, Figure 1)

form is becoming especially common in 'labour-intensive industries, where vertical disaggregation is less costly and easier to administer'. They point to developments in the US college textbook publishing industry in which many of the leading firms have now contracted out not just printing and binding functions (often overseas) but also many editing operations, artwork, graphics and design.

Two more detailed examples of dynamic network organizations can be given to provide some idea of their actual, as opposed to their theoretical, characteristics:

- The first example is a US toy manufacturer, Lewis Galoob Toys, Inc., which was, in the mid-1980s, a multimillion-dollar company, yet, in the words of a *Business Week* (3 March 1986, p. 61) report, 'is hardly a company at all' –

  A mere 115 employees run the entire operation. Independent inventors and entertainment companies dream up most of Galoob's products, while outside specialists do most of the design and engineering. Galoob farms out manufacturing and packaging to a dozen or so contractors in Hong Kong, and they, in turn, pass on the most labour-intensive work to factories in China. When the toys land in the US, they're distributed by commissioned manufacturers' representatives. Galoob doesn't even collect its accounts. It sells its receivables to Commercial Credit Corp., a factoring company that also sets Galoob's credit policy. In short, says Executive Vice President Robert Galoob, 'our business is one of relationships'. Galoob and his brother, David, the company's president, spend their time making all the pieces of the toy company fit together, with their phones, facsimile machines, and telexes working overtime.

- The second example, less extreme in form, is the Nike athletic footwear company described by Donaghu and Barff (1990) and Korzeniewicz (1994). Like other athletic footwear firms, Nike does not wholly own any integrated production facilities but is characterized by 'the large-scale vertical disintegration of functions and a high level of subcontracting activity' (Donaghu and Barff, 1990, p. 539). Its development displays great flexibility in adapting to changing competitive circumstances. As

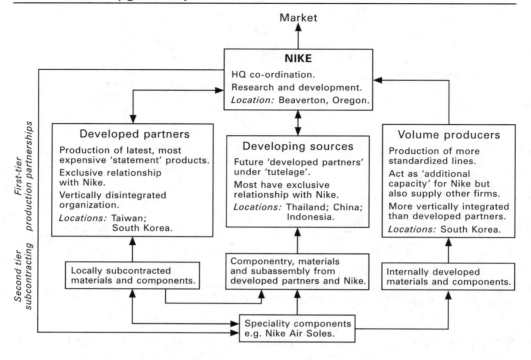

Figure 7.15   The Nike network
Source: Based on Donaghu and Barff (1990, Figure 4 and pp. 542–44)

Figure 7.15 shows, Nike consists of a complex tiered network of subcontractors which perform specialist roles:

> Nike's 'in-house' production may be thought of as production from its exclusive partners. Nike develops and produces all high-end products with exclusive partners, while volume producers manufacture more standardized footwear that experience larger fluctuations in demand . . . Nike acts as the production co-ordinator and three categories of primary production alliance form the first tier of subcontractors. A second tier of material and component subcontractors supports production in the first tier . . . [all] . . . production takes place in South East Asia while the headquarters in Beaverton, Oregon, houses Nike's research facilities.
>
> (Donaghu and Barff, 1990, p. 544)

> In the case of Nike Corporation, marketing and advertising have driven the rest of the commodity chain. Marketing, advertising, and consumption trends dictate what will be manufactured, how it will be manufactured, and where it will be manufactured. In explaining Nike Corporation's success, manufacturing processes are secondary to the control over the symbolic nature and status of athletic shoes.
>
> (Korzeniewicz, 1994, p. 263)

These two examples are consistent with Gereffi's *buyer-driven commodity chain*, one of whose basic characteristics is that

> frequently these businesses do not own any production facilities. They are not 'manufacturers' because they have no factories. Rather, these companies are 'merchandisers' that design and/or market, but do not make, the branded products they sell. These firms rely on complex tiered networks of contractors that perform almost all their specialized tasks . . . The main job of the core company in buyer-driven commodity chains is to manage

these production and trade networks and make sure all the pieces of the business come together as an integrated whole. Profits in buyer-driven chains thus derive not from scale economies and technological advances as in producer-driven chains, but rather from unique combinations of high-value research, design, sales, marketing, and financial services that allow the buyers and branded merchandisers to act as strategic brokers in linking overseas factories and traders with evolving product niches in their main consumer markets.

(Gereffi, 1994, p. 99)

We seem to be getting closer to the 'virtual firm' or, less dramatically, to the 'hollow corporation'. At the very least, it is clear that the boundaries of firms are becoming increasingly blurred. The 'new map . . . [of the firm] . . . displays no sharp dividing line separating the inside of the firm from the outside. Rather, it shows the firm as a dense network at the centre of a web of relationships' (Badaracco, 1991, p. 314).

Connecting the organizational and geographical dimensions of business networks

Figure 7.16 summarizes the primary modes of TNC operations and relates them to the major contributory influences. As we have shown in the two previous sections of this chapter, these modes of operation have become both increasingly diverse and increasingly complex. The idea that TNC activity can be captured effectively through the single lens of foreign direct investment is clearly totally inadequate. Strategic

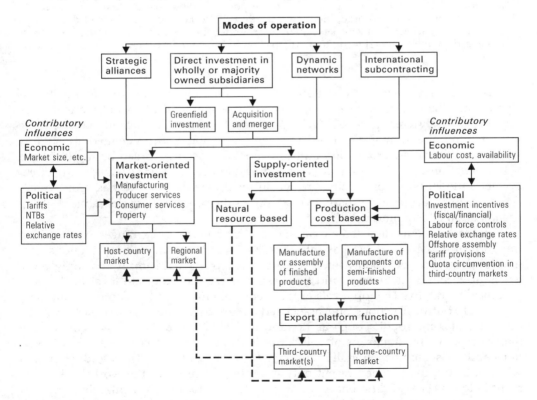

Figure 7.16   Different modes of TNC operation and their contributory influences

alliances, international subcontracting and sourcing, and the embryonic development, at least, of dynamic and flexible network forms add enormously to the organizational complexity of the system. The question we pose now is: how are these diverse organizational forms expressed 'on the ground' in organizational-geographical space?

The situation is complicated enough even if we take a static picture. It is, of course, far more complex than this because all these networks of relationships are in a continuous state of flux. The ways in which the production chain is organized, and the boundary between which functions are internalized within a firm and which are externalized and performed as a division of labour between firms, are extremely fluid. As we have seen, there are signs of increasingly flexible network forms of organizational relationship in the global economy. The combination of these various networks of relationships, both within TNCs and between independent and quasi-independent firms, creates a highly complex *geographical* structure.

The earliest attempt to explore this link between the organizational structures of TNCs and their geographical form was made by Stephen Hymer in 1972. Hymer focused solely on the internal organization of TNCs posing the question 'does the internal division of labour within the TNC correspond to an international division of labour?' He drew upon the theories of organizational hierarchies devised by Alfred Chandler and the theory of location of Alfred Weber to argue that such an organizational-geographical correspondence did, indeed, exist:

> the [trans]national corporation tends to create a world in its own image by creating a division of labour between countries that corresponds to the division of labour between various levels of the corporate hierarchy. It will tend to centralize high-level decision-making occupations in a few key cities (surrounded by regional sub-capitals) in the advanced countries, thereby confining the rest of the world to lower levels of activity and income.
>
> (Hymer, 1972, p. 59)

There is obviously some validity in Hymer's view, as our discussion in this chapter has demonstrated. There *are* recognizable hierarchical spatial tendencies in the location of different parts of TNCs. Corporate headquarters *do* tend to concentrate in a small number of major metropolitan centres; regional offices *do* favour a slightly wider range of cities; production units *are* more extensively spread both within and between nations, in developed and developing countries. Not surprisingly, therefore, Hymer's model of a world being created in the image of the TNC has been widely accepted. But, of course, it grossly oversimplifies the complexity of the modern global organization of economic activity.

Hymer's scheme depicted a clear and distinct hierarchical arrangement in which the vertical division of labour within the TNC is reflected in an unambiguous geographical division of labour. Interpreted literally this would imply that only those levels of corporate activity would be present at each appropriate level in the geographical hierarchy. The top levels of the metropolitan hierarchy would contain only high-level control functions and associated occupations. But this is clearly not the case. A major metropolis, such as London or New York does, indeed, contain the major corporate headquarters of TNCs. But it also contains lower-level control functions and, most significant of all in this context, many of the kinds of production unit which are supposed to be present only in peripheral areas of the world. It is more realistic to conceive of different types of geographical location as containing different *mixes* of corporate units in which the actual proportions vary.

Hymer, quite reasonably given the circumstances prevailing at the time he was writing, conceived of the TNC as a very simple hierarchically organized structure. But he did not consider how the various levels of the firm interact with other firms, other than to note the agglomerative tendencies of TNC headquarters. In fact, as we have seen in this chapter, the global economy is made up of a variety of *complex intra-organizational and interorganizational networks* – the internal networks of TNCs, the networks of strategic alliances, of subcontracting relationships and of other, newer, organizational forms.

These intersect with *geographical networks* structured particularly around linked agglomerations or clusters of activities. As we emphasized in Chapter 1, not only is all economic activity grounded in specific places but also geographical clustering is the norm. The key issue, then, is what form do such clusters take and how are they interconnected? In Chapter 5 we noted the claim of writers in the flexible specialization school of post-Fordism that the development of technical and organizational flexibility has led to the emergence of so-called 'new industrial districts' based upon very dense localized transactions between small independent firms. We also discussed in Chapter 5 the strong tendency for innovative activity to be geographically clustered into innovative milieux or technology districts which constitute the key growth foci of the global economy.

There undoubtedly are 'new industrial districts' of the kind postulated in the post-Fordist literature but the extent to which they are made up of relatively self-contained, interlinked webs of small entrepreneurial firms which generate greater local economic autonomy has been exaggerated.[11] More generally, localized clusters of economic activity consist of a differentiated mixture of independent firms of various sizes and of the branch plants and affiliates of multiplant firms, many of which are TNCs. These different types of establishment are connnected into much larger organizational and geographical structures. In the case of the branches and affiliates they are obviously part of a specific corporate structure and will be constrained in their autonomy by parent company policy. The extent to which they are functionally connected into the local economy will be enormously variable. The 'independent' firms in a local economy may, in fact be rather less independent than they appear at first sight. Some, at least, will be integrated into the supply networks of larger firms, again including TNCs whose decision-making functions are very distant. As we saw in our discussion of subcontracting there are both benefits and costs associated with such a role for subcontracting firms. Other firms may be linked together through strategic alliances or they may be a part of the flexible business networks co-ordinated by key 'broker' firms.

In other words, it is very difficult indeed to generalize about the precise forms and relationships involved. What can be said is that the global economy is made up of intricately interconnected *localized clusters* of activity which are embedded in various ways into different forms of corporate network which, themselves, vary greatly in their geographical extent. Some TNCs are globally extensive, others have a more restricted geographical span. Either way, however, firms in specific places – and, therefore, the places themselves – are increasingly connected into international and global networks. The precise role played by firms in these networks will have very significant implications for the communities in which they are based. Hymer's model captured part of this issue in terms of the uneven geographical distribution of 'high' level and 'low' level corporate functions. Places which consist of substantial concentrations of high-level functions are obviously substantially better off than those which are allocated entirely low-level functions. Similarly, the kind of supply role played by firms within TNC networks will be critical to local well-being. As we have seen, customer–supplier relationships increasingly involve a greater emphasis on long-term, closer relationships based upon a high degree of mutual trust. This may encourage closer geographical relationships between customers and suppliers.

However, TNCs are increasingly operating an upper tier of preferred suppliers which are closely integrated at all stages of the production process, from design through to final production. For any one firm, such preferred suppliers will be relatively few in number and unevenly distributed geographically. TNCs still have a strong propensity to source over extensive geographical distances, including globally. Not every local economy, therefore, can hope to participate in these new supply networks:[12]

> First, the most likely scenario is that a smaller number of places will become the hosts to more integrated multinational corporate investments. For these favored places, the prognosis is relatively good . . . the stability of the investment, implicating as it does, multiple-linked firms, is likely to be significantly higher. Yet, while many local firms will no doubt be drawn into the production complex, it is perhaps less likely that they will become core members of the collaborative partnership, remaining rather in a subordinate position to it. Secondly, as these investments become more concentrated in particular regions, the excluded regions are likely to become that much more excluded. Rather than a general embracing of more and more territory into the productive orbit of multinational networks, the degree of geographical differentiation will tend to increase.
>
> (Schoenberger, 1991, quoted in Dicken, 1994, pp. 122–23)

In exploring the complex relationships between the organizational and the geographical dimensions of transnational corporate networks we have implicitly taken a 'top-down' approach. In fact, it is quite wrong to suggest that the map of economic activity, at whatever geographical scale, is merely the result of decisions made by business firms or any other organization simply projected on to the earth's surface in a top-down manner. As Walker (1988, p. 385) has rightly observed, 'It is impossible to separate out the organizational from the geographical, much less to treat industrial location as the simple outcome of organizational forms or decisions . . . space is . . . basic to human action . . . all forms of organization are inherently spatial to some degree'. In fact, the very character – the history, culture, institutional structures – of particular places itself exerts a considerable influence on the processes and networks we have been discussing. Specifically, place and spatial relationships within and between territorial complexes of economic activity are, in themselves, an *intrinsic* part of the production system as a whole. Organizational and geographical processes constitute a mutually interactive dynamic.

Conclusion

The aim of this chapter has been to explain how economic activities are organized and reorganized through dynamic networks of relationships within and between transnational corporations and other types of firm. We have emphasized, in particular, the immense variety and diversity of processes and outcomes, both organizational and geographical. The production chain – the basic building block of the economic system – can be articulated through many different combinations of organizational structures and geographical configurations. In this chapter we focused on three specific aspects of transnational corporate activity.

First, we focused on TNCs as networks of internalized relationships. Consistent with the view that TNCs are not converging towards a single form we identified a variety of organizational structures. The particular type employed appears to be influenced both by the firm's specific history and by the nature and complexity of the industry environment(s) in which it operates. Strategically, all TNCs have to resolve the basic tension between globalizing pressures on the one hand and localizing pressures on the other. In pursuing their specific strategies, TNCs create not just organizational structures but also have to configure their component parts geographically. Because different parts of a firm have different locational needs, and because these may be satisfied in different types of location, each tends to take on distinctive geographical characteristics. We illustrated this by looking at three functions: corporate and regional headquarters, R&D facilities, and production units. But, of course, these structures are dynamic, not static. Restructuring and reorganization are endemic within TNCs.

Secondly, we explored the networks of externalized interfirm relationships within which TNCs are embedded. Again, these take on a bewildering variety of forms. There has been a strong propensity for western firms to engage in strategic alliances and other forms of relationships with other firms, including firms which are their direct competitors. The nature and extent of the relationship between firms and their suppliers have also become increasingly complex with, in general, a tendency for major firms to develop longer-term relationships with key suppliers. The geography of these customer–supplier relationships also appears to be variable. On the one hand, there is still a tendency to seek out supplies across vast geographical distances; on the other hand, the adoption of just-in-time systems creates some pressures for closer spatial links between customers and suppliers. In fact, although closer network relationships have been heralded as something new for western firms they are a long-established feature of east Asian firms, as our brief account of the Japanese *keiretsu* and Overseas Chinese family businesses revealed. What does seem to be new, however, is the emergence of vertically disintegrated flexible network organizations. Here most production chain functions are performed by specialist firms within a flexible network co-ordinated by a key firm.

Finally, we connected up the organizational dimension of TNC networks (both internal and external) with the geographical dimension. Stephen Hymer's early study of the relationships between the division of labour within a TNC and the international division of labour between places retains some validity. However, not only have the internal structures of TNCs become more complex but also Hymer did not investigate the geographical implications of their externalized networks. Combining both sets of organizational networks with the localized geographical clusters which make up the global economic map creates an immensely complex system. The global economy is made up of intricately interconnected localized clusters of activity which are embedded in various ways into different forms of corporate network which, in turn vary greatly in their geographical extent. Some TNCs are globally extensive, others have a more restricted geographical span. Either way, however, firms in specific places – and, therefore, the places themselves – are increasingly connected into international and global networks.

Notes for further reading

1. For a discussion of some of these influences and how they affect firm structure and behaviour see Heenan and Perlmutter (1979), Bartlett (1986), Bartlett and Ghoshal (1989), Schoenberger (1997).

2. Rather confusingly, from our point of view, Bartlett and Ghoshal use the term *transnational* to describe this kind of emergent organization. In this book, we use the term 'transnational corporation' in a generic sense to encompass all kinds of international business organization; for this reason I have used the term 'complex global organization' to describe Bartlett and Ghoshal's fourth ideal type.

3. See Howells (1990), Patel and Pavitt (1991), Pearce and Singh (1992), Office of Technology Assessment (1994), Patel (1995)..

4. Stephen Hymer who, as we have seen, was a pioneer in developing theoretical explanations of the TNC, was probably the first to use the term 'new international division of labour' (NIDL) to explain the shift of production from the industrialized economies of the 'core' to the economies of the 'global periphery'. However, it was the publication of Fröbel, Heinrichs and Kreye's book, *The New International Division of Labour*, in English in 1980 which stimulated a considerable body of work in this area. For a critique of the NIDL concept see Jenkins (1984), Elson (1988), Schoenberger (1988b).

5. See Clark (1994), Clark and Wrigley (1995), Schoenberger (1997).

6. See Kindleberger (1988).

7. Hamilton (1991), Whitley (1992a), Castells (1996, Chapter 3) provide broad reviews of Asian business organizations and networks.

8. The Japanese *keiretsu* are discussed by Aoki (1984), Helou (1991), Fruin (1992), Gerlach (1992), Odagiri (1992). A popular treatment is provided by Miyashita and Russell (1994).

9. See Redding (1991), Whitley (1992a), Kao (1993), Yeung (1997).

10. There is a growing literature on strategic alliances. Useful contributions are by Business International (1987), Morris and Hergert (1987), Hagedoorn and Schakenraad (1990), Office of Technology Assessment (1993, Chapter 5), Anderson (1995).

11. See the literature cited in Note for Further Reading 11, Chapter 5.

12. Amin (1992), Amin and Malmberg (1992), Schoenberger (1997) explore this issue.

CHAPTER 8

Dynamics of conflict and collaboration: '*both* transnational corporations *and* states matter'

TNCs and states/states and TNCs: the ties that bind

> It is perhaps most useful . . . to view the relationship between [trans]nationals and governments as both cooperative and competing, both supportive and conflictual. They operate in a fully dialectical relationship, locked into unified but contradictory roles and positions, neither the one nor the other partner clearly or completely able to dominate.
>
> (Gordon, 1988, p. 61)

This quotation captures the essence of the intricate relationships between TNCs and states as containing elements of both rivalry and collusion (Pitelis, 1991). In a sense, each needs the other, even though their relationships may be conflictual in certain circumstances. States need firms to generate material wealth and provide jobs for their citizens. They might prefer such firms to be domestically bounded in their allegiance but that is not an option in a capitalist market economy. Conversely, firms need states to provide the infrastructural basis for their continued existence; both physical infrastructure in the form of the built environment and social infrastructures in the form of legal protection of private property, institutional mechanisms to provide a continuing supply of educated workers and the like.

As we showed in Chapter 3, states constitute both *containers* of distinctive business practices and cultures – within which firms are embedded – and *regulators* of business activity. National boundaries, therefore, create significant differentials on the global political-economic surface; they constitute one of the most important ways in which location-specific factors are 'packaged'. They create discontinuities in the flow of economic activities which are extremely important to the ways in which TNCs can operate. In particular, states have the potential to determine two factors of fundamental importance to TNCs and which may force them to modify their strategic behaviour:[1]

- The terms on which TNCs may have *access* to markets and/or resources.
- The *rules of operation* with which TNCs must comply when operating within a specific national territory.

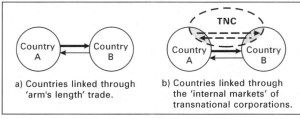

a) Countries linked through 'arm's length' trade.

b) Countries linked through the 'internal markets' of transnational corporations.

Figure 8.1 The 'incorporation' of parts of a nation's territory by TNCs

At the same time the fact that, by their very nature, TNCs not only span national boundaries but also, in effect, incorporate *parts* of national economies within their own firm boundaries (Figure 8.1) creates major potential problems for states. The nature and the magnitude of the problem vary considerably according to the kinds of strategies pursued by TNCs. Most important is the extent to which TNCs pursue globally integrated strategies within which the roles and functions of individual units are related to that overall global strategy.[2] As we saw in Chapter 7, integrated TNCs do indeed geographically fragment their operations in pursuit of global profits. Some types of TNC operation create higher value for the communities in which they are located than other types of operation. More generally, states tend to be fearful about the autonomy and stability of those TNC units located within their national territory as well as concerned about the leakage of tax revenues.

At first sight it may seem obvious that TNCs will invariably seek the removal of all regulatory barriers that act as constraints and impede their ability to locate wherever, and to behave however, they wish. The ultimate preference for TNCs, it would seem, is the removal of all barriers to entry, whether to imports or to direct presence; freedom to export capital and profits from local operations; freedom to import materials, components and corporate services; freedom to operate unhindered in local labour markets. Certainly, given the existence of differential regulatory structures in the global economy, TNCs will seek to overcome, circumvent or subvert them. Regulatory mechanisms are, indeed, constraints on a TNC's strategic and operational behaviour.

Yet it is not quite as simple as this. TNCs may perceive the very existence of regulatory structures as an *opportunity*, enabling them to take advantage of regulatory differences between states by shifting activities between locations according to differentials in the regulatory surface – that is, to engage in *regulatory arbitrage* (Leyshon, 1992). One aspect of this is the propensity of TNCs to stimulate competitive bidding for their mobile investments by playing off one state against another as states themselves strive to outbid their rivals to capture or retain a particular TNC activity. More generally, TNCs seem to have a rather ambivalent attitude to state regulatory policies:[3]

> TNCs have favoured minimal international coordination while strongly supporting the national state, since they can take advantage of regulatory differences and loopholes . . . While TNCs have pressed for an adequate coordination of national regulation, they have generally resisted any strengthening of international state structures . . . Having secured the minimalist principles of national treatment for foreign-owned capital, TNCs have been the staunchest defenders of the *national state*. It is their ability to exploit national differences, both politically and economically, that gives them their competitive advantage.
>
> (Picciotto, 1991, pp. 43, 46)

More specifically, it has been argued that TNCs will increasingly tend to support a strategic trade policy in their home country, with the expectation that this will open up market access in foreign countries and enable them to benefit from large-scale economies and learning curve effects (Yoffie and Milner, 1989).

It is clear, therefore, that the relationships between TNCs and states are exceedingly complex. In this chapter we focus on two major aspects of these relationships:

- The costs and benefits of transnational corporations to both host economies and home economies.
- The nature of the bargaining relationship between transnational corporations and states.

TNCs and states: evaluating the costs and the benefits

According to viewpoint, TNCs either expand national economies or exploit them; they are either a dynamic force in economic development or a distorting influence; they either create jobs or destroy them; they either spread new technology or pre-empt its wider use and so on. At one extreme the charge is one of political interference in national affairs or of bribery of national officials. At the other extreme the TNC is regarded as a greater force for international economic well-being than the parochially bounded nation-state. Virtually every aspect of the TNC's operations – economic, political, cultural – has been judged in diametrically opposed ways by its proponents and its opponents. In this section, the major ways in which TNCs may affect various aspects of national economies are outlined as a framework within which specific cases can be assessed. Their effects will differ, therefore, according to whether the country involved is *host* to the operations of a foreign firm or is the *home* base from which a firm extends its operations overseas. We look at each of these two perspectives in turn.

Transnational corporations and host economies

The establishment of an overseas facility by a TNC incorporates a package of qualities – financial, technological, managerial, marketing – which, together, have far-reaching implications for the host economy. As Figure 8.2 shows, a number of variable factors are involved:

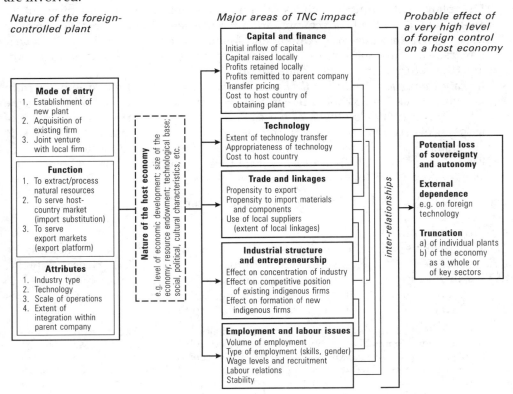

Figure 8.2 Effects of transnational corporations on host economies

- *The nature of the foreign-controlled plant*:
  1) The *mode of entry* – how the foreign plant is established (by setting up a completely new facility, by acquiring an existing indigenous enterprise or by forming a joint venture with local capital). A completely new 'greenfield' plant is generally regarded with greater favour by host countries than acquisition of existing capacity. A new plant obviously adds initially to the host country's stock of productive capacity whereas acquisition merely transfers ownership of existing capacity to a foreign firm. Of course, the outcome is never as clear cut as this. Much depends on the effect of a new plant on other firms and, in the case of acquisition, on whether the acquired plant's performance is enhanced to the benefit of the domestic economy. The mode of entry obviously depends on the opportunities available. Foreign acquisitions have been more prevalent in developed host economies, simply because there are many more suitable candidates for take-over, but they are by no means absent in developing countries.
  2) The *function* of the plant. Figure 8.2 reminds us that foreign branch plants tend to be established for one of three reasons: to exploit a localized material resource, to serve the host market itself by substituting for imports or to use the host country location as a platform for exporting either finished products or components.
  3) The *operational attributes* of the plant, including industry type, technology employed, scale of operations and the extent to which the plant is integrated into the parent company structure.

Thus, plants with different modes of entry, different functions and different attributes will affect the host economy in different ways. Each of the major areas of impact shown in Figure 8.2 – capital and finance, technology, trade and linkages, industrial structure and employment – may be influenced in various ways according to the nature of the foreign plant involved.

- The *nature and characteristics of the host economy*. Most TNC activity originates from the highly industrialized and affluent developed market economies. In so far as the bulk of this activity also flows to these same developed economies the degree of 'dissonance' between a foreign-controlled operation and the host economy is likely to be small. For less industrialized host countries, however, and for those with very different socio-cultural characteristics from those of the foreign firm's home country, the 'shock waves' may be much greater.

Let us now look at each of the major areas of TNC impact.[4]

Capital and finance

The most obvious and immediate impact of foreign investment on a host economy is the inflow of capital. It is especially important for those countries suffering from a shortage of investment capital. TNCs have certainly been responsible for injecting capital into host economies, both developed and developing. But not all new overseas ventures undertaken by TNCs involve the actual transfer of capital into the host economy:

> About a half of all funds obtained externally by majority-owned foreign affiliates (non-bank) of United States TNCs (non-bank) were raised in host countries in 1992 . . . In the case of Japan, funds raised externally financed 58 per cent of total overseas investments in 1992, with funds obtained from local (host-country) financial institutions accounting for 35 per cent of that total . . . Externally raised capital appears to finance a larger share of investment funds in developed than in developing countries.
>
> (UNCTAD, 1995, pp. 142–43)

Thus, at least some of the capital employed by TNCs may either be borrowed on the host-country capital markets or arise from the reinvestment of retained earnings from the foreign affiliate. Local firms may be bought with local money. Local firms may even be squeezed out of local capital markets by the perceived greater attractiveness of TNCs as a use for local savings.

Even where capital inflow does occur, or where earnings are retained, there will, eventually, be a reverse flow as the foreign plant remits earnings and profits back to its parent company. This reverse flow may, in time, exceed the inflow of capital. Any net financial gain to the host country also depends on the trading practices of the TNC. A host country's balance of payments will be improved to the extent that the foreign plant exports its output and reduced by its propensity to import. A vital issue, therefore, is the extent to which financial 'leakage' occurs from host economies through the conduit of the TNC. This raises the question of the ability of host-country governments to obtain a 'fair' tax yield from foreign firms, many of which are capable of manipulating the terms of their intracorporate transactions through transfer pricing.

The problem of transfer pricing

In external markets, prices are charged on an 'arm's-length' basis between independent sellers and buyers. In the internal 'market' operated by TNCs, however, transactions are between *related* parties – units of the *same* organization. The rules of the external market do not apply. The TNC itself sets the transfer prices of its goods and services within its own organizational boundaries and, therefore, has very considerable flexibility in setting its transfer prices to help achieve its overall goals. The ability to set its own internal prices – within the limits imposed by the vigilance of the tax authorities – enables the TNC to adjust transfer prices either upwards or downwards and, therefore, to influence the amount of tax or duties payable to national governments. For example, it would be in a TNC's interest to charge more for the goods and services supplied to its subsidiaries located in countries with high tax levels and *vice versa*. A similar incentive exists where governments restrict the amount of a subsidiary's profits which can be remitted out of the country.

In general, the greater the differences in levels of corporate taxes, tariffs, duties, exchange rates, the greater will be the incentive for the TNC to manipulate its internal transfer prices. As Lall (1973, p. 180) points out, 'transfer pricing can be used most effectively by very large corporations with tightly exercised central control, sophisticated computational facilities and a wide experience of world conditions and of dealing with governments, and not by investors with limited overseas operations and a great deal of autonomy between different units'.

TNCs have a strong incentive to engage in transfer pricing. The very large, highly centralized, global TNC has the greatest potential for doing so. But it has proved extremely difficult for governments (and researchers) to gather hard evidence on its actual extent. A United States House of Representatives study in 1990 claimed that more than half of almost forty foreign companies surveyed had paid virtually no taxes over a ten-year period. It was estimated that some $35 billion is being lost through the transfer pricing mechanism. In the United Kingdom, a study of 210 TNCs showed that 83 per cent had been involved in a transfer pricing dispute.[5] But even in advanced economies like the United States or the United Kingdom it is extremely difficult for the Internal Revenue Service to assess the actual extent of transfer pricing. It is even more difficult for developing countries to do so.

The financial gain (or loss) to host countries from foreign investment, therefore, consists primarily of the net balance of capital flows plus the net earnings from trade. But there is also the cost of actually obtaining the foreign investment to consider. Most states, national and subnational, offer very considerable financial and fiscal incentives to attract internationally mobile investment. These costs, together with the provision of the necessary physical infrastructure may be quite substantial. We will return to this issue in the context of the bargaining process between TNCs and states later in this chapter.

Technology

Technology may be embodied in the form of capital goods, such as machinery, equipment and physical structures; or it may be disembodied in such forms as industrial property rights, unpatented know-how, management and organisation, and design and operating instructions for production systems. *Foreign direct investment has traditionally been one of the most important channels of technology transfer as it involves the physical relocation of entire production systems*, combining in a single package capital goods and a number of the forms of disembodied technology.

(UNCTC, 1983, pp. 162–63, emphasis added)

Three issues are especially important in evaluating the technological impact of foreign enterprises on host economies:

- The extent to which technology is *transferred* both within the bounds of the TNC and to other – domestically based – enterprises.
- The *appropriateness* of the technology transferred – both processes and products.
- The *costs* to the host economy of acquiring the technology.

Transfer of technology

Each of the modes of foreign involvement shown in Figure 7.16 – foreign direct investment, collaborative ventures, international subcontracting, dynamic and flexible business networks – is a potential channel for technology transfer. Simply by locating some of its operations outside its home country the TNC engages in the geographical transfer of technology. In this respect, technologies are, indeed, spread more widely. In so far as a foreign affiliate employs local labour there will be a degree of technology transfer to elements of the local population through training in specific skills and techniques. But the mere existence of a particular technology within a foreign-controlled plant does not guarantee that its benefits will be widely diffused through a host economy. The critical factor here is the extent to which the technology is made available to potential users outside the firm either directly, through linkages with indigenous firms, or indirectly via the 'demonstration' effect.[6]

However, the very nature of the TNC may inhibit the spread of its proprietary technology beyond its own organizational boundaries. Possession and exploitation of technology are a diagnostic feature of the TNC. Such technology is not lightly handed over to other firms. Control over its use is jealously guarded: the terms under which the technology is transferred are dictated primarily by the TNC itself in the light of its own overall interests.

A major tendency, as we saw in Chapter 7, is for TNCs to locate the bulk of their technology-creating activities – research and development – either in their home country or in the more advanced industrialized countries. So far, relatively little R&D has been relocated to the developing countries. Thus, 'even if it is admitted that TNCs transfer the best production technology, they do not transfer the capability to generate new technology to affiliates in the Third World. They transfer "know-how" (production engineering) and not "know-why" (basic design, research and development)' (Lall, 1984, p. 10). Nevertheless, some R&D – mostly lower-level support laboratories – has been located in a few of the newly industrializing economies. In some cases this is a direct result of host-government pressure on TNCs to establish R&D facilities in return for entry. Such leverage is probably greatest where the TNC wishes to establish a branch plant to serve the host-country market itself.

Appropriateness of technology

In the case of developing countries, in particular, a major issue is the appropriateness of the technology transferred via the TNC. Do the processes and products being introduced match local conditions and local needs? New technologies are invariably introduced by TNCs first of all in their home country or in other industrialized countries. Since they reflect the prevailing cost and availability of factors in those countries the technologies tend to be capital- rather than labour-intensive. But in most developing countries the abundant factor tends to be low-skilled labour while capital is relatively scarce. Hence, there is much disagreement about the extent to which TNCs adapt their *process* technology for use in developing countries to make it more appropriate.

Lall and Streeten (1977, p. 72) make a distinction between the *adaptability* of technology and its actual *adaptation*:

> (a) As far as *adaptability* goes, much of modern 'high' technology cannot be changed to suit LDCs' endowments: the demands of precision, continuity, scale and complexity are too great. However, some 'low' technologies (for instance, in simple industries such as textiles) and 'peripheral' or 'ancillary' technology (for instance, transport or handling) are more adaptable. The scope of adaptability can be extended, but the cost, in terms of R&D and organisational requirements, may be quite high.
> (b) As far as *adaptation* of foreign technology goes, the bulk of *basic or core production technology* transferred by TNCs, both directly and by licensing, is not adapted in any significant way to low-wage conditions, though some *scaling down* of technology seems to be undertaken to adjust to smaller runs than would be appropriate in developed countries.

On the other hand Lall (1984, p. 9) claims that 'at the micro-level, every new application of a technology entails considerable adaptive effort. The core process may not be significantly altered, but changes in scale, inputs, outputs, automation etc., may constitute between 10–60 percent of total project costs'. In fact, research shows great variation in the relative capital intensity of production by foreign and domestic firms operating in the same industry in developing countries. In some cases, there is no significant difference; in others, foreign firms are significantly more capital-intensive in their operations than domestic firms.[7]

The second major issue in the appropriateness debate relates to the kinds of *products* transferred by TNCs to developing countries. This has become an especially contentious issue largely because of some highly publicized cases. In their drive to create global markets, TNCs have sought to introduce and sell their products throughout the developing world. The creation of particular types of demand and the shaping of consumer tastes and preferences are an intrinsic part of the TNC system. The problem is that

> the use of scarce resources for the production of goods which are over-differentiated, over-packaged, over-promoted, over-specified and within the reach of only a small elite,

or, if bought by the poor, at the expense of more essential products, is not conducive to 'national welfare'. This is not to say that all TNC technology is unnecessary in LDCs – clearly, that would be absurd. But the free import of foreign capital and of the sort of technology many TNCs excel in would reproduce the pattern of developed countries and would be undesirable. In other words, a definition of welfare based on meeting *basic social needs* would lead to a *fairly small proportion of TNC technology being regarded as beneficial.*

(Lall and Streeten, 1977, p. 71)

Against those products which are clearly 'inappropriate' to Third World countries, however, there are undoubtedly others that have brought great benefit, including those in agriculture (seeds, fertilizers) and health care.

A final aspect of the appropriateness issue relates to the *environmental and safety dimension* of TNC activity. Do TNCs export technologies to developing countries which are environmentally objectionable or which are less safe than they should be?[8] There have been claims that TNCs systematically shift some of their environmentally noxious or more hazardous operations to developing countries with less stringent environmental and safety standards. Although this may indeed occur in specific cases, there is no evidence to suggest that this is the general practice. Leonard's (1988) study of US companies did not support the claim of US firms engaging in industrial flight to pollution havens. The years following the introduction of stringent environmental regulations in the United States were not characterized by the widespread relocation of pollution-intensive industries to countries with lower regulatory standards.

A rather different aspect of the problem relates to safety and environmental management. Industrial disasters, such as the one at the Union Carbide plant at Bhopal, India, in 1994, focused attention on the safety practices of TNCs. A frequent claim was that TNCs tend to adopt less stringent safety practices in their developing country plants than in their home plants. The recent conflict in Nigeria involving Shell's environmental practices in Ogoniland raised both environmental and political issues. These cases, and others, are serious in the extreme. But it is dangerous to generalize from them to produce universal statements on the environmental behaviour of all TNCs. UNCTAD has carried out comprehensive surveys which give a more balanced picture. While not minimizing the seriousness of specific cases, the UNCTAD findings do not support the view that TNCs in general are environmentally irresponsible.

As far as 'industrial accidents' are concerned,

> while the number of industrial accidents appears to have risen over the last fifteen years, available evidence indicates that transnational corporations have been involved in less than half of them. Many accidents have occurred in purely national firms or in State-owned enterprises. The vast majority of industrial accidents have occurred in the home countries of transnational corporations, or in other developed market economies.
>
> (UNCTC, 1988, p. 228)

In terms of environmental management, the UNCTAD survey found that

> Transnational corporations are increasingly taking a more strategic approach towards environmental management issues, tending to view the costs associated with environmental management as long-term investment central to successful business ventures . . . TNCs are increasingly establishing targets with respect to the environmental performance of their operations . . . North American TNCs, with their generally longer experience of FDI, are relatively more sensitive to or aware of the international aspects of their activities than TNCs from other regions.
>
> (UNCTAD, 1995, pp. 176–77)

Questions about the appropriateness of technology tend to be less relevant in the developed economies for obvious reasons. But one issue of growing importance in such economies is the influence of foreign TNCs on *business organization and practice.* Japanese investment in Britain is a case in point. One argument for encouraging such investment has been that the introduction of highly efficient Japanese business methods will rub off on British industry in general and raise the level of efficiency in the economy. In other words, it is the *demonstration* of the effect of the 'social' innovations of work organization, labour relations, relationships with suppliers and so on which are suggested to be highly appropriate to the needs of the UK economy. Whether such a process of diffusion is occurring, and to what extent, is a matter of some debate.

Cost of technology transfer via the TNC
Finally, we need to consider the question of the cost to the host country involved in technology transfer through the TNC. But precisely what that cost is, and whether it is a 'reasonable' price to pay, are extremely difficult to determine. First, technology is only one part of the overall package of attributes which the TNC brings to a host country; it is difficult to separate out. Secondly, assessment of the cost involved assumes that it can be measured against alternative ways of acquiring the same technology. The two major alternatives are 1) to buy or license the technology alone from its owner (the TNC); that is, to 'unbundle' the TNC package; or 2) to produce the technology domestically.

The parallel usually drawn is with Japan, which rebuilt its postwar economy without the introduction of direct foreign investment, mainly by *licensing* technology from western firms. Although a great deal of technology is licensed by developing (and developed) countries from TNCs it is not always a feasible alternative. A TNC may be unwilling to license the technology or it may charge an exorbitant price. It may be a question of the host country accepting the entire TNC package or getting nothing at all. The possibility of producing the technology domestically may be feasible for some of the more advanced industrial nations but rarely so for developing countries.

Trade and linkages
Two of the most important questions surrounding the impact of TNCs on all host economies, whether developed or developing, are

- their role in the host country's *trade* with the outside world. Exports and import-substituting production by TNCs contribute towards a positive trade balance, imports by TNCs contribute towards a negative trade balance, although this may be offset if the imports are essential for export-producing activities;
- the extent to which their operations are *integrated* into the local economy through *linkages with domestic firms.* This is the most significant mechanism through which technology is transferred, additional employment created and opportunities increased for the formation of new local enterprises.

How far TNCs actually contribute to the host economy's *trade balance* depends on a number of factors. The most obvious is the primary purpose for which the foreign plant is established (Figure 8.2): that is, whether its role is to serve the host market itself or to serve third-country markets using the host economy as an export platform. In general, TNCs tend both to export and to import more than local firms but that merely reflects their greater overall size and the geographical extensiveness of their operations. Plants set up to serve the local market, by definition, will not be large exporters. How far they are large importers of intermediate products will depend upon corporate sourcing policy, local supply capabilities and host-government policy towards local content. Plants set up as export platforms, on the other hand, are bound to be major exporters – that is their entire *raison d'être*. But, again, the *net* effect on the trade balance will be influenced by the extent to which its export-producing activities depend upon imported components.

Interfirm linkages are the most important channel through which technological change is transmitted. By placing orders with indigenous suppliers for materials or components which must meet stringent specifications technical expertise is raised. The experience gained in new technologies by local firms enables them to compete more effectively in broader markets, provided, of course, that they are not tied exclusively to a specific customer. The sourcing of materials locally may lead to the emergence of new domestic firms to meet the demand created, thus increasing the pool of local entrepreneurs. The expanded activities of supplying firms, and of ancillary firms involved in such activities as transport and distribution, will result in the creation of additional employment. But such beneficial spin-off effects will occur only *if* the foreign affiliates of TNCs *do* become linked to local firms. Where TNCs do not create such linkages they remain essentially as foreign enclaves within a host economy, contributing little other than some direct employment.

As far as local linkages are concerned the most significant are *backward* or *supply* linkages. Here, the crucial issue is the extent to which TNCs either import materials and components or procure them from local suppliers. The actual incidence of local linkage formation by foreign-controlled plants depends upon three major influences:

● The particular *strategy* being followed by the TNC and the *role* played by the foreign plant in that strategy.

The influence of TNC strategy on the development of local linkages within host economies has a number of aspects. One is the general corporate policy towards the sourcing of inputs, which will determine the degree of sourcing autonomy granted to individual plants. Those TNCs which are strongly vertically integrated at a global scale are less likely to develop local supply linkages than firms with a lower degree of corporate integration. But even where vertical integration is low the existence of strong links with independent suppliers in the TNC's home country or elsewhere in the corporate network may inhibit the development of local linkages in the host economy. Familiarity with existing supply relationships may well discourage the development of new ones, particularly where the latter are perceived to be potentially less reliable or of lower quality.

A particularly important factor is the role of the foreign plant itself in the TNC's overall strategy: whether it is oriented to the host market and is, therefore, import substituting, or whether it is an export platform activity. Foreign plants which serve the host market are more likely to develop local supply linkages than are export platform plants. Lall (1978) identifies four types of export-oriented activity in TNCs and suggests that each will have rather different implications for the creation of local supply linkages, particularly in developing countries:

1) *TNCs which began as local market-oriented operations but which have subsequently developed strong export orientation.* These generally utilize relatively stable and

unsophisticated technologies and are located in areas where the labour force is skilled but cheap and where there is an established indigenous industrial sector. Such firms may have established significant local linkages during their import-substituting phase which then form the basis of broader export-oriented operations.

2) *TNCs involved in more traditional industries* (e.g. textiles, food processing, sports goods) which employ standardized technology but where product differentiation, marketing or product innovations are important considerations. These activities have a high potential for the creation of local linkages with domestic suppliers to manufacture either components or complete products.

3) *New TNC investments in 'modern' industries which are located in developing country export platforms.* The technologies used are fairly complex and production is directed towards world markets. In general, these activities are very closely controlled from corporate headquarters in the TNCs' home countries and often utilize established supply links. The potential for developing local linkages is fairly limited but not entirely absent.

4) *TNC investments which are, essentially, merely 'sourcing' operations* in which only a particular process (usually highly labour intensive) is located in developing countries. The clearest example is probably the semiconductor industry (Chapter 11). The potential for developing local linkages is low because of the very tight product specification, the need to keep production costs as low as possible and the dynamic nature of the technology involved.

- The characteristics of the host economy

In general, we would expect to find denser and more extensive networks of linkages between TNCs and domestic enterprises in the developed economies than in the developing economies. Within developing countries such linkages are likely to be greatest in the larger and more industrialized countries than in others. In addition, host-country governments may well play a very important role in stimulating local linkages by insisting that TNCs utilize a certain level of locally sourced materials and components. Such local content policies have become increasingly widespread in both developing and developed countries. Indeed, the ILO's (1981a, p. 94) view is that, at least in developing countries,

> government intervention in the sourcing choices of MNEs appears to have been the single most powerful determinant for the creation of local linkages of MNEs . . . Without such government intervention it is likely that, despite some market pressure, local MNE linkages would be much less developed than they are today in various countries and industries.

But much depends on the relative strength of the host country's bargaining power *vis-à-vis* the TNC, and on the extent to which local supplies are of an appropriate quantity and quality. Again, it tends to be in the larger and the more industrialized developing countries that such local content policies have the greatest impact, and also in those TNC activities serving the local market. Indeed, it could be that the export-oriented industrialization strategies of developing countries actually inhibit the development of local supply linkages.

- Time

The third influence on the establishment of TNC linkages with local suppliers is time. Particularly in view of the closer relationships between firms and their suppliers which have been emerging (see Chapter 7) it should not be expected that a foreign plant, newly established in a particular host country, will immediately develop local supplier linkages. Not only do appropriate suppliers have to be identified but also it takes time for supplier firms to 'tune in' to a new customer's needs.

Empirical evidence of local linkage formation by TNCs is extremely variable. Examples can be found at all positions on the spectrum of possibilities, from virtually no local linkages on the one hand to a high level of local linkages on the other. Studies within the smaller developing countries, particularly those with a short history of industrial development, tell a fairly uniform story of shallow and poorly developed supply linkages between local firms and foreign-controlled plants. A common observation is that foreign plants located in export-processing zones (EPZs) are particularly unlikely to develop supplier linkages with the wider economy.[9] In the case of the Mexican *maquiladora* plants, for example, it was found that

> Very little Mexican content goes into *maquila* production. In fact, it has not amounted to 2 per cent of total purchases of material inputs during the 1980s and actually dipped below 1 per cent in 1985 . . . Most products sourced by *maquiladoras* in Mexico are low value-added products and, as opposed to many of the *maquila* products themselves, require very little technological sophistication. The picture is bleaker when we concentrate on the border region where products supplied were virtually devoid of any requirement for modern technology or design skills. *Maquiladoras* are much less likely to source high or middle-tech items in Mexico . . . and even less likely to do so in the northern border area.
>
> (Fuentes *et al.*, 1993, pp. 172, 174)

The cause appears to be the reluctance of corporate (rather than local) purchasing managers to buy Mexican inputs.

In Indonesia, the pattern of local purchasing was more varied:

> only the firms in the automotive industry . . . have been increasing the local content of the final goods through the mandatory increase of purchases of locally made parts and components. In contrast, in the case of pharmaceuticals, virtually all basic raw materials had to be purchased from the [TNCs] . . . In the case of food processing and chemicals companies, various raw materials were procured locally, but in general these linkages did not involve an appreciable increase in the technical capabilities of local suppliers.
>
> (Thee, quoted in Hill, 1993, p. 210)

Hill (1993, p. 210) suggests that although

> this aspect of [TNCs'] contribution to Indonesian development is rather small . . . it can be expected to develop over time as the quality of the potential pool of subcontractors improves and on the assumption that the Government continues with its programme of trade policy reform. TNCs have essentially responded to market signals rather than being the primary cause of the problem.

However, there are circumstances where there may be quite a high level of *local sourcing* but which does not involve *local firms*. Research in the Johor Bahru region of southern Malaysia, for example, has shown that although the new foreign manufacturing plants established there 'are sourcing a large part of their inputs in Johor . . . the regional effect is confined to foreign, mainly Japanese and Singaporean, suppliers. As a result, the linkages of the new manufacturing plants are only in part beneficial to the local economy' (van Grunsven, Egeraat and Meijsen, 1995, p. 3).

The situation in the older-industrialized countries is rather different. Often, of course, foreign firms have been in operation for so long that they are an almost indistinguishable part of the landscape and have developed dense networks of local supplier linkages over a long period of time. IBM, for example, claims to have some 11,000 suppliers in the United Kingdom. Even some of the more recent arrivals have gradually built up local supply networks. But, as the experience of Sony in the United Kingdom shows, it may take up to ten years to do so. One reason for the generally low level of local supply linkages of Japanese firms in Europe as a whole, therefore, is that many of them are very recent arrivals. On the other hand, it is by no means inevitable that TNCs in industrialized countries will develop strong local linkages. In Canada there have been persistent criticisms of the failure of many US firms to source their major inputs from Canadian suppliers. The extent of local supply links of foreign electronics plants in Scotland is also small (Turok, 1993).

Apart from the extent of linkages created by foreign firms in a host economy there is also the question of their *quality* and the degree to which they involve a beneficial transfer of technology (either production or organizational) to supplier firms. A common criticism is that many TNCs tend to procure only 'low-level' inputs from local sources, for example cleaning services and the like. This may be because of deliberate company policy to keep to established suppliers of higher-level inputs or because such inputs are simply not available locally (or are perceived not to be so). Where development of higher-level supply linkages occurs there does seem to be a positive effect on supplier firms. For example, the rigorous specifications laid down by Japanese electronics firms and by American firms such as IBM have undoubtedly had a beneficial technological spin-off on their UK suppliers. In these cases, at least, there is evidence of technological linkages increasing over time. Table 8.1 summarizes the differences between 'dependent' and 'developmental' scenarios. Clearly, from a host-country perspective the aim must be to achieve a linkage structure which is developmental. Much may depend upon its bargaining power (see pp. 270–5).

Table 8.1   Dependent and developmental linkage structures

Attribute	Dependent structure	Developmental structure
Form of local linkages	Unequal trading relationships Conventional subcontracting Emphasis on cost saving	Collaborative, mutual learning Basis in technology and trust Emphasis on added value
Duration and nature of local linkages	Short-term contracts	Long-term partnerships
Degree of local embeddedness of inward investors	Weakly embedded Branch plants restricted to final assembly operations	Deeply embedded High level of investment in decentralized, multifunctional operations
Benefits to local firms	Markets for local firms to make standard, low-technology components Subcontracting restricts independent growth	Markets for local firms to develop and produce their own products Transfer of technology and expertise from inward investor strengthens local firms
Prospects for the local economy	Vulnerable to external forces and 'distant' corporate decision-making	Self-sustaining growth through cumulative expansion of linked firms
Quality of jobs created	Predominantly low-skilled, low-paid. May be high level of temporary and casual employment	Diverse, including high-skilled, high-income employment

Source: Based on Turok (1993, Table 1).

Industrial structure and entrepreneurship

Although not all TNCs are giant corporations, it is often the case that foreign plants are larger than their domestic competitors. Hence, the entry of a foreign plant into a host economy may have a number of repercussions on the structure of domestic industry, particularly on the competitiveness, survival and birth of domestic enterprises. But, as in other aspects of TNC impact, there is no inevitability about such structural effects. Much will depend upon the specific domestic context itself and on the relative size and market power of the TNC affiliate in that context. In general, the difference between a foreign plant and a domestic enterprise in developed market economies, especially those which are themselves sources of TNCs, will be far less than that in developing countries. The industrial structure of most developing economies tends to be much more strongly *dualistic*, with a small, technologically advanced sector (relatively speaking) which is oriented to the more modern urban market and a technologically less advanced sector characterized by traditional production and attitudes. TNCs in developing countries are, by definition, part of the technologically advanced sector of the host economy.

A major long-term effect of the entry of TNCs into a host economy – both developed and developing – is likely to be an increase in the level of *industrial concentration*. The number of firms is likely to be reduced and the dominance of very large firms increased. Lall (1979, p. 328) suggests two reasons:

> First, regardless of the TNC's market conduct, the *attributes* of these enterprises can raise barriers to entry for local firms: TNCs often introduce advanced, usually larger-scale and more capital-intensive technology; they generally produce a wider, more differentiated and better marketed range of products; they utilise newer managerial and organisational skills; they have better access to financial, technical and marketing resources abroad; and they may be more prepared to challenge 'live-and-let-live' rules of the game observed by local oligopolists than local entrants.
>
> Second, their *conduct* may speed up the process of concentration. TNCs are generally the leading forces in the developed countries in diversifying across industries, in effecting takeovers and in lobbying policy makers, and they may be expected to transfer these highly developed strategies to all the host countries in which they operate. Thus, in LDCs, TNCs may purchase local firms on especially favourable terms because of their strong hold over technology or input markets . . . ; they may be able to outlast local competitors in price-cutting wars because of financial staying power; or they may be able to win more favourable concessions from host governments. The market power and tactics of TNCs may also cause higher concentration by inducing defensive mergers among local firms.

Thus, two important *negative* effects of TNCs on host economies are

- the possible squeezing out of existing domestic firms
- the suppression of new indigenous enterprises.

Both these fears have been voiced especially strongly in the case of developing countries but there is no reason to believe that they do not also apply to particular sectors in developed economies if high TNC penetration has occurred. But, clearly, the less developed the indigenous sector the greater is the likelihood of its being swamped by foreign entry and of local entrepreneurship being suppressed. But we should beware of assuming that the involvement of TNCs in a particular host economy will inevitably destroy or suppress domestic enterprise. There may well be *positive* effects:

- Where substantial local linkages are forged by foreign plants, particularly on the supply side, opportunities for local businesses may well be enhanced. Existing firms may receive a boost to their fortunes or new firms be created in response to the stimulus of demand for materials or components.

- The formation of new enterprises may be stimulated through the 'spin-off' of managerial staff who set up their own businesses on the basis of experience and skills gained in employment with the foreign firm.

Employment and labour issues[10]

For most ordinary people, as well as for many governments, the most important issue in the debate over the TNC is its effect on jobs:

- Does the entry of a foreign-controlled plant create new jobs?
- What kinds of jobs are they?
- Do TNCs pay higher or lower wages than domestic firms?
- Do TNCs operate an acceptable system of labour relations?

Aggregate levels of TNC employment

TNCs employ very large numbers of workers in both developed and developing countries:

> Overall, TNCs are estimated to account directly for a total of over 73 million jobs worldwide, of which over 60 per cent are in parent companies, primarily based in developed countries, and 40 per cent in their foreign affiliates. About 12 million – more than a half of foreign affiliates' total – are directly employed in foreign affiliates in developing countries. However, TNC employment constitutes only a negligible proportion – about 3 per cent – of the world's labour force. Given that TNCs are primarily engaged in capital- and technology-intensive activities, their relatively modest direct contribution to overall employment levels in home and host countries is not surprising. At the same time, TNCs account for about one-fifth of paid employment in non-agricultural activities in developed countries and some developing countries, suggesting that their direct contribution to employment in manufacturing and services is far from negligible.
>
> (UNCTAD, 1994, pp. 163–64)

To these we must add the *indirect* employment in linked firms which are, themselves, not TNCs. UNCTAD estimates that in developing countries between one and two jobs are created indirectly for each direct employee in a TNC. But it is extremely difficult to estimate accurately the level of indirect employment creation because, as Figure 8.3 shows, there are many ways in which such employment may be generated. The number of jobs created *directly* in a particular TNC plant will depend upon

- the *size* of its activities and
- the *technological* nature of the operation; particularly on whether it is capital intensive or labour intensive.

The number of *indirect* jobs created will also depend upon two major factors:

- The extent of *local linkages* forged by the TNC with domestic firms.
- The *amount of income generated* by the TNC and *retained* within the host economy. In particular, the wages and salaries of TNC employees and of those in linked firms will, if spent on locally produced goods and services, increase employment elsewhere in the domestic economy.

A further reason for treating aggregate employment estimates with some caution is they do not show the *net* employment contribution of TNCs to a host economy. In

Employment effects	Definition or illustration
Direct employment effects	Total number of people employed within the TNC subsidiary.
Indirect employment effects	All types of employment indirectly generated throughout the local economy by the TNC subsidiary.
1. *Macro-economic effects*	Employment indirectly generated throughout the local economy as a result of spending by the TNC subsidiary's workers or shareholders.
2. *Horizontal effects*	Employment indirectly generated among other local enterprises as a result of competition with the TNC subsidiary.
a) *Narrow horizontal effects*	Employment indirectly generated among local enterprises competing in the same industry as the TNC subsidiary.
b) *Broad horizontal effects*	Employment indirectly generated among local enterprises active in other industries than the TNC subsidiary.
3. *Vertical effects*	Employment indirectly generated by the TNC subsidiary among its local suppliers and customers.
a) *Backward effects (or linkages)*	Employment indirectly generated by the TNC subsidiary among its local suppliers (of raw materials, parts, components, services, etc.).
b) *Forward effects (or linkages)*	Employment indirectly generated by the TNC subsidiary among its local customers (e.g. distributors, service agents, etc.).
Note:	The above employment effects, if they could be measured, should be calculated in net terms (i.e. gross employment directly or indirectly generated, minus total employment displacement).

Figure 8.3   Direct and indirect employment-generating effects of TNCs
Source: ILO (1984, p. 39)

other words, against the number of jobs *created* in and by TNCs we need to set the number of jobs *displaced* by any possible adverse effects of TNCs on indigenous enterprises. For reasons outlined earlier, domestic enterprises may be squeezed out by the size and strength of foreign branch plants while new firm formation may be inhibited. In these respects, the effect of the TNC may be to displace existing or potential jobs in domestic enterprises. Hence, the overall employment effect of TNCs in host economies depends upon the balance between job-creating and job-displacing forces: net employment contribution of TNC to host economy = $(DJ + IJ - JD)$,
where DJ = number of direct jobs created in TNC;
IJ = number of indirect jobs in firms linked to TNC and in other sectors;
JD = number of jobs displaced in other firms.

Types of employment

The number of jobs created by TNCs in host economies is only part of the story. What kind of jobs are they? Do they provide employment opportunities which are appropriate for the skills and needs of the local labour force? The answer to these questions depends very much on the attributes of the foreign plant (Figure 8.2): the type of industry, the nature of the technology used, the scale of operations and the extent to which the foreign plant is integrated into its parent organization. In particular, where the plant 'fits' into the TNC's overall structure and how much decision-making autonomy it has are key factors. In general, the fact that TNCs tend to concentrate their higher-order decision-making functions and their R&D facilities in the developed economies produces a major geographical bias in the pattern of types of employment at the global scale.

In developing countries, the overwhelming majority of jobs in TNC plants are *production* jobs. The ILO estimates that production workers make up between 60 and 75 per cent of the total employment in TNCs in developing countries. Such figures should cause no surprise in view of the nature of most TNC investment in developing countries, particularly in export-platform activities. In export-processing zones, of course, low-level production jobs, especially for young females, are the rule, although this partly reflects the types of industry which dominate in EPZs. More generally, the ILO suggests that the proportion of higher-skilled workers employed by TNCs in developing countries has tended to increase over time as has the proportion of local professional and managerial staff. Such changes have progressed furthest in the more advanced industrial countries of Latin America and southeast Asia. Even so, the TNC labour force in developing countries remains concentrated in low-skill production and assembly occupations.

Geographical segmentation of high- and low-order occupations (white-collar and blue-collar jobs) in TNCs is not confined solely to that between developed and developing countries. As we saw in Chapter 7, a similar geographical segmentation of TNC functions exists within developed economies. This dichotomy is clearly illustrated in the United Kingdom where the higher-order occupations are disproportionately concentrated in southeast England in and around London. A major criticism of the foreign branch plants located in Scotland, Wales, northern England and Northern Ireland, therefore, is that they employ primarily lower-skilled workers and that they generally lack higher-level employment opportunities. Whether or not better types of job would be available in such areas in the absence of the foreign plants is, of course, an open question.

Wage and salary levels

Related to the type of employment offered by foreign plants is the question of wages and salary levels. In so far as TNCs take advantage of geographical differences in prevailing wage rates between countries they do, in fact, 'exploit' certain groups of workers. The exploitation of cheap labour in developing countries at the expense of workers in developed countries is one of the major charges levelled at the TNC by labour unions in western countries (an issue we will return to in Chapter 13). The major problem here seems to be in relation to conditions in subcontracting companies rather than in the TNCs themselves and there is much controversy over this issue. For example, in the mid-1990s there was much criticism of the exceptionally low wages and poor working conditions in the Asian subcontracting plants making athletic footwear for Nike. The general response of TNCs facing such allegations is that they do not have complete control over what goes on in independent factories although, in the face of these criticisms, many TNCs are now implementing codes of practice to which their subcontractors must conform. However, as far as their *directly owned affiliates* are concerned, the general consensus seems to be that TNCs generally pay either at or above the 'going rate' in the host economy. UNCTAD's (1994, pp. 197, 198) conclusion is that

> Generally speaking, at the aggregate and industry levels, the workforce directly employed by foreign affiliates enjoys superior wages, conditions of work and social security benefits relative to the conditions prevailing in domestic firms. In developed countries, for instance, the average level of wages and salaries in foreign affiliates has been found without exception to be above that in domestic firms, and the gap between affiliates and domestic companies continued to grow during the 1980s . . . [in developing countries] . . . In Indonesia, Malaysia, Peru and Thailand, on average, TNCs pay generally higher wages than local companies.

Where TNCs do pay above the local rate, of course, they may well 'cream off' workers from domestic firms and possibly threaten their survival. This point relates to the kinds of *recruitment policies* used by TNCs. TNCs tend to operate very careful screening procedures when hiring workers. This may well mean that employees for a newly established foreign plant are drawn from existing firms rather than from the ranks of the unemployed. Another aspect of recruitment, at least in some industries, is the extent to which TNCs recruit particular types of workers to keep labour costs low. In the textiles, clothing, and electronics industries, for example, there is a very strong tendency to prefer females to males in assembly processes and, in some cases, to employ members of minority groups as a means of holding down wage costs and for ease of dismissal (see Chapters 9 and 11). But such practices would appear to be specific to particular industries and should not necessarily be regarded as universally applicable to all TNCs in all industries.

Labour relations

In most developing countries labour is either weakly organized or labour unions are strictly controlled (or even banned) by the state. Even in developed economies some major TNCs simply do not recognize labour unions in their operations. But most TNCs, however reluctantly, do accept labour unions where national or local circumstances make this difficult to avoid. Whether labour unions are involved or not, the question of the nature of labour relations within TNCs focuses on whether they are 'good' or 'bad', that is, harmonious or discordant. Some studies suggest that TNCs tend to have better labour relations in their plants than domestic firms; others point to a higher incidence of strikes and internal disputes in TNCs. But it is often difficult to separate out the 'transnational' element. In the case of strikes, for example, it may be plant or firm size which is the most important influence rather than nationality of ownership.

One of the most acute concerns of organized labour is that decision-making within TNCs is too remote: that decisions affecting work practices and work conditions, pay and other labour issues are made in some far-distant corporate headquarters which has little understanding or even awareness of local circumstances. Hamill's (1984) study of labour relations in a sample of thirty foreign TNCs operating in the United Kingdom is a good example which provides some valuable insights into the issues involved. Some labour relations decisions made by TNCs were far more centralized than others in that they were either made directly at corporate headquarters or required its approval. These areas mainly related to the operating costs of the subsidiary and reflected the parent company's concern to control financial and labour costs. However, Hamill also found considerable variation between TNCs in their degree of headquarters' involvement in labour relations. Such variety in TNC labour relations policies can be explained by several factors:

- *Degree of intersubsidiary production integration.* Highly integrated production systems require a greater degree of control and co-ordination:

    If the overseas subsidiary is a vital part of an integrated production system – a system which could be widely disrupted if this particular key plant was to close down – then labour relations problems at this level could severely affect not only the operational efficiency of the local subsidiary itself, but also that of the whole worldwide network. In such circumstances, therefore, the parent company will become closely involved in labour relations developments at this level.

    (Hamill, 1984, p. 33)

- *Nationality of ownership.* Firms tend to transfer at least some of their domestic labour relations practices to their overseas operations.
- *Method of establishment.* Labour relations were more highly centralized in newly established (greenfield) foreign plants than in plants acquired from domestic firms.
- *Performance of the subsidiary.* Poor performance tended to result in a significant increase in corporate involvement in labour relations at the plant. This was especially so where such poor performance was perceived to be caused by labour problems.
- *Relative importance of the parent company as a source of funds for the subsidiary.* A heavy dependence on corporate finance was generally associated with a fairly centralized labour relations policy.

Hence,

general statements cannot be applied to the organisation of the labour relations function within such firms. Rather, different TNCs adopt different labour relations strategies in relation to the environmental factors peculiar to each firm. In other words, *it is the type of transnational under consideration which is important rather than transnationality itself.*

(Hamill, 1993, p. 34)

Table 8.2   Subsidiary strategies and the employment effects of TNCs on host countries

| Subsidiary strategy | Spatial division of labour | Employment effects | | | Labour relations |
		Level of employment	Type of employment	Degree of job security	
'Miniature replica'	Located in close proximity to major markets. Adds to regional imbalances	Employment creation through import substitution and multiplier effects. Restricted long-term employment growth prospects due to limited product/ market role of subsidiary	Low–medium skill content; limited R&D	Depends on market share and market growth. In the longer term, job security is threatened by shift to global/ regional strategies	Host-country industrial relations practices
'Rationalized manufacturer'	Reinforces international division of labour through international sourcing in low-cost countries and location of final assembly close to major markets	Employment-creating effects limited by restricted plant role and high import content in final assembly operations	Low skill content; routine assembly work, 'screwdriver' operations; limited functional responsibility or R&D; limited workforce training	Plant status dependent on international factor cost movements; but possible upgrading of assembly plants over time through widening product/market responsibilies and increased local content	Anti-union practices to reduce threats to co-ordinated and integrated global production system
'Product specialist'	Located mainly in highly developed countries or regions; access to skilled labour	Major positive effect through wide product/ market role of plant	High skill content and decentralization of R&D and other functional activities	Long-term job security due to enhanced plant status	Transfer of parent company industrial relations practices

Source: Based on Hamill (1993, Figure 3.5).

An especially important factor in the employment effects of TNCs on host countries is the particular role played by the foreign subsidiary in the parent company's overall strategy. Table 8.2 provides one perspective on this by relating various aspects of employment impact – level, type, job security, labour relations – to three types of subsidiary role: the miniature replica, the rationalized manufacturer and the product specialist. 'Clearly, from a host country perspective, "product specialists" with enhanced product and market responsibilities are preferable to "miniature replicas" or "rationalized manufacturers"' (Hamill, 1993, p. 72).

Dependence, truncation and hollowing out

It should now be clear from our discussion of individual aspects of TNC impact on host economies that no unequivocal general evaluation can be made. Whether a foreign plant creates net costs or net benefits will depend on the *specific context*: the interaction between the attributes and functions of the plant itself within its corporate system and the nature and characteristics of the host economy. It also depends, critically, on the alternatives realistically available. But what if a host economy – or an important sector within the economy – develops a high level of foreign TNC involvement? Does the presence of a large foreign-controlled sector tip the balance of evaluation? In a long-term sense, the answer to these questions would appear to be 'yes'. Whereas the involvement of some foreign plants in a host economy will have beneficial effects – not only in creating employment but also in introducing new technologies and business practices – overall dominance by foreign firms is almost certainly undesirable from a host-country viewpoint. There are real dangers in acquiring the status of a *branch plant economy*.

Precisely what constitutes an undesirable level of foreign penetration is open to debate. It is made more complex by the fact that a country may be dominated by, and dependent upon, external forces even where there is very little direct foreign investment in the economy. This may occur, for example, where a large segment of an economy is engaged in subcontracting work for foreign customers. Here, however, our concern is with the effects of a large direct foreign presence. Until recently, most of the debate was focused upon developing countries and formed part of the broader dependency debate. But such concern is no longer confined to developing countries. Most obviously, those developed economies which have a very high level of foreign direct investment also share the same kinds of problem. Even in those developed economies where, overall, the degree of foreign penetration is relatively low the dominance of specific sectors by foreign firms is beginning to raise concern. The current controversy over foreign (especially Japanese) penetration of the US economy is a good example of this.

The potential problems of a high level of foreign penetration of a host economy can be summarized under two closely related headings:

● dependence
● truncation or 'hollowing out'.

A major consequence of a high level of dependence on foreign enterprises is a reduction in the host country's sovereignty and autonomy: its ability to make its own decisions and to implement them. At the heart of this issue are the different – often conflicting – goals pursued by nation-states on the one hand and TNCs on the other. Each is concerned to maximize its own 'welfare' (in the broadest sense). Where much of a host country's economic activity is effectively controlled by foreign firms, non-national goals may well become dominant. It may be extremely difficult for the host government to pursue a particular economic policy if it has insufficient leverage over the dominant firms. The tighter the degree of control exercised by TNCs within their own corporate hierarchies and the lower the degree of autonomy of individual plants the greater this loss of host-country sovereignty is likely to be. In the *individual* case this may not matter greatly but where such firms *collectively* dominate a host economy or a key economic sector it most certainly does matter.

Perhaps the most significant aspect of dependence upon a high level of foreign direct investment is that of *technological dependence*. This is

> the continuing inability of a . . . country to generate the knowledge, inventions, and innovations necessary to propel self-sustaining growth . . . If a country does not produce its own technology in at least some industries, it is argued, it will suffer slower growth and more disadvantageous terms of trade in the long run . . . Technological dependence may mean slower or 'distorted' growth and reduced economic sovereignty.
>
> (Newfarmer, 1983, pp. 177–79)

A second consequence of a very high level of foreign dominance is the likely truncation, or 'hollowing out', of various facets of the economy. This can be interpreted at two levels. The most obvious level is that of the foreign-controlled *plant* itself. A truncated plant is one which performs only some of the firm's total functions. Truncation is implicit in the very nature of the large, geographically extensive TNC. As we saw in Chapter 7, TNCs characteristically subdivide their internal operations and locate specialized units in different types of location. Essentially, a truncated plant is one which concentrates mainly or exclusively on production activities and which, therefore, lacks the higher-level administrative and R&D functions. Again, the problem arises not from this phenomenon in itself but where *most* foreign-controlled plants in a host economy are truncated. One result will be a highly skewed set of occupational opportunities with relatively few openings for higher-skilled and professional, scientific and technical workers. In addition, truncated plants are likely to be deficient in technological dynamism, relying upon their parent company for innovative activity.

The second level at which truncating effects of foreign dominance may apply is that of an entire *industry* or even of a *host economy as a whole*. 'As the proportion under foreign control rises, an industry becomes a shell. In terms of its products, the industry seems to be complete and comprehensive, but large elements of the production system are missing or deficient' (Britton and Gilmour, 1978, p. 98). The result is a 'hollowed out' economy. More generally, a high level of foreign penetration will tend to exacerbate the negative repercussions of foreign plants discussed earlier. In particular it will tend to inhibit or suppress the development of indigenous firms either because foreign plants create few local linkages or because indigenous firms are squeezed out by the competitive strength of foreign plants which are backed by much larger corporate resources.

To repeat, dependence and truncation of a host economy do not inevitably arise from the entry of a TNC. But they do seem to be a most likely outcome of a high level of foreign penetration. However, none of this is to argue that foreign investment should be avoided completely. What should be avoided by host economies is an excessive degree of foreign penetration. The major need is to avoid *technological dependence*, because it is technology which is the seed corn of future economic development. As Lall (1984, p. 11) points out, the correct strategy then must be a judicious and careful blend of permitting TNC entry, licensing and stimulation of local technological effort. The stress must always be – as it was in Japan – to keep up with the best practice technology and to achieve production efficiency which enables local producers (regardless of their origin) to compete in world markets. This objective will necessitate TNC presence in some cases but not in others.

Making the connections

The organizing framework for our discussion of the impact of TNCs on host economies, shown in Figure 8.2, simply provided a list of the major processes involved. No attempt was made to indicate the interconnections between each of the individual processes themselves. Yet it is most important to stress that they are all *interlinked* in highly complex ways. Figure 8.4 shows some of the major interconnections. It helps us to appreciate just how complex the relationship can be between the operation of a foreign-controlled unit and the host economy in which it is placed. Most studies of the impact of TNCs on host economies tend to focus on just a small segment of these relationships. There is nothing wrong in this; it is often necessary to do so in order to make the research task feasible. But it is important always to set such partial investigations within the overall picture. Ideally, of course, one would wish to be able to obtain empirical data on every one of the linked processes shown in Figure 8.4 so that an estimate of the *net* impact of a TNC on a host economy can be established. Sadly, that remains a Utopian goal, not least because the whole process is *dynamic*, not static. As we observed in Chapter 7, TNCs are continuously adjusting, reorganizing and restructuring their operations. Transnational corporate networks are, by their very nature, in a continuous state of flux.

In this section we have concentrated overwhelmingly on the economic impact of TNCs. However, they are not only extremely significant transmitters of economic change but also of *social, cultural and political* change. This is particularly the case for developing countries. A person employed in a TNC factory or office acquires not just work experience but also a new set of attitudes and expectations. The effect of the employment of women is often to transform, not always favourably, family structures and practices. TNCs introduce patterns of consumption which reflect the preferences of industrialized country consumers. In this respect, the transnational advertising agencies are especially important. They 'are not simply trying to sell specific products in the Third World, but are engineering social, political and cultural change in order to ensure a level of consumption that is "the material basis for the promotion of a standardized global culture"' (Sklair, 1991, p. 149). But it is not just TNCs in advertising and the 'cultural' industries which are important channels for the transmission of a global culture; the production of goods is part of the same process because TNC goods embody particular cultural attributes.

Transnational corporations and home economies

Most of the arguments over the possible costs and benefits of TNCs have been concerned with their effects on host economies. This is not surprising: the geographical destinations of TNC investments are far more diverse and numerous than their origins. As we saw in Chapter 2, most TNCs originate from a relatively small number of developed countries. Of these, the United States has been until recently by far the most significant and it was there that widespread concern about the possible domestic impact of outward investment first surfaced. But the issue is one which faces an

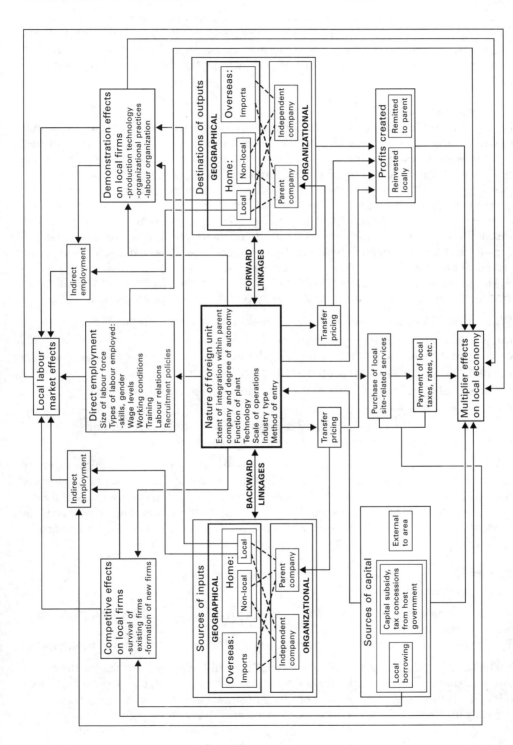

Figure 8.4   Making the connections: the direct and indirect effects of a TNC on a host economy

increasing number of countries as more and more firms have extended their operations across national boundaries. As we have seen at several points throughout this book, a general tendency of the past two decades has been for large firms in particular to expand their overseas operations at a much faster rate than their domestic activities.

Questions and assumptions

What are the implications for the firm's home country of such increased international production? Does it adversely affect the country's economic welfare by, for example, drawing away investment capital, displacing exports or destroying jobs? Or is it an inevitable feature of today's highly competitive global economy which forces firms to expand overseas in order to remain competitive? *Proponents* of overseas investment argue that the overall effects on the domestic economy will be positive, raising the level of exports and of domestic activity to a level above that which would prevail if overseas investment did not occur. *Opponents* of overseas investment, on the other hand, argue that the major effect will be to divert capital which could have been invested at home and to displace domestic exports.

The problem, as Musgrave (1975) pointed out in the US context, is that these opposing viewpoints depend in a critical way on two initial assumptions:

- What would have happened if the investment had not been made abroad?
  - (a) would that investment have been made at home or
  - (b) would the resources which went into the foreign investment have been used in higher levels of consumption and/or public services?
- What would have been the effect of foreign investment on domestic exports?
  - (a) would the foreign sales of the product of the investment have been filled by exports from the home economy in the absence of the investment or
  - (b) would they have been taken over by foreign competitors?

The fundamental problem in assessing the impact of TNCs is, once again, a counterfactual one. The need is to establish what would have happened if such investment had not occurred. But, since the alternative situation cannot be established empirically, we have to make assumptions about that alternative. The critical issue is the extent to which domestic investment could realistically be *substituted* for overseas investment.

Estimating the employment impact of outward investment

Two key processes are involved:

- The *displacement effect*: the loss of domestic activity or employment through overseas investment.
- The *stimulus* which overseas investment has on domestic activity and employment.

The act of establishing an overseas operation will have implications for the home country's balance of payments, through its influence on capital and financial flows and its effects on trade. But the most obvious implication for the average citizen is the effect on employment.[11] Overseas investment may have very significant repercussions on domestic employment. It is this issue which has received most attention in the United States, which has become increasingly contentious in those European

Table 8.3 The potential effects of outward investment on home-country employment

Area of impact	Direct		Indirect	
	Positive	Negative	Positive	Negative
Quantity	Creates or preserves jobs in home location, e.g. those serving the needs of affiliates abroad	Relocation or 'job export' if foreign affiliates substitute for production at home	Creates or preserves jobs in supplier/service industries at home that cater to foreign affiliates	Loss of jobs in firms/ industries linked to production/activities that are relocated
Quality	Skills are upgraded with higher-value production as industry restructures	'Give backs' or lower wages to keep jobs at home	Boosts sophisticated industries	Downward pressure on wages and standards flows on to suppliers
Location	Some jobs may depart from the community, but may be replaced by higher-skilled positions, upgrading local labour market conditions	The 'export' of jobs can aggravate regional/local labour market conditions	The loss of 'blue collar' jobs can be offset by greater demand in local labour markets for high value-added jobs relating to exports or inter-national production	Demand spiral in local labour market triggered by layoffs can lead to employment reduction in home-country plant locations

Source: UNCTAD (1994, Table IV–1).

countries which are also the sources of significant numbers of TNCs and, more recently, in Japan with the acceleration of Japanese overseas investment. But the interpretations of the employment effects of outward investment are diametrically opposed because they are based on totally contrasting assumptions:

- Labour interests are adamant that the overseas activity of TNCs dramatically increase unemployment at home because it is frequently assumed that *all* the activity undertaken overseas could realistically have been retained at home.
- Most business interests, on the other hand, are equally adamant that overseas investment increases, or at least preserves, jobs at home. The business-interest estimates tend to make the opposite assumption: that *none* of the overseas activity could realistically have been retained at home.

Quite clearly, more sophisticated approaches are needed to estimate even the approximate impact of overseas investment on domestic employment.

Table 8.3 summarizes the positive and negative aspects of both direct and indirect employment effects of outward investment in terms of three attributes: quantity of jobs, quality of jobs and the location of jobs. In interpreting this table we need to bear in mind that the precise effects of outward investment on home-country employment are highly contingent on the specific circumstances involved. Unfortunately, there have been relatively few attempts to calculate actual home-country employment changes associated with outward investment.

The pioneer study of this issue was made by Hawkins (1972). He subdivided the possible direct employment effects into four categories:

- The *production-displacement effect* (DE): employment losses arising from the diversion of production to overseas locations and the serving of foreign markets by these overseas plants rather than by home-country plants, that is, the displacement of exports.

- The *export-stimulus effect* (XE): employment gains from the production of goods for export created by the foreign investment which would not have occurred in the absence of such investment.
- The *home-office effect* (HE): employment gains in non-production categories at the company's headquarters made necessary by the expansion of overseas activities.
- The *supporting-firm effect* (SE): employment gains in other domestic firms supplying goods and services to the investing firm in connection with its overseas activities.

Thus, the *net employment effect* (NE) of overseas investment on the home economy is:

$$NE = -DE + XE + HE + SE$$

Hawkins calculated the employment-displacement effect for three differing assumptions of the amount of production which could have remained in the United States if the foreign investment had not occurred for the year 1968. His estimates ranged from a possible gain of 279,000 jobs to a possible loss of 322,000 jobs.

Hawkins's three alternative scenarios were based upon somewhat arbitrary substitution values. A more sophisticated attempt to estimate the employment effects of overseas investment, again in the case of the United States, was made by Glickman and Woodward (1989). They used complex formulae to calculate the displacement and stimulus effects of overseas investment by US firms for the period 1977–86. They estimated that

- 3.3 million jobs were *displaced* by overseas investment;
- 588,000 jobs were *stimulated* by such investment;
- the result was a *net loss* of 2.7 million domestic jobs (i.e. jobs which, in the absence of the overseas investment, would have been created in the United States).

Clearly, within the aggregate job loss figures there were many particular cases of employment growth through the stimulus effect of overseas investment. Often, however, the jobs stimulated are not necessarily in the same industries, occupational groups or geographical locations as the jobs displaced. The adverse employment impact varied substantially from one *industry* to another, being greatest in non-electrical machinery and chemicals, industries in which US overseas investment was particularly high. There was also a differential impact on *different groups of workers*. In general, women workers were harder hit than male workers; black and Hispanic workers more than white workers; blue-collar workers more than higher-skilled white-collar workers.

In other words, *the winners and the losers are not the same*. Two quotations make this point very clearly:

> It becomes a problem of a *structural mismatch* between the jobs eliminated and the jobs created, even if the latter dominate. One of the obvious results is that the jobs that are eliminated will be almost exclusively production jobs . . . Other staff, and occasionally managerial, jobs may be eliminated, but this is less frequent and the reabsorption of such workers in similar jobs in the same firm is much more likely than in the case of production workers . . . The skill mix [of the jobs created] obviously does not conform closely to the jobs eliminated through substitution of foreign production for exports. Thus one tendency, on a priori grounds, is for an expansion of foreign investment to relatively reduce the demand for production workers and to expand the number of clerical, professional, skilled, and managerial workers within the same industry.
>
> (Hawkins and Jedel, 1975, cited in ILO, 1981b, p. 60, emphasis added)

On computer tapes, jobs may be interchangeable. In the real world they are not. A total of 250,000 new jobs gained in corporate headquarters does not, in any political or human sense, offset 250,000 old jobs lost on the production line. When Lynn, Massachusetts becomes a ghost of its former self, its jobless citizens find little satisfaction in reading about the new headquarters building on Park Avenue and all the secretaries it will employ. The changing composition of the workforce and its changing geographical location brought about by the globalization of US industry are affecting lives of millions of Americans in serious and largely unfortunate ways.

(Barnet and Muller, 1975, p. 302)

Again, however, it must be emphasized that the volume, nature and location of employment change will not necessarily be the same in other circumstances. The examples quoted above were time and place specific.

Is overseas investment discretionary or obligatory?

It is virtually impossible to say, with certainty, that overseas investment could equally as well have been made in the firm's home country. We can make various assumptions about what might have happened, but that is all. Ultimately, the key lies with the *motivations* which underlie specific investment decisions. As profit-seeking organizations, firms invest overseas for a whole variety of reasons, for example, to

- gain access to new markets
- defend positions in existing markets
- circumvent trade barriers
- diversify the firm's production base
- reduce production costs.

It might be argued that foreign investment which is undertaken for *defensive* reasons – to protect a firm's existing markets, for example – is less open to criticism than *aggressive* overseas investment. The argument in the case of defensive investment would be that in the absence of overseas investment the firm would lose its markets and that domestic jobs would be lost anyway. Such investment might be made necessary by the erection of trade barriers by national governments, by their insistence on local production, or by the appearance of competitors in the firm's overseas markets. But, presumably, defensive investment might also include the relocation of production to low-wage countries in order to remain competitive in cost terms. Here, the alternative might be the introduction of automated technology in the domestic plant which would also lead to a loss of jobs.

In this context, the product life-cycle model has obviously formed the basis of much of the argument regarding the probable effects of overseas investment on domestic employment.[12] As a product moves through the stages of the cycle the relocation of production to overseas locations, first, to maintain access to markets and, subsequently, to hold down production costs is depicted as being inevitable (see Chapter 6). But, as we have seen, the reasons for international production are far more varied and complex than the product-cycle model suggests. It is, therefore, a matter of fine judgement to draw the line between defensive and aggressive foreign investment. Even the product cycle can be interpreted in both ways.

Although there may well be some clear-cut cases – particularly where access to markets is obviously threatened or where proximity to a localized material is mandatory – there will inevitably be many instances where there is substantial disagreement over the need to locate overseas rather than at home. The various interest groups will have different perceptions of the situation. Thus,

> Multinational management tends to explain investments abroad as reflecting imperative requirements of markets and competition, i.e. global change factors calling for such adjustment by multinational enterprises in the interest of efficient production and ultimate survival. Its critics see the search for profits as the driving factor.
>
> The available evidence is mixed as to the extent to which multinational enterprises do, or can, in fact exercise discretion in decisions to transfer production abroad on a large scale, especially to low-wage developing countries. Instances can be found of multinational enterprises apparently having set up off-shore sourcing subsidiaries, in Hong Kong, Singapore or elsewhere, before they were obliged to do so in self-defence, by competition in home or export markets, either from local firms or from multinational enterprises based in other home countries. On the other hand, examples can be found where such foreign subsidiaries were set up only after substantial competition, for instance from national enterprises in NICs, had been encountered in home or third markets. Still other cases exist – notably in the US television industry – where some TNCs resisted production transfers abroad until they were thrust upon the firm by imminent bankruptcy. The problem for the researcher is that, in all such cases, observable events must be compared with hypothetical, non-observable alternatives. *Thus a clear-cut general answer as to the degree to which TNCs have been wilful discretionary agents of production transfers abroad is not possible. For this purpose the actual decision-making processes of individual multinationals . . . would need to be investigated through detailed case studies.*
>
> (ILO, 1981b, pp. 69–70, emphasis added)

It is this kind of complexity that makes the adoption of national policies towards outward investment so problematical. A comprehensive restriction on overseas investment by domestic firms might make a national economy worse, rather than better, off. On the other hand, wholesale outflow of investment must surely be detrimental to home-country interests. If there is to be a national policy in this area, therefore, it must be both well informed and selective. It is worth recalling that Japanese policy changed from that of tight restriction on outward investment to one of carefully encouraging certain types of overseas investment as a matter of national policy.

The bargaining relationship between TNCs and states[13]

The relationship between TNCs and states is especially complex where TNCs are pursuing a strategy of international integration of their activities in which individual units in a specific host country form only a part of the firm's international or even global operations. Such integration 'not only creates a need to shift the host government–TNC relationship from regulation to negotiation, it also forces a much sharper recognition of relative bargaining power and of the evolution of relative power over time' (Doz, 1986a, p. 233). In the final analysis, therefore, the relationship between TNCs and host countries (actual or potential) revolves around their relative bargaining power.

The bargaining process is immensely complex and varies enormously from one instance to another. However, it is important to understand some of the general features of the process. For the sake of simplicity we assume that a host country can be regarded as a single entity in the bargaining relationship. In fact, of course, many competing interest groups are involved, including domestic business interests which may have varying attitudes towards TNCs.

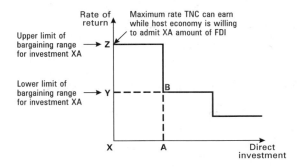

Figure 8.5   A simplified model of the bargaining relationship between a TNC and a host country
Source: Based on Nixson (1988, Figure 1)

Figure 8.5 is a highly simplified, hypothetical example. The vertical axis of the graph shows the rate of return which a TNC may seek for a given level of investment (XA) on the horizontal axis. The bargaining range for this level of TNC investment is shown to vary between

1) a lower limit (XY), which is the minimum rate of return that the TNC is prepared to accept for the amount of investment XA;
2) an upper limit (XZ), which is determined by the cost to the host economy of either developing its own operation, finding an alternative investor or managing without the particular advantages provided by the TNC. XZ is the maximum return the TNC can make for the amount of direct investment (XA) permitted by the host economy. It is in the interests of the TNC to try to raise the upper limit (XZ); conversely, it is in the interests of the host economy to try to lower that upper limit: 'The higher is the cost to the host economy of losing the proposed [investment], the greater are the possibilities for the TNC of setting the bargain near the maximum point' (Nixson, 1988, p. 379).

On the other hand, the more possibilities the host economy has of finding alternatives the greater are its chances of lowering that upper limit. The greater the competition between TNCs for the particular investment opportunity the greater are the opportunities for the host country to reduce both the upper and lower limits: 'In addition, the host economy has an interest in lowering the lower limit, through the creation of an advantageous "investment climate" (political stability, constitutional guarantees against appropriation, etc.) which might persuade the TNC to accept a lower rate of return' (Nixson, 1988, p. 380).

Competitive bidding for investment
The greater the competition between potential host countries for a specific investment the weaker will be any one country's bargaining position because countries will tend to bid against one another to capture the investment. Indeed, one of the most striking developments of the last two or three decades has been the enormous intensification in *competitive bidding* between states (and between communities within the same state) for the relatively limited amount of internationally mobile investment. Such 'cut-throat' bidding undoubtedly allows TNCs to play off one state against another to gain the highest return for their investment. In fact, much of the actual financial investment may be provided by the host government itself in the form of various kinds of financial and fiscal (tax) deals as well as in the form of physical and social infrastructure. A recent study by UNCTAD found that only 4 countries out of 103 did not offer some kind of fiscal incentive to inward investors during the early 1990s while financial incentives were offered in 59 out of 83 countries surveyed. Table 8.4 shows just a few examples of the kinds of deals that have been made by states in their scramble to capture major investments. It is no accident that most of the examples are

Table 8.4 The costs of attracting foreign investment: some examples of incentive packages

Location	Company	Company investment ($ millions)	State investment ($ millions)	State's financial investment per employee ($)
Smyrna, TN (USA) [1983]	Nissan (Japan)	745–848	22 Road access 7.3 Training 33 Total	25,384
Flat Rock, MI (USA) [1984]	Mazda (Japan)	745–750	19 Training 5 Road improvement 3 On-site works 21 Economic development grant/loan 5 Water system 48.5 Total	13,857
Georgetown, KT (USA) [1985]	Toyota (Japan)	823.9	12.5 Land purchase 20 Site preparation 47 Road improvement 65 Training 5.2 Toyota families' education 149.7 Total	49,900
Tuscaloosa, AL (USA) [1993]	Mercedes-Benz (Germany)	300	68 Site development 77 Infrastructure 15 Private sector/goodwill 90 Training 250 Total	166,667
Spartenburg, SC (USA) [1994]	BMW (Germany)	450	130 Total	108,333
Setubal (Portugal) [1991]	Auto Europa (Ford/VW) (USA/Germany)	2,603	483.5	254,451
West midlands (UK) [1995]	Ford/Jaguar	767	128.72	128,720
Northeast England [1994/95]	Samsung (Korea)	690.3	89	29,675
Lorraine (France) [1995]	Mercedes-Benz (Germany)	370	111	56,923

Source: Based on UNCTAD (1995, Table VI.3).

drawn from the automobile industry which, as we shall see in Chapter 10, is such a major job provider.

Table 8.4 shows both the size of the company's investment and the amount of financial incentive provided by the state. Most interestingly, in the final column, it shows the state investment per job created. The sheer size of the numbers in the final column is staggering. It is not surprising that there was widespread consternation in neighbouring southern states when Alabama offered the equivalant of $167,000 per job created to attract the Mercedes-Benz plant for which virtually all the southern states had been competing. These figures provide a graphic demonstration of the bargaining power of major TNCs to offset some of their investment costs on to the state. But it is not only new TNC investments which are involved in this process. It has become increasingly common for TNCs for attempt to lever various kinds of state subsidies in order to persuade them to keep a plant in a particular location.

Two recent examples from the European automobile industry illustrate this practice. During early 1997, Ford announced plans greatly to reduce the labour force at its Halewood plant in Merseyside, England and made it clear that the actual future of the plant, even at a reduced scale, might be in doubt unless the UK government came up with an appropriate deal. The government duly did just that. At around the same time, the French firm Renault announced plans to close down its Belgian plant (with the loss of 3,100 jobs) but also tried to persuade the Spanish government to provide state funds to modernize its plant at Valladolid. Not surprisingly, the European Union did not approve of such an attempt to use state aid to switch production between different European countries. Efforts are now being made at the European Union level to prevent firms from exploiting differences in the map of investment incentives and subsidies.

Like firms, therefore, states engage in *price competition* in their attempts to capture a share of the market for mobile investment. Like firms, states also engage in *product differentiation* by creating particular 'images' of themselves[14] such as the strategic nature of their location (it is amazing how many places project themselves as being at the 'centre' of the world), the attractiveness of the business environment, the quality of the labour force and so on. Hence, the figures shown in Table 8.4 do not reveal the full extent of states' incentives to attract and retain TNC investment.

States undoubtedly face a major dilemma. If they do not join the bidding battle they face the probability of being left out of TNCs' investment plans:

> incentives are not among the main determinants of FDI locational decisions. Nevertheless, competition among countries to attract and keep investment through incentives is strong and pervasive. This is partly so because, other things being equal, incentives can induce foreign investors towards making a particular locational decision by sweetening the overall package of benefits and hence tilting the balance in investors' locational choices. Incentives can be justified if they are intended to cover the wedge between the social and private rates of return for FDI undertakings that create positive spillovers. However, incentives also have the potential to introduce economic distortions (especially when they are more than marginal) . . . It is not in the public interest that the cost of incentives granted exceeds the value of the benefits to the public. But, as governments compete to attract FDI, they may be tempted to offer more and larger incentives than would be justified, sometimes under pressure from firms that demand incentives to remain in a country.
>
> (UNCTAD, 1995, p. 299)

Relative bargaining power

The extent to which a state feels the need to offer large incentives will depend on its relative bargaining strength in any specific case; conversely, the extent to which a TNC is able to obtain such incentives will depend on its relative bargaining strength. The outcome will depend on a number of factors which have been neatly summarized by Gabriel (1966). The price which the *host country* will ultimately pay is a function of

- the number of foreign firms independently competing for the investment opportunity
- the recognized measure of uniqueness of the foreign contribution (as against its possible provision by local entrepreneurship, public or private)
- the perceived degree of domestic need for the contribution.

The terms the *TNC* will accept are a function of

- the firm's general need for an investment outlet
- the attractiveness of the specific investment opportunity offered by the host country, compared to similar or other opportunities in other countries

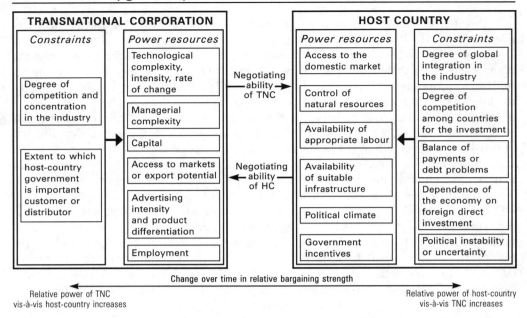

Figure 8.6 Components of the bargaining relationship between TNCs and host countries
Source: Based on material in Kobrin (1987)

- the extent of prior commitment to the country concerned (e.g. an established market position).

Figure 8.6 sets out the major components of the bargaining relationship between TNCs and host countries. Both possess a range of 'power resources' which are their major bargaining strengths. Both operate within certain constraints which will restrict the extent to which these power resources can be exercised. The relative bargaining power of TNCs and host countries is a function of three related elements:

- The *relative demand* by each of the two participants for resources which the other controls.
- The *constraints* on each which affect the translation of potential bargaining power into control over outcomes.
- The *negotiating status* of the participants.

Figure 8.6 suggests that, in general, host countries are subject to a greater variety of constraints than are TNCs, a reflection of the latter's greater potential flexibility to switch its operations between alternative locations. Nevertheless, the extent to which a TNC can implement a globally integrated strategy is constrained by nation-state behaviour. Where a company particularly needs access to a given location and where the host country does have leverage, then the bargain which is eventually struck may involve the TNC in making concessions. In general, the scarcer the resource being sought (whether by a TNC or a host country) the greater the relative bargaining power of the controller of access to that resource and *vice versa*.

For example, states which control access to large, affluent domestic markets have greater relative bargaining power over TNCs pursuing a market-oriented strategy than states whose domestic markets are small. It is in this kind of situation that the host country's ability to impose performance requirements – such as local content levels – on foreign firms is greatest. It may also give the host country sufficient leverage to persuade the inward investor to establish higher-level functions such as research and development facilities. On the other hand, the nature of the domestic market may not be a consideration for a TNC pursuing the kinds of integrated production strategy shown in Figure 7.7. Where, for example, the TNC's need is for access to low-cost labour which is very widely available then an individual country's bargaining power will be limited. On the whole, cheap labour is not a scarce resource at a global scale.

On the other hand, states may well have the ability to inhibit the achievement of globally integrated strategies by TNCs: 'host countries wield the power to limit the extent of, or even to dismantle, the MNC integrated manufacturing and trade networks with more regulations and restrictions on foreign investments and market access' (Doz, 1986b, p. 39). At the extreme, of course, both institutions – TNCs and governments – possess sanctions which one may exercise over the other. The TNC's ultimate sanction is not to invest in a particular location or to pull out of an existing investment. A nation-state's ultimate sanction against the TNC is to exclude a particular foreign investment or to appropriate an existing investment.

The problem, of course, is that the whole process is *dynamic*: the bargaining relationship changes over time as the bottom section of Figure 8.6 suggests. In most studies of TNC–state bargaining the conventional wisdom is that of the so-called 'obsolescing bargain' in which

> once invested, fixed capital becomes 'sunk', a hostage and a source of bargaining strength. The high risk associated with exploration and development diminishes when production begins. Technology, once arcane and proprietary, matures over time and becomes available on the open market. Through development and transfers from FDI the host country gains technical and managerial skills that reduce the value of those possessed by the foreigner.
>
> (Kobrin, 1987, pp. 611–12)

In this view, after the initial investment has been made, the balance of bargaining power is believed to shift from the TNC to the host country, in other words, to move to the right in Figure 8.6. But although this may well be the case in natural resource-based industries it is far less certain that this applies in those manufacturing sectors in which technological change is frequent and/or where global integration of operations is common. In such circumstances, 'the bargain will obsolesce slowly, if at all, and the relative power of MNCs may even increase over time' (Kobrin, 1987, p. 636). In so far as such industries are becoming increasingly important in the world economy, the associated shift in bargaining power towards the TNC will pose major problems for host countries.

Conclusion

The aim of this chapter has been to explore the inter-relationships between TNCs and states which form such an important nexus within the global economy. We have emphasized, in particular, the fact that TNC–state relationships may be both conflictual and co-operative; that TNCs and states may be rivals but, at the same time, they may collude with one another. In a real sense, states need firms to help in the process of material wealth creation while firms need states to provide the necessary supportive infrastructures, both physical and institutional, on the basis of which they can pursue their strategic objectives. Within this general framework, we focused in this chapter on two major aspects of TNC–state relationships.

First, we explored the question of the costs and benefits for both host and home countries of TNC activity. In both cases, the analytical problem is the same: it is a counterfactual problem. We cannot unambiguously establish what the situation would be like *if* the TNC were *not* involved in a particular case. Assumptions have to be made as to what the alternative situation would realistically be. The 'bottom line' is the *net effect* which takes into account the opportunities forgone by the presence or absence of the TNC. But the situation has become vastly more complex in today's highly inter-connected global economy. From a host economy's viewpoint, could the particular item of technology, the particular level of employment and so on, be created without the involvement of the TNC? In some cases, the answer will undoubtedly be 'yes': in others, the answer will just as undoubtedly be 'no'. Similarly, from a home economy's view-point, what would be the effect if the country's firms were not involved in overseas activities? Can a firm opt out of international operations in today's increasingly global production environment? Would jobs at home be lost anyway even if the firm did not invest overseas? Would the economy be better or worse off? Everything will depend, as we have seen, on whether the overseas investment is obligatory or discretionary, but even here differences of viewpoint will exist. What is quite clear is that *TNCs tie national and local economies more closely into the global economy.*

Secondly, we explored the bargaining processes between TNCs and states as each tries to gain maximum advantage from the other in pursuit of their own strategic objectives. States have become increasingly locked into a cut-throat competitive bidding process for investments; a process which provides TNCs with the opportunity to play off one bidder against another. But the extent to which this happens depends upon the specific *relative bargaining power* of TNCs and states. There is little doubt that there has been a shift in the relative power of TNCs and states but the position is far less straightforward than has often been supposed. Each bargaining process is different and highly contingent on the specific circumstances involved. That is why it is so difficult to make broad generalizations about the relative 'balance of power' between TNCs and states. Stopford and Strange's (1991, pp. 215–16, emphasis added) view is that

> governments *as a group* have indeed lost bargaining power to the multinationals . . . Intensifying competition among states seems to have been a more important force for weakening their bargaining power than have the changes in global competition among firms. This is not to deny that governments can maintain considerable power in their dealings with any one foreign firm. The reasons lie in the competition for world market shares. It seems to us that the changes in the production structure . . . have altered the relative importance of those factors over which states had most control, as compared with those over which firms had most control . . . [however] . . . does it follow that firms *as a group* have increased their bargaining power over the factors of production? Here, the argument becomes complex, for the power of the individual firm may be regarded as having also fallen as competition has intensified. New entrants have altered the rules and offer governments new bargaining advantage. *One needs to separate the power to influence general policy from the power to insist on specific bargains.*

It is this question of the 'specific bargain' which needs to form the basis of the much-needed research in this area.

Notes for further reading

1. See Reich (1989).
2. Doz (1986b) provides an extensive discussion of these issues.
3. See, for example, Yoffie and Milner (1989), Picciotto (1991), Rugman and Verbeke (1992).
4. The annual *World Investment Report* published by UNCTAD devotes a major section each year to specific aspects of TNC impact. For example, the 1994 report was concerned specifically with employment issues. These reports provide an excellent source of up-to-date material.
5. The study was conducted by Ernst & Young and reported in *The Financial Times* (23 November 1995).
6. See UNCTAD (1995, Chapter III).
7. Reviews of empirical studies from a variety of developing countries are provided by UNCTC (1988), Hill (1990).
8. The research literature on the environmental dimension of the TNC is not large. See, for example, Pearson (1987), Leonard (1988), UNCTAD (1993a). However, specific issues have attracted much media attention (e.g. the major disasters at the Hoffman LaRoche plant at Seveso, Italy in 1976, at the Union Carbide plant in Bhopal, India in 1984, the environmental conflict involving Shell in Nigeria in 1995).
9. Relevant studies include Warr (1987), ILO (1988), Sklair (1989), Fuentes *et al.* (1993).
10. The most comprehensive reviews of the employment impact of TNCs can be found in studies by the ILO (Bailey, Parisotto and Renshaw, 1993) and by UNCTAD (1994).
11. See ILO (1981a; 1981b), UNCTAD (1994).
12. The influence of product-cycle thinking on views about the employment-displacing effects of overseas investment is discussed by Enderwick (1982).
13. The literature on the bargaining relationships between TNCs and states is fairly limited. Useful contributions are provided by Gabriel (1966), Guisinger (1985), Poynter (1985), Doz (1986a), Encarnation and Wells (1986), Kobrin (1987), Behrman and Grosse (1990), Stopford and Strange (1991), Grosse and Behrman (1992), UNCTAD (1995, Chapter VI).
14. This is a point discussed by Guisinger (1985) and Encarnation and Wells (1986).

PART III

Global shift: the picture in different sectors

INTRODUCTION

The choice of case-study sectors

The six chapters of Part II set out the general ways in which the major institutions and processes of change shape and reshape the global economic map. The central theme was that the globalization and internationalization of economic activity arise from the dynamic interplay between three sets of processes: the strategies of TNCs, the strategies of national governments and the character and direction and nature of technological change. Particular emphasis was placed upon the interconnections and interdependencies between economic activities and upon the networks of organizational relationships which form the fabric of the global economy.

Part II, therefore, was concerned with the *general* operation of these complex and dynamic processes. But precisely how they operate and the *specific* outcomes produced vary substantially between different types of economic activity. TNCs are more directly involved and influential in some industries than in others. Their strategic orientations may vary both within and between industries. The importance of strategic alliances and the nature of materials and components sourcing may differ from one sector to another. Similarly, the involvement of national governments is not uniform across all industries. Some industries are regarded as being more important to governments than others. The precise form of government involvement may differ by sector. The nature and the intensity of the interactions between TNCs and governments may well vary from one industry to another. Technology and technological change are also far from uniform across economic sectors. Indeed, one of the major distinguishing features of an industry is its specific product and process technology. Even the general technologies of transport and communication, which affect all economic activities, may have differential effects.

For all these reasons it is important to look at *specific* cases to see just how the general processes operate and interact to produce particular outcomes. The globalizing of economic activity is not a uniform process. In Part III, therefore, we examine four sectors – textiles and clothing, automobiles, electronics, services – each of which has experienced major global shifts in recent decades. It is not suggested that these four sectors are in any way 'typical' or 'representative' of all types of industry. Rather, the purpose is to show how the processes of change combine to create particular organizational and geographical forms at the global scale. Each sectoral study throws a slightly different light on the processes of change.

The treatment of each case study follows a broadly similar pattern:

- general significance of the sector and the nature of its particular production chain
- global shifts in production and trade
- the changing pattern of demand
- technology, technological change and the production process
- government policies
- corporate strategies
- jobs.

In this way it becomes possible to see more clearly the similarities and differences between the four sectors. It also makes it easier to compare any one particular characteristic (for example, the importance of strategic alliances or the nature of government policy) across all four sectors.

CHAPTER 9

'Fabric-ating fashion': the textiles and clothing industries

Introduction[1]

The textiles and clothing industries were the first manufacturing industries to take on a global dimension. They are the most geographically dispersed of all industries across both developed and developing countries. They are organizationally very complex, containing elements of both very new and very old organizational practices. They are changing very rapidly in their organization, their technology and in their geography. Indeed, global shifts in the textiles and clothing industries exemplify many of the intractable issues facing today's world economy, particularly the trade tensions between developed and developing economies. These changes continue to cause intense political friction.

The textiles industry was the archetypal industry of the first industrial revolution of the eighteenth and nineteenth centuries in Britain. To a considerable extent, the first industrial revolution was a textiles revolution (see Figure 5.1): the cotton textiles industry was the primary engine of growth. Its major geographical centre of production – Lancashire – became the exemplar of the nineteenth-century industrial landscape with its oft-described 'dark satanic mills'. Its marketing capital, Manchester, became the first global industrial city – the 'Cottonopolis' – of an industrial system whose tentacles spread across the globe. All the 'newly industrializing countries' of the nineteenth century – the United States, Germany, France, The Netherlands – also developed large textiles industries employing many hundreds of thousands of workers, often in strongly localized geographical clusters. A similarly concentrated pattern occurred in the rather later development of a factory clothing industry in the second half of the nineteenth century.

The sheer strength – both economic and political – of the British textiles industry in the nineteenth century effectively strangled the development of an indigenous textiles industry in the major colonies, especially India.[2] But such dominance could not last, particularly in an industry which was so ideally suited to the early stages of industrialization. It was possible to produce textiles and clothing using relatively simple technologies and low-skill labour. The traditional craft skills of hand spinning, weaving and sewing were a ready basis for larger-scale industrial application. The capital investment required was relatively modest compared with many other types of industry. Where local supplies of the raw materials were also available there was an even more obvious case for the development of a textiles industry. But materials availability is not a prerequisite; cotton and similar materials are not expensive or difficult to transport. Lack of indigenous supplies has not inhibited the development of highly successful textiles industries in many parts of the world.

Globally the textiles and clothing industries are very large-scale employers of labour. Some 13 million workers are directly employed in the textiles industry worldwide and a further 6 million in clothing manufacture. But these figures grossly understate the actual numbers involved. Countless unregistered workers, employed both in factories and at home, need to be added to reach a true picture, especially in the clothing industry. Despite the changes wrought by new technologies, corporate rationalization and competition from new producers, the textiles and clothing industries continue to be important sources of employment in the developed economies. In particular they employ many of the more 'sensitive' segments of the labour force: females and immigrants, often in tightly localized communities. In the developing countries the industries employ predominantly young female workers in conditions often leaving a great deal to be desired and which recall those of the sweatshops and mills of nineteenth-century cities in Europe and North America. This became a major political issue in the late 1990s as pressure built up for clothing firms to ensure decent working conditions and wage levels in their factories.

The importance of textiles and clothing as a basis for today's newly industrializing and less industrialized countries, together with their continued, though much diminished, importance in the older industrial economies, have made these industries into an international political football. They are the subject of fierce political controversy between developed and developing countries and, increasingly, between the developed economies themselves. Textiles and clothing are the only industries in the world economy to which special international trade restrictions apply through the *Multi-Fibre Arrangement (MFA)*.

The textiles–clothing production chain

The textiles and clothing industries form part of a larger production chain,[3] as Figure 9.1 shows. Each stage has its own specific technological and organizational characteristics and particular geographical configuration. Each has been changing very substantially in recent decades. The textiles industry itself consists of two major operations: the preparation of yarn and the manufacture of fabric. Both stages are performed by firms of all sizes, from the very small domestic enterprise to the very large subsidiary of the transnational corporation. The general trend, however, has

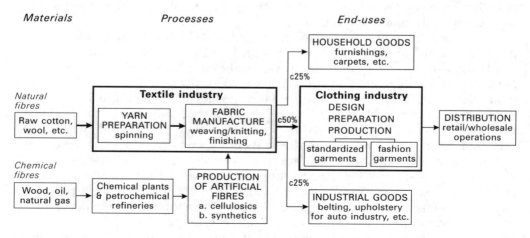

Figure 9.1   The textiles–clothing production chain

been for textile manufacturing to become more and more capital intensive and for large firms to be increasingly important. The output of the textiles industry goes to three types of end-use, of which the clothing industry is by far the most important. Approximately 50 per cent of all textiles production goes into the manufacture of garments.

Despite some recent changes, the clothing industry remains far more fragmented organizationally than the textiles industry and is less sophisticated technologically. It is also an industry in which subcontracting is especially prominent. Very often the design and even the cutting processes are performed quite separately from the sewing process, the latter being particularly amenable to international subcontracting. The clothing industry itself produces an enormous variety of often rapidly changing products. However, the most important distinction is between *standardized garments* on the one hand and *fashion garments* on the other. Finally, although not part of the production sequence itself, the role of the distributors of garments – particularly the retailers – is of considerable and growing importance. The increasing dominance of much retail trade by very large firms has enormous implications for the organization and for the global geography of clothing manufacture. It is in the clothing industry, in fact, that *buyer-driven* production chains are especially prominent.

A development sequence

Figure 9.2 suggests a generalized sequence of development through which individual producing countries appear to have passed. Six stages are shown, beginning with the

	Type of production	**Trade characteristics**	**Examples of countries**
Stage 1	Simple fabrics and garments manufactured from natural fibres.	Production oriented to domestic market. Net importers of fibre, fabric and clothing.	Least developed.
Stage 2	Production of clothing for export. Mostly standard items or those requiring elaborate 'craft' techniques.	Export of clothing to developed country markets on basis of low price.	Less advanced Asian, African, Latin American.
Stage 3	Increase in quantity, quality and sophistication of domestic fabric production. Expansion of clothing sector with upgrading of quality. Development of domestic fibre manufacturing.	Much increased international involvement in export of fabric, clothing and even of synthetic fibres.	More advanced ASEAN, and eastern European. China starting to enter this stage.
Stage 4	Further development and sophistication of fibre, fabric and clothing production.	Full-scale participation in international trading system. Substantial trade surpluses.	Taiwan, South Korea, Hong Kong.
Stage 5	Output of textiles and clothing continues to increase but employment declines. Increased capital intensity and specialization.	Facing increased international competition.	Japan, United States, Italy.
Stage 6	Substantial reduction in employment and number of production units. Decline both relative and, in some sectors, absolute.	Severe problems of competition. Substantial trade deficits.	United Kingdom, Germany, France, Belgium, The Netherlands.

Figure 9.2   An idealized sequence of development for the textiles and clothing industries
Source: Based, in part, on Toyne *et al.* (1984, pp. 20–21)

embryonic stage typical of the least developed countries through to the maturity and decline of the older industrial countries. Figure 9.2 shows the type of production likely to be characteristic of each stage and the related kinds of trade. It also gives examples of countries which are characteristic of each stage. The sequence is a very useful summary of what has happened so far but, like all such sequential models, it should not be regarded necessarily as being predictive of what will happen in the future – as charting the inevitable course of individual countries. Although many countries have passed through some or all of these stages, the precise path of development depends upon a number of factors which, together, produce the specific geographical patterns of the textiles and clothing industries.

Global shifts in the textiles and clothing industries

Textiles

Figure 9.3 maps the world distribution of textiles manufacturing employment in 1993, the latest data available at the time of writing. Although widely spread geographically, some clear concentrations stand out. Excluding the former Soviet Union, China has by far the largest level of textile employment in the world with more than 6 million employed in the industry. India, with almost 1.5 million is a distant second. Textiles employment in the United States in 1993 was around 800,000, that in Japan just under 600,000. Most of the other major textiles activity is in Europe, and in Asia. Since the 1960s the pattern of change in textiles production has varied enormously between individual nations. In general, the industry grew most rapidly in the developing market economies and in the economies of eastern Europe and either stagnated or declined in the developed economies.

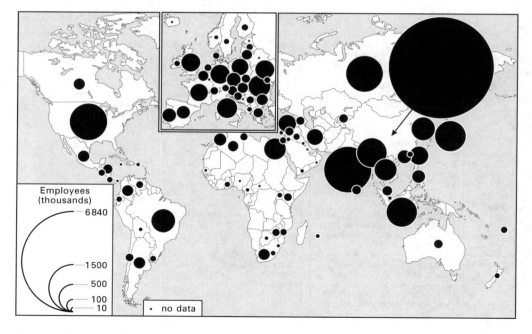

Figure 9.3   Global distribution of employment in the textiles industry
Source: ILO (1996b), UNIDO (1996, *International Yearbook of Industrial Statistics, 1996*)

Table 9.1   Annual growth of textiles production, 1980–93

Geographical area	Annual growth in value-added (%)	
	1980–89	1990–93
Industrialized countries (excl. eastern Europe and former USSR)	0.4	–2.3
Eastern Europe and former USSR	1.2	–
All developing countries	2.3	1.2
First-generation NIEs	1.9	1.0
Second-generation NIEs	5.6	0.9
Other developing countries	1.1	1.3

Source: UNIDO (1996, *International Yearbook of Industrial Statistics, 1996*, Table 1.10).

These broad trends are shown very clearly in Table 9.1 for the period 1980–93. In the developed market economies textiles production grew by only 0.4 per cent per year during the 1980s and then declined steeply in the early 1990s (by 2.3 per cent annually). The major growth of textiles production in the world economy during the 1980s took place in the developing market economies and, especially in the 'second generation' NIEs. But although growth rates remained positive in the early l990s the general level was low. With some exceptions (notably China) the rate of growth of textiles production in the developing countries has slackened (in this sense, the textiles sector in these countries has been growing more slowly than the clothing sector, as we shall see).

Substantial global shifts have occurred in textiles production in the last few decades, with the decline or stagnation of traditionally dominant producers and the emergence of new centres of production. Changes in the pattern of *international trade* in textiles reflect these shifts in production, although not in an identical manner. For example, despite the undoubted growth of textiles manufacturing in the developing countries and the relative decline as textiles producers of the older industrialized

Table 9.2   The world's leading textile exporting countries, 1995

	Country	Share of world exports		Annual percentage change			
		1980	1995	1992	1993	1994	1995
1	Germany	11.4	9.3	5	–	9	12
2	China	4.6	9.1	7	1	36	18
3	Hong Kong	–	–	12	2	12	10
	domestic exports	1.7	1.2	–2	–6	–7	–7
	re-exports	–	–	17	4	17	13
4	Italy	7.6	8.3	8	–	15	17
5	Korea	4.0	8.1	12	9	19	15
6	Taiwan	3.2	7.8	3	8	25	16
7	Belguim–Luxembourg	6.5	5.1	2	–	9	14
8	France	6.2	4.9	8	–	15	20
9	USA	6.8	4.8	5	2	9	12
10	Japan	9.3	4.7	8	–5	1	6
11	United Kingdom	5.7	3.4	4	–	15	15
12	Pakistan	1.6	2.8	12	–2	14	7
13	India	2.1	2.9	16	–1	31	–
14	The Netherlands	4.1	2.3	3	–	9	31
15	Spain	1.3	1.7	8	–	24	33
	Above total	76.1	76.0				

Source: WTO (1996, *Annual Report, 1996*, Table IV.51).

nations, the leading textiles-exporting country is still Germany although it seems likely to be overtaken by China in the very near future. As Table 9.2 shows, eight of the fifteen leading textiles-exporting countries are older industrialized countries (including Japan). The fifteen countries as a group generated 76 per cent of world textile exports in 1995, the same proportion as in 1980. However, comparison of the 1980 and 1995 rankings, and of the annual growth rates, indicates substantial volatility in the relative importance of the leading textiles exporters. The major gainers of export shares were the east Asian countries (China, Hong Kong, South Korea, Taiwan) – although not Japan, whose share of world textiles exports declined from 9.3 per cent in 1980 to 4.7 per cent in 1995. All the developed country exporters, apart from Italy, had a smaller share of the world total in 1995 than in 1980.

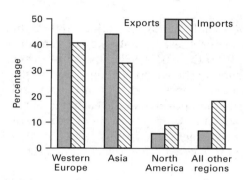

Figure 9.4   Regional shares of world trade in textiles, 1995
Source: WTO (1996, *Annual Report 1996,* Chart IV.10)

Figure 9.4 shows that world textiles exports are dominated by western Europe and Asia (primarily east Asia), each of which accounted for more than 40 per cent of the total in 1995. The most striking feature is that around four-fifths of western Europe's textiles exports are intraregional.

Despite the major shifts which have been occurring in the global pattern of the textiles industry, almost half of all world textiles exports still originates from the developed economies. But, as Table 9.3 shows, an increasing number of them have substantial trade deficits in textiles. Only Italy has a textile trade surplus which approaches that of the leading developing country producers. The EU as a whole had a positive balance of $3.7 billion with the rest of the world, compared with the United States which had a substantial trade deficit in textiles, of $3.1 billion in 1995. Japan's former textiles trade surplus has been transformed into a deficit.

Among developing countries, the most spectacular export performance has been by China. Although China has long had the largest textiles labour force in the world, its production was, until recently, mostly for its own domestic market. Within a decade, however, China has emerged as a leading textiles exporter in the developing world and is now second in the world as a whole after Germany. As Table 9.3 shows,

Table 9.3   Trade balances in textiles, 1995

Developed economies	$ millions	Developing economies	$ millions
Belgium–Luxembourg	+3,672	China	+3,004
Canada	−1,831	Hong Kong	−3,044
France	−24	Korea	+8,345
Germany	+2,036	Taiwan	+10,116
Italy	+6,356	Pakistan	+4,134
Japan	−1,193	India	+3,504
The Netherlands	+24		
United Kingdom	−2,536		
United States	−3,069		

Source: Calculated from WTO (1996, *Annual Report, 1996,* Table IV.53).

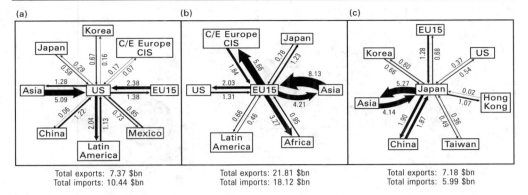

Figure 9.5   The textile trade networks of the United States, the EU and Japan, 1995
Source: WTO (1996, *Annual Report, 1996*, Table A15)

Taiwan has a huge textiles trade surplus as does Korea (both of them larger than the combined surplus of Germany and Italy). Only Hong Kong of the leading Asian textiles producers has a trade deficit.

Figure 9.5 shows the major components of the textile trade network for the world's three leading economies: the United States, the European Union and Japan. The three networks show both a common Asian element and a distinctive regional element which differs in each case. Almost 50 per cent of the United States' imports of textiles (Figure 9.5a) originates from Asia, of which around 12 per cent of all imports comes from China. One-quarter comes from the EU together with significant flows from Latin America. Figure 9.5b reveals that 45 per cent of the EU's textile imports originates from Asia but with a further 10 per cent from central and eastern Europe and the CIS. (These figures exclude intra-EU trade which accounts for two-thirds of the EU's total textile trade.) In the case of Japan (Figure 9.5c) the local region is clearly dominant: almost 70 per cent of Japan's textile imports originate from Asia. But Japan is also, itself, a significant exporter of textiles to Asia; almost three-quarters of Japan's textile exports go to other Asian countries.

Clothing

Like textiles, the manufacture of clothing is very widely spread geographically, as Figure 9.6 shows. The largest concentration of clothing employment is in China, with 1.6 million followed, a long way behind, by the United States (770,000), the Russian Federation (630,000) and then Japan (450,000). Clothing manufacture remains extremely important in western Europe – the United Kingdom, Germany, France and Italy each had between 115,000 and 170,000 employed in the industry in 1993. Romania (200,000) and Poland (145,000) are the leading clothing employers in eastern Europe. Among developing countries, east and southeast Asia is overwhelmingly dominant. Apart from China, there are major concentrations of clothing workers in Indonesia (317,000), Thailand (227,000), the Philippines (183,000), South Korea (197,000) and Hong Kong (170,000).

The current geographical distribution of clothing production is the result of the substantial global shifts that have occurred since the 1960s and 1970s. The major changes are summarized in Table 9.4. As a comparison with Table 9.1 shows, the

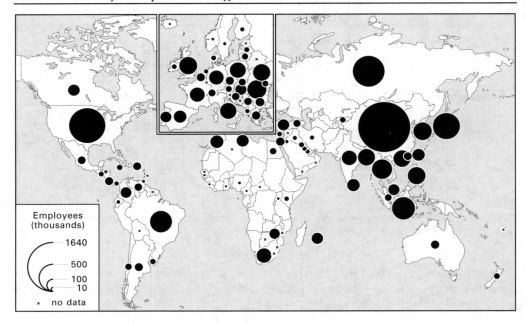

Figure 9.6   Global distribution of employment in the clothing industry
Source: ILO (1996b), UNIDO (1996, *International Yearbook of Industrial Statistics, 1996*)

developed economies as a group fared far worse in clothing than in textiles. Their average annual growth in clothing production was negative throughout the period from 1980 to 1993. Expressed in relation to a 1980 production level of 100, clothing production in France had fallen to 57, in Italy to 82, in Germany to 53. Only Belgium fared especially well (a 1993 index of 131) while the United Kingdom did relatively well with its production index falling by only one point to 99. The United States' index had declined to 95 and Canada's to 85.

In comparison, clothing production increased very substantially in the developing countries as a whole during the 1980s (Table 9.4) but the most notable feature was the shift towards the 'second generation' NIEs. As a group they had an annual growth rate of 5.6 per cent in the 1980s (compared with 3.3 per cent for the 'first generation' NIEs) and of 7.1 per cent in the early 1990s (compared with a negative rate in the 'first

Table 9.4   Annual growth of clothing production, 1980–93

Geographical area	Annual growth in value-added (%)	
	1980–89	1990–93
Industrialized countries (excl. eastern Europe and former USSR)	−0.3	−3.7
Eastern Europe and former USSR	2.0	–
All developing countries	3.1	−1.5
First-generation NIEs	3.3	−5.0
Second-generation NIEs	5.6	7.1
Other developing countries	–	–

Source: UNIDO (1996, *International Yearbook of Industrial Statistics, 1996*, Table 1.10).

Table 9.5   The world's leading clothing exporting countries, 1995

Country	Share of world exports		Annual percentage change			
	1980	1995	1992	1993	1994	1995
1 China	4.0	15.2	36	10	29	1
2 Hong Kong						
domestic exports	11.5	6.0	12	5	2	−1
re-exports	−	−	23	16	2	1
3 Italy	11.3	8.9	4	−	12	12
4 Germany	7.1	4.7	12	−	1	11
5 United States	3.1	4.2	27	18	13	18
6 Turkey	0.3	3.9	20	4	6	34
7 France	5.7	3.6	10	−	9	13
8 Korea	7.3	3.1	−9	−9	−8	−12
9 United Kingdom	4.6	2.9	8	−	20	13
10 Thailand	0.7	2.9	3	11	8	2
11 India	1.5	2.6	23	4	25	−
12 Portugal	1.6	2.3	12	−	5	11
13 Indonesia	0.2	2.1	40	11	−8	5
14 Taiwan	6.0	2.1	−8	−9	−8	−6
15 The Netherlands	2.2	1.8	10	−	11	6
Above total	66.9	66.1				

Source: WTO (1996, *Annual Report, 1996*, Table IV.58).

generation NIEs of −5 per cent per annum). Particularly strong growth of production volumes occurred in the Philippines (a 1993 index of 723 over 1980), Malaysia (239) and Indonesia. In comparison, the growth of South Korea and Hong Kong clothing production was far more modest. In Taiwan, there was a substantial decline while that of Singapore was static between 1980 and 1993. Outside Asia, strong growth occurred in North Africa, especially in Morocco and in central America, notably Honduras. The most rapid growth in clothing production in eastern Europe occurred in Bulgaria and Romania, but at much lower rates than in Asia.

These shifts in production are reflected in the trade figures. Compared with textiles there is a greater dispersion of *clothing-exporting* countries. A comparison of Tables 9.2 and 9.5 shows that the leading fifteen clothing-exporting countries accounted for 66 per cent of the world total, whereas in textiles the top fifteen accounted for 76 per cent. Although there are elements in common between the two tables, there are also some very substantial differences. China has replaced Hong Kong as the world's leading exporter of clothing, followed by Italy and Germany. Both Italy and Germany had become substantially less important as clothing exporters by 1995. On the other hand, China's growth as a clothing exporter has been dramatic. In 1980, China generated just 4 per cent of world exports; by 1995, its share had increased fourfold. Four of the fifteen leading clothing exporters are not on the textiles list at all: Turkey, Portugal, Thailand and Indonesia. The increasing significance of the 'second generation' NIEs is readily apparent. Japan is no longer one of the world's leading clothing exporters although it is the tenth most important textiles exporter. On the other hand, the United States' relative importance as a clothing exporter increased from 3.1 per cent in 1980 to 4.2 per cent in 1995.

Figure 9.7 shows the regional pattern of clothing trade in 1995. As in the case of textiles (Figure 9.4) the export picture is dominated by western Europe and Asia although Asia's share is now very significantly greater than that of western Europe. The picture for clothing imports is different. The dominant destinations are western Europe followed by North America. It is very clear that, at the broad regional scale,

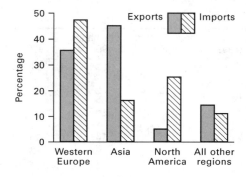

Figure 9.7   Regional shares of world trade in clothing, 1995
Source: WTO (1996, *Annual Report, 1996,* Chart IV.11)

Asia has a very large trade surplus in clothing and North America a very large deficit, with western Europe occupying an intermediate position. Within Asia itself, there has been a very considerable shift in the importance of individual countries:

> as the NIEs in East Asia were shifting into higher-value-added production in the 1980s and 1990s, clothing exports became a growth pole for other low-wage countries in the region. The Southeast Asian nations of Thailand, Indonesia, and Malaysia increased their share of global apparel exports more than fivefold . . . In Bangladesh, Sri Lanka and Mauritius, apparel exports climbed to one-half or more of each economy's merchandise exports . . .

Textiles and clothing, the preeminent export sector in the East Asian NIEs in the 1960s and 1970s, actually shrank as a proportion of these economies' total exports between 1980 and 1993.

(Gereffi, 1996a, pp. 91, 93)

The pattern of trade surpluses and deficits in clothing has long been a particularly sensitive issue in the politics of international trade. Table 9.6 shows the situation in 1995. It is very different from that in textiles (Table 9.3). Apart from Italy every one of the leading developed economies has a big trade deficit in clothing. Most obvious is the enormous US deficit of $34.7 billion, two-fifths the total clothing deficit of the developed countries listed. Japan's and Germany's deficits are also very large, each at about 20 per cent of the group total. In contrast, all the developing countries listed had a trade surplus in clothing.

The major components of the clothing trade networks of the United States, the EU and Japan are shown in Figure 9.8. As in the case of textiles (Figure 9. 5) each network has a strong Asian element together with a distinctive regional element. Almost two-thirds of US clothing imports emanate from Asia (Figure 9.8a), of which China accounts for almost a quarter. A further 22 per cent originates from Latin America and a small share (5 per cent) from the EU. In the case of the EU (Figure 9.8b) exactly half of its clothing imports are from Asia but the distinctive feature of the EU's clothing imports is the importance of countries on its immediate periphery: central and eastern Europe (15 per cent of EU clothing imports) and North Africa (13 per cent). As in the

Table 9.6   Trade balances in clothing, 1995

Developed economies	$ millions	Developing economies	$ millions
Belgium–Luxembourg	−1,724	Hong Kong	+8,634
Canada	−1,674	Korea	+3,884
France	−4,664	Malaysia	+2,118
Germany	−16,845	Taiwan	+2,374
Italy	+9,438	Portugal	+2,827
Japan	−18,228		
The Netherlands	−2,238		
United Kingdom	−3,695		
United States	−34,716		

Source: Calculated from WTO (1996, *Annual Report, 1996*, Table IV.60).

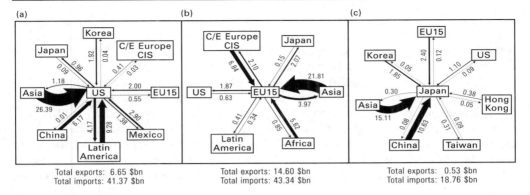

Figure 9.8   The clothing trade networks of the United States, the EU and Japan, 1995
Source: WTO (1996, *Annual Report, 1996*, Table A15)

case of textiles, a large proportion of the EU's clothing trade is intra-EU trade (70 per cent of the total). Not surprisingly, around 80 per cent of Japan's clothing imports originate from Asia (Figure 9.8c). Almost three-quarters of these imports come from China. Interestingly, however, the EU accounts for 13 per cent and the United States for 6 per cent of Japan's clothing imports.

Thus, a particularly interesting aspect of international trade in clothing is the existence of several strong regional patterns. For example, a growing proportion of clothing imports into the United States is from Mexico and from the Caribbean. An increasing share of the EU's clothing imports originates from the Mediterranean rim, including some north African countries; most of Japan's clothing trade is with neighbouring Asian countries. Such propinquitous trade patterns reflect a number of influences, not least geographical distance but also the operation of corporate and political actions. As we shall see below, it is in the clothing sector that the greatest degree of penetration of developed country markets by developing country producers has occurred. At the same time, a substantial share of the world's clothing trade still occurs between the industrialized countries themselves. There is a very high degree of interpenetration of developed country markets, especially within the EU.

The changing pattern of demand

Demand is a fundamental influence on the level and location of the textiles and clothing industries. Each sector in the textiles–clothing production sequence shown in Figure 9.1 relates to rather different markets. However, since some 50 per cent of all textiles production goes to the clothing industry, the major influence on the demand for textiles is the demand for clothing.

Demand for clothing is perhaps the easiest to identify since it is final consumer demand. The major general determinant is the level and distribution of personal income. Since personal incomes are so very unevenly distributed geographically (see Figure 6.3) it is the affluent parts of the world which largely determine the level and the nature of the demand for clothing. The generally low incomes in developing countries clearly restrict the size of their domestic clothing markets and this has undoubtedly had an important influence on their adoption of export-oriented policies.

But the relationship with income is not total: beyond a certain level of income, demand for clothing tends not to increase at the same rate as incomes increase. In other words, clothing tends to have an elasticity of demand of less than one. As personal incomes rise a relatively smaller proportion is spent on basic clothing.

This poses a major problem for clothing manufacturers and retailers: they need to stimulate demand through *fashion change*. Only the innocent would regard fashion as the spontaneous expression of consumer demand. Fashion – and the resulting demand – are largely induced, not least by the clothing and textiles manufacturers and, especially, the retailers themselves.[4] Fashion change has been especially important to the textiles and clothing industries of the developed market economies. Enormous efforts (and expenditures) have gone into promoting fashion products and creating 'designer' labels. Indeed, one of the most interesting developments in the clothing industry during the past few years has been the rapid proliferation of designer-labelled garments. Such a practice covers a very broad spectrum of consumer income levels from the exceptionally expensive to the relatively cheap. Designer labelling is basically a device to *differentiate* what are often relatively similar products and to cater to – and to encourage – the segmentation of market demand for clothing.

The growing power of the retailing chains

Within the clothing industry, in particular, demand is becoming increasingly dominated by the purchasing policies of the major multiple retailing chains. A smaller and smaller percentage of clothing sales is being channelled through independent retailers. In the United States, big companies such as Sears, J.C. Penney, Dayton Hudson, K-Mart, Wal-Mart, account for a very large proportion of clothing sales, as do Daiei, Mitsukoshi, Daimaru and Ito Yokado in Japan. In Germany, the leading clothing retailers include Karstadt, Kaufhof, Schickendanz; in The Netherlands, C&A; in France, Carrefour; in Britain, Marks & Spencer. However, the extent and the nature of multiple retailer dominance over the clothing market varies a good deal from one country to another.

There has certainly been a 'retailing revolution' in both the United States and the United Kingdom since the 1960s. In the United States, the retail industry

> became more oligopolistic during the 1960s and 1970s as giant department stores swallowed up many once-prominent independent retailers . . . The growth of large firms at the expense of small retail outlets was encouraged by several forces, including economies of scale, the advanced technology and mass advertising available to retail giants, government regulation, and the financial backing of large corporate parent firms. In the 1980s, the department store in turn came under siege . . . While this format typically met the needs of the suburban married couple with two children and one income, by 1990 less than 10 per cent of American households fitted that description. Today the generalist strategy no longer works. The one shopper of yesterday has become many different shoppers, with each member of the family constituting a separate buying unit.
>
> (Gereffi, 1994, p. 105)

The result has been the rapid emergence of specialist clothing retailers – like The Gap, The Limited, Liz Claiborne – serving niche markets.

A similar process has occurred in Europe. In the United Kingdom, for example, in the early 1980s almost three-quarters of clothing sales were channelled through the multiple retailers. In 1986 the four leading clothing retailers were responsible for almost 30 per cent of total clothing sales in the United Kingdom (Elson, 1990). However, the nature of clothing retailing changed substantially during the 1980s as retailers began to focus on segmented markets, such as particular age and income groups. As in the United States, specialist chains emerged such as Next, Principles, Laura Ashley, together with local branches of some of the United States firms (notably The Gap). During the 1990s, this trend has intensified further with the spread of such firms as Jigsaw, DKNY and the like, catering to affluent young consumers.

These kinds of development, which are occurring to a greater or lesser degree in all developed market economies, have very profound implications for textiles and clothing manufacturers. The highly concentrated purchasing power of the large retail chains gives them enormous leverage over textiles and clothing manufacturers. When the market was dominated largely by the mass-market retailers, the demand was for long production runs of standardized garments at low cost. As the market has become more differentiated and more frequent fashion changes have become the rule, manufacturers are having to respond far more rapidly to retailer demands and specifications. Under such circumstances, the *time* involved in meeting orders becomes as important as the cost. This basic shift in the structure of marketing is having repercussions throughout the textiles and clothing industries and influencing both the adoption of new technologies and also corporate strategies. Although relatively few retail chains are themselves manufacturers of clothing, they are very heavily involved in subcontracting arrangements. The production chain in these industries is becoming transformed into a *buyer-driven* chain.

Production costs and technology

The production characteristics of each sector of the textiles–clothing production sequence vary considerably (Figure 9.9), although much depends on *where* production takes place. A process which is relatively capital intensive in one country may be relatively labour intensive in another, depending upon relative factor endowments. Similarly, there is considerable variation in the size of production units between one country and another. In clothing manufacture capital intensity is generally low, labour intensity is high, the average plant size is small and the technology relatively unsophisticated.

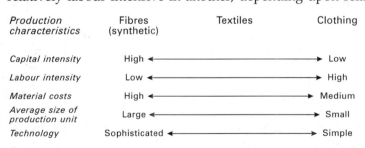

Figure 9.9 Variations in production characteristics between major components of the textiles–clothing production chain

Variations in labour costs

From the viewpoint of the changing global distribution of these industries the most important consideration is the extent to which the individual production factors are geographically variable in either their cost or their availability. For textiles and clothing manufacture there is no doubt that labour costs are the most significant production factor. Not only are textiles and clothing two of the most labour-intensive

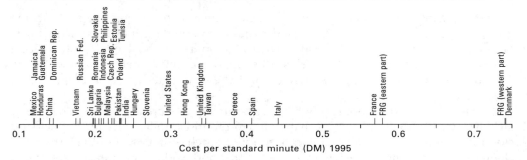

Figure 9.10 Labour costs per standard minute in the clothing industry, 1995
Source: Based on material in KSA (1995)

industries in modern economies but also labour costs are the most geographically variable of the production costs of the industries. Figure 9.10 shows just how wide the labour cost gap is between different countries. The measure used is the 'cost per standard minute'.[5] The spread is enormous, from DM0.742 in Denmark and DM0.741 in (western) Germany to less than DM0.120 in Yucatan, Mexico. There is substantial labour cost differentiation within Asia between the first and second-generation NIEs, with very low labour costs in China. But a significant feature of Figure 9.10 is that the labour cost levels in eastern Europe and the Russian Federation are comparable with those in low-cost Asia. It is also noteworthy that the United States' cost per standard minute is relatively low.

The difficulty of competing in labour cost terms with other producers was a major reason for the large-scale shift of the US textile industry from New England and of the garment industry from New York to the southern states. However, in terms of unit labour costs, which allow for levels of productivity, the United States is in a far better position *vis-à-vis* its industrialized competitors. But even allowing for productivity differences the developing countries have an enormous labour cost advantage over the developed market economies, particularly in the production of clothing. The major advantage of low labour-cost producers lies in the production of standardized items of clothing, which sell largely on the basis of price, rather than in fashion garments in which style is more important. The difference between the two is one of *rate of product turnover*. Fashion clothing has a rapid rate of turnover which tends to reflect the idiosyncracies of particular markets. Proximity to such markets is important and this helps to explain the survival of many developed country clothing manufacturers. This also partly explains the relative advantage of low-cost countries located close to the major consumer markets of the United States (e.g. Mexico, the Caribbean), Europe (e.g. central and eastern Europe, the Mediterranean rim) and Japan (the Asian countries).

Technological change

Both the cost of production and the speed of response to changes in demand are greatly influenced by the technologies used. Technological innovation may reduce the time involved in the manufacturing process and make possible an increased level of output with the same size – or even smaller – labour force. As international competition has intensified in the textiles and clothing industries the search for new, labour-saving technologies has increased, especially among developed country producers. However, the potential for such innovation varies very considerably between different manufacturing operations within the production sequence.[6] Two kinds of technological change are especially important:

- Those which increase the speed with which a particular process can be carried out.
- Those which replace manual with mechanized and automated operation.

On both counts, technological change has been far more extensive in the textiles manufacturing sector than in the manufacture of clothing.

Technological changes in textiles production

One of the most important technological innovations in the spinning of yarn was the introduction of open-ended spinning, which combines what were formerly three separate processes into a single process using rotors instead of spindles. Spinning speeds have been increased at least fourfold and labour requirements reduced by approximately 40 per cent. Textile weaving has been revolutionized by the introduction of the shuttleless loom in which the shuttle is replaced by a variety of alternative devices (for example, 'rapier' looms, 'projectile' looms, 'waterjet' and 'air jet' looms). Again, the major result has been a spectacular increase in the speed of the weaving process and this, together with wider loom capacity, has raised productivity by a huge factor. Parallel developments in knitting technology and in finishing – the latter now being especially highly automated – have contributed further to the increased speed of textiles operations and the consequent reduction in the number of workers needed.

Technological changes in clothing production

In contrast, technological developments in the clothing industry have been far less extensive. There was relatively little change in clothing technology between the industry's initial emergence in the late nineteenth century and the early 1970s. The basic sewing machine was not so very different from that in use fifty years earlier. The manufacture of clothing remains a complex of related *manual* operations, especially in those goods in which production runs are short. The problem is that,

> with the exception of the simplest articles . . . a garment is still a relatively complicated product for its price. The materials from which it is made, being limp and relatively delicate, do not lend themselves readily to mechanical handling. Many thicknesses of cloth may be cut out at a time, but each garment has to be put together individually.
>
> (Plant, 1981, p. 63)

Most of the recent technological developments in the industry, including those based on microelectronic technology, have been in the non-sewing operations: grading, laying out and cutting material in the preassembly stage, and warehouse management and distribution in the postassembly stage (Mody and Wheeler, 1987). The application of electronically controlled technology to these operations can achieve enormous savings on materials wastage and greatly increase the speed of the process. Hoffman and Rush (1988) point out that the grading process may be reduced from four days to one hour; computer-controlled cutting can reduce the time taken to cut out a suit from one hour to four minutes. But these developments do not reach the core of the problem. The sewing and assembly of garments account for four-fifths of all labour costs in clothing manufacture. So far only very limited success has been achieved in mechanizing and automating the sewing process itself.

Current technological developments in the manufacture of clothing are focused on three areas.[7] To

- increase the *flexibility* of machines: robots are being designed which can recognize oddly shaped pieces of material, pick the pieces up in a systematic manner and align the pieces on the machine correctly whilst also being able to sense the need to make adjustments during the sewing process itself
- address the problem of *sequential operations*, particularly the difficulty of transferring semi-finished garments from one workstation to the next while retaining the shape of the limp material
- develop the *unit production system* which will deliver individual pieces of work to the operator on a conveyor-belt system. This will greatly reduce the large amount of (wasted) production time spent by the operator on unbundling and rebundling work pieces. The handling process has been estimated to take up to 60 per cent of the operator's total time.

The drive to introduce new technologies in textiles and clothing has been stimulated mainly by the need for developed country producers to be cost competitive in the face of the very low labour costs in developing countries. But cost reduction is not the only benefit derived from the new technologies. At least as important, if not more so, are the *time savings* that result from automated manufacture. This has two major benefits:

- First, speeding up the production cycle reduces the cost of working capital by increasing the velocity of its use.
- Secondly, it becomes possible for the manufacturer to respond more quickly to customer demand. This is especially important in an environment in which there is strong retail dominance and where the retailer increasingly pushes for frequent deliveries in order to avoid holding too much stock.

Electronic point-of-sale (EPOS) technologies permit a direct, real-time link between sales, reordering and production. The classic example of this is the Italian clothing manufacturer, Benetton:

> Micro-electronics technology is vital to the Benetton operation . . . At the Italian head-quarters is a computer that is linked to an electronic cash register in every Benetton shop; those which are far away, like Tokyo and Washington are linked via satellite. Every outlet transmits detailed information on sales daily, and production is continuously and flexibly adjusted to meet the preferences revealed in the market. Benetton produces entirely to the orders received from the shops.
>
> (Elson, 1989b, p. 103)

The jeans manufacturer Levi Strauss also operates a computerized system both to keep track of the work in progress in each factory (and payments to workers on the basis of completed work) and to control the overall volume of production. As the production chain has become increasingly buyer driven, these IT-based innovations have become extremely important. Not only do they permit very rapid response to sales and demand at the point of sale but also they enable the buyer firm to pass on the costs of producing and holding inventory to the manufacturer.

Government policies towards the textiles and clothing industries

In developing economies, textiles and clothing manufacture have occupied a key position in national industrialization strategies. Hence, the kinds of import-substitution and export-oriented measures outlined in Chapter 4 have been applied extensively to the encouragement of the industries in most developing countries. But it is in the older-established producing countries of Europe and North America and, more recently, Japan, faced with increasingly severe competition from low-cost or more efficient producers, that government intervention has been especially marked. The political sensitivity of these industries has forced governments to intervene. The policies adopted by governments have been of two kinds:

- Those aimed at encouraging the *restructuring and rationalization* of the country's textiles and clothing industries.
- Those aimed at *stimulating* the industries through offshore assembly provisions and preferential trading agreements.

Let us look at each of these in turn.

Policies to encourage restructuring

A major policy pursued by a number of developed country governments – most notably in Europe – has been the encouragement of restructuring in their textiles and clothing industries through the use of subsidies and investment assistance programmes. For example, in the United Kingdom as early as the 1950s the textiles industry once again found itself faced with severe competition from low-cost producers in Asia. In addition to the imposition of 'voluntary' trade restraints, the United Kingdom government implemented its Cotton Industry Act, whose aim was to remove half of the industry's capacity and to stimulate re-equipment and modernization in the surviving firms. Firms were encouraged to engage in such 'voluntary euthanasia' by generous financial compensation for scrapped capacity and grants for re-equipment. Even though the 50 per cent scrapping target was broadly achieved the problem of overcapacity did not disappear and the subsequent decades have seen a continuation of rationalization within the industry. Large-scale financial assistance to the textiles and clothing industries has also been common throughout Europe.

Government involvement in restructuring domestic textiles and clothing industries has not been confined solely to the older producing nations, however. From the late 1960s the Japanese government actively sought to restructure its textiles industry by reducing capacity, modernizing plant and moving into higher-value products. Even the newer producers such as South Korea and Taiwan have found it necessary to intervene to maintain their competitive position through the encouragement of new investment and the scrapping of old capacity. In both cases, the hand of the government has been firmly involved.

Offshore assembly provisions and preferential trading agreements

A rather different kind of national policy which has had important effects on the global geography of the textiles and clothing industries is the use of offshore assembly provisions. Such provisions, part of a country's customs and excise regulations, have been especially common in the United States and West Germany. As explained in Chapter 4, offshore assembly provisions permit a company to export materials, have them made up into garments in another country (invariably one with low labour costs) and then reimport them into the company's domestic market paying duty only on the value-added in offshore processing. In the United States, the use of a specific item of the US tariff code has been used extensively for this purpose. It also established the Caribbean Basin Initiative (CBI) which allowed access to the US market, under specific conditions, of garments produced in the Caribbean countries.

Both the tariff provision and the CBI were devices to regulate the offshore practices of US manufacturers:

> The 807 and CBI programmes were sold to industry and labour as a means of retaining high value operations (thus high wage jobs) in the US. In reality, benefits flowed primarily to larger apparel firms that have established low cost production facilities throughout the Caribbean and especially along the maquila corridor of Northern Mexico. Hence, the State and larger industrial players colluded for their respective objectives and sacrificed the increasingly marginalized smaller firms of the US textile and apparel industry which did not have links to low cost supply.
>
> (Glasmeier, Thompson and Kays, 1993, p. 29)

More recently, the signing of the North American Free Trade Agreement will have a substantial effect on the textile and clothing industries within North America. Under the NAFTA, Mexican, US and Canadian tariffs on textiles and clothing are to be phased out over a ten-year period. However, for clothing to be eligible for tariff concessions it must be sewn with fabric which has been woven in North America. The major fear amongst US and Canadian textile and clothing producers is that the huge labour cost differential between Mexico and the United States/Canada will pull more and more production across the border:

> In large measure a firm's size and the market segment in which it competes determine the competitive options afforded by the . . . trading agreement. While for most large firms NAFTA simply affords new markets to exploit or makes non-domestic sourcing options more attractive, for smaller domestic firms the new trade regime may force decisions that require fundamental restructuring of operations or market orientation.
>
> (Glasmeier, Thompson and Kays, 1993, p. 29)

A similar scenario could be played out in Europe as the low-cost countries to the east of the EU and on its southern periphery become more integrated. The European Union has preferential agreements with certain Mediterranean countries which permit market access (and, therefore, offshore production by European and other firms). But the biggest threat is from central and eastern Europe which are already major supply sources for German clothing manufacturers in particular.

The international regulatory framework: the Multi-Fibre Arrangement[8]

Individual national policies to stimulate domestic textiles and clothing industries or to facilitate their rationalization and restructuring have been extremely important in helping to reshape the industry globally. Of far greater significance, however, is the *international* regulatory framework within which the industries have operated for the past quarter of a century. The textiles and clothing industries are unique in that they are the only industries to which special international trade regulations apply, under the *Multi-Fibre Arrangement* (MFA). Today, the majority of all world trade in textiles and clothing is covered by this agreement. Its provisions and their implementation – and avoidance – have been a major factor in the changing global pattern of production and trade. Although protectionism in these industries is not solely a postwar phenomenon, the origins of its modern variant, the MFA, are to be found in the problems which faced developed country producers, particularly the United States and the United Kingdom, in the 1950s.

Faced with a massive inflow of low-price imports from Japan, Hong Kong and some other Asian producers, both the United States and the United Kingdom negotiated separate 'voluntary' agreements with the Asian exporters to restrict imports for a limited period. By 1962 such arrangements had become broadened into the Long-Term Arrangement (LTA), within GATT, which regulated international trade in cotton textiles. The aim of the LTA was to encourage orderly development of the international cotton textiles market to allow the developed countries to restructure their cotton textiles industry. It allowed an importing country to limit shipments from any source in any cotton textile category which would 'cause or threaten to cause disruption in the market of the importing country'.

The LTA allowed for a gradual increase in imports from developing country signatories of 5 per cent a year. It remained in force for eleven years. During that time, however, the world picture became far more complex. First, there was the massive growth of artificial fibres which were not covered by the LTA. Secondly, an increasing number of developing countries emerged as important exporters of textiles and, especially, of clothing. The precipitous decline of the industries in the developed economies continued. In 1973 a much broader trade agreement, which included the European countries and also covered artificial and other non-cotton fibres, was negotiated. This was the first Multi-Fibre Arrangement. By Article 1(2), its principal aim was

> to achieve the expansion of trade, the reduction of barriers to such a trade and the progressive liberalization of world trade in textiles products, while at the same time ensuring the orderly and equitable development of this trade and avoidance of disruptive effects in individual markets and on individual lines of production in both importing and exporting countries.

Like the LTA, the MFA was initially negotiated for a limited period: four years from January 1974. Like the LTA, too, its aim was to create an 'orderly' development of trade in textiles and clothing which would benefit *both* developed and developing countries. Access to developed country markets was to increase at an annual average rate of 6 per cent, though this was far below the 15 per cent sought by the developing countries. At the same time, the developed countries were to have safeguards to protect the 'disruption' of their domestic markets. Within the MFA, individual quotas were negotiated which set precise limits on the quantity of textiles and clothing products which could be exported from one country to another. For every single product a quota was specified. When that quota was reached no further imports were permitted.

In practice it has been the disruptive, rather than the liberalizing, aspect which has been at the forefront of trading relationships. Since 1974 the MFA has been renegotiated or extended four times (in 1977, 1982, 1986, 1991). In general, the MFA has become more, rather than less, restrictive. Both the EU and the United States have negotiated much tighter import quotas on a bilateral basis with most of the leading developing country exporters, and in several cases have also invoked anti-dumping procedures.

The effects of the MFA on world trade in textiles and clothing have been immense. Without doubt, it has greatly restricted the rate of growth of exports from developing countries, which have been far lower than would have been the case in the absence of the MFA. A major initial beneficiary of this dampening of the relative growth of developing country exports was the United States, which greatly increased its penetration of European textiles and clothing markets during the 1970s. During the early 1980s, however, it was the European producers which greatly increased their penetration of the United States market.

An inevitable consequence of the increased restrictiveness on developing country exports of textiles and clothing has been a parallel increase in efforts to circumvent the restrictions. Such evasive action may take a variety of forms:

- For a producing country which has reached its quota ceiling in one product to switch to another item.
- To use false labelling to change the apparent country of origin (an illegal act).
- For firms to relocate some of their production to countries which either are not signatories to the MFA or whose quota is not fully used by indigenous producers.

There is no doubt that the Multi-Fibre Arrangement has greatly influenced global trade and investment in textiles and clothing. There have been winners and losers from what is, in many respects, a distortion of the GATT principles.

A major task of the Uruguay Round of the GATT negotiations in the late 1980s/early 1990s was to integrate the MFA into the GATT. As we noted in Chapter 3, the basic principle at the heart of GATT rules on international trade is the most favoured nation principle, which implies non-discrimination between all trading partners. The MFA clearly contravenes this principle. From the standpoint of the industrialized countries, 'the MFA by its structure discriminates among supplier countries, most notably between industrial countries, which enjoy free access, and developing countries (and Japan), which face controls' (Cline, 1987, p. 232). On the other hand, the developing countries are indicted by the industrialized countries for restricting access to their domestic markets through both very high tariffs and non-tariff barriers.

As a result of the Uruguay Agreement of the GATT, trade in textiles and clothing has now been incorporated into the WTO. It was agreed that the MFA would be phased out over a ten-year period (1995–2004) but in three stages:

- Stage 1 (1995–98): 16 per cent of tariff lines to be integrated into the GATT.
- Stage 2 (1998–2002): 17 per cent of tariff lines to be integrated.
- Stage 3 (2002–2004): 18 per cent to be integrated after 2002 followed by the final 49 per cent by the end of 2004.

Thus, 'integration is . . . heavily backloaded, putting most of the difficult liberalization off to the future' (Hoekman and Kostecki, 1995, p. 209). In fact, both the US and the EU are to 'integrate' first those products which already enter their markets freely. The United States' ten-year liberalization schedule in effect leaves the integration of 70 per cent of imports by value to the very end of the transition period' (*The Financial Times*, 10 January 1996).

Not surprisingly, developing countries are unhappy with what they regard as a deliberate dragging of feet by the world's two largest textile and clothing markets. In response, European and United States' producers argue that developing countries need to be more positive in increasing access to imports into their own domestic markets. Quite clearly, full liberalization of the textiles and clothing industries is far from assured. Meanwhile, bilateral deals continue to be the basis of trade with the world's leading clothing producer, China, which does not yet belong to the WTO. Particularly acrimonious trade relationships have occurred between the United States and China, although a new agreement on textiles and clothing was signed by the two countries in early 1997.

Corporate strategies in the textiles and clothing industries[9]

As we have seen in this chapter, the manufacture of textiles and clothing is geographically very widespread. They are relatively rare instances of globally significant industries which are extensively present in many developing countries, rather than in just a few. Vast numbers of, mostly small, developing country firms are involved in textiles and clothing production. Nevertheless, the globalization of these industries has been driven primarily by developed country firms. Indeed, it is paradoxical that a significant proportion of the textiles, and especially the clothing, imports, which are the focus of such concern in developed countries are, in fact, organized by the international activities of those very countries' own firms. But the processes and strategies involved are both complex and dynamic.

In examining the development of these strategies two basic points need to be made:

- First, the globalization of the textiles and clothing industries cannot be explained simply as a relocation of production from developed to developing countries in search of low labour costs. This was the position taken in theories of the new international division of labour.[10] As Elson (1988) points out, other factors are involved including, in particular, orientation to specific markets.
- Second, where firms have internationalized their production operations they have used a variety of methods, notably international subcontracting and licensing, which do not necessarily involve equity participation. We discussed such strategies in general terms in Chapter 7. Foreign direct investment has been relatively unimportant in textiles and clothing although it does, of course, occur.

Major factors in how and where the various modes of international involvement have been used are the existence of the Multi-Fibre Arrangement, with its complex system of national quotas, and the volatility of currency exchange rates. Technological innovations in production and distribution processes are also part of the strategic equation in so far as they influence both the costs of producing textiles and clothing and also the speed of response to changing customer demands. These, in turn, are heavily affected by the dominance of the major retail chains.

Within this multiplicity of influences on the strategic behaviour of textiles and clothing firms it is the need for *flexibility* which is increasingly the dominant consideration. Whether a firm pursues a strategy of cost leadership or product differentiation and whatever its focus (product or geographic) it must be able to respond quickly and flexibly to changing circumstances. Consequently, the internationalization strategies of textiles and clothing companies can best be understood in these terms.

Corporate strategies in the textile industry

Although textiles firms of all sizes continue to exist, the textiles industry is increasingly an industry of large firms. In the world as a whole, some thirty or so textiles corporations are especially important. These form what Clairmonte and Cavanagh (1981) termed the 'world textile oligopoly': the group mainly responsible for reshaping the global textiles industry. They include such companies as Burlington Mills in the United States, Toray in Japan, Coats Viyella in the United Kingdom and the Marzotto Group in Italy.

Textile firms have tended to pursue one of three major strategies in attempting to remain competitive.[11] To

- produce standardized goods for large markets using economies of scale to reduce costs and enable them to compete on price
- supply large markets on the basis of utilizing low-cost labour in offshore locations

- produce small quantities of specialized goods for specific market niches. This presupposes high-quality products which can be sold at a premium price to offset the additional costs of switching specifications.

Although the British and European textiles companies have a lengthy history of international involvement in textiles production, the first really major wave of such activity occurred in the early 1960s and was led by the Japanese textiles firms and the general trading companies (the *sogo shosha*). The Japanese textiles firms were already strongly vertically integrated in Japan itself and operated a dualistic-hierarchical network of domestic subcontractors. The *sogo shosha* were responsible for organizing a huge proportion of Japanese imports and exports and already had an intricate international distribution system. When the United States introduced the LTA in 1962 to protect its domestic market from Japanese cotton textiles imports, it triggered the first surge of overseas involvement by Japanese firms. To avoid the problem of quotas both the textiles firms and the *sogo shosha* set up international subcontracting links in other east and southeast Asian countries. Very often, as in Japan itself, the principal firms took a small equity share in the subcontractors.

Relatively quickly, therefore, the Japanese textile industry became an international, vertically integrated operation which incorporated an extensive network of local Asian producers. According to Oman (1989), by 1980 Japan's nine leading *sogo shosha* were involved in 150 textiles ventures outside Japan. The leading vertically integrated textiles companies (Toray, Asahi, Teijin, Toyo, Mitsubishi) were also involved in a vast array of international operations, primarily in Asia but with some in Latin America and Africa. Some of these took the form of direct investment but most were international subcontracting arrangements.

Compared with the Japanese companies, US and European textiles firms are less internationalized. In the United States the tendency has been to increase the degree of domestic concentration in the industry through acquisition and merger and to upgrade domestic productivity through heavy investment in new technology. In Europe, too, these strategies have been pursued, although some of the very large textiles companies have also become increasingly internationalized. Major examples are the two leading British companies: Coats Viyella and Courtaulds. These two companies have been engaged in massive rationalization and restructuring programmes since the 1970s. Such programmes have involved a varying mixture of product rationalization and focus; technological innovation to reduce costs and increase flexibility; reduction of production capacity particularly in the United Kingdom but also overseas; and the use of international subcontracting, licensing and other forms of relationship with local firms in developed and developing countries.

For example, Coats Viyella, the largest textile company in Europe, was created through a series of acquisitions of British textile and clothing companies during the 1970s and 1980s.[12] Coats Viyella has gradually transformed itself from a 'production-driven firm' with an immense range of products to a 'market-driven firm with five core businesses: threads, clothing, homewares (e.g. bed linen), 'fashion retail' and precision engineering. Locationally, Coats Viyella has adopted a two-pronged strategy:

- Shifting low value-added activities, such as undyed thread manufacture and zip production, to low-cost countries, in the eastern and southern European periphery and in Asia.
- Locating dyeing factories, which are less labour intensive, closer to its main markets.

In addition, Coats Viyella has made major investments in information technology, both to improve its production processes and more importantly to connect more directly to its retailers (both its own 'fashion' outlets, such as Jaeger, and independents). It can now fill orders in two weeks compared with six weeks under the old system.

However, Coats Viyella faces the major problem common to most textile firms: its customers themselves are continuously shifting their locations to remain competitive. This is a particular problem for its main business – threads:

> Coats Viyella is being forced to follow the textiles industry round the world . . . its main preoccupation is the relocation of other companies' stitching operations. Threads account for nearly half the group's sales. And it is threads that absorbs 60 per cent of its reorganiz-ation costs, as it expands in eastern Europe, closes down elsewhere on the continent and refocuses in Asia. These shifts will only end when the world's clothing producers stop moving – and that may be another 10 years, as the implications of phasing out the Multi-Fibre Arrangement are worked through.
>
> (The Financial Times, 12 September 1996)

Corporate strategies in the clothing industry

Of all the major parts of the textiles–clothing production chain, the manufacture of clothing is by far the most fragmented and least dominated by large firms. In part this is explained by the relatively low technological sophistication of the clothing process and the low barriers to entry to the industry. In part it reflects the vagaries of the market for clothing, which restrict long production runs and high-volume production to staple items. Even in this archetypal small-firm industry, however, large firms are becoming increasingly important: only they can afford to invest in the new technologies, or to build a worldwide brand image based on mass advertising. Thus, although the clothing industry of most countries is made up of a myriad of very small firms, many of which operate as subcontractors, there is an undoubted trend towards increased concentration.

As in the case of textile firms, three broad strategies can be identified among clothing producers:[13]

- Production of standardized goods for large markets utilizing economies of scale to lower costs and to be price competitive. Examples in the United States include Levi Strauss, VF Corporation, Fruit of the Loom. However, as we saw earlier, the clothing segment is far less amenable to automation than the textiles segment and there are lower limits to efficient scale.
- Operation of small workshops – often in the form of 'sweatshops' in large cities – using immigrant and sometimes undocumented labour. Such firms generally work as short-term subcontractors producing lower-quality garments.
- Production of short orders to fill manufacturers' production gaps, often for very specific segments of the market.

However, in the case of clothing there is an additional organizational component of great significance:

> firms . . . which have limited or no manufacturing facilities, but organize systems of production from design and manufacture of raw fabrics to delivery of goods to retail establishments. Combining highly sophisticated design capability with state-of-the-art telecommunications and distribution structures, new competitors market garments that range from very high fashion to very low quality. These operations have emerged in the

1980s to include some of the nation's largest and most successful apparel companies such as Liz Claiborne and The Gap. The most dynamic part of this segment is in the value niche in which fashion, quality, and price are optimized and capture high sales volume. These hybrids are major players beyond the apparel industry alone, in some cases they have integrated forward into retail distribution ... The success of this consumer preference-driven strategy with its self-optimizing production system is evident in the fact that Liz Claiborne and The Gap are each $2 billion corporations with contract production facilities in at least 45 countries and sales in the US, Canada, and, increasingly, in Europe.

(Glasmeier, Thompson and Kays, 1993, p. 24)

Although some of these firms, notably the ones which have integrated forward into retailing, are now well known there are many others which are not. For example, in the United States,

Frederick Atkins (a co-operative) apparel packager, produces private label clothing from textile design to final product for the nation's largest department stores. The company currently orchestrates production of approximately 8 per cent of the nation's apparel supply. Products arrive from all over the world via intermediaries who negotiate contracts with textile producers and garment manufacturers.

(Glasmeier, Thompson and Kays, 1993, p. 23)

These major international *retail chains* and *buying groups* exert enormous purchasing power and leverage over clothing manufacturers. Although the production and retailing of clothing may be fragmented in individual markets, international buying operations are highly concentrated.

International subcontracting, licensing and other forms of non-equity international investment have been even more pervasive and influential in the clothing sector than in textiles. In some cases – such as the Japanese and British firms discussed in the previous section – clothing is part of a firm's vertically integrated activities. Again, as in the case of textiles, it was Japanese firms that initiated the extensive use of international subcontracting in clothing. During the 1960s and 1970s Japanese companies established subcontracting arrangements in Hong Kong, Taiwan, South Korea and Singapore. Their production was mostly exported to the United States and not to Japan's own domestic market. Again, the Japanese *sogo shosha* were at the leading edge of these international subcontracting developments, often using minority investments in local firms. Probably 90 per cent of Japanese overseas clothing operations are still located in east and southeast Asia and most were set up in the 1970s. Overall, however, the Japanese clothing industry is far less internationalized than its textiles industry. The opposite is true of the US and European clothing industries, which are far more internationalized than their textiles industries.

The strategies developed by US clothing firms to cope with the intensified competition of the 1970s and 1980s were 'two-pronged':

On the one hand, companies concentrated on product and marketing strategies on the leading edge of the fashion market ... On the other hand, manufacturers focused on investments to cut costs and raise productivity. As a complement, the larger US apparel firms significantly increased offshore processing via subcontracting arrangements with producers in developing countries.

(Oman, 1989, p. 228)

The geographical pattern of this activity has changed over time; an initial emphasis on Mexico and the Philippines has been superseded by increasing subcontracting activity in the Caribbean region. Some of the larger US clothing companies are also involved in direct investment in overseas operations through wholly or majority-owned subsidiaries. For the most part, such plants are oriented towards local markets rather than third-country exports and are mainly in developed countries.

Some of the most interesting strategic mixes are used by those clothing firms which compete in mass markets but on the basis of brand names supported by extensive advertising. The best examples are probably the jeans manufacturers, Levi Strauss and VF Corporation. Even by the late 1970s Levi Strauss was spending $50 million a year on worldwide advertising. Its current problem is how to adapt to the demographic change which is altering the size of its traditional market segment of fifteen- to twenty-four-year-olds without moving too far from its core product, the denim jean. One of its major responses was to develop the Dockers range of cotton trousers aimed at the postjeans generation.

Levi Strauss's international production strategy has been to develop its own branch factories in western and eastern Europe, Latin America and Asia, although it now also has licensing agreements. It employs some 40,000 workers worldwide, of which 28,000 are in North America, 7,000 in Europe and 2,000 in Asia. Levi Strauss is the market leader in this sector; its nearest rival is VF Corporation, which manufactures another heavily branded product, Wrangler Jeans, which it acquired from Blue Bell in 1987. According to Oman, Blue Bell's international strategy was the classic one of the market follower. Instead of developing an international network of directly owned manufacturing plants, Blue Bell opted for the less exposed and less risky strategy of licensing its product to independent manufacturers in more than forty countries.

The adoption of offshore production strategies by European clothing firms has been most pronounced among German and British companies. German clothing firms have been especially heavily involved in international subcontracting arrangements, as the detailed survey by Fröbel, Heinrichs and Kreye (1980) revealed. They calculated that in the 1970s around 70 per cent of all German clothing firms, including some quite small ones, were involved in some kind of offshore production. Roughly 45 per cent of the arrangements involved international subcontracting and a further 40 per cent involved varying degrees of equity involvement by German firms in local partners. German firms tended to concentrate their offshore activities in eastern Europe but with a growing presence in countries of north Africa and the Mediterranean (notably Greece, Malta and Tunisia). Overall, Fröbel, Heinrichs and Kreye calculated that more than four-fifths of all West German clothing imports manufactured under subcontracting arrangements came from East Germany, Poland, Hungary, Romania, Bulgaria and Yugoslavia.

The case of the German fashion company, Hugo Boss, provides a good example of current trends. Faced with high domestic production costs, Hugo Boss has long used offshore subcontractors, primarily in Romania, the Ukraine, Poland and the Czech Republic where wages are 50 per cent lower than in Germany. In 1989, Boss acquired an American clothing producer, Joseph & Feiss of Cleveland, Ohio. In 1991, Hugo Boss was itself acquired by the Italian textile and clothing group, Marzotto. In addition to sourcing an increasing proportion of its garments overseas, the company has also moved strongly into retailing through franchising its brand name in around 200 stores worldwide. According to its (German) chairman, 'we are no longer a production-oriented company. Today, we are a company with a strong emphasis on creativity and design, marketing and logistics' (quoted in (*The Financial Times*, 9 January 1996).

Similarly, the two leading British clothing companies – Coats Viyella and Courtaulds – have developed a mixture of international strategies. Courtaulds has established clothing factories in Portugal, Morocco and Tunisia. Coats Viyella is in the process of relocating a substantial proportion of its clothing production from the UK to Poland, Hungary, Morocco, Sri Lanka, China, India and Indonesia. In one specific instance in 1996, a Coats Viyells shirt factory on Merseyside, which manufactures for Marks & Spencer, was closed down and its production transferred to another group factory in Mauritius.

Of all the European clothing producers the Italians have developed most differently. We noted earlier that Italy is the only major European country whose clothing industry has continued to perform well in the teeth of intensive global competition. We noted, too, the innovative methods used by the Italian company Benetton to link together its production and distribution networks. Benetton has a highly sophisticated and complex global sales and distribution system based upon a network of franchised stores (more than 7,000 shops in 120 countries). But unlike the usual franchise arrangement, Benetton receives neither royalties nor takes back unsold stock. In fact, each shop is totally independent and carries the risk. Benetton, in contrast, has an assured set of high-profile outlets which it supports mainly through its controversial advertising campaigns. Benetton sees itself not as a manufacturer or retailer of clothing but as a 'clothing services company'. Whereas most European firms have shifted much of their production to Asia, around four-fifths of Benetton's garments are still manufactured in Europe, mostly in Italy. But not by Benetton itself:

> the company handles in-house only those bits of manufacturing – mainly design, cutting, dyeing and packing – that it considers crucial to maintain quality and cost-efficiency. It contracts out the rest to local suppliers. This devolved system reduces Benetton's risk and confers legendary flexibility. Last-minute dyeing means that the company can respond swiftly to sales trends.
>
> (*The Economist*, 23 April 1994)

So, instead of following the lead of most other European clothing firms and relocating production to low-cost countries, Benetton constructed two high-technology plants at its headquarters in Italy. One plant is a state-of-the-art cutting and dyeing plant; the other is a highly computerized warehouse. These two plants employ very few workers. Benetton's actual production of garments is carried out by a labour force of 250,000 employed in subcontractors' factories in Italy (*The Sunday Times*, 14 April 1997).

In general, the Italian producers have pursued a strategy of product specialization and fashion orientation with the aim of avoiding dependence upon those types of good most strongly affected by low-cost competition. This involves mainly small firms in a decentralized production system:

> Decentralised production has occurred in those areas of textiles where the fashion element (hence the need for greater risk and flexibility) is important. Italy's unique strength in Western Europe is to have created a kind of price- competitive mass-market in fashion, where certain products often enjoy an area 'trademark' (Como silk ties, Prato wool fabrics, and so on).
>
> (Shepherd, 1983, p. 42)

More recently, however, some Italian firms have established international licensing agreements for production of high-fashion and designer-label products.

Table 9.7 Global sourcing patterns of clothing retailers

Type of retailer	Types of order	Main sourcing countries
Fashion-oriented companies (e.g. Armani, Donna Karan, Boss, Gucci, Polo/Ralph Lauren)	Expensive 'designer' products High level of skill Orders in small quantities	Italy, France, UK, Japan. South Korea, Taiwan, Hong Kong, Singapore
Department stores, specialty stores, brand-name companies (e.g. Bloomingdales, Saks Fifth Avenue, Neiman-Marcus, Macy's, The Gap, The Limited, Liz Claiborne, Calvin Klein)	Top-quality, high-priced garments sold under variety of national brands and private labels (store brands) Medium to large-sized orders, often co-ordinated by store buying groups	South Korea, Taiwan, Hong Kong, Singapore Malaysia, Indonesia, Philippines, southern China, India, Turkey, Egypt, Brazil, Mexico, Thailand Dominican Republic, Jamaica, Haiti, Guatemala, Honduras, Costa Rica, Colombia, Chile, Poland, Hungary, Czech Republic, Bulgaria, Kenya, Zimbabwe, Mauritius, Macao, Pakistan, Sri Lanka, Bangladesh, interior China, Tunisia, Morocco, UAE, Oman
Mass merchandisers (e.g. Sears, Montgomery Ward, J.C. Penney, Woolworth)	Good-quality, medium-priced goods Mostly sold under private labels Large orders	South Korea, Taiwan, Hong Kong, Singapore Malaysia, Indonesia, Philippines, southern China, India, Turkey, Egypt, Brazil, Mexico, Thailand Dominican Republic, Jamaica, Haiti, Guatemala, Honduras, Costa Rica, Colombia, Chile, Poland, Hungary, Czech Republic, Bulgaria, Kenya, Zimbabwe, Mauritius, Macao, Pakistan, Sri Lanka, Bangladesh, interior China, Tunisia, Morocco, UAE, Oman
Discount chains (e.g. Wal-Mart, Kmart, Target)	Low-priced goods Store brand names Very large orders	Malaysia, Indonesia, Philippines, southern China, India, Turkey, Egypt, Brazil, Mexico, Thailand Dominican Republic, Jamaica, Haiti, Guatemala, Honduras, Costa Rica, Colombia, Chile, Poland, Hungary, Czech Republic, Bulgaria, Kenya, Zimbabwe, Mauritius, Macao, Pakistan, Sri Lanka, Bangladesh, interior China, Tunisia, Morocco, UAE, Oman Qatar, Peru, Bolivia, El Salvador, Nicaragua, Vietnam, Russia, Lesotho, Madagascar North Korea, Myanmar, Cambodia, Laos, Yap, Maldives, Fiji, Cyprus, Bahrain
Small importers	Pilot purchase and special items Sourcing done for retailers by small importers acting as 'industry scouts' in seeking out new sources of supply Relatively small orders initially but could grow rapidly	Dominican Republic, Jamaica, Haiti, Guatemala, Honduras, Costa Rica, Colombia, Chile, Poland, Hungary, Czech Republic, Bulgaria, Kenya, Zimbabwe, Mauritius, Macao, Pakistan, Sri Lanka, Bangladesh, interior China, Tunisia, Morocco, UAE, Oman Qatar, Peru, Bolivia, El Salvador, Nicaragua, Vietnam, Russia, Lesotho, Madagascar North Korea, Myanmar, Cambodia, Laos, Yap, Maldives, Fiji, Cyprus, Bahrain

Source: Based on Gereffi (1994, Figure 5.2 and Table 5.3).

In the clothing industry, therefore, the boundary between production and retailing is becoming increasingly blurred as the power within the production chain shifts further towards the buyers (including the department stores, mass merchandisers, discount chains and fashion-oriented firms). As Gereffi shows, precisely how they organize the sourcing of their garments varies according to the position they occupy within the market. Table 9.7 summarizes the situation. It shows a close association between the type of retailer and the location from which garments are sourced. For example, in general the fashion-oriented companies source within Europe and the first-generation NIEs in Asia while the discount stores source from lower-cost countries.

Strategies of developing country firms

The entire emphasis in this discussion of international corporate strategies in the textiles and clothing industries has been on firms from the industrialized countries. But a growing number of producers from the leading Asian NIEs in particular, are themselves increasingly involved in a variety of internationalization strategies. Khanna (1993) calls this the 'second migration of production'. Firms in Hong Kong, Singapore, South Korea and Taiwan face increasing competitive pressure from the newer wave of Asian producers (notably China, but also Malaysia, Thailand and Indonesia) as well as restrictions on their trade with North America and Europe through the MFA. In fact Hong Kong firms began to shift clothing production to other Asian countries as early as the mid-1960s. With the tightening grip of the MFA in the 1970s, Hong Kong firms set up plants in the Philippines, Thailand, Malaysia and Mauritius and, subsequently, in Indonesia and Sri Lanka to get round quota restrictions. In the past decade, a huge number of investments have been made in China. East Asian firms have also begun to establish plants in Europe and North America (including the Caribbean) to serve developed country markets directly. Significantly, also, a specific emphasis on product specialization and on higher-value and fashion goods is now being adopted by leading manufacturers in the more advanced of the developing country producers, particularly those in Hong Kong, Singapore, South Korea and Taiwan.

Gereffi (1996a, pp. 97–98) sees the changing position within east Asia as a process of *triangle manufacturing*:

> The essence of triangle manufacturing, which was initiated by the East Asian NIEs in the 1970s and 1980s, is that US (or other overseas) buyers place their orders with the NIE manufacturers they have sourced from in the past, who in turn shift some or all of the requested production to affiliated offshore factories in low-wage countries (e.g. China, Indonesia, or Guatemala). These offshore factories can be wholly owned subsidiaries of the NIE manufacturers, joint-venture partners, or simply independent overseas contractors. The triangle is completed when the finished goods are shipped directly to the overseas buyer . . . Triangle manufacturing thus changes the status of NIE manufacturers from established suppliers for US retailers and designers to 'middlemen' in buyer-driven commodity chains that can include as many as 50 to 60 exporting countries. Triangle manufacturing is socially embedded. Each of the East Asian NIEs has a different set of preferred countries where they set up their new factories . . . These production networks are explained in part by social and cultural factors (e.g. ethnic or familial ties, common language), as well as by unique features of a country's historical legacy (e.g. Hong Kong's British colonial ties gave it an inside track on investments in Mauritius and Jamaica).

Corporate strategies in textiles and clothing: a summary

The strategies adopted by textiles and clothing producers, therefore, are extremely varied and complex. The combinations of technological innovation, different types of internationalization strategy, the relationship with retailers and the constraints of the Multi-Fibre Arrangement combine to produce a more complex global map of production and trade than a simple explanation based on labour-cost differences would suggest. Although firms in these industries do engage in foreign direct investment this is a far less significant practice than other forms of international involvement, especially international subcontracting and licensing arrangements, often orchestrated by the large retailers and buyer groups. These are the dominant force of international production in these industries, particularly in the manufacture of clothing. In this industry, the influence of transnational corporations tends to be more indirect than direct and the involvement of local capital and entrepreneurship greater than in many other industries. The manufacture of clothing is an ideal candidate for international subcontracting. It is highly labour intensive in the developed countries, it uses low-skill or easily trained labour, the process can be fragmented and geographically separated, with design and often cutting being performed in one location (usually a developed country) and sewing and garment assembly in another location (usually a developing country).

Although international subcontracting in clothing manufacture knows no geographical bounds – with designs and fabrics flowing from the United States and Europe to the far corners of Asia and finished garments flowing in the opposite direction – there are some strong geographical biases in the relationships. For example, most of the clothing manufactured in the Caribbean region and in central America is organized by and for the US market and operates within the US government's tariff provisions on offshore assembly. Preferential access to the EU market, together with geographical proximity, have also been important in the development of offshore clothing production in the Mediterranean and north African countries. For example, the clothing industries of countries such as Malta and Tunisia both depend heavily on their links with the EU. So, too, do the transitional market economies of central and eastern Europe and the former Soviet Union.

Jobs in the textiles and clothing industries[14]

Job losses in the older centres of production

Within the developed market economies the past thirty years have witnessed a massive decline in employment in the textiles and clothing industries. Altogether, the five leading EU countries lost almost 1.2 million jobs in textiles and around 700,000 jobs in clothing between 1970 and 1993. The early 1970s were a watershed for textiles and clothing employment in most of the leading industrialized countries. In the United Kingdom, a total of 420,000 disappeared in textiles and 180,000 in clothing between 1970 and 1993. Germany lost 290,000 in textiles and 250,000 in clothing; France lost 225,000 textile jobs and 170,000 clothing jobs. Italy lost 185,000 jobs in textiles and 42,000 jobs in clothing. Almost 320,000 jobs disappeared in the United States textiles industry and 380,000 in the clothing industries.

The popular view – and, indeed, the political view as expressed through such measures as the Multi-Fibre Arrangement – is that these job losses have been caused by the wholesale geographical shift of production to cheap-labour locations and to the resulting high levels of import penetration of the domestic market. There is no doubt at all that such imports have adversely affected employment in the textiles and clothing industries of the industrialized economies. But it is misleading to attribute all – or even, in some cases, most – of the blame directly to this single cause.

Employment change within a nation's industry is the result of the operation of several inter-related forces. The most important are

- changes in domestic demand
- changes in productivity (output per worker)
- changes in exports and imports.

It is extremely difficult to calculate the effects of each of these on employment change. Cline's (1987) study of the US textiles and clothing industries between 1962 and 1985 claimed that the effect of imports on employment change in textiles was negligible compared with the effect of productivity growth and the growth of domestic demand. The picture in clothing was a little less clear but still in the same direction. Earlier studies of the components of employment change in the United Kingdom and West Germany during the 1970s reached broadly similar conclusions: that the biggest source of employment loss in the textiles and clothing industries was productivity growth.

These are, of course, aggregate figures and it is quite likely that import penetration has a more significant employment effect in some types of textiles and clothing product than others. But there is a further complication in trying to calculate the employment effects of imports in the textiles and clothing industries. The method of calculation is based upon the assumption that each factor operates independently of the others. This is clearly not the case. At least some of the increase in productivity – through technological change in the process and organization of production – is stimulated by the pressures of competition from overseas, particularly from lower-cost producers. Much of the competition has come from other developed countries, however, and only in certain types of textiles and clothing is the direct impact of developing country imports the major influence on employment loss.

All the emphasis tends to be on job losses in the older industrialized countries. However, the first-generation NIEs have also experienced substantial employment decline in these industries. Between 1987 and 1991, textile and clothing employment in South Korea declined from 784,000 to 534,000; in Taiwan from 460,000 to 311,000 (Khanna, 1993).

Characteristics of the labour force and conditions of work

Whatever the precise cause of employment loss in the textiles and clothing industries of the developed economies its actual impact is extremely uneven, not only geographically but also in terms of the kinds of worker affected. Textiles and clothing manufacture tends to be strongly concentrated in particular locations within individual countries. In the United Kingdom four regions – the north west, Yorkshire and Humberside, the east Midlands and Scotland – contain most of the country's textiles and clothing employment. In the United States the major concentration is in the southern states. In Canada the province of Quebec is the major location.

In western Europe, Lorraine in France, the Wallonian region of Belgium and the Mezzogiorno of Italy have a large proportion of their employment in the industries. The manufacture of clothing has also tended to be an inner-city industry, clustering into congested areas in old buildings: the garment districts of Los Angeles, New York City, London, Manchester and Montreal typify such tight geographical concentration. Hence, the large-scale employment losses in the industries have hit specific areas very hard indeed. In many cases, these are areas with very limited alternative employment opportunities.

The problem is made more acute by the nature of the labour force itself. Some four-fifths of the workers in the clothing industry and more than half of those in textiles manufacture are female. A substantial proportion of the labour force is relatively unskilled or semi-skilled, with no easily transferable skills. The specific sociocultural role of women, in particular their family and domestic responsibilities, also makes them relatively immobile geographically. A further characteristic of the textiles and clothing workforce in the older-industrialized countries is that a large number tend to be immigrants or members of ethnic minority groups. This is a continuation of a very long tradition, especially in the clothing industry. The early industry of New York, London, Manchester and Leeds was a major focus for Jewish immigrants. Subsequent migrants have also seen the industry as a key point of entry into the labour market. The participation of Italians and eastern Europeans in both the United States and the United Kingdom has been followed more recently by the large-scale employment of blacks, Hispanics and Asians in the United States and by non-white Commonwealth immigrants in the United Kingdom. All these 'sensitive' segments of the labour market experience real problems of alternative employment when their jobs in textiles and clothing disappear.

The history of these industries – and especially of the clothing industry – is one of appalling working conditions in sweatshop premises. At least in the textiles industries of the developed economies such conditions are now relatively rare; factory and employment legislation have seen to this. But the sweatshop has certainly not disappeared from the clothing industries of the big cities of North America and Europe. The highly fragmented and often transitory nature of much of the industry makes the regulation of such establishments extremely difficult. Some argue that there has been a major resurgence of clothing sweatshops in some big western cities. Taplin (1994, pp. 211–12), for example, argues that in the United States

> since the late 1970s the institutional regulation of labour markets has been subordinated to meeting the needs of employers, in particular giving them a freer hand in their utilization of labour. Poorly regulated labour markets, particularly weak government enforcement of workplace health and safety and few sanctions on employers for illegal employment practices have resulted, since the 1980s, in a rebirth of sweatshops in regions where immigrant labour is available. In regions such as southern California, Miami and New York/New Jersey many contractors have been able to use a Third World labour force in what amounts to de facto Third World labour market conditions . . . Meanwhile, the southeastern states continue to provide an institutional environment hostile to unions, with minimal government interference, low taxes and the availability of a surplus black labour force.

A survey of clothing establishments in San Francisco and Oakland in 1994 found 'more than half of them in violation of minimum wage standards. Sewing jobs for Esprit, Liz Claiborne, Izumi and other glittering names were being done by underpaid workers' (*The Economist*, 12 February 1994). Similar problems have been uncovered in the United Kingdom in 1996 in a series of investigations of clothing workshops in the big cities. 'Workers earn less than £2 an hour for a 50-hour week. Yet some of the UK's best-known high-street retail chains buy from these manufacturers, even though they appear to break their own guidelines' (*The Financial Times*, 3 October 1996).

In the rapidly growing textiles and clothing industries of the developing countries the labour force is similarly distinctive. Employment in the industries tends to be spatially concentrated in the large, burgeoning cities and in the export-processing zones. The labour force is overwhelmingly female and predominantly young. Many workers are first-generation factory workers employed on extremely low wages and for very long hours: a seven-day week and a twelve to fourteen-hour day are not uncommon. Employment in the clothing industry in particular tends to fluctuate very markedly in response to variations in demand. Hence, a high proportion of out-workers are employed; women working as machinists or hand-sewers at home on low, piecework, rates of pay. Such workers are easily hired and fired and have no protection over their working conditions. Many are employed in contravention of government employment regulations. Yet there is no shortage of candidates for jobs in the fast-growing clothing industries of the developing countries. Factory employment is often regarded as preferable to un- or underemployment in a poverty-stricken rural environment. A factory job does provide otherwise unattainable income and some degree of individual freedom. Often the wages earned are a crucial part of the family's income and there is much family pressure on young daughters to seek work in the city factories or in the EPZs:

> The conditions of employment in many (not necessarily all) textiles and clothing factories in the developing countries hark back to those found in the nineteenth century in Europe and North America. This tradition continues in many developing countries today: Workers hesitate to stay away from their work even when sick, due to the fear of losing their employment. Illness is quite widespread, however, not least due to insufficient ventilation, high workshop temperatures, overcrowded workrooms, and inadequate safety provisions. Complaints about health hazards are reported, such as fatigue and backstrain caused by working long hours at machines in the clothing industry. Poor lighting and demanding work often cause eyestrain, while certain materials cause skin allergies. Although the risk of industrial accidents in the clothing industry is lower than in some others, health risks due to overcrowding in poorly ventilated rooms are the main causes of widespread premature invalidity.
>
> (Robert, 1983, pp. 31–32)

As a result of pressure from groups such as Oxfam, the major clothing companies are now giving undertakings to monitor the operations of their suppliers and sub-contractors in both developing and developed countries to remove illegal practices, especially of child labour. The major UK retailers have promised to end contracts with firms which contravene their guidelines. Similarly in the United States in April 1997, the leading garment firms (including Nike, Liz Claiborne, Nicole Miller, L.L. Bean and Reebok) formally announced a code of conduct to eliminate domestic and overseas sweatshop conditions. But the process of monitoring and detection is difficult. It is even more difficult to monitor the practice of homeworking which, again, tends to be highly exploitative of the most disadvantaged groups who work at home for minimal rates of pay and no benefits. But in an industry as fragmented and organizationally complex as clothing this is an immense task.

Thus, despite the enormous global shifts which have occurred in the textiles and clothing industries and their rapid and widespread growth in Third World countries, these industries continue to be significant and sensitive sectors in the older indus-trialized countries.

Notes for further reading

1. Dickerson (1991), Scheffer (1992), ILO (1996b), provide a comprehensive treatment of the textiles and clothing industries. Earlier studies by Clairmonte and Cavanagh (1981), Toyne *et al.* (1984), Cline (1987) remain very useful general sources. Several of the chapters in Gereffi and Korzeniewicz (1994) deal with specific aspects of the industries (e.g. those by Gereffi; Appelbaum, Smith and Christerson; Taplin).

2. The irony of this is that 'the cotton revolution . . . began by imitating Indian industry, went on to take revenge by catching up with it, and finally outstripped it. The aim was to produce fabrics of comparable quality at cheaper prices. The only way to do so was to introduce machines – which alone could effectively compete with Indian textile workers' (Braudel, 1984, p. 572).

3. See Clairmonte and Cavanagh (1981), Toyne *et al.* (1984), Gereffi (1994), Taplin (1994).

4. The increasingly significant role of the retailers in these industries is discussed by Clairmonte and Cavanagh (1981), Elson (1990), Dickerson (1991), Scheffer (1992), Gereffi (1994), Taplin (1994).

5. This calculation, made by the industry consultant KSA, is based upon a 'model factory' with specific characteristics which enables comparison across countries. It is more reliable than the usual hourly labour cost comparisons. I am indebted to Markus Hassler for providing this information.

6. Discussion of technological change in these industries is provided by Hoffman (1985), Mody and Wheeler (1987), Toyne *et al.* (1984), Hoffman and Rush (1988), Dickerson (1991).

7. Mody and Wheeler (1987).

8. Hoekman and Kostecki (1995, Chapter 8) provide details of the MFA and its incorporation into the GATT agreement in the Uruguay Round.

9. Valuable studies of corporate strategies in the textiles and clothing industries are provided by Elson (1989; 1990), Oman (1989), Scheffer (1992), Gereffi (1994; 1996b).

10. See Fröbel, Heinrichs and Kreye (1980).

11. Glasmeier, Thompson and Kays (1993, p. 23).

12. See Peck and Dicken (1996) for a case study of Coats Viyella's acquisition of Tootal, then Britain's third largest textiles company. *The Economist* (18 February 1995) provides a good account of Coats Viyella's current restructuring activities.

13. Glasmeier, Thompson and Kays (1993, pp. 23–24).

14. The ILO (1996b) analyses employment issues in these industries.

CHAPTER 10

'Wheels of change': the automobile industry

Introduction[1]

The automobile industry was the key manufacturing industry for most of the middle decades of the twentieth century. Peter Drucker (1946, p. 149) captured its significance at that time: 'the automobile industry stands for modern industry all over the globe. It is to the twentieth century what the Lancashire cotton mills were to the early nineteenth century: the industry of industries.' The internal combustion engine was, quite literally, the major engine of growth for most of the developed market economies until the middle 1970s. The significance of the industry lay not only in its sheer scale but also in its immense spin-off effects through its linkages with numerous other industries. The motor vehicle industry came to be regarded as a vital ingredient in national economic development strategies.

For a while it seemed that the industry was maturing and ageing gracefully and that it had lost its propulsive influence on the industrialized economies. Not so. In the opinion of the researchers on the MIT International Motor Vehicle Program,

> 'the auto industry is even more important to us than it appears. Twice in this century it has changed our most fundamental ideas of how we make things. And how we make things dictates not only how we work but what we buy, how we think, and the way we live.'
>
> (Womack, Jones and Roos, 1990, p. 11)

Some 3–4 million workers are employed directly in the manufacture of automobiles throughout the world and a further 9–10 million in the manufacture of materials and components. If we add the numbers involved in selling and servicing the vehicles, we reach a total of around 20 million.

Organizationally the automobile industry is one of the most global of all manufacturing industries. It is an industry of giant corporations, many of which are increasingly organizing their activities on internationally integrated lines. In contrast to the textiles and clothing industries discussed in the previous chapter, the world automobile industry is predominantly an industry of transnational corporations. The ten leading automobile producers account for no less than 71 per cent of world production. Almost 90 per cent of the world total is produced by a mere twenty companies. All these either have fully fledged manufacturing operations in different countries or, at the very least, foreign assembly operations.

However, the global pattern of automobile production is not solely the reflection of the locational whims of these transnational firms. Throughout its history, the international location of the industry has been strongly influenced by the policies of national governments. Tariff, and especially non-tariff, barriers continue to exert an extremely important influence in both developing and developed economies. At the same time, national governments have struggled to outbid one another in their efforts to secure the large manufacturing plants of the major automobile manufacturers. Indeed, the automobile industry is a particularly good example of the competitive bidding process for internationally mobile investment which we discussed in Chapter 8. Not surprisingly, the giant global corporations of the industry have developed consummate skills in playing one government off against another to secure the maximum advantage from the situation.

Most national government involvement in the industry is motivated by the desire either to build a new automobile industry or – as in the traditional producing countries – to retain their existing industry. The large scale of automobile production creates very large, strongly localized concentrations of employment. In addition, the high linkage intensity of the automobile industry means that the closure or contraction of a large assembly plant will almost inevitably have very serious 'knock-on' effects on employment in the components industries.

The automobile production chain

The automobile industry is essentially an *assembly* industry. It brings together an immense number and variety of components, many of which are manufactured by independent firms in other industries. It is a prime example of a producer-driven production chain. As Figure 10.1 shows, there are three major processes prior to final assembly: the manufacture of bodies, of components and of engines and transmissions. The nature of the industry offers the possibility of organizational and geographical separation of the individual processes. How far vehicle manufacturers carry out the separate parts of the production sequence themselves varies considerably. In fact, the production of automobile components comprises a specialized set of industries in which the leading firms have themselves become increasingly transnational as the global character of the automobile industry itself has evolved. In this chapter we are concerned primarily with automobile manufacture rather than with component manufacture and with the production of passenger cars.

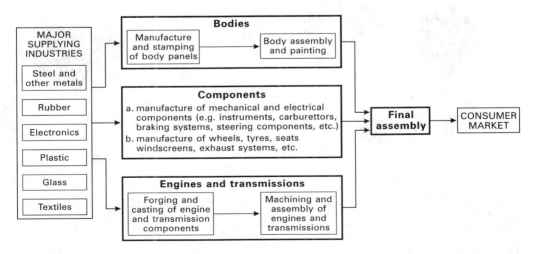

Figure 10.1   The automobile production chain

A development sequence[2]

It is possible to identify a series of stages in the development of a country's automobile industry Although it is by no means inevitable that all countries will actually pass through the sequence, it does provide a useful classification of the world automobile industry:

- *Stage 1:   Import of completely built-up (CBU) vehicles by local distributors.* This tends to be limited in scale because of high transport costs and possibly by government import restrictions.
- *Stage 2   Assembly of completely knocked-down (CKD) vehicles* imported from the home plants of world manufacturers. Permits transport cost savings and provides the opportunity to make minor modifications for the local market.
- *Stage 3   Assembly of CKD vehicles but with increasing local content.* This both depends upon, and encourages, the development of a local components industry. Strongly favoured by national governments.
- *Stage 4   Full-scale manufacture of automobiles.* Tends to be restricted to a smaller number of countries than stages 2 and 3. It is by no means inevitable that countries which have reached stage 3 will then move to full-scale local manufacture. It is even possible that a country might regress from the status of full-scale local manufacturer to that of mere assembler.

Global shifts in the automobile industry

Production of automobiles

Between 1960 and 1995 world production of passenger cars increased by 185 per cent: from 13 million to 37 million vehicles. During those three and a half decades major

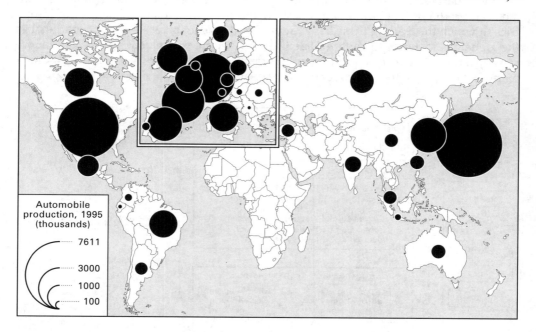

Figure 10.2   World production and assembly of automobiles
Source: AAMA (1996, *World Motor Vehicle Data, 1996*); SMMT (1996, *World Automotive Statistics, 1996*)

Table 10.1 Growth of automobile production by major countries, 1960–95

Country	1960 Production (000 units)	1960 World share (%)	1989 Production (000 units)	1989 World share (%)	1995 Production (000 units)	1995 World share (%)
France	1,175	9.0	3,409	9.6	3,050	8.2
Germany	1,817	14.0	4,564	12.9	4,360	11.8
Italy	596	4.6	1,972	5.6	1,423	3.8
Spain	43	0.3	1,639	4.6	1,959	5.3
Sweden	108	0.8	384	1.1	388	1.1
UK	1,353	10.4	1,299	3.7	1,532	4.1
Canada	323	2.5	984	2.8	1,339	3.6
USA	6,675	51.4	6,823	19.2	6,350	17.1
Japan	165	1.3	9,052	25.5	7,611	20.6
Korea	–	–	872	2.5	2,003	5.4
Malaysia	–	–	94	0.3	195	0.5
Taiwan	–	–	–	–	282	0.8
Argentina	30	0.2	112	0.3	227	0.6
Brazil	38	0.3	731	2.1	1,303	3.5
Mexico	28	0.2	439	1.2	699	1.9
Australia	–	–	357	1.0	292	0.8
Czech Rep.	–	–	184	0.5	228	0.6
Poland	–	–	289	0.8	392	1.1
World	12,999	100.0	35,455	100.0	37,045	100.0

Note:
– data unavailable.

Source: OECD (1983, *Long-Term Outlook for the World Automobile Industry*); AAMA (1996, *World Motor Vehicle Data*); SMMT (1996, *World Automotive Statistics, 1996*).

changes occurred in the global distribution of the industry. Figure 10.2 is the world automobile map for 1995, while Table 10.1 shows the major changes which occurred in the output of the major manufacturing nations. Automobile production is very strongly concentrated in the developed market economies, particularly western Europe, Japan and North America. In 1995 four-fifths of world automobile output was produced in this global triad.

By far the most dramatic development of the 1970s and 1980s was the spectacular growth of Japan as an automobile producer. In 1960 Japan produced a mere 165,000 cars, 1.3 per cent of the world total. In 1989 Japan produced 9 million autos, 26 per cent of the world total. By 1980, in fact, Japan had already overtaken the United States as the world's leading producer of passenger cars. By 1982 one car in every four produced in the world was made in Japan. By 1995, Japan's domestic automobile production had fallen back to 7.6 million cars (21 per cent of the world total), but this largely reflected the fact that an increasing proportion of Japanese vehicle production is now carried out in overseas Japanese plants.

In 1960 the United States produced more than half of total automobile output in the world; by 1995 its share had fallen to 17 per cent. Less dramatic, though nevertheless very significant, was the decline of the United Kingdom's automobile industry. In 1960 the United Kingdom produced 10.4 per cent of the world total, more than eight times more cars than Japan; in 1995 the United Kingdom's share was a mere 4.1 per cent, although that represented a substantial revival over the 1980s. Within the EU, Germany and France remain the dominant producers of passenger cars, with 12 per cent and 8 per cent of the world total respectively. The most impressive growth in automobile production in Europe occurred in Spain. In 1960 Spain produced a mere 43,000 cars; by 1995 its output was almost 2 millions.

Outside the core areas of Japan, the United States and western Europe there are few important concentrations of automobile production. One is Latin America, notably Brazil, Mexico and Argentina which, together produced 2.2 million cars in 1995. A second, more recent, centre of automobile production is in east and southeast Asia where South Korea, in particular, has very suddenly emerged as an important producer. As recently as the early 1980s, Korea was producing only 20,000 automobiles; in 1995 Korean output was 2 *million*. Malaysia and Taiwan, together, produce only half a million cars while China produces 240,000. In most of the other developing countries, however, most automobile production is simple assembly of imported components. Prior to the collapse of the state socialist economies at the end of the 1980s, there was a third significant automobile concentration in the Soviet Union and eastern Europe (notably in the USSR itself, together with Czechoslovakia, Poland and East Germany). Today, however, these automobile production complexes are in some disarray. This is especially true of the former Soviet Union. The East German industry has, of course, been absorbed into that of the former West Germany while the former state-owned automobile industries in Poland, the Czech Republic, Hungary and Romania are in various stages of transition into new forms, commonly through joint ventures with foreign manufacturers.

International trade in automobiles

The high level of geographical concentration of automobile production in the global economy is reflected in the structure of international trade. As Figure 10.3 shows, this trade is essentially *triangular* between (and within) the three major producing regions: western Europe, North America and east Asia. Approximately 94 per cent of world

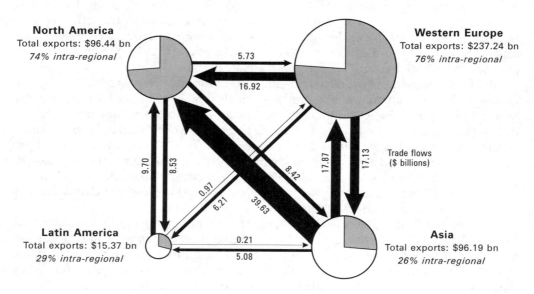

Figure 10.3 The world automobile trade network, 1995
Source: WTO (1996, *Annual Report, 1996*, Table A7)

Table 10.2   The world's leading exporting countries of automotive products, 1995

Country	Share of world exports		Annual percentage change			
	1980	1995	1992	1993	1994	1995
Germany	21.0	18.6	15	–	17	20
Japan	19.8	17.7	11	2	3	–2
USA	12.7	11.5	15	10	12	6
Canada	6.9	9.6	11	18	14	7
France	9.9	7.3	10	–	17	18
Belgium–Luxembourg	4.9	5.3	4	–	20	12
Spain	1.8	4.9	13	–	29	26
UK	5.8	4.4	4	–	16	26
Italy	4.5	4.0	–4	–	23	34
Mexico	0.3	3.1	20	23	21	39
Sweden	2.8	–	2	–15	34	–
Korea	0.1	2.0	22	51	18	57
The Netherlands	1.1	1.5	–1	–24	31	54
Austria	0.5	–	25	–14	13	–
Brazil	1.1	0.6	49	2	5	–9
Above total	93.2	94.9	11	–4	14	14

Source: WTO (1996, *Annual Report, 1996*, Table IV.44).

automobile trade is accounted for by these three regions. Western Europe alone is responsible for 52 per cent of the world total. But, as Figure 10.3 shows, three-quarters of western European automobile trade is intraregional. Table 10.2 shows the fifteen leading automobile exporting countries. Together, these accounted for 95 per cent of total automobile exports in 1995, with Germany and Japan the clear leaders, accounting for more than one-third of world exports. As in the case of production, it was the growth of Japan which was the most dynamic and transforming element in world trade in automobiles during the 1970s and 1980s in particular. In 1963 only 7.6 per cent of Japan's car production was exported; in 1995 well over 50 per cent of Japanese automobile production was exported. This, combined with a low level of imports, has given Japan a huge trade surplus in automobiles.

At the global scale, therefore, Japan came to dominate world automobile trade. It was during the 1970s, in particular, that Japanese car exports began to penetrate most world markets, including the domestic markets of the major automobile-producing nations. Import penetration increased rapidly in all cases throughout the 1970s although in the United States and some EC countries this penetration was restricted through the operation of 'voluntary export restraints'. Even so, in 1987 almost one-quarter of domestic car sales in the United States was of Japanese cars. A similar degree of Japanese penetration was evident in Belgium and The Netherlands. In the United Kingdom and West Germany the level of penetration was 10 and 15 per cent respectively, but it was much lower in France (3 per cent) and negligible in Italy. By far the highest degree of Japanese import penetration into countries with their own domestic vehicle industry was in Australia, where virtually one car in every two sold in 1987 was Japanese made. As we shall see later, the threat and subsequently the reality of Japanese competition in the automobile industry was the single most important force stimulating change among competing firms in the other major producing countries.

Of all the major automobile-producing countries, Japan has the most geographically extensive pattern of exports. As Figure 10.4 shows, its tentacles spread throughout the world although with particular emphasis on North America (46 per

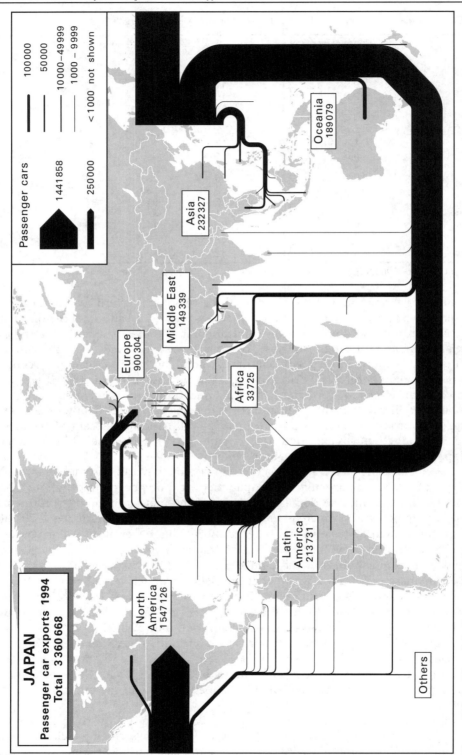

Figure 10.4 The global pattern of Japanese automobile exports, 1994
Source: AAMA (1996, *World Motor Vehicle Data*, 1996)

Table 10.3 Geographical composition of automobile exports, 1994

	Europe	North America	Latin America	Africa	Asia/Middle East	Oceania
			Share by destination (%)			
USA	12.0	55.0	8.0	0.2	24.0	1.0
Canada	0.8	98.9	0.1		0.3	
Germany	77.0	9.0	1.0	0.5	11.0	1.0
France	86.0	0.2	5.6	2.0	3.0	1.0
Italy	91.0	0.2	2.8	1.0	5.0	0.2
Spain	93.1	2.0			1.0	
UK	85.5	4.9	0.5	0.4	6.2	2.0

Source: Calculated from AAMA (1996, *World Motor Vehicle Data, 1996*, AAMA, Washington DC).

cent of the total) and Europe (27 per cent). In comparison, the trade patterns of the other leading producer countries are far less extensive than that of Japan. Table 10.3 summarizes the broad patterns. In the case of the leading European producers, in particular, automobile exports are overwhelmingly focused on Europe itself. This is most apparent in the case of Spain and Italy, but it is notable that even Germany sends more than three-quarters of its automobile exports to European markets. The world's third leading automobile exporter, the United States, sends two-thirds of its cars to Canada (55 per cent) and Latin America (8 per cent); 24 per cent to Asia (including the Middle East); and 12 per cent to Europe. Canada, of course, is an integral part of the United States automobile industry so it is not surprising to find that 99 per cent of Canada's automobile exports flow to the United States.

Outside the dominant producers of the developed market economies there has been considerable growth in exports from some of the developing countries. Although their overall share of total world automobile exports is still modest, their emergence as vehicle and component exporters is a significant development. For the most part, however, there is a strong geographical bias to these exports as Table 10.4 reveals. Virtually all car exports from Brazil, for example, go to other parts of Latin America, while Mexico is strongly tied in to the US market. The major exception is South Korea which, as we have seen, is the only major new indigenous automobile producer to have emerged in the past few years. The geographical composition of Korea's automobile exports shown in Table 10.4 is not dissimilar from that of Japan. It is striking that Korea's automobile exports are not dominated by Asian countries but by North America and Europe. From virtually a standing start in the 1980s, the Korean automobile industry has become a world player. Thus, although both production and trade in the automobile industry remain strongly dominated by the core

Table 10.4 Geographical composition of automobile exports from leading developing countries, 1994

	Europe	North America	Latin America	Africa	Asia/Middle East	Oceania
			Share by destination (%)			
Brazil	7.3		92.7			
Mexico		87.7	8.9		2.5	
South Korea	22.1	36.2	14.0	2.3	18.4	7.0

Source: Calculated from AAMA (1996, *World Motor Vehicle Data, 1996*, AAMA, Washington DC).

producers of Japan, North America and western Europe, it is clear that considerable changes have occurred and that some new producers are emerging.

The changing pattern of demand for automobiles

Changes in the level, the composition and the geography of demand have all played a key role in the evolving global map of the automobile industry and in the problems facing many of the traditional producing nations and firms. Automobile production is strongly *market oriented*. Thus, historically, production has developed within large, affluent consumer markets in which high levels of demand have permitted the achievement of economies of scale. The world's three major vehicle-producing regions – North America, western Europe and Japan – are also the world's most developed consumer markets. The changing demand for automobiles has three major characteristics:

- it is highly cyclical
- there are long-term (secular) changes in demand
- there are signs of increasing market segmentation and fragmentation.

Demand for automobiles is very sensitive to changes in the level of economic activity as a whole. Periods of high and increasing demand are separated by periods of stagnant or declining demand. However, demand trends vary substantially between different parts of the world. Demand for automobiles is growing most slowly in western Europe and North America – what one industry observer called 'glacial growth' in these mature markets – and fastest in some of the east and southeast Asian countries. There is an expectation of a major surge in demand in eastern Europe in the 1990s but this will clearly depend upon the rate of economic reconstruction there. The major automobile manufacturers, therefore, are pinning their hopes on continuing high levels of growth in demand outside North America and Europe. In both these mature producer regions, there is very substantial excess capacity amounting to several million vehicles, hence the continuing process of rationalization involving plant closures and contractions. However, recent reports suggest that the growth potential of Asia may be exaggerated. As we show in a later section of this chapter, the massive growth of automobile production in Asia may well exceed the ability of the market to absorb it by the year 2000.

The slow growth in demand for automobiles in the western European and North American markets reflects more than just cyclical forces. There are deeper *secular* or structural characteristics in these markets which limit future growth in car sales. Demand for automobiles is broadly of two kinds: new demand and replacement demand. Rapid growth in demand is associated with new demand. But as a market 'matures' and vehicle ownership levels approach 'saturation', the balance between new and replacement demand alters. More and more car purchases become replacement purchases. In the mature automobile markets today some 85 per cent of total demand for automobiles is replacement demand. Replacement demand is generally slower growing and also more variable because it can be postponed. In such mature markets, manufacturers have long adopted strategies similar to those of clothing manufacturers: they regularly introduce 'new' models to entice existing owners to replace their vehicles. Often, the changes are little more than cosmetic although promoted by massive advertising campaigns as being both significant and highly desirable. New production technologies are, however, making possible a far greater variety of vehicle types.

Until relatively recently there were considerable differences between the major automobile markets, each tending to favour particular types of vehicle. The greatest difference was between North America on the one hand and western Europe and Japan on the other. Demand in the highly affluent, highly mobile, cheap energy-based North American market was overwhelmingly for very large cars. In contrast, in western Europe and Japan generally lower incomes and higher energy costs plus more congested driving conditions were expressed in a demand for smaller, more fuel-efficient cars. But even within Europe, individual national markets tended to be served separately by domestically based producers. During the past twenty years, and especially since the oil crises of 1973 and 1979, which reduced the attractiveness of large cars, these circumstances have changed dramatically.

In the case of the automobile industry, of course, the impact of oil price rises was especially severe. Virtually at a stroke, the cost to the consumer of owning and running a car accelerated to unprecedented levels. The impact was especially severe in North America: demand for large gas-guzzling cars plummeted (though it later recovered somewhat). The large US manufacturers moved to 'downsize' their range, a trend reinforced by US government regulations to reduce the average gasoline consumption of automobiles. A major reason for the increased import penetration of the US market in the 1970s, therefore, was the inability of the domestic manufacturers to produce small, fuel-efficient cars to compete with European and, especially, Japanese vehicles.

One result of all these changes in the structure of demand has been to reduce some of the geographical differences between individual national markets. Demands have begun to *converge*, at least in the mass-market sector. Regional and even global markets have become apparent. Within such markets, however, there are signs of greater market *segmentation* and *fragmentation*: demand for particular types of car for particular uses (e.g. four-wheel-drive recreation vehicles) or for a customized version of a general model. At least in the affluent consumer markets this is leading to more consumer-driven choice, which can, in turn, be satisfied because of the dramatic changes taking place in the way automobiles are made.

Production costs, technological change and the changing organization of production[3]

Mass production

The basic method of producing automobiles changed very little between 1913, when Henry Ford first introduced the moving assembly line, and the 1970s, when a radically new system of production began to emerge in Japan.[4] The automobile industry was *the* mass-production industry. It produced a limited range of standardized products for mass-market customers. It produced in very large volumes in massive assembly plants using very rigid methods in which each assembly worker performed a highly specialized and narrow task very quickly and with endless repetition. The automobile industry appeared to have abolished craft production for ever, apart from the small number of firms manufacturing for the luxury car market.

As developed by Ford and General Motors, and adopted by all the other western manufacturers, this 'Fordist' method of production certainly brought the automobile as a commodity into the reach of millions of consumers. But, although very efficient in many respects, it contained one major limitation: its *rigidity*. In order to reduce costs a particular model had to be produced in huge volumes. Assembly lines were 'dedicated' to a specific model and to change models took a great deal of time and money. Individual assembly workers were mere small cogs in the wheel of production, with no responsibilities beyond their very narrow single task.

The automobile industry is an assembly industry requiring hundreds of thousands of components – ranging from entire engines and bodies down to small pieces of cosmetic trim. The big American and European manufacturers developed a particular kind of relationship with their suppliers, based on short-term, cost-minimizing contracts. The supplier–customer relationship was distant not just in functional terms but also, increasingly, in geographical terms. The major producers scoured the world for low-cost component suppliers. The close geographical proximity of customer and supplier, which had been a feature of the early years of the industry, began to break down as technological developments in transport and communication made long-distance transactions feasible. The increased geographical distance between the assemblers and their suppliers made it necessary for the assemblers to hold huge inventories of components at their assembly plants. In this way, the possibility of the assembly line being disrupted by a temporary shortage of components (or by faulty batches) was reduced. This was the 'just-in-case' system, described in Chapter 5.

Production of an automobile, from its initial design stage through to its appearance in the sales rooms, is an immensely complex and expensive process. A five-year development period has been the norm; costs of up to $5 billion have been common for the very large-scale and complex projects. Such enormous costs of development forced the major producers to seek very large production volumes to achieve economies of scale in production. It was generally agreed that a vehicle manufacturer needed to produce approximately 2 million vehicles per year to reap the maximum benefits from scale economies, although the minimum efficient scale of production varied according to the particular manufacturing process involved. In locational terms,

> body assembly, paint trim and final assembly on a particular model should be carried out together, because of the high costs involved in transporting assembled bodies. Transport of unassembled bodies is less expensive and the location of the pressing plant is therefore less crucial. Engine and power train (engine and gearbox) components are easier and cheaper to transport, and the production technology involved is also very different, so there is no particular need to have these plants on the assembly sites. The same would apply to gearbox and axle production.
>
> (Bhaskar, 1980, p. 55)

For the automobile industry as a whole, labour costs may account for between a quarter and a third of total production costs. The importance of labour costs in motor vehicle manufacture, like scale economies, varies between different processes. Final assembly is the most labour-intensive stage. For example, the labour involved in body-making and final assembly may account for about 23 per cent of total vehicle production costs, whereas the labour costs involved in engine and transmission production are only 8 per cent of the total. Thus, certain parts of the automobile production process are more sensitive to labour cost differentials than others. As always, of course, it is not simply the cost of labour which is important but, rather, the overall quality and reliability of the labour force. Such considerations have played an important part in the investment-location decisions of most major transnational vehicle manufacturers. Certainly they have sought out areas of surplus labour and, especially, those in which labour relations seem less likely to be problematical, both within and between individual countries. The importance of geographical differences in labour costs is, however, declining as the industry becomes more capital intensive.

A major factor here is, quite obviously, *technological change*. The industry's production cost structure has not been static; production cost relations have been altered by developments in the technology of both products and processes. Until recently, technological change in the industry seemed to be creating two distinct – and opposed – tendencies. One was towards even greater scale of production of *standardized* vehicles, the other was towards smaller-scale, more specialized production of vehicles aimed at particular *niches* in the market. Some automobile manufacturers have favoured one path of development while some favoured the other. However, this distinction is becoming increasingly blurred.

Product innovation in the motor vehicle industry has generally been incremental. However, very rapid and far-reaching changes have occurred since 1973 in the aftermath of the oil crisis. The most important innovations have been those directed towards reducing fuel consumption, for example by making more efficient engines, by reducing the weight of materials used (such as substituting plastics and non-ferrous metals for steel), by reducing the size of cars (especially in North America) and by increased use of electronics to control engine performance. In addition, government safety and anti-pollution regulations in many countries have created pressures for change in car design. As yet, however, the development of cars powered by alternative fuels has been very limited.

Lean production

Many of the characteristics of the automobile industry just described remain very important. Since the 1970s, however, and with accelerating speed, major – even revolutionary – changes in the ways in which automobiles are developed and manufactured have been occurring. The source of such changes, which are dramatically reshaping the industry, is the leading Japanese producers, led, initially, by Toyota. Womack, Jones and Roos (1990) use the term *lean production* to contrast with the mass-production techniques which have pervaded the industry. In their view, 'lean production combines the best features of both craft production and mass production – the ability to reduce costs per unit and dramatically improve quality while, at the same time, providing an even wider range of products and even more challenging work' (Womack, Jones and Roos, 1990, p. 277). The major differences between craft production, mass production and lean production were discussed in Chapter 5 and are summarized in Table 5.2.

However, such flexible systems do not, as is so often stated, mean the reduced importance of economies of scale:

> What the new lean production techniques allow is the production of a variety of cars but within a large annual volume. Hence, an assembly plant is still optimum at around 250,000 units a year, although the model specific optimum can be lower . . . What the new techniques do is to make it easier for large companies to make a variety of products, but they do not make it easier for small companies to survive.
>
> (Rhys, 1990)[5]

What they also appear to do is make possible shorter design periods per model.

A major requirement of the flexible production of automobiles is ease of assembly. The problem is that

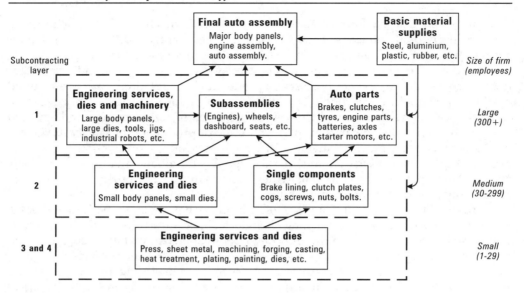

Figure 10.5   Functional tiers in the Japanese automobile supplier system
Source: Based on Sheard (1983, Figure 2)

before you can make things flexibly, you must first make them simple. Simplifying is difficult because each new car is more complex and more costly to develop than the last. So the trick is to get somebody else to develop parts of the car for you – and ideally to build them as well . . . Modular manufacturing involves designing and assembling the entire car as a series of sub-assemblies or modules. New modules can be developed directly to replace an existing one, allowing cars to be changed easily.

(*The Economist*, 29 July 1989)

Such modules may be made by the automobile assembler but, in many cases, they are made by outside suppliers.

The relationship between customer and supplier in such a flexible production system must, however, be very different from that in the mass-production system. The relationship has to be extremely close in functional terms, with design and production of components being carried out in very close consultation. Long-term relationships are preferable to short-term relationships. The system adopted by companies like Toyota is one of tiered suppliers, each tier having different tasks. The first-tier suppliers relate directly to the customer, second-tier suppliers to first-tier suppliers and so on (Figure 10.5). The use of first-tier (or preferred) suppliers tends to involve a smaller number of suppliers than is common in the more remote, cost-driven, procurement systems. There is also a clear trend for the number of first-tier suppliers to be decreasing. Not only are there close functional relationships between customers and suppliers in this system but also there are closer geographical relationships. The use of just-in-time methods, which we discussed in Chapter 5, encourages geographical proximity although the distances involved may be very variable, ranging from suppliers located literally next door to assemblers (e.g. in 'supplier parks') to those located within the broad region (e.g. within Europe).

Automobile component suppliers, in fact, are now increasingly involved in different kinds of functional relationships with the automobile assemblers as the latter

Table 10.5   Types of supplier responsibilities within the automobile industry

Traditional subcontracting	Provision of component system	Parallel development	Codevelopment
Production of components from detailed specifications provided by automobile manufacturers	Production and assembly of complete component systems based upon technical directives of automobile manufacturers	The supplier is involved in the manufacture, assembly and development of components	The supplier has complete responsibility for the design and production of components
The supplier has full responsibility for performance and quality	The supplier controls the quality and cost of parts procured from their suppliers	The supplier has the capability of making technical and cost adjustments during the design process	Application of simultaneous engineering of product and process
The supplier delivers the components to the assembly plant	The supplier controls the logistics chain of the component system	The automobile manufacturer retains control of development, design and prototype testing	The automobile manufacturer manages the technical interface between components

Source: Based on Laigle (1996, Table 1).

shift more responsibility (and risk) on to the supplier firms. The evolving relationships can take a number of different forms as Table 10.5 shows. Each type involves rather different kinds of supplier responsibility. The most significant trend is for suppliers to manufacture entire 'subassemblies' rather than individual components. Indeed, VW is introducing an even more revolutionary system at a new plant at Resende in Brazil:

> Instead of buying parts from outside, VW will ask component makers to fit their products directly on the assembly line . . . Suppliers will assume responsibility for putting together and installing four 'modules' in all: the chassis; axles and suspension; engines and transmissions; and driver's controls . . . Resende's move to bring suppliers directly on to the assembly line is unprecedented . . . The way the factory will be managed is similarly mould-breaking. Responsibilities will be shared between VW and its main partner-suppliers, forming . . . a 'consortium' . . . Together with VW they will run the plant and divide the profits.
>
> (*The Financial Times*, 29 March 1996)

Whether or not this 'assembly line of the future' becomes accepted practice elsewhere within VW or, more widely, amongst other automobile manufacturers is a matter of debate. More important, however, is the fact that new forms of assembler–supplier relationships are being forged throughout the automobile industry. They are part of the series of profound technological and organizational changes which are sweeping through the automobile industry. From being an apparently 'mature' industry in which technologies were relatively stable, the industry has 'dematured'. As we shall see in a later section of this chapter, these developments are having profound implications for the corporate strategies of the automobile manufacturers. Before reaching that point, however, we need to consider the influence of government policies on the industry.

The key role of the state[6]

The world automobile industry is dominated by a small number of global corporations, yet today, as in the past, their investment-location decisions are greatly affected by the policies of national governments towards the industry. Before looking at the strategies pursued by the major manufacturers themselves in their efforts to remain afloat in the turbulent seas of international competition we should look, therefore, at the general ways in which national governments affect the automobile industry. As Reich (1989) has shown, in an industry like automobiles, which is dominated by transnational producers, there are two key aspects of state policy. The first is the *degree of access* to its domestic market which the state allows to foreign firms to establish production plants there. The second important aspect of state policy is the kind of *support provided by the state* to its domestic firms and the extent to which the state discriminates against foreign firms.

Reich suggests that each of the major automobile-producing countries has pursued a different mix of these two policies:

- *France*: limited access granted to foreign firms and discriminatory intervention in favour of domestic producers.
- *Britain*: unlimited access granted to foreign firms and equal treatment of both foreign and domestic producers.
- *Germany*: unlimited access granted to foreign firms and discriminatory intervention in favour of domestic producers.
- *United States*: unlimited access to foreign firms and equal treatment of both foreign and domestic producers.

To this list we should add:

- *Japan*: limited access granted to foreign firms and discriminatory intervention in favour of domestic producers.
- *Korea*: no access to foreign firms as producers; foreign automobile imports highly restricted by non-tariff barriers.

The shift from tariff to non-tariff barriers

Historically, the existence of national tariff barriers around sizeable consumer markets explains much of the early geographical spread of the automobile industry. Most governments in Europe and elsewhere (including Japan, Canada and Australia, for example) in the period before the Second World War had erected tariffs against automobile imports. These provided an obvious stimulus both to the setting up of foreign branch plants by the major vehicle producers and also to the growth of domestic manufacturers. Today, few of the developed market economies levy particularly high import tariffs against automobile imports although they remain high in some cases. The common EU tariff is 11 per cent, the US automobile tariff is only 3 per cent. Japan now has no import tariffs on motor vehicles although it has been much criticized for its slowness in removing certain non-tariff barriers, as the dispute with the United States exemplifies.

Although tariffs in the major developed country markets are now generally low, there has been a strong upsurge in other forms of protection, primarily against Japanese imports. We saw earlier just how rapidly and deeply Japanese import penetration has developed since the 1970s. In response to the growing domestic outcry from major industry pressure groups both the United States and several western European countries negotiated 'voluntary' export restraints with Japan. In 1981 the US government established such a bilateral agreement as an allegedly temporary measure to last for three years. Although the import ceiling has been raised since then the agreement still exists. The overall level was negotiated between the US and Japanese governments. The actual share of the cake between the Japanese firms themselves has been determined by the Japanese government.

In Europe, the completion of the Single European Market in 1992 replaced the individual bilateral agreements which had previously existed between several countries and Japan. After lengthy and acrimonious negotiations, the European Commission negotiated an agreement with Japan which limited Japanese automobile imports to a given percentage of total EU vehicle sales for a transitional period of seven years. After that, there will be no restrictions on Japanese automobile imports into the EU. Apart from internal disagreements over the length of the transitional period (with the French, Italians and Spanish pushing especially hard for a lengthy transitional period before voluntary export restraints were phased out), the biggest area of controversy was over the kind of treatment to be accorded to the vehicles actually manufactured within Europe itself (i.e. in the UK where, at the time, all the Japanese transplants were located).

There was particularly bad feeling between the French and Italians on the one hand and the British on the other over the treatment of cars manufactured in Japanese plants in Britain. Are they sufficiently 'European' or are they really Japanese? In the continuing political argument, Britain has been variously described as a 'Trojan Horse' infiltrating Japanese cars into the European market or as an 'aircraft carrier' moored offshore and launching Japanese cars into Europe. Given Toyota's recently announced proposal to consider building its second European plant in northern France it will be interesting to see whether the attitude of the French government changes, although that of the French automobile producers will not. What is clear, however, is that

> The EC's 'consensus' with Japan . . . appeared only to have papered over the cracks of internal dissent, let alone establish a secure framework for international trade. It was subject to the differing interpretations of member states over what represented an appropriate industrial policy, and the EC's neo-liberal line – in this as in other sectors – was sharply discordant with that in at least some European capitals.
>
> (Sadler, 1995, p. 35)

Local content and export requirements

The increased fear of Japanese imports into both North America and western Europe has stimulated additional policy measures. One is to encourage Japanese vehicle manufacturers to build *overseas production plants* in their major markets to displace imports. The other is to insist on specific levels of *local content* in vehicle manufacture by foreign producers. Again, it is in Europe that the greatest controversy has developed over this issue, both over precisely what local content means and also how it should be measured. For a specific EU member state 'local' refers to all EU countries and not just its own domestic territory. But how much local (i.e. EU) content must an automobile contain in order to count as a domestic product? The Treaty of Rome does not lay down any quantitative threshold of local content; it simply states that a product is regarded as originating in the European Community 'if the last substantive manufacturing process takes place in a Community country'. As far as Japanese firms in Europe are concerned, an eventual level of 80 per cent local content seems to be regarded as acceptable (although the French would prefer 100 per cent).

Such measures, though relatively new in the developed country context, have long been important elements in the national policies of developing countries towards the automobile industry. Virtually all developing economies with any kind of motor vehicle industry operate both local content requirements and various types of tariff and non-tariff import restrictions. The use of both local content and other import restrictions together with very high tariffs on vehicle imports formed the basis of the strong import-substitution policies followed by a number of developing countries, particularly during the 1950s and 1960s. The aim was to build a domestic automobile industry to serve the domestic market. Each of the major industrial countries in the developing world – India, Brazil, Argentina, Mexico – as well as countries such as Spain adopted this kind of strategy. In most cases local content requirements were set at between 50 and 90 per cent, usually on a progressively rising scale over a period of several years.

Although domestic market protection remains a prominent feature of most developing countries' automobile industry the general shift of policy towards export promotion (see Chapter 4) has also been reflected in the automobile industry. Apart from the usual battery of financial and tax incentives and concessions, each of the major Latin American countries has export requirements under which manufacturers must export a specified proportion of their output. The increased involvement of Brazil, Mexico and Spain, in particular, in the international automobile trade is a reflection of these efforts and of the response of transnational automobile manufacturers. In the late 1980s Mexico made explicit moves to increase its involvement in the internationally integrated operations of the major transnational firms by substantially modifying its local content requirements and reducing import restrictions on vehicles but still insisting on a proportionate level of exports by the TNCs operating within Mexico.

General government involvement in the automobile industry

Apart from questions of trade regulation the major developed country governments have also intervened in their automobile industries in a number of ways which vary considerably from one country to another. In general, the western European governments, especially France, the United Kingdom and Italy, have been more extensively involved in their domestic vehicle industries than the governments of the United States and Japan, and have also gone furthest in attempting to restructure their industries. In the past, of course, Japanese government 'guidance' of the country's growing automobile industry was indeed considerable.

Japanese government involvement was especially marked in the 1950s and 1960s and took a number of forms. First, very tight protective barriers were placed around the domestic industry: strict import quotas and extremely high tariffs were imposed, which prevailed until relatively recently. Second, the direct involvement of foreign manufacturers in the Japanese industry was prohibited for a long period. The much-needed automobile technology was acquired through licensing agreements with foreign manufacturers. Third, MITI attempted to encourage rationalization among the major Japanese vehicle producers although with little success. Fourth, the Japanese government was heavily involved in assisting overseas marketing and exports, through financial and other assistance. These measures are no longer in force; instead the Japanese government's involvement in the automobile industry is now primarily in negotiating voluntary export restraint agreements with its trading partners.

Until recently, the governments of France, the United Kingdom and Italy were directly involved both in direct ownership of vehicle producers and in large-scale financial support. (Only in France does a residue of state ownership remain in the case of Renault where the government retains a stake.) Each of the state-owned enterprises had needed enormous financial subsidies. European governments have also used their regional development policies to influence the location of automobile manufacturers and to persuade both domestic and foreign firms to build plants in depressed regions of the country. Massive financial incentives and restrictions on development elsewhere have been used to establish huge assembly plants in areas of high unemployment. Although the US government has not pursued such a spatially discriminatory policy, huge incentive packages have been used by individual states to attract Japanese and German automobile assembly and component plants to specific locations. As we saw in Table 8.4, the sums involved were enormous.

The implications of regional integration

A particularly significant aspect of state policy towards the automobile industry is the emergence, or strengthening, of regionally integrated trade areas in North America, Europe and Latin America.

The Single European Market

The establishment of the European Economic Community in 1957 and its subsequent enlargement in 1973, 1981, 1986 and 1995 to the present fifteen member states have had a dramatic influence on the shape of the automobile industry within Europe.[7] Indeed, it is in Europe that the greatest progress has been made by the automobile TNCs, especially Ford and General Motors, in creating international integrated production networks. The completion of the single market in 1992 merely added the finishing touches by removing the remaining technical and physical barriers to the flows of vehicles and components. Technical barriers were especially significant for the achievement of economies of scale in the automobile industry. The practice whereby an automobile manufacturer had to obtain different type-approval certification for every member state, involving compliance with often minute variations in vehicle specification, has now been abolished. This greatly reduces the costs of detailed design and manufacturing variations. In general, the major beneficiaries of the single market have been those automobile producers which already had the most highly integrated operations across national boundaries within Europe.

The European automobile industry has undoubtedly changed in the aftermath of '1992'. More significant in the long run, however, will be the opening up of the eastern European region. Suddenly, a huge contiguous region is being created with the potential of being both a large consumer market and also a low-cost production location for sourcing both components and finished vehicles. The political developments of the late 1980s and early 1990s are presenting major strategic opportunities for automobile manufacturers operating in Europe as we shall see in a later section. At the same time, the political transformation of the entire European automobile production space poses serious problems for countries such as Spain and Portugal. As the countries of eastern Europe become increasingly integrated into the European Union then the EU states on the western periphery will be even further removed from the centre of gravity of the European automobile industry.

Regional integration in North America

In North America, too, political agreements on the integration of national markets have had profound repercussions on the structure of the automobile industry. The

1965 Automobile Pact between the United States and Canada reshaped the industry, producing an integrated structure in which production by the major manufacturers was rationalized and reorganized on a continental, rather than a national, basis.[8] By the early 1970s, in fact, the automobile industry in North America was fully integrated; Canadian plants performed specific functions within the larger continental production and marketing system.

The 1988 Canada–United States Free Trade Agreement (CUSFTA) also contained important provisions for the automobile industry. In particular, it redefined the level of 'North American content' necessary for a firm to be able to claim duty-free movement within the North American market. The CUSFTA, by making distinctions between different types of automobile manufacturer based on their current status, in effect created a 'two-tier' automobile industry in Canada:

> On the one hand, General Motors, Ford, Chrysler . . . so long as they comply with the provisions of the Auto Pact, will be able to continue to bring parts and vehicles into Canada duty-free from any country (including such low cost countries as Brazil, Korea, Mexico, Taiwan and Thailand). On the other hand, Honda, Hyundai, Toyota and any future new producer, operating in Canada in the same way as their commercial rivals, will have to pay duty on anything they import from countries other than the United States'.

(Holmes, 1990, pp. 173–74)

The North American Free Trade Agreement (NAFTA) has even more far-reaching implications for the automobile industry because of the lower production cost characteristics of the Mexican auto industry and the fact that it was already becoming increasingly integrated into the North American market anyway through the strategies of the American, Japanese and European producers and the changing attitude of the Mexican government. As we showed in Box 3.4, automobile tariffs are to be removed within NAFTA over a period of ten years and vehicles will have to meet a 62.5 per cent local content requirement in order to be free of tariffs.

Regional integration in Latin America: Mercosur

Until recently, the major Latin American automobile producing countries, Brazil and Argentina, operated strongly protective, nationally based import-substitution policies. Such a stance has not disappeared but the creation of the Mercosur regional bloc, involving Argentina, Brazil, Paraguay and Uruguay has produced a substantial shift in policy emphasis. One result of Mercosur is to create an expanded regional market of some 2 million vehicles a year, access to which is limited through tariffs which are to fall to 20 per cent by the year 2000. Consequently, as we shall see in a later section, there has been an enormous upsurge of investment in both Argentina and Brazil by American and European automobile manufacturers.

Environmental and safety legislation

A final aspect of state policy to be mentioned is that of environmental, fuel-efficiency and safety legislation. All governments have some legislation covering these areas, each of which has profound implications for the design, technology and materials used in cars and, therefore, in their cost. Complying with changes in legislation can be especially problematical where it involves fundamental design changes. For US producers, one of the major effects of the 1973 oil crisis on the industry was the introduction of fuel economy standards whose aim was to lower the average fuel consumption level. Legislation to control noxious emissions from automobile engines has been particularly stringent for many years in states such as California. Recent heightened concern with the environment has led universally to intensified measures to clean up engine emissions through the introduction of catalytic converters and the like. The 1990 US Clean Air Act's fifty new requirements were estimated to cost the industry between $8 billion and $10 billion a year to implement.

Corporate strategies in the automobile industry

Increasing global concentration; increasing global connectedness
In the early, pioneer days of the automobile industry in North America and western Europe there were scores of manufacturers each producing a limited range of vehicles for individual national markets. In 1920, for example, there were more than 80 automobile manufacturers operating in the United States, more than 150 in France, 40 in the United Kingdom and more than 30 in Italy. Today, after decades of acquisitions and mergers each major national market is dominated by a very small number – literally a handful in most cases – of massive corporations. The top four firms account for between 88 and 100 per cent of national automobile production in each of the major producing countries. In France and Sweden the entire national output of cars is produced by only two companies; in Italy one firm, Fiat, is totally dominant. But such high levels of concentration are evident not only at the national scale but also internationally: the global automobile industry is in the hands of a small number of very large firms.

Table 10.6 lists the world's leading automobile manufacturers. It shows that the fifteen firms produced four-fifths of world automobile output in 1994. The two leading firms – General Motors and Ford – produced one-quarter of the world total between them. The top five produced 49 per cent and the top ten 71 per cent of the world total. GM and Ford have remained the world's leading automobile producers throughout the postwar period although their share of the world total has declined. Chrysler, for long the world number three, plummeted to twelfth position, shedding its European operations as part of its efforts to stay in business. Other changes in relative position occurred through acquisitions and mergers.

Table 10.6 The world league table of automobile producers, 1994

Rank	Company	Country of origin	Automobile production	Share of world total (%)	Percentage produced abroad 1994	1989
1	GM	USA	5,183,238	14.5	49.7	41.8
2	Ford	USA	3,854,149	10.8	56.9	58.7
3	Toyota	Japan	3,357,337	9.4	17.5	8.3
4	VAG	Germany	2,693,861	7.5	43.7	30.5
5	PSA	France	2,243,280	6.3	21.1	15.4
6	Nissan	Japan	2,016,276	5.7	33.5	13.8
7	Renault	France	1,850,394	5.2	24.6	17.6
8	Fiat	Italy	1,658,130	4.6	25.7	7.1
9	Honda	Japan	1,456,817	4.1	41.7	28.0
10	Mitsubishi	Japan	1,116,075	3.1	20.1	–
11	Mazda	Japan	1,068,469	3.0	23.1	18.3
12	Chrysler	USA	972,187	2.7	43.3	13.0
13	Hyundai	Korea	896,592	2.5	–	–
14	Daimler-Benz	Germany	583,833	1.6	–	–
15	BMW	Germany	550,636	1.5	0.1	–

Source: Calculated from AAMA (1996, *World Motor Vehicle Data, 1996*).

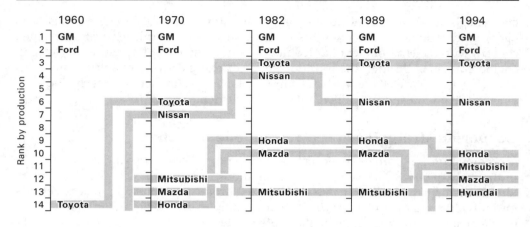

Figure 10.6   The rise and rise of Japanese (and Korean) automobile manufacturers

By far the most dramatic shifts in relative position – not surprisingly – involved the leading Japanese firms, particularly Toyota and Nissan (Figure 10.6). Before 1960 no Japanese manufacturer ranked among the world's top fifteen vehicle producers. In 1960 Toyota appeared in the top league for the first time – in fourteenth position. By 1965 Toyota had moved up to ninth position and had been joined by Nissan in eleventh place and Toyo Kogyo (Mazda) in thirteenth place. Five years later, Mitsubishi and Honda had also entered the first division while Toyota and Nissan had moved into sixth and seventh position respectively. By 1989 Toyota was challenging Ford for second place in the world league, Nissan ranked sixth, Honda ninth and Mazda tenth. In 1994, Honda had slipped down one place to tenth, Mitsubishi had climbed two places to eleventh while Mazda fell from tenth to twelfth. The biggest change, however, was the emergence of the Korean automobile firm, Hyundai, in thirteenth position.

Table 10.6 also indicates the extent to which the leading automobile manufacturers are *transnational* in their operations. Quite clearly a good deal of variety exists. Ford is, by a considerable margin, the most transnational producer of all: in 1994 almost 60 per cent of its passenger car production was located outside the United States. GM's degree of transnationality, however, though substantially lower – 50 per cent in 1994 – has been increasing rapidly in recent years (in 1989, for example, only 42 per cent of GM's production was outside the United States). Of the other very large automobile manufacturers (those producing more than 1 million cars) Volkswagen has the largest share of its operations located overseas (44 per cent). But if the big US producers are the most transnational in their activities the big Japanese producers are clearly the least transnational of the major producers. Apart from some local assembly operations which build cars from imported knock-down kits for local markets, Toyota had no overseas production facilities for passenger cars in 1982, while less than 3 per cent of Nissan's total production was located outside Japan. Among the smaller Japanese companies only Honda had overseas production facilities (in the United States).

However, the transnationality of the Japanese producers changed dramatically during the 1980s, although Honda still remains the most internationalized of the Japanese companies. In 1989, 28 per cent of Honda's output was produced outside

Japan; in 1994, the proportion was 42 per cent. Toyota more than doubled its share of its overseas production (from 8 per cent to 18 per cent); Nissan produced one-third of its cars outside Japan in 1994 compared with 14 per cent in 1989.

The statistics shown in Table 10.6 do not reflect one of the most important developments of recent years: the fact that most of the world's automobile manufacturers are deeply embedded in *collaborative agreements* with other manufacturers and in complex transnational sourcing arrangements for components with the major component firms. The very high level of concentration in the world automobile industry is largely the outcome of factors discussed earlier in this chapter, in particular the drive to achieve efficiency in design, production and marketing in an increasingly competitive global market. A major problem, however, is that the development of new models, which themselves have a shrinking life, requires massive investment not only in machinery and other equipment but also in research and development. Consequently, even the very largest firms are involved in collaborative ventures with other manufacturers, while the very survival of smaller firms seems to depend increasingly on interfirm agreements to supply parts, to produce jointly under licence and to engage in joint R&D.

Consequently, a veritable transnational spider's web of strategic alliances has developed, a web which stretches across the globe. All three US automobile firms – Ford, GM and Chrysler – have forged collaborative links with Japanese and Korean companies. For example, GM has an arrangement with Isuzu (of which GM owns 39 per cent) for the Japanese firm to supply transmissions and axles to GM plants and to manufacture a small car. GM has a joint venture with Toyota to build a small car at Fremont, California (the NUMMI plant) and one with Suzuki in Canada. Ford owns 33 per cent of Mazda, which has, for some time, supplied transmissions for Ford cars and kits for assembly of vehicles in Asia. The Ford Escort in North America was jointly developed with Mazda, which also sold a version under its own badge in other markets.

The Ford–Mazda link was significantly strengthened in 1997 when it was announced that the two firms are to 'synchronize production cycles of a large number of vehicles and to share platforms and powertrains (engines and transmissions) . . . The measures would allow the two to share design and engineering expertise and free resources to create a variety of products' (*The Financial Times*, 18 April 1997). Ford also has an agreement with Yamaha for the Japanese company to develop a new engine for some Ford cars. Within Japan itself there is an exceptionally complex network of relationships between the major automobile manufacturers. Japanese firms also have close links with Korean automobile manufacturers which, in turn, have close links with United States firms. For example, the Korean firms Daewoo and Kia have manufactured cars which have been sold in the United States under GM and Ford brand names respectively.

Within Europe an especially complex collaborative network has developed over the years. For example, Honda had a long-standing arrangement with Rover to produce cars in the United Kingdom. Virtually all the major European manufacturers are involved in collaborative deals among themselves. For example, Fiat and Peugeot collaborate on the production of engines and steering components; six European firms – Fiat, Peugeot, Renault, Rover, VW and Volvo – have had a co-operative research programme. Volkswagen and Ford operated a joint venture (Autolatina) for production in Argentina and Brazil.

Transnational collaborative arrangements also occurred within the former USSR and eastern Europe. Fiat was particularly active in coproduction and licensing agreements in the USSR, Poland and Yugoslavia, as was Renault in Romania and Citroen in Romania, East Germany and Yugoslavia. In fact, much of the former Soviet Union's and Poland's postwar automobile industry was built in conjunction with Fiat, including the massive assembly plant at Toglliattigrad. The recent political changes in eastern Europe have produced a new wave of such arrangements between western automobile manufacturers and local firms in the production of both vehicles and components.

However, such is the volatility of the global automobile industry that a number of these alliances and joint ventures have unravelled. For example, the Honda–Rover alliance was terminated by BMW's take-over of Rover in 1994; the close link between Renault and Volvo ended; the Ford–VW collaboration in the Autolatina project has been dissolved. But there is no reason to believe that the use of alliances and joint ventures will not continue; it is just that the particular partnerships may change.

Types of competitive strategy

Although the automobile industry is undoubtedly a global industry, there is a good deal of *variety* in the specific corporate and competitive strategies adopted by the leading companies. During the late 1970s and early 1980s two clearly defined strategic options appeared to be available for the world's major automobile producers:

- One option was the so-called *world car* strategy: the manufacture of mass-produced cars for a world market. The world car strategy was based explicitly upon transnationally integrated production between the overseas affiliates of the parent company. Component manufacture was to be located in the most favourable – least-cost – locations at a global scale. These highly specialized plants would then supply a strategically located network of assembly plants which served specific, large-scale geographical markets. The emphasis was on standardization with minor, often cosmetic, modifications to suit particular markets. Hence, economies of scale were a key consideration both for component manufacture and for the assembly of finished vehicles. The world car, therefore, was not merely to be sold to a world market: more significantly, it was to be *manufactured* on a world scale.
- The alternative strategy was seen to be one confined to the *luxury car* manufacturers, producing in relatively small volumes for a small market in which premium prices could be charged. Such a strategy was regarded as being less likely to involve transnational production.

For a while it certainly seemed to be the case that the world automobile industry would segment according to these two strategic models. Certainly the two leading US manufacturers, GM and Ford, moved a considerable way along the world car path while firms like BMW, Daimler-Benz and Volvo continued to operate as specialist manufacturers in the luxury car niche. In fact, by the late 1980s the picture had changed dramatically. The stark distinction between mass-produced and lower-volume cars began to disappear in the face of the Japanese development of alternative methods of production. During the 1990s, different variants of a new world car strategy emerged among some, though not all, of the major producers. The dominant trend is towards some form of *glocalization* in which the leading producers are establishing systems of automobile production in each of the major global markets of North

America, Europe and east and southeast Asia. Since the Japanese have been the main force driving changes in the world automobile industry it makes sense to begin with the Japanese companies.

The strategies of the Japanese automobile manufacturers

The dramatic emergence of Japan as the world's leading automobile producer and exporter (Tables 10.1 and 10.2) and the spectacular rise of the leading Japanese companies up the world league table (Figure 10.6) were achieved almost entirely without any actual overseas production. Japanese firms had established local *assembly* operations in a number of countries, such as Australia and a number of developing countries, to put together finished vehicles from imported kits. But before 1982 there was not a single Japanese automobile *production* plant outside Japan. Japanese vehicle manufacturers had shown themselves perfectly capable of serving the North American and European markets by exports from Japan.

The price competitiveness of Japanese vehicle exports was based upon the extremely large-scale, flexibly organized and highly automated production plants in Japan. Japan's massive technological investment in motor vehicle production during the 1950s and 1960s, aided by government measures, resulted in extremely efficient vehicle production. The very high degree of vertical integration with Japanese component suppliers added to this remarkably efficient system. These production cost efficiencies more than offset any transport cost penalties which arose from Japan's geographical distance from major markets.

From the early 1980s, however, the Japanese companies began to change their global strategy by locating major production plants in their major markets. The primary stimulus for this change was the increasingly powerful political opposition in both North America and western Europe towards the growth of Japanese imports. The establishment of 'voluntary' restraint agreements between Japan and a number of countries was the visible sign of the level of trade friction being engendered by Japanese import penetration. Certainly in the United States, one of the objectives of

Table 10.7   Japanese automobile plants in North America

Company	Location	Date established	Minimum capacity	Employment
Honda	Marysville, OH	1982	360,000 cars	2,600
	East Liberty, OH	1989	150,000 cars	1,800
	Alliston, Ontario	1987	100,000 cars	1,200
Nissan	Smyrna, TN	1983	480,000 cars	3,300
	Decherd, TN	1986	300,000 engines	1,000
Toyota	Fremont, CA (NUMMI with GM)	1984	340,000 cars	2,500
	Georgetown, KY	1988	400,000 cars	3,000
	Cambridge, Ontario	1988	50,000 cars	1,000
Mazda	Flat Rock, MI	1987	240,000 cars	3,500
Mitsubishi	Normal, IL (with Chrysler)	1988	240,000 cars	2,900
Subaru/Isuzu	Lafayette	1989	120,000 cars	1,700
Suzuki	Ingersoll, Ontario (CAMI with GM)	1989	200,000 cars	2,000

Source: Press and company reports.

the trade protection policy was to persuade Japanese automobile manufacturers to locate production facilities there. In so doing, it was believed the Japanese companies would have to operate under the same conditions as the US companies: the playing field would be levelled. But trade protectionist measures, although the most important, were not the only stimulus to Japanese automobile investment in the United States. Given the increasing possibilities of tailoring vehicles to variations in customer demand, it was becoming more important for Japanese firms to locate close to their major markets. The same story was subsequently repeated in western Europe.[9]

Having been extremely cautious overseas investors, the Japanese automobile manufacturers have all established full production facilities in North America. As Table 10.7 shows, by 1991 there were more than a dozen Japanese automobile 'transplants' in place in the United States and Canada. In fact, the pioneer was neither of the two largest Japanese producers, Toyota or Nissan, but Honda, which established a manufacturing plant at Marysville in Ohio in 1982. This was soon followed by the Nissan plant at Smyrna, Tennessee in 1983. The leading Japanese manufacturer, Toyota, entered North America as a producer in a very cautious manner: by establishing a 50/50 joint venture with GM to produce cars at GM's Fremont, California, plant. This NUMMI plant began production in 1984; four years later Toyota began production at wholly owned plants in Georgetown, Kentucky, and Cambridge, Ontario. Honda continued to develop its American operations by opening a further plant in Ohio and one in Ontario. Mazda began production in Michigan in 1987 and Mitsubishi entered a joint venture with Chrysler – named Diamond Star – in Illinois.

Each of the major Japanese firms has continued to increase its planned capacity and to make major investments in engine, transmission and components plants. By the early 1990s the Japanese transplant factories in North America had a planned production capacity of 2.7 million vehicles and some 25,000 employees. In other words, during the period of less than a decade an entirely new Japanese-controlled automobile industry has been created in North America in fierce, direct competition with domestic manufacturers. The leading Japanese firms are now exporting some of their output not only to third countries like Europe or Taiwan but also to Japan itself.

As the Japanese plants in the United States progressively increased their North American content they were followed by a continuing wave of Japanese component manufacturers. There are now more than 300 such plants in operation. The locational pattern of both assemblers and suppliers in North America is quite distinctive; perhaps not surprisingly since, as newcomers, the Japanese have no existing plants or allegiance to specific areas. Mair, Florida and Kenney (1988, p. 361) identify a 'transplant corridor', a

> region stretching from Southern Ontario south through Michigan, Illinois, Indiana, Ohio, and Kentucky to Tennessee. This 'transplant corridor' is organized principally along several interstate highways . . . The single exception to the pattern of regional concentration is the NUMMI joint venture between Toyota and GM, which reopened a previously closed GM plant in Fremont, California.

Suppliers have generally followed the assemblers because of the use of just-in-time delivery systems. At the finer geographical scale the strong preference of Japanese automobile firms has been for greenfield sites near to small towns in rural areas. The old-established automobile industry centres are not favoured. The major exceptions to this rural orientation are the Toyota joint venture in California and the Mazda plant in Michigan. Choice of greenfield, rural locations has been determined mainly by the desire of the Japanese producers to minimize the influence of the labour unions. Recruitment has been primarily of young workers with little factory experience but with the 'right attitudes' towards the kinds of flexible labour practices employed in the plants. The advantages of being able to start from scratch, on greenfield sites, with newly designed plants and with a hand-picked workforce prepared to accept new working conditions and practices are enormous.

Table 10.8   Japanese automobile plants in Europe

Company	Location	Date established	Capacity	Employment
Nissan	Sunderland, northeast England	1986	300,000 cars and engines	4,600
Toyota	Burnaston, east Midlands	1992	200,000 cars	3,000
	Shotton, north Wales	1992	200,000 engines	300
Honda	Swindon, Wiltshire	1991	200,000 engines	600
	Swindon, Wiltshire	1992	100,000 cars	1,400
Mitsubishi	Born, The Netherlands (with Volvo)	1995	200,000 cars	6,800

Source: Press and company reports.

Compared with their entry into North America, Japanese automobile manufacturers were slower to establish production facilities in Europe. However, the momentum increased markedly at the end of the 1980s (Table 10.8). As far as passenger cars are concerned – as in the United States – Honda was the pioneer. But Honda merely signed a joint venture agreement in 1979 with BL (the predecessor of Rover and then state-owned). Nissan was the first to announce plans for an actual manufacturing plant in the United Kingdom, in 1981, but it was not until 1986 that production actually began. In 1989 both Toyota and Honda announced plans for both car and engine plants, also to be located in the United Kingdom. In 1995, Mitsubishi established a joint venture with Volvo and the Dutch government to produce cars in The Netherlands. In early 1997, Toyota implied that it might establish a large production plant at Lens in northern France. Although there has not been the same following inflow of Japanese components suppliers as yet, a number have indeed arrived.

The three leading Japanese automobile firms have adopted similar locational criteria as in North America. They have avoided traditional automobile manufacturing areas, they have opted for greenfield sites and they have adopted very specific recruitment policies. All three have insisted on single-union agreements. The Japanese plants in the United Kingdom are specifically oriented towards the European market and this, as we have seen, has led to political friction within the EU. Again, both market-access/political factors and market-proximity factors have been influential in the Japanese decisions to locate production in Europe.

Of course, it goes without saying that Japanese automobile manufacturers are deeply and extensively involved within Asia itself. Through a network of assembly plants and joint ventures with domestic firms, Japanese cars are assembled in Thailand, Malaysia, the Philippines, Indonesia, Taiwan and China. In several of these countries, Japanese manufacturers totally dominate the automobile market. In Thailand, for example, Japanese firms have a market share of more than 90 per cent;

Toyota alone controls almost 30 per cent of the Thai vehicle market. Faced with increasingly difficult circumstances in the Japanese market itself (e.g. the problems created by the high value of the yen, the slowdown in demand) Japanese firms are putting increased emphasis on raising their penetration of the Asian market by beginning to develop cars specifically tailored to that market and not just versions of existing models. In early 1997, the Honda City and the Toyota Soluna were introduced. Both were based upon a very different approach to producing cost-efficient cars for a low-income market:

> no south-east Asian market is large enough to support a highly-automated car production site on its own. Because import tariffs on finished vehicles are high throughout the region, Asian cars had to be cost-effective even when assembled manually in small lots of less than 10,000 units per month. To solve these problems, both Toyota and Honda used a new design technique that reversed the traditional process of designing a car and then reducing costs by squeezing parts suppliers. Attempting to localize production as much as possible, engineers at both Toyota and Honda focused first on what parts local companies could produce cheaply and then designed a car with those components in mind . . . What Honda and Toyota came up with was a car priced several thousands dollars lower but only slightly smaller than the companies' previous bottom-of-the-range passenger car models . . . Seventy per cent of the parts in both the Soluna and City come from within south-east Asia, against only about half for the Corolla and Civic.
>
> (*The Financial Times*, 11 February 1997)

Thus, each of the leading Japanese automobile manufacturers is moving towards the greater transnationalization of their production. Nissan has greatly exceeded its plan to manufacture 25 per cent of its production outside Japan by the early 1990s. Toyota's objective is to double its overseas production capacity by the year 2000 in order to become the world's number one automobile manufacturer. But as Figure 10.7 shows, there is a long way still to go. Only Honda has a really large proportion of its output outside Japan and virtually all of that is in North America. Nissan has a major production facility in Mexico whose capacity is about to be doubled. It is clear that, as yet, the degree of transnationalization among Japanese automobile producers is very limited, particularly when compared with the leading US firms, GM and Ford. But it is equally clear that the primary objective of the Japanese producers is to develop a major direct presence in each of the world's major automobile markets – the global triad – of Asia, North America and western Europe.

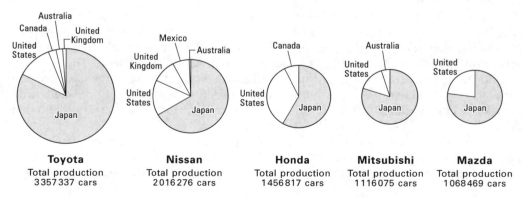

Figure 10.7   International automobile production by the leading Japanese firms, 1994
Source: AAMA (1996, *World Motor Vehicle Data, 1996*)

Strategic responses to Japanese competition

For two decades the Japanese automobile producers have been perceived as the major competitive threat by the American, and most of the European, producers. In facing up to the intensification of global competition American and European firms have pursued rather different competitive strategies and created different geographical configurations in their production networks. In so far as both US and European producers face a shared competitive threat from the Japanese, however, they have all adopted a varying mixture of the following measures:

- Introduction of *new technology*, notably flexible manufacturing systems.
- Introduction of *new work practices* with, again, a particular emphasis on flexibility, multiskilling and the reduction of the multifarious task divisions which have characterized all American and European automobile plants.
- Modification of *component sourcing procedures* to move towards a closer functional relationship with key suppliers and a just-in-time system.
- Development of *strategic alliances* with other firms.
- Continued *reorganization and rationalization* of their production activities both domestically and internationally. It is in this latter respect that the strategies of the non-Japanese companies differ most of all.

The strategies of the United States automobile manufacturers

Ford and GM are the longest-established transnational vehicle producers in the world. In general, Ford has expanded internationally mainly by opening new plants; GM primarily by acquiring existing foreign manufacturers. Ford established a manufacturing plant in Canada in 1904, just across the Detroit River in Windsor, Ontario; GM acquired the Canadian vehicle company, McLaughlin, in 1918. Ford opened its first European car assembly plant at Trafford Park, Manchester in 1911 (subsequently replaced by a massive integrated operation at Dagenham in 1931), followed by a plant at Bordeaux in 1913. Ford began to assemble cars in Berlin in 1926 and in Cologne in 1931. GM's entry into Europe began in 1925 with the acquisition of Vauxhall Motors in England, followed, in 1929, by the purchase of the Adam Opel Company in Germany. During the 1920s both firms also established assembly operations in Latin America as well as in Australia and Japan.

These early transnational ventures were triggered primarily by the existence of protective barriers around major national markets as well as by the high cost of transporting assembled vehicles from the United States. In recent years, however, both Ford's and GM's transnational strategies have been concerned initially with expanding and integrating and subsequently rationalizing their operations on a global scale. As Table 10.6 showed, Ford is the more transnational of the two leading companies, although GM has been investing heavily overseas in recent years as part of its intensified global strategy.

Figure 10.8 shows the broad pattern of Ford's and GM's international production. It can be seen immediately how extensively transnational the two firms are. Although there are slight differences of detail, their geographical similarities are very considerable. Both companies have heavily rationalized and restructured their domestic operations with extensive closure and contraction of plants. GM, in particular, has undertaken a massive rationalization programme in the United States, including the closure of numerous plants. On the other hand, GM gambled heavily

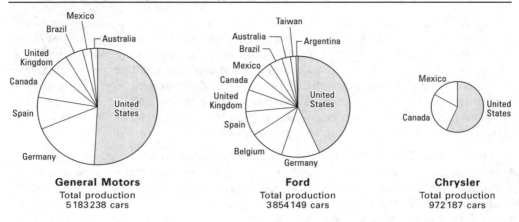

Figure 10.8   International automobile production by Ford, GM and Chrysler, 1994
Source: AAMA (1996, *World Motor Vehicle Data, 1996*)

on a totally new car project, Saturn, located at Spring Hill, Tennessee, whose objective was to produce a totally new range of small (subcompact) cars to compete with the Japanese. It was the first fully integrated automobile production facility to be built in the United States since 1927, incorporating engine, gearbox and finished car production on the same site. Its success has led GM to announce a second Saturn model to be produced in 1998 at its Wilmington, Delaware, plant. In addition, as we have seen, GM operates a highly successful joint venture (NUMMI) with Toyota at Fremont in California.

Figure 10.8 breaks down the world distribution of Ford's and GM's automobile production. Two-thirds of Ford's overseas car production is located in Europe. Germany is the major production focus, with 21 per cent of the overseas total, followed by Belgium (19 per cent), Spain (14 per cent) and the United Kingdom (12 per cent). Outside Europe, Ford's foreign production operations are primarily in Latin America (Brazil, Argentina), Mexico and Australia. It was within Europe, however, that Ford's strategy of transnational integration developed most fully and earliest.

Ford was the first automobile producer to take full advantage of the development of the European Community. In 1967 Ford reorganized its entire European operations previously separately focused on the United Kingdom, West Germany and Belgium into a single organization, Ford Europe. The separate national operations were transformed into a transnationally integrated operation, with each plant performing a specified, often specialist, role within the corporation to achieve economies of scale. In the early stages of the process Ford simply reallocated the production of certain models to specific European plants. However, the first real product of Ford's transnationally integrated strategy in Europe was the Fiesta in the 1970s. Key elements in Ford's European strategy from the early 1970s were the desire to gain a foothold in the heavily protected Spanish market and to produce a small car.

The Spanish operation became, from the beginning, a key node in the transnational network of production for the Ford Fiesta. It became one of three final assembly locations for the Fiesta, the others being at Dagenham and Saarlouis. By locating assembly at three separate locations Ford could not only serve specific geographical markets but also make up for shortfalls in production at one location by shipping cars from one of the others. The three assembly plants were, in turn, locked into a highly complex network of component suppliers, both inside and outside Ford.

In effect, Ford created a production system which was, in many ways, a microcosm of a globally integrated system on a European scale. But there was little real connection between Ford Europe and the rest of Ford's global operations, including Ford US. As the 1990s wore on, Ford – like all automobile producers – found itself facing intensified competition in all its major markets. Its response, in early 1995, was to embark on a hugely ambitious process of global reorganization ('Ford 2000'). The North American and European operations have been merged into a single organization, Ford Automotive Operations, with the intention of ultimately integrating its Asian, Latin American and African operations as well. The philosophy underlying this global structure is as follows:

> We can't allow human and financial resources to be wasted duplicating vehicle platforms, powertrains (engines and transmissions) and other basic components that serve nearly identical customer needs in different markets . . . In the future we will have one small engine family in Europe and North America instead of two separate families that power the same car for the same kind of customer – yet are completely different and used duplicate resources in their development.
> (Alex Trotman, Chairman of Ford, quoted in *The Financial Times*, 3 April 1995)

Ford has created one worldwide product development organization out of its previously separate North American and European design and engineering operations and created five *vehicle centres*, each of which has responsibility for the global development of specific products. Four of the five vehicle centres are located in the United States and one – responsible for small and medium-sized, front-wheel-drive cars – is located in Europe (jointly in the UK and Germany). In other words, the 'world car' is back on the agenda at Ford. It will certainly involve further rationalization of existing facilities as Ford's decision to slim down its Halewood, Merseyside, plant demonstrates.

GM has a global production system much like Ford's in its geographical structure but the nature, the timing and the organization of its development have been different. Figure 10.8 shows GM's global manufacturing operations. Like Ford, two-thirds of GM's overseas car production is in Europe but with a much greater emphasis on Germany, where GM's Opel subsidiary is responsible for 58 per cent of the firm's European production. In contrast, the United Kingdom is far less important to GM than it is to Ford. For both companies, however, Spain occupies an especially significant position as a location for the production of small cars. Outside Europe, Canada is the most significant area of operations, followed by Brazil, Mexico and Australia. Mexico, as with Ford, is an increasingly important sourcing point for the North American market. Again, however, it is in Europe that the greatest degree of transnational integration has occurred. Until the early 1980s GM was far weaker in Europe than Ford; now it produces more cars there than its rival and has very radically reshaped and rationalized its European operations.

GM's European operations had long been based upon two separate national subsidiaries: Vauxhall in the United Kingdom and Opel in Germany. During the 1970s Vauxhall's performance became progressively weaker as the investment emphasis shifted towards Opel. In 1979 GM drastically redrew the production responsibilities of the two European subsidiaries. Vauxhall was relegated to being a domestic supplier only and much of this production was simple assembly of imported components and subassemblies from Opel. This situation has only recently begun to change again. Like Ford, GM set up a major new European engine plant (in Vienna) to supply its global operations, together with a new automatic transmission plant at Strasbourg. More significantly, GM also built a major new manufacturing plant in Spain (at Zaragossa).

But, although faced with exactly the same competitive pressures as Ford, GM has not chosen to pursue a wholesale global reorganization of its business. It has established a global strategy board which consists of the heads of GM's North American and international operations and its component division but this has not created a radically new structure. Although GM wants to 'globalize' its operations 'we want to do it within our existing structure' (Jack Smith, GM Chairman, quoted in *The Financial Times*, 15 March 1996).

Both Ford and GM are making strenuous efforts to improve their position in the Asian market, although they are a very long way behind the Japanese in this respect. The US automobile firms in total have only about 1 per cent of the southeast Asian car market compared with the Japanese share of 75 per cent. Both Ford and GM have limited assembly and component operations in Asia. Both have been selling through their alliances with Mazda and Isuzu respectively. In 1996, however, GM announced a $750 million investment in a new manufacturing plant in Thailand. Similarly, both US companies are endeavouring to build bases in the eastern European countries. GM is building a major car assembly plant near Katowice in Poland to employ 2,000 workers, it has an engine plant in Hungary and a highly advanced plant at Eisenach in the former East Germany. Ford has been far more cautious than GM in committing itself to eastern European manufacturing; it opened a modest-sized plant in Poland in 1995 and has component activities in Hungary.

The strategies of the European automobile manufacturers[10]

The major European automobile producers are significantly more parochial locationally than either GM or Ford. Figure 10.9 summarizes their production geography. Of the four leading European companies, VW has pursued by far the most extensive and systematic international strategy. Outside Europe, VW is a major producer in Brazil where it has 40 per cent of the market. Its production facilities in Mexico have effectively replaced much of the production formerly located at VW's Pennsylvania plant, which was closed in 1988. Mexico is now the sole production location for the VW Beetle. Within Europe, prior to the opening up of eastern Europe, VW concentrated its production in two countries in a clear strategy of spatial segmentation. High-

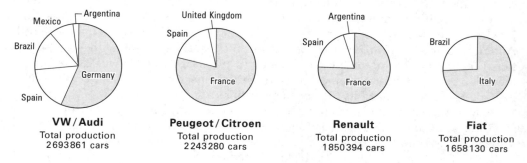

Figure 10.9 International automobile production by the major European firms, 1994
Source: AAMA (1996, *World Motor Vehicle Data, 1996*)

value, technologically advanced cars are produced in the former West Germany; low-cost, small cars are produced in Spain where VW undertook a massive investment programme in the former SEAT company. During 1990, however, VW moved very rapidly to establish production of small cars in eastern Germany and to take a 70 per cent stake in the Czech firm, Skoda. VW has embarked on a major restructuring programme which involves reducing the number of vehicle 'platforms' used by its four component companies (VW, Audi, SEAT, Skoda) to ensure the maximum sharing of basic structures on which differentiated vehicles can be based. As we saw earlier, VW is also introducing radically new sourcing arrangements with its key suppliers although, as yet, these are being introduced outside the sensitive home base.

While VW was expanding its European production base to incorporate Spain, the Italian automobile firm, Fiat, initially moved in the opposite direction and recon-centrated production in its home market. Like the French producers, Fiat is heavily dependent on its domestic market. Unlike the French firms, however, Fiat currently manufactures its entire EU output in Italy; 70 per cent of its EU sales are also in its home market. Such geographical reconcentration was the outcome of a major strate-gic shift which began in the early 1980s. But such parochialism has not lasted. A major element in Fiat's current strategy is to create an extensive production network in the former Soviet Union and eastern Europe where it has the longest-established links of any automobile producer in the world. Fiat's unique history of collaborative ventures in those regions is now the basis for new developments in the changed political environment. Fiat's 'grand European design' is to build a manufacturing network extending from the Mediterranean to the Urals. At the same time, Fiat is strengthening its position in Brazil. Both Brazil and Poland are seen as the major bases for Fiat's small 'world car', the Palio, which began production in Brazil early in 1996. Fiat's current aim is to produce more than 50 per cent of its cars outside Italy by the year 2000.

The two French automobile companies, Peugeot-Citroen (PSA) and Renault, fall between these two positions. Both are strongly home-country oriented in their pro-duction. Peugeot-Citroen was formed by a state-induced merger of the separate Peugeot and Citroen companies in 1975, followed by the purchase of Chrysler's European operations in 1978. It has faced major problems in digesting the results of these mergers and has become the leading lobbyist in favour of a strong political stand against Japanese automobile companies in Europe. Seventy-nine per cent of Peugeot-Citroen's production is located in France with a further 18 per cent in Spain and 3 per cent in the United Kingdom. It has no production plants outside Europe, although there is a possibility it might return to the United States.

Renault[11] has been, for more than forty years, the French government's national champion, supported by massive state aid, which served to constrain its activities. State control is in the process of being removed and Renault, like Peugeot-Citroen, has been involved in major restructuring. An important strategic development of the early 1990s was its agreement with Volvo which, it was believed, would eventually result in full-scale merger of the two companies. But in late 1993, Volvo pulled out, leaving Renault in a difficult position. Like Peugeot, Renault has a substantial presence in Spain which is increasingly integrated with its French operations. It also has some production in Argentina. But Renault sold its major non-European operations – its involvement with the American Motors Corporation (AMC) in North America – to Chrysler and closed its Mexico plant.

Even allowing for the case of VW, the most internationally oriented European automobile producer, the major European companies are far narrower in global terms than the US and, increasingly, the Japanese firms. Apart from some limited involvement in Latin America (and excluding simply local assembly plants in various countries), they are entirely European in their production networks. It is especially notable that, while the Japanese have been busy building large manufacturing plants in North America during the 1980s, the Europeans have not been doing this. Indeed, both VW and Renault have pulled out of North America.

New centres of automobile production: east and southeast Asia

The global automobile companies based in Japan, the United States and Europe exert such market dominance that there have been virtually no new entrants in the industry during the past twenty-five years. As we have seen, the automobile industry in Argentina, Brazil and Mexico – though quantitatively large – is entirely dominated by foreign firms. The major exception is South Korea which, in the space of a just a few years, has emerged as a significant international force.[12] As we noted in Chapter 4, the development of a Korean automobile industry has been strongly influenced by the policies of the Korean government:

> Autos were identified as one of the priority industries in the Heavy and Chemical Industry Plan of 1973. In 1974 an industry-specific plan for automobiles was published covering the next ten years. The objectives were to achieve a 90 per cent domestic content for small passenger cars by the end of the 1970s and to turn the industry into a major exporter by the early 1980s.
>
> (Wade, 1990b, p. 310)

To meet these ambitious objectives, the Korean government took the following steps:

- It specified who the three primary producers were to be: Hyundai, Kia and what was later to become Daewoo (it was originally called GM Korea and then Saehan).
- The government had to approve the plans of each producer, having stipulated their minimum size and the maximum permitted size of engines to be manufactured.
- A promotional plan for the components industry was initiated which required the three automobile producers to co-operate in producing standard parts and to meet local content requirements.
- The three producers were later required to set export targets on a consecutive basis: southeast Asia, Latin America and the Middle East, Canada 'all by way of preparation for a big push into the US market' (Wade, 1990b, p. 310).
- The producers were encouraged to set export prices below production costs.
- The government provided them with substantial export subsidies (i.e. credit).

During the early 1980s, in response to difficult economic conditions, the Korean government effectively made Hyundai the leading producer in the industry, giving it an enormous relative advantage which the firm used to grow at a very rapid rate. Today, the Korean automobile industry consists of five firms: Hyundai, Daewoo, Kia, Ssangyong and Samsung (which was recently given permission to produce passenger cars from 1998). The first three producers have all depended in their early development on close technological and marketing relationships with US and Japanese firms.

Hyundai developed on the basis of a close relationship with Mitsubishi, although its first car, the Pony, produced in the 1970s, drew on technology and design from a variety of foreign sources, including Italy, the United Kingdom, Japan and the United

States. The Pony was succeeded by a new car, the Excel, based on Mitsubishi designs, and built in a huge new plant at Ulsan. Both the Pony and the Excel were exported in large volumes, especially to North America. Hyundai's strategy was to compete with the Japanese in a very narrow product range and entirely on price. Initially, Hyundai's export success was remarkable. In 1986, 300,000 cars were exported to North America, a level more or less maintained for the following two years. On the strength of this success, Hyundai built a second new plant in Korea. More ambitiously still it built a plant in Canada at Bromont near Montreal, with the capacity to produce 120,000 cars and to employ 1,200 workers directly, but this plant was not a success. Currently, Hyundai plans to double its production to 2.4 million vehicles by 2000, with plans to manufacture in Turkey, eastern Europe, India and Vietnam. As we have seen, by 1994, it was already the thirteenth largest automobile producer in the world.

Daewoo developed out of a joint venture with General Motors (hence its original name, GM Korea). For a time, it was 50 per cent owned by GM and was quite closely integrated into GM's global operations. In particular, Daewoo manufactured cars sold in North America by GM's Pontiac division. This relationship no longer exists. Daewoo has become an especially aggressive entrant into the European market, using innovative methods of direct distribution. It has also been exceptionally active in setting up production plants in eastern Europe. It has purchased a substantial share in a Romanian manufacturer and in FSO, the major Polish automobile manufacturer (much to the distress of its own former partner, GM). Of the other Korean automobile manufacturers, Kia[13] retains a close link with Ford (which owns 10 per cent), Samsung has a close technical relationship with Nissan. Each of the Korean automobile firms, all of which are members of the giant industrial conglomerates, the *chaebol*, is making huge investments both in Korea and overseas. Hyundai, Daewoo and Samsung each has the ambition to be one of the ten leading automobile producers in the world.

Elsewhere in Asia, one country – Malaysia – is building an indigenous automobile industry on the basis of a mixture of foreign assemblers and a domestic producer of a national car. This, the Proton project, was launched by the government in 1985 and was based upon a close relationship with Mitsubishi. The early Protons were, essentially, Mitsubishis; the Japanese company still retains a 16 per cent stake in Proton. But the Malaysian firm is now growing rapidly in its own right. In addition to its original factory, the company is building a massive modular car factory near Kuala Lumpur which is designed to produce 1 million cars a year eventually. Europe has been targeted as a key export market for Proton which also bought the British specialist producer, Lotus, in 1996.

The other major automobile production foci in east and southeast Asia are Taiwan, Thailand, China and, to a lesser extent, Indonesia and the Philippines. Thailand is currently regarded as the 'car capital' of southeast Asia with virtually all the major foreign producers having a presence there. China has experienced rapid growth of automobile production based on joint ventures involving such firms as VW, but it has substantial overcapacity and the Chinese government has placed restrictions on further joint ventures. It is clear that the very rapid growth of automobile production throughout east and southeast Asia in the past decade is in danger of creating severe overcapacity.

Corporate strategies in the automobile industry: a summary
In summary, the strategies of the major automobile producers are more diverse than is often realized. They are also in a condition of flux as major technological and

organizational changes sweep through the industry, as 'mass' production is replaced by 'lean' production. There is, undoubtedly, a global battle raging in the automobile industry – 'car wars' – in which there will certainly be casualties. During the late 1980s and early 1990s, the major battleground was Europe, as Japanese penetration increased the pressure on local manufacturers. This battle had both an economic and a political dimension as governments, as well as firms, strove to protect and enhance their positions. The strategy of the leading Japanese companies was clearly to establish a major integrated production system in each of the three global regions: Asia, North America, western Europe. GM and Ford are moving in the same direction and have the advantage of already operating a sophisticated integrated network within Europe. In comparison, the European manufacturers remain far more limited geographically. Only VW has much of an internationally integrated system. Its withdrawal from production in the United States is being replaced by its growing involvement in Mexico, which could give VW increasing access to the North American market. The European picture is being changed by the new developments in eastern Europe. However, whilst intense competition will continue in both Europe and North America (where NAFTA is changing the automobile production map) it seems likely that the site of the next car wars could well shift to Asia. Although Japanese producers currently have a dominant position they are being threatened by the Koreans and by other low-cost regional producers, as well as by the intensified efforts being made by GM and Ford to increase their market penetration.

Jobs in the automobile industry

Changes in demand for automobiles in the developed market economies, technological and organizational changes in the production process itself and the increased tendency for the larger TNCs to expand overseas, to integrate their production operations transnationally and to engage in international sourcing of vehicles and components have all combined to produce major changes in employment in the automobile industries of the older-industrialized countries. High levels of import penetration, particularly by Japanese manufacturers, have also adversely affected the domestic motor industry of individual countries, reducing the size of their labour forces.

All the major American and European vehicle manufacturers have been restructuring their operations and, in so doing, have greatly reduced the size of their labour force. During the 1980s, for example, employment in the US automobile industry fell by 24 per cent, from 470,000 to 355,000. Plant closures and contractions continue and the effects are felt not only by blue-collar workers but also by white-collar workers. GM had cut its white-collar staff in the United States by 15,000 by 1993. Plant closures have been under way in the United States since the late 1970s. During 1979 and 1980 alone,

> domestic automobile manufacturers closed or announced the imminent shutdown of twenty facilities employing over 50,000 workers. As a consequence of these closings and of the output reductions in other auto plants, suppliers of materials, parts, and components to the automotive industry closed nearly 100 plants, eliminating the jobs of about 80,000 additional workers . . . among the major permanent closings in the last years of the decade were thirteen Chrysler plants employing nearly 31,000 workers, five Ford plants including the huge facility at Mahwah, New Jersey, and seven plants in the General Motors system. Of these twenty-five shutdown, eleven were located in Michigan and six more were in other midwestern states. But even the Sunbelt lost some of its automobile

capacity. Ford shut down its large Los Angeles assembly plant in 1980, while Chrysler closed a small facility in Florida.

(Bluestone and Harrison, 1982, p. 36)

Between 1987 and 1990 GM and Chrysler closed a further ten assembly plants in the United States. Set against these closures, of course, are the newly opened Japanese transplants and the new GM Saturn plant. Even so, there has certainly been a major net job loss in the US automobile industry.

In Europe, Ford cut its blue-collar labour force by 26 per cent (104,000 to 77,500) and its white-collar labour force by 34 per cent (18,300 to 12,000) during the 1980s. Ford's Dagenham plant employed 29,000 workers in the late 1960s; today it employs a mere 7,500. Recent restructuring by Ford will reduce the jobs at its Merseyside plant by a further 1,300. Renault's French workforce fell during the 1980s from 89,000 to 77,000. In early 1997, Renault closed its Belgian plant with the loss of 3,100 jobs. Peugeot-Citroen cut its Spanish workforce by 22 per cent. Saab (now part of GM) reduced its labour force from 17,000 to 11,500 and closed its Malmö plant – only three years after it was opened. The German automobile producers have forecast a loss of some 100,000 jobs in the German car and components industry by 2000 unless the government agrees on measures to reduce business costs. Some of these predictions are deliberately aimed at putting pressure on governments to provide various forms of subsidy. As we noted in Chapter 8, automobile companies have proved to be extremely adept at such bargaining (or blackmail, as some would see it). So far, the Japanese automobile firms have not shed labour on a large scale although, for the first time since 1945, some plant closures have occurred. For example, Nissan closed its Zama plant in Tokyo and an engine plant in Kyushu in 1995. It is in the process of shedding one-quarter of its domestic labour force although, in the Japanese style, none of these will be compulsory redundancies.

The large size of automobile plants and the historical tendency for the industry to be geographically concentrated into specific localities pose particularly severe problems. The employment ramifications of the closure of an automobile assembly plant are far more extensive than those caused by the closure of a textile mill or a clothing factory. The very nature of the automobile industry – its use of a myriad of materials and components from many different industries – means that the knock-on effects are far greater than in almost any other industry.

In Europe additional problems have arisen because of the large numbers of immigrant workers employed in the motor vehicle industry. In Germany and France, in particular, thousands of migrant workers from Turkey, North Africa and southern Europe flocked to the automobile assembly plants during the boom years of the 1960s. For example, more than half of the 17,000 workers employed at the Peugeot/Talbot plant at Roissy in France in the early 1980s were African immigrants. With the drastic downturn in the fortunes of the European automobile industry in the 1980s and the severe rationalization programmes being pursued by VW, Peugeot and Renault the problems of these so-called 'guestworkers' have become acute. Minority groups in the United States have also been seriously affected by contraction in the motor vehicle industry. For example, 'In August 1979 virtually 30 percent of Chrysler's national employment was made up of black, Hispanic, and other minorities while over half of its Detroit work force was nonwhite' (Bluestone and Harrison, 1982, p. 54).

Each of the traditional motor vehicle-producing nations is, therefore, experiencing major problems of employment change and employment adjustment. In each case, the strong geographical localization of the industry and its tentacular reach into many other industries make such problems especially serious in particular places. Hardly any of these job losses can be attributed to a relative shift of automobile production to developing countries for such geographical relocation has been very limited. They are the result of a combination of circumstances: a profound change in the structure of demand for automobiles in the developed market economies; the adoption of new technologies in the production process; and the international production and sourcing strategies of the major TNCs as they seek to survive and flourish in an increasingly competitive global market. The problem is that

> the current Western automobile workforce is in precisely the opposite position of craft workers in 1913. The introduction of mass production created new jobs for craft workers – these workers made the production tools needed by the new system. By contrast, lean production displaces armies of mass-production workers who by the nature of this system have no skills and no place to go.
>
> (Womack, Jones and Roos, 1990, pp. 235–36)

For those still employed in automobile plants there has been a major change in the nature of their jobs. As we saw in Chapter 5, there is controversy over the effects of the Japanese-inspired work practices: 'Lean' may well imply 'mean' (Harrison, 1994).

Notes for further reading

1. The automobile industry is one of the most heavily reported of all industries as a regular reading of *The Financial Times*, *The Wall Street Journal*, *The Economist* or *Business Week* will show. The most comprehensive analyses are those conducted as part of the MIT International Automobile Program (see, for example, Womack, Jones and Roos, 1990). A recent study of the European automobile industry is provided by Hudson and Schamp (1995).
2. The sequence was suggested by Bloomfield (1978).
3. Womack, Jones and Roos (1990) provide a very readable account of the history of technological change in the automobile industry although their specific approach is based upon their claim that traditional mass production is being replaced by one particular form of production: 'lean' production. Williams et al (1992) are strongly critical of their claims. Kaplinsky (1988) interprets technological and organizational change within the industry in terms of the restructuring of the labour process.
4. See the discussion in Chapter 5.
5. Letter by Garel Rhys to *The Financial Times* (18 October 1990).
6. Reich (1989) presents a detailed historical analysis of the evolution of government policy towards automobile producers in France, Germany, Britain and the United States. Sadler (1995) discusses the development of state and EC policy towards the European automobile industry.
7. See Dicken (1992).
8. See Holmes (1990; 1992), Eden and Molot (1993).
9. The development of Japanese automobile investment in North America is analysed by Mair, Florida and Kenney (1988), Reid (1990), Womack, Jones and Roos (1990), Drache (1996). The European case is discussed by Dicken (1987; 1992), Hudson (1995).
10. The various chapters in Hudson and Schamp (1995) analyse the strategies of the major European automobile manufacturers.
11. Savary (1995) provides a detailed analysis of Renault.
12. The development of the Korean automobile industry is analysed by Amsden (1989), Wade (1990a), Kim and Lee (1994), Lee and Cason (1994).
13. Kia experienced severe financial problems in 1997 and had to be bailed out by the Korean government.

CHAPTER 11

'Chips and screens': the electronics industries

Introduction

Of all the manufacturing industries discussed in this book the electronics industry is both the youngest and also the one with the most far-reaching implications for the future economic evolution of developed and developing economies alike. Microelectronics has emerged as the dominant technology of the last three decades, extending its transformative influence into all branches of the economy and into many aspects of society at large. Although the early stirrings of an electronics industry became apparent in 1901 with the introduction of radio, its really spectacular development is entirely a product of the last few decades.

The first step in creating the modern electronics industry was the development of the *transistor* in the Bell Telephone Laboratories in the United States in 1948. The transistor replaced the thermionic valve or vacuum tube and made possible the development of a *micro*electronics industry. Unlike the valve or tube, the transistor is a solid-state device which uses materials such as silicon which, with the addition of chemical impurities, can be made to act as a *semiconductor* of electric current. Within a few years of the initial development of the transistor, semiconductors were being manufactured commercially in the United States. The end of the 1950s saw the emergence of a second major innovation – the *integrated circuit* – which consists of a number of transistors connected together on a single piece or 'chip' of silicon. By the early 1970s it had become possible to incorporate a number of very sophisticated solid-state circuits on to a single chip the size of a finger-nail to create a *microprocessor* which was able to perform the functions which only two decades earlier had taken a whole roomful of valve computers.

The progressive refinement of these basic innovations over a very short period dramatically increased the power of electronic components and also spectacularly decreased their size. It is now possible to pack millions of individual circuits on to a single chip of silicon less than one square centimetre in size. Increased *miniaturization* has been a fundamental development, for it permits the incorporation of electronic components into a vast range of products, from pocket electronic products (like calculators, televisions, telephones) to highly complex computers, industrial robots and aircraft guidance systems.

In fact, it is the *increasingly pervasive application* of electronics, rather than its absolute size, which makes the industry of such very great significance. Indeed, the microelectronics industry is more aptly described as today's 'industry of industries' than the automobile industry. It is, without doubt, 'the engine of the digital age'.[1] The extensive ramifications of the electronics industry, not only for other sectors of the economy but also for telecommunications and national defence, have made all governments increasingly anxious to avoid being left out or left behind in what is a rapidly moving technological scene. The electronics industry, like textiles, steel and automobiles before it, has come to be regarded as the touchstone of industrial success. Hence, all governments in the developed market economies, as well as those in the more industrialized developing countries, operate substantial support programmes for the electronics industry, particularly microprocessors and computers. At the same time, shifts in the global pattern of production of consumer electronics – particularly television and audio products – have led to considerable trading frictions and the implementation of various protectionist measures against imports, especially from Asia to North America and western Europe.

The electronics production chain

In this chapter we are concerned with just two branches of the electronics industry: semiconductors and consumer electronics. Figure 11.1 shows where these two subsectors fit into the electronics production chain as a whole. The core of the electronics industry is the *components* sector. Its most important elements are the active components, based on the *semiconductor*, which control the flow of electrical current. Semiconductors themselves can be divided into two major categories:

- *memory chips* which contain preprogrammed information
- *microprocessors* which are, in effect, 'computers on a chip'.

Semiconductors and related components are employed in a bewildering variety of *applications* which, for convenience, can be classified into two categories:

- electronic equipment
- consumer electronics.

Within the electronic equipment sector there is obviously much overlap, particularly because the computer – the biggest single user of semiconductors – is universally involved in the other equipment sectors. The recent spectacular growth of the personal computer, with its increasing multimedia applications, has put what was

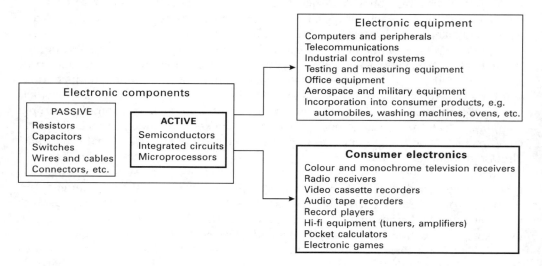

Figure 11.1   The electronics production chain

formerly a product used only by industry, commerce or the public sector into individual households. In this sense, the computer has begun to invade the consumer electronics sector as well. Of course, other types of electronic equipment have also been incorporated into various personal and household products, such as automobiles, washing machines, ovens and similar items. But consumer electronics as a sector is usually defined in terms of 'complete' electronic products such as radios, televisions, video and audio recorders, hi-fi equipment and so on.

Both the sectors examined in this chapter – semiconductors and consumer electronics – are highly capital-intensive industries in which very large transnational firms tend to dominate. However, certain activities have also been open to new, dynamic and initially small, electronics firms, particularly in the United States. As we shall see, the semiconductor and consumer electronics industries differ somewhat in their global distribution and, therefore, in the degree to which developing countries are involved. The key to such variation is the extent to which parts of the production chain – those which are highly labour intensive – can be geographically separated from the other stages in the sequence. Finally, the nature of employment in the electronics industry in both developed and developing countries poses a number of significant problems.

Global shifts in the electronics industries

The semiconductor industry

The growth of *production* of semiconductors has been truly phenomenal since their commercial introduction in the 1950s. In terms of volume of production, output virtually doubled every year throughout the 1970s. Geographically, the commercial production of semiconductors began in the United States during the 1950s and for

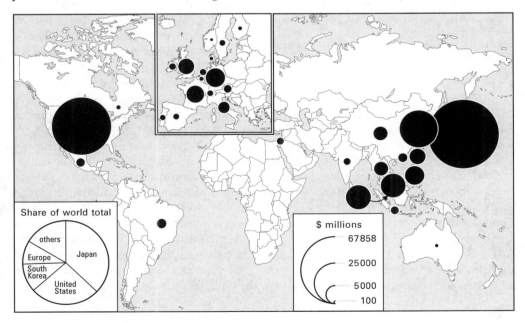

Figure 11.2   World production of active electronic components, 1996
Source: Based on data in Elsevier Advanced Technology (*Yearbook of Electronic Data, 1996*)

nearly two decades the United States dominated world production. During the 1980s, however, Japan overtook the United States to become the world's leading producer. Figure 11.2 shows the geographical distribution of the production of 'active' electronic components, which include semiconductors, integrated circuits and microprocessors. In 1996 Japan accounted for 37 per cent of the world total, the United States for 26 per cent, South Korea for 11 per cent. In comparison, the whole of Europe accounted for only 9 per cent of total world production. Within Europe the leading producers are Germany (29 per cent of the European total), France (24 per cent) and the United Kingdom (19 per cent). Outside these three core areas of production the major centres are in east and southeast Asia, notably Singapore, Malaysia, Taiwan, the Philippines and Thailand. The most notable development since the late 1980s has been the emergence of South Korea as a major world player in the semiconductor industry from a modest share of world production to the third leading producer in 1996.

The global pattern of *trade* in semiconductors has been described as a 'war of supremacy' between the United States and Japan, with western Europe striving to maintain a presence. Increasingly, however, we have to include South Korea and the other east and southeast Asian NIEs in this scenario. Figure 11.3 shows that Japan had a massive trade surplus in active electronic components of $18 billion in 1994 while South Korea had a surplus of almost $8 billion. Conversely, the United States had a deficit of $8 billion and western Europe a deficit of almost $10 billion. The group of east and southeast Asian NIEs also had a substantial trade surplus in active electronic components of $4.5 billion.

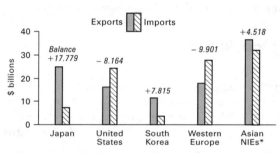

* Hong Kong, Indonesia, Malaysia, Philippines, Singapore, Taiwan, Thailand

Figure 11.3 Trade balances in active electronic components, 1994
Source: Based on data in Elsevier Advanced Technology (Yearbook of Electronic Data, 1996)

A number of significant changes have been occurring in the global geography of semiconductor production and trade in recent years, however, changes affecting both the established production areas of the United States and Europe and also the newer centres within east and southeast Asia. Within the United States, the locational core of the semiconductor industry remains the Santa Clara valley of California – 'Silicon Valley'.[2] But Silicon Valley's dominance is less marked than it was as new centres of production have emerged in such states as Colorado, Oregon and Utah. Within western Europe there has been considerable development of semiconductor production in certain more peripheral regions, most notably in Ireland, Wales and Scotland. In the latter case, the Central Valley of Scotland has acquired a substantial semiconductor industry, giving rise to the label 'Silicon Glen' and its description as 'the largest "chipshop" in Europe'.[3] Within east and southeast Asia, too, locational change is becoming evident as the 'gang of four' move into more sophisticated products and processes and new centres of production emerge in such countries as China, Malaysia, the Philippines, Thailand and Indonesia.[4]

The consumer electronics industries

The *production* of consumer electronics products is generally much more widely spread globally than that of semiconductors. It also displays far more obvious global

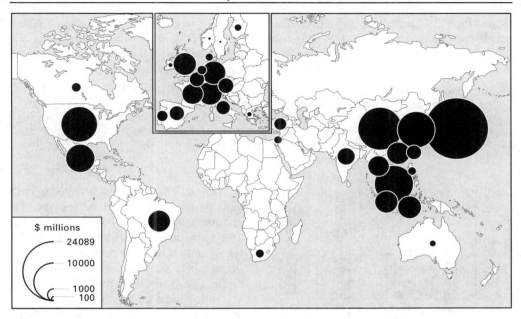

Figure 11.4   World production of consumer electronics, 1996
Source: Based on data in Elsevier Advanced Technology (Yearbook of Electronic Data, 1996)

shifts and, therefore, a greater involvement of developing countries. It is in consumer electronics, in particular, that the emergence of east and southeast Asian countries as major foci of production and exports is most clearly seen. Figure 11.4 portrays the global pattern of consumer electronics production in 1996. Compared with the production of semiconductors, consumer electronics manufacture is rather less concentrated geographically. Whereas four-fifths of semiconductor production is located in the United States, Japan, South Korea and western Europe, these account for only 60 per cent of consumer electronics production. East and southeast Asia (including Japan) is clearly the leading region of consumer electronics production today, with two-thirds of total world production.

Global shifts in consumer electronics production are shown especially clearly in the case of television receivers because this is one product which originated in the older industrialized countries of North America and Europe. The global pattern of television receiver production changed markedly between the late 1970s and late 1980s in particular. During that period, world production grew by 68 per cent but in Europe growth was only 9 per cent and in the United States 21 per cent. The major growth region was Asia (+168 per cent). But, within Asia there was very substantial differential growth. Japan, which, in 1978, was the world's leading TV manufacturer, increased its output by a mere 13 per cent. China, on the other hand, came from nowhere in 1978 to become the world's largest producer by 1987. Malaysia's TV output increased by more than 725 per cent, South Korea's by 204 per cent and Singapore's by 193 per cent. The global shift of TV manufacture to Asia was very marked indeed. Those trends continued into the 1990s.

The dominant position of the east and southeast Asian countries in the global consumer electronics industries is shown even more clearly in the pattern of *trade* flows. Figure 11.5 indicates the huge trade surpluses of the east and southeast Asian producers compared with the large trade deficits of Europe and, especially, the United States. Without doubt, there has been deep import penetration of the North American and European consumer electronics markets. For example,

> from 1948 till 1962, the USA was totally self-sufficient in television manufacture. In 1962 the first Japanese imports developed in the black-and-white sector. The first imported colour television sets were introduced into the US market in 1967 and within three years, imports were making up 17 percent of the US market for colour sets and about 50 percent for black-and white . . . Beginning in late 1975, there was a marked acceleration of this import penetration.
>
> (Turner, 1982, p. 55)

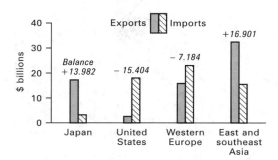

Figure 11.5   Trade balance in consumer electronics, 1994
Source: Based on data in Elsevier Advanced Technology (Yearbook of Electronic Data, 1996)

In fact, the dominance of Asian consumer electronics producers is far greater than country-based production and trade figures suggest. Such figures tell us only where the sets originate; they do not tell us who actually produces them. In fact, the vast majority of all the TV sets produced in the United States and Europe are manufactured by Asian firms which have either established new factories there or acquired domestic producers. The United States – which formerly dominated the world TV industry – no longer has an indigenous TV manufacturer. The last survivor, Zenith, has been acquired by a South Korean firm, LG Electronics.

These shifting global patterns of production and trade in semiconductors and consumer electronics products reflect a complexity of causal factors. Perhaps more than in any other industry the global geography of the electronics industry is highly volatile, in response to both technological and organizational developments within the corporate strategies of the major producers and also to the actions of national governments. It is to these factors that we now turn.

The changing pattern of demand

Semiconductors

Demand for semiconductors is a *derived* demand: it depends upon the growth of demand for products and processes in which they are incorporated. In the United States, at least initially, most of the stimulus came from the defence-aerospace sector and this explains much of the initial growth of the semiconductor industry in the United States. In Europe, and especially in Japan, defence-related demand was far less important than industrial and, particularly, consumer electronics applications. Nevertheless, governments have often been dominant customers, either directly or indirectly, throughout the semiconductor industry's short history. The microelectronics industry, and especially the semiconductor, is at the heart of the information technology revolution which is transforming both products and processes throughout the global economy.

In 1995, world semiconductor sales were more than $150 billion, some three times greater than in 1985 (*The Economist*, 23 March 1996). It has been estimated that the global market for semiconductors is in the process of tripling every five years. Overall, the biggest end-user is the computer industry and, especially, the personal computer industry. Possibly 60 per cent of semiconductor revenues are derived from sales to computer manufacturers. The two industries, PCs and semiconductors, are symbiotically locked together. On the one hand, demand by PC manufacturers drives much of the demand for semiconductors whilst, on the other hand, the leading semiconductor manufacturers (notably Intel and AMD) create demand for more powerful PCs by introducing new generations of increasingly powerful chips. Such a cumulative process is reinforced by the dominant position of Microsoft, the world's leading software company, whose new generations of software require increasingly powerful microprocessors. The introduction of Windows 95 provided an especially strong example of this process.

A major reason for the spectacular growth in demand for semiconductors has been the phenomenal decline in their selling price: the fall has been vertiginous. Partly because of technological developments (discussed in the next section) and partly because of fierce price competition between producers, the price of semiconductors is a tiny fraction of that prevailing only a short time ago. Between the mid-1960s and the late 1970s, for example, 'the cost per electronic function . . . [fell] . . . by a factor of 100,000 in less than two decades' (Siegel, 1980, p. 3). However, the memory capacity of semiconductors has grown phenomenally since the 1970s. Higher-capacity chips are appearing so quickly that each affects the price of its predecessors.

But growth in demand for, and the price of, semiconductors has been far from continuous and uninterrupted. The industry is subject to spectacular fluctuations in demand; gluts and shortages are endemic. Supply gluts further intensify price competition; at other times, severe shortages of chips have created major problems and raised prices for end-users. For example, during the mid-1990s the price of memory chips (the basic 'commodities' of the semiconductor industry) plunged from $13 in November 1995 to $9 in March 1996:

> The price of the 4-megabyte memory chip used in many personal computers has dropped from $125 to $25. Each 4MB chip is made up of eight 4-megabit D-Ram chips – and these basic chips can now be picked up for $3 each; they cost $13.50 last year. Spot prices from commodity brokers go even lower . . . Richard Gordon, semiconductor analyst at US research consultancy Dataquest, said: . . . 'It would be cheaper for a burglar to ring a computer broker and buy the chips over the phone than to steal them'. Indeed, the price of buying these chips is now lower than the cost of making them.
>
> (*The Observer*, 21 June 1996)

It was predicted that the slump in demand for semiconductors could last up to two years before the next generation of memory chips appeared to boost demand. As a result, virtually all the major semiconductor manufacturers delayed their plans to build massive new production capacity (plans which had been announced only a short time earlier) until the upturn occurs.

The basic problem for the semiconductor industry results from a combination of factors: the massive fixed costs of production facilities (see next section), the fact that they take at least a year to build, and the fact that demand for chips can change rapidly. 'The long lead times needed to build a plant mean that the supply of chips tends to be sticky; that is, it does not adjust smoothly to price changes or to changes in demand' (*The Economist*, 23 March 1996). The situation is made more complex by the strongly *segmented* character of demand for semiconductors. There are six major types of semiconductor, each of which has rather different patterns of demand:

- *Standard devices* which are technically indifferent to their final use.
- *Exclusive devices* which, like standard devices, are technically indifferent to final use but which are produced by one or a few producers with a technological monopoly.
- *Specific devices* which are mass produced but which can be used only in specific applications.
- *Custom devices* which are manufactured to very particular requirements of a specific user.
- *Microprocessors* which can be mass produced and used for multiple purposes but which can be programmed for particular applications.
- *Semicustom devices* in which certain parts are mass produced but which can then be tailored to a particular use by leaving the final connections to be decided in the light of the user's requirements. Such devices are also known as application-specific integrated circuits (ASICs). Demand for them has been growing exceptionally rapidly.

In general, the geography of demand for semiconductors correlates closely with the geography of production. It is overwhelmingly concentrated in the three major regions of the United States, western Europe and east Asia.

Consumer electronics

The pattern of demand for consumer electronics products such as television receivers differs considerably from that for semiconductors. First, of course, they are *final demand products*. Secondly, they are products for which demand is income-elastic, whereas this is less so for semiconductors in which industrial and government demand is especially important. In many ways, in fact, demand for television receivers is similar to that for automobiles. Demand for sets grew especially rapidly throughout the 1960s and well into the 1970s, with a strong geographical concentration of demand in the higher-income countries. But by the end of the 1970s, in addition to the effects of economic recession on consumer purchasing, demand for television sets was approaching saturation in the developed market economies. Replacement demand had become the major characteristic of the industry, creating severe problems for producers because, as we saw in the case of automobiles, the growth of demand in a saturated, replacement market is inevitably very much slower and more volatile than in a less developed market.

Hence, the most likely increases in demand for television sets in the coming decades will be in some developing countries and in the transitional economies of eastern Europe, assuming that incomes in these countries continue to grow. In the traditional markets of North America, Europe and Japan, however, television manufacturers face real difficulties in adjusting to prevailing market conditions. It is for such reasons that the consumer electronics companies are constantly seeking ways of developing new, related products. During the 1980s the major development was the video cassette recorder (VCR) and the compact disc player but those markets, which grew very rapidly, are now also reaching saturation in the industrialized countries.

In the mid-1990s, expectations were based upon a number of developments, including digital TV, digital video discs and, in particular, a variety of multimedia and multifunctional products. For example, it is predicted that TV screens will be transformed from their current box shape, which is determined by the size and shape of the conventional TV tube, to a wall-hung, flat format which will become the control centre for all the consumer electronics products used within the home, including not only radio, audio and TV but also personal computers. It is too early to say whether such ambitious developments will actually occur at the scale needed to sustain the growth of the consumer electronics market. What such developments do point to, however, is the increasing integration between the semiconductor, computer and consumer electronics sectors.

Production costs and technological change[5]

In the fiercely competitive environment of the global electronics industries the drive to push down production costs is a paramount consideration. The industries have become increasingly capital intensive whilst retaining a considerable degree of labour intensity in certain parts of their production chain. The urge to minimize production costs is reflected both in the nature and the rapid rate of technological change and in the changing geography of production at various spatial scales.

Semiconductors

The pace of technological change has been especially rapid and far-reaching in semiconductor production, where it is now necessary for production equipment to be replaced or updated every two or three years. The cost of establishing a semiconductor plant has escalated accordingly. In the 1960s such a plant could be set up for roughly $2 million; by the early 1970s between $15 million and $20 million was needed; ten years later, in the early 1980s, an investment of between $50 million and $75 million was necessary. By the late 1980s a new chip production facility was costing $150 million. As newer generations of chips have been developed, the costs have escalated even further:

> New fabrication plants ('fabs') can cost well over $1 billion, and the biggest will cost more than $2 billion within a year or two. State-of-the-tech fabs become obsolete in three to five years; staying ahead in such a business requires a chip maker to spend vast sums simply to keep up with technology. Last year semiconductor firms spent $30 billion on new fabrication capacity. Intel, America's largest chip manufacturer, spent $3 billion alone.
>
> (*The Economist*, 23 March 1996)

Even allowing for inflation it has obviously become a much more costly operation in which the *capital barriers to entry* have increased very substantially. Not surprisingly, therefore, the industry is one of the most capital- and research-intensive of all manufacturing industries.

Two considerations are especially important in the manufacture of semiconductors:

- The reliability of the product itself. A great deal of effort and expenditure is devoted to improving the 'yield' of the manufacturing process and reducing the number of rejected circuits. One important element here is the cleanliness of the environment in which wafer fabrication occurs. The very high cost of creating 'pure' environments is one of the reasons for the very high capital costs of semiconductor fabrication plants. Currently, one approach is to create 'mini-environments' which are, in effect, sealed containers to ensure that 'as the wafer travels through the factory from one processing step to another during its production cycle, it never comes into contact with the "dirty" factory environment' (*The Financial Times*, 2 July 1996).

- The need to pack as many circuits as possible on to a single chip. One of the most remarkable features of the semiconductor industry is the fact that the number of components per chip has increased phenomenally. By the mid-1990s, a microprocessor could contain as many as 5.5 million individual transistors compared with just 3,500 on Intel's first commercial microprocessor produced in 1971. Texas Instruments claims to have developed a technology able to pack 125 million transistors on to a single chip. Each transistor measures only 18 hundred millionths of a metre across which is 30 per cent smaller than the size of the most powerful chip currently avaliable (*The Financial Times*, 29 May 1996).

It is the combination of these two needs – to increase both the yield and the number of circuits per chip – which accounts for the steep increase in the cost of setting up a new semiconductor plant. Every major increase in the number of components per chip demands far more sophisticated and expensive manufacturing equipment which, in turn, becomes obsolete more quickly. At the same time, the trend is towards larger, more capital-intensive and research-intensive production. But to understand the changing global structure of the semiconductor industry we need to look more closely at the production process itself. We also need to take into account the different types of semiconductor being produced, because the potential for mass production varies greatly between the standard device and the custom device. Production of standard devices demands large-scale mass production; manufacture of custom chips demands much smaller runs. However, the development of semi-custom and application-specific (ASIC) devices has extended the use of mass-production techniques into new areas.

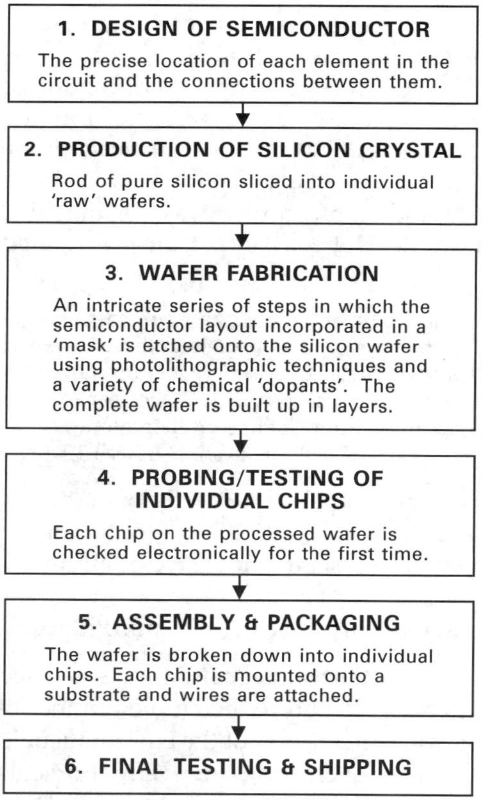

The general sequence of semiconductor production is shown in Figure 11.6. The differing characteristics of each stage have very important implications for the spatial organization of the industry at a global scale. The process begins with the *design* of a new circuit (1). Its precise form will obviously depend upon the function it is to perform. Production of complex circuits involves the superimposition of a series of separate layers, each one being produced initially as a pattern or mask, from which the actual circuits will eventually be made. The production of the silicon (2) from which the chip wafers will be made is a process whereby the silicon crystal is drawn out and formed into a

1. DESIGN OF SEMICONDUCTOR

The precise location of each element in the circuit and the connections between them.

↓

2. PRODUCTION OF SILICON CRYSTAL

Rod of pure silicon sliced into individual 'raw' wafers.

↓

3. WAFER FABRICATION

An intricate series of steps in which the semiconductor layout incorporated in a 'mask' is etched onto the silicon wafer using photolithographic techniques and a variety of chemical 'dopants'. The complete wafer is built up in layers.

↓

4. PROBING/TESTING OF INDIVIDUAL CHIPS

Each chip on the processed wafer is checked electronically for the first time.

↓

5. ASSEMBLY & PACKAGING

The wafer is broken down into individual chips. Each chip is mounted onto a substrate and wires are attached.

↓

6. FINAL TESTING & SHIPPING

Figure 11.6   Stages in semiconductor production
Source: Based on UNIDO (1981, pp. 69–71); Ó hUallacháin (1997, Figure 2)

cylindrical rod roughly 100 mm in diameter. Generally, this is done by specialist firms, although a few of the very large semiconductor manufacturers produce their own silicon. The silicon rods are then sliced into individual wafers 0.5 mm thick. The wafer fabrication stage (3) consists of a number of intricate and highly precise processes in which the circuits are etched on to the wafers, layer by layer, using the masks in a photolithographic-chemical process. Each wafer contains large numbers of identical circuits. The next stage (4) is that in which each chip on the wafer is tested electrically before being sent for assembly (5). Here, the individual wafers are broken down into separate chips, assembled into the final integrated circuit or microprocessor using a bonding/wiring process and individually packed. They are then subjected to final testing (6) and shipped to the customer for use in the final product.

A particularly important distinction exists between the design and wafer fabrication stage on the one hand and the assembly stage on the other. Each has different production characteristics and neither needs to be located in close geographical proximity to the other. Design and fabrication require high-level scientific, technical and engineering personnel while the fabrication stage itself requires an extremely pure production environment and the availability of suitable utilities (pure water supplies, waste disposal facilities for noxious chemical wastes). In contrast, the assembly of semiconductors is carried out using low-skill labour, notably female, and although there is still a need for a 'clean' production environment this is less critical than for wafer fabrication. The low-weight/high-value characteristics of semiconductors permit their transportation over virtually any geographical distance. Hence, it is the assembly stage of the production sequence which has been most susceptible to relocation to low-labour cost areas of the world.

Developments in the technology of semiconductor production – new lithographic techniques, new methods of wafer fabrication and the introduction of automation at all stages of the process – have created substantial changes in the organization and structure of the industry. In particular, the relative importance of labour costs has been greatly reduced especially in the higher-value products. At the same time, there have been increased pressures to enlarge the scale of production plants. A wafer fabrication plant, in order to be profitable, needs to produce at very high volumes.

Overall, therefore, the manufacture of semiconductors is a highly capital- and research-intensive industry but one in which there are distinct 'breaks' in the production sequence. Such breaks are used by producers, as we shall see, to create a complex global geography of production.

Consumer electronics

In general, the manufacture of consumer electronics products, such as television receivers, is less advanced technologically than the production of semiconductors. In product cycle terms, television sets have come to be regarded as mature products in which the technology has become standardized and in which the emphasis is on reducing production costs through economies of scale and the use of less skilled labour. There is a good deal of validity in this view. But colour television production in recent years has shown signs of technological rejuvenation; a 'dematuring' similar to that evident in the automobile industry. This is reflected both in new process technology and also in the development of new functions for television receivers and of new, related, products. These are part of broader developments in communications systems in which the domestic television receiver is seen, potentially, as the centre of a sophisticated entertainment, educational, commercial and general information system.

The production process itself consists of three related stages. As in semiconductor manufacture each stage has rather different characteristics which have important organizational and locational implications. The design stage is highly research intensive, particularly as efforts are made to develop new functions for television sets and to improve the efficiency of the production process. The manufacture of components, particularly that of television screen tubes, is heavily capital intensive. It is a stage in which economies of scale are especially significant: the optimal scale of production for television tubes is approximately 1 million per year compared with roughly 400,000 for complete sets. Again, it is the assembly stage which is the most highly labour intensive, employing large numbers of low-skilled, mainly female, workers. The potential advantage of low labour-cost areas is clear.

There have been two particularly important technological developments in television manufacture in recent years which have affected the production process very considerably. One is the progressive simplification of set engineering, as reflected in the number of components used. During the 1970s alone, the number of components in the average television set declined from 1,400 to 400. The other major development is increased automation of assembly, particularly the use of automatic insertion techniques, which were pioneered by Japanese firms but which are now widely spread throughout the industry. Reductions in the number of components used and increased automation of the production process have together altered the needs for different types of labour. In general, fewer skilled workers are now needed to manufacture a television set.

The role of governments in the electronics industries

Semiconductors[6]

National governments throughout the world involve themselves in the electronics industry to a greater or lesser degree. Such intervention has important implications for the kinds of strategies pursued by producing firms, which are examined in the following section. Government involvement is most extensive and fundamental in the semiconductor sector as well as in computers and telecommunications. In these sectors, government interest dates from the industry's earliest days. It is easy to appreciate why governments have attempted to intervene in the development of the semiconductor industry. Semiconductor production is recognized as a key technology with enormous ramifications throughout the economy. If a country is to benefit fully from it – or if it is to avoid being left behind – it must have access to what is an expensive and rapidly changing technology.

This access may be achieved in several ways. One is to build an indigenous production capacity based upon domestically owned firms. Another is to attract foreign semiconductor firms to establish production units. A third is to purchase semiconductors on the open market and concentrate on developing the end-uses. Problems are inherent in all three options. Setting up a viable domestic industry may be beyond the means of many countries. On the other hand, relying on foreign investment or the open market may lead to problems of dependency of supplies on foreign sources. Such potential vulnerability may be important not only for industrial applications but especially for defence. Semiconductor technology is at the heart of all modern defence systems, hence the sensitivity of most national governments to developments in this industry.

The particular policies pursued by governments reflect their specific national circumstances, including the country's relative position in the global semiconductor industry. In the United States, the country in which the semiconductor industry originated, the dominant influences have been the federal defence and aerospace sectors. In the early days of the semiconductor industry these set the direction and nature of the industry's development because they were its dominant customers. Although defence-related forces are now less important to the American semiconductor industry their influence persists. In recent years, as the United States' lead in semiconductor production has been eroded, there has been much broader criticism of the direction (even the existence) of US policy towards the microelectronics industry. The sharp deterioration in the country's global position as a semiconductor producer led to strong political and industry lobbying for targeted policies. The relative decline of the industry is seen as a threat to national security, both economic and military.

Apart from some Defense Department support for Sematech, a consortium of US semiconductor firms established in 1987, however, no specific US government policy towards the industry has yet emerged except in the area of *trade*. In 1986 the US government signed a 'semiconductor pact' with Japan. This arose out of allegations by the United States that Japanese producers were 'dumping' chips at excessively low prices in the United States (and, therefore, undercutting US producers) and also that the Japanese were unfairly restricting access to their domestic market. The initial pact expired in 1991 and was renegotiated for a further five years with a particular US emphasis on access for American semiconductor manufactures to the Japanese market. On the expiry of the semiconductor pact in 1996, the US and Japan reached agreement to set up two new international industry bodies.

Advocates of an explicit US government strategy towards the semiconductor industry point, in particular, to the case of Japan. In Japan, policy has been directed to the industrial and consumer applications of semiconductors. We have already discussed the general nature of Japanese industrial policy (Chapter 4) but we should recall that microelectronics and the related computer and information technologies form the central focus of the drive to develop knowledge-intensive industries. Japanese policy in this sector, as in others, has been one of general guidance through MITI and the establishment of a suitable framework in which the private sector firms can operate competitively. The initial aim was to avert technological dominance by the United States by discouraging direct foreign participation in the Japanese semiconductor industry and acquiring foreign technology through other means. Protection of the industry was extremely tight until the end of the 1970s, both through import controls and restrictions on inward investment.

Since that time Japanese policy has become more liberal. But the Japanese determination to get ahead of the competition in specific types of microelectronics remains very strong. A clear example was the very large-scale integration (VLSI) project begun in 1976 on the initiative of MITI. This involves research collaboration between the five major Japanese companies to develop highly sophisticated integrated circuits for the next generation of computers. The Japanese government funded some 40 per cent of the total cost. Although MITI's involvement has certainly been important, Fransman (1990) argues that its actual contribution has been less significant than is often believed.

However, there is no doubt that government involvement in the South Korean semiconductor industry has been highly significant. As Wade (1990b) has shown, the Korean government used a variety of devices – including its control of the telecommunications industry – to foster the entry of large Korean firms into the semiconductor industry. In 1976, the government established the Korea Institute of Electronics Technology (KIET) whose responsibility was 'planning and coordinating semiconductor R&D, importing, assimilating, and disseminating foreign technologies, providing technical assistance to Korean firms, and undertaking market research' (Wade, 1990b, p. 313). In 1982, the Long-Term Semiconductor Industry Promotion Plan was announced whereby substantial fiscal incentives were made available to the four leading Korean firms.

Turning to western Europe, the third major focus of semiconductor production, we find a very uneven policy picture. Not only does the European semiconductor industry lag a good way behind those of the United States and Japan but also much of its production capacity is in American-, Japanese- and, more recently, Korean-owned plants. In general, European governments have not only welcomed such investments but also, in some cases, assiduously courted the foreign semiconductor firms. According to Dosi (1983), western European policies towards the semiconductor industry have evolved in three stages. First, the period to the mid-1960s was largely one of non-intervention (apart from defence-related R&D and some bias towards national producers in government purchases). Second, between the mid-1960s and mid-1970s government focus on the computer industry gave some stimulus to semiconductor research. But in neither period was government involvement particularly influential. The third period, from the mid-1970s onwards, marked a major intensification in government involvement, this time focused on the information technologies including microelectronics.

Since the early 1980s European governments have been involved in supporting major collaborative ventures in information technologies in general (the ESPRIT programme) and specifically in semiconductors. The $4 billion (£2.4 million) JESSI programme – the Joint European Submicron Silicon Initiative – was concerned with developing advanced microchip technology. There has also been increasing pressure on non-European semiconductor manufacturers to locate more than merely assembly operations in Europe: the objective has been to persuade US and Japanese companies to locate more of their design and fabrication plants in Europe. The aims of these European initiatives were threefold: to protect European capacity in the core technologies of microelectronics; to accelerate innovation; and to encourage crossborder links between national electronics firms within Europe. A major problem, however, has been whether or not to allow non-European firms with major European operations to participate in the collaborative programmes.

The European Commission has also been increasingly active in protecting the semiconductor industry against what are seen to be unfair trading practices. In 1990 it secured a voluntary agreement with the Japanese on the minimum prices at which standard memory chips would be sold in the EC. Subsequently, the Commission initiated anti-dumping measures against Korean semiconductor manufacturers who were alleged to be dumping chips at below acceptable prices. In effect, this was the price agreed with the Japanese, who were regarded as the lowest-cost producers. These minimum prices were reimposed in 1997 on the biggest selling semiconductor chips from 14 Japanese and Korean manufacturers (*The Financial Times*, 1 April 1997).

Consumer electronics

In general, government involvement in the consumer electronics industry has been far less extensive than its involvement in semiconductors. The involvement has also been primarily defensive.[7] Until the 1970s, of the major industrial nations only Japan and, to a lesser extent, France had adopted specific policies towards consumer electronics. In Japan consumer electronics, and especially television receivers, were regarded as an important sector for national development not only in itself but also as a significant complement to the semiconductor and electronic components sectors. Thus, consumer electronics in Japan were given the 'usual treatment' reserved for targeted growth sectors. Emphasis was placed on creating a highly efficient and productive sector able to achieve world leadership in products such as colour television sets. The policy was both protective and stimulative. Of the European countries, only France had adopted a long-term strategy explicitly for consumer electronics, mainly by encouraging Thomson as the major television manufacturer and by protective measures.

In western Europe as a whole, in fact, protection of the television market has operated since the early 1960s through the existence of exclusive transmission systems (notably the PAL and SECAM systems), which, until they began to expire in the 1980s, had been used to keep out Japanese competition. Before 1970 licences were not granted at all to Japanese manufacturers; after 1970 some limited access was allowed. But protection was mainly against the larger sets and did little to prevent the inflow of smaller sets, for which demand was growing most rapidly.

The rapid increase in import penetration led to most western countries adopting quantitative trade restrictions against television imports. In 1977 a 'voluntary export restraint' agreement was reached between Japan and the United States which substantially reduced the volume of Japanese imports. Restrictions were subsequently extended to South Korean and Taiwanese producers. At the same time some European governments, notably the British, set out to attract Japanese consumer electronics firms to set up production plants. The French were far more reluctant although they have subsequently softened their attitude. France, however, continued to follow a 'national champion' policy in its support of Thomson.

Among developing countries it is the three Asian NIEs – South Korea, Taiwan and Singapore – which have adopted the most positive policies towards the consumer electronics sector as part of their export-oriented industrialization strategy.[8] Singapore and Taiwan have been especially open to investment by TNCs, particularly from Japan. In contrast, South Korea's strategy has been one of developing an indigenous industry and providing far less scope for the direct involvement of foreign firms. For example, the South Korean government has had a major influence on the development of the country's electronics industry using a mix of policies, including:

- Protectionist trade barriers, both tariffs and a ban on the import of foreign-made electronics.
- Provision of low-interest capital for companies in targeted sectors, which has given the state an influence on industry decision-making.
- Heavy government investment in electronics R&D, particularly in the design of semiconductors, the results of which are made available to Korean producers. 'The result of these co-ordinated initiatives by capital and the state is that . . . South Korea is the only one of Asia's newly industrializing countries (NICs) that has developed a manufacturing base to produce high value-added goods sufficient to sustain high growth rates over the next decade' (Henderson, 1989, pp. 66–67).

A broadly similar state electronics strategy has been followed in Taiwan.

Faced with increasing competition in the global electronics industries, the South Korean government has been urging Korean electronics companies to adopt a multi-pronged strategy, including:

- Diversifying their geographical markets to reduce dependence upon the United States.
- Establishing overseas production facilities either to achieve lower production costs or to circumvent trade barriers.
- Developing their own brand names to reduce their dependence as original equipment manufacturers.
- Raising expenditure on R&D in order to lessen dependence on foreign technology.
- Moving their product emphasis towards higher value-added products.

Corporate strategies in the electronics industries

In such a volatile technological and competitive industry as electronics, firms inevitably employ a whole variety of strategies to ensure their survival and in pursuit of growth. However, corporate strategies – whether offensive or defensive in nature – increasingly are being implemented at a *global*, rather than a purely national, scale. Indeed, in the face of fierce global competition, firms have been systematically *rationalizing* and *reorganizing* both their domestic and overseas operations.

It is the inexorable pursuit of such strategies – set within the context of rapidly changing market, technological and political forces – which does most of all to shape and reshape the global map of production and trade in the electronics industry. Of course, many factors influence the particular mix of strategies employed. One important variable is *size of firm*: large firms tend to operate in rather different ways from smaller firms. Another influence is, undoubtedly, a *firm's geographical origins*: the domestic context in which it has developed. There are substantial differences in behaviour between American, Japanese and European electronics firms.

A strategy common among some firms has been to *specialize* in specific market segments, to pursue a 'niche' strategy such as the manufacture of high-value microcircuits for very specific applications. Among other firms, however, the preferred strategy has been to increase the degree of *vertical integration*. Among television manufacturers, particularly American and European firms, a common strategy has been that of product diversification to reduce their dependence upon a product in which competition from Japanese and other Asian producers is especially fierce. Cutting across these strategies are those of increasing *internationalization of production*, of *automation* and, overall, of *rationalization and reorganization on a global scale*.

One characteristic common to each major region of production and to both semiconductor and television manufacture is the very high degree of firm concentration. Increasingly, a relatively small number of very large transnational corporations dominates production, a reflection of the technological and production characteristics of these sectors (the accelerating necessity of very large capital investments brought about by the rapid and highly expensive nature of technological change). In such circumstances, small-scale operations become less and less viable.

Strategies in the semiconductor industry

The transformation from a fragmented industry structure, in which small and medium-sized firms were the norm, to one of large-firm dominance and to a more restricted role for the small firm, has been especially dramatic in the semiconductor industry of the United States. During the late 1950s and through much of the 1960s, entry into the semiconductor industry was relatively easy as the proliferation of new small firms particularly in Silicon Valley demonstrated. New firms sprang up overnight, often spinning off directly from established firms (many of which were, themselves, relatively new). The starting point was the departure of William Shockley, one of the pioneers of transistor technology, from Bell Telephones to set up on his own in Palo Alto, California.

But this was only the beginning:

> In 1957 eight of his young assistants . . . quit the company to form, with the backing of Fairchild Camera and Instruments, Fairchild Semiconductor. Although Fairchild soon became one of the top merchant semiconductor houses and introduced a number of key technological advances, its Eastern management was unable to satisfy the innovators. In clusters of twos and threes, a large number of Fairchild's researchers and executives, including the eight founders, quit to found or re-organize other semiconductor firms in the area. The first group formed Rheem Semiconductor (Raytheon) in 1959, and two other groups . . . formed Signetics and Amelco (now Teledyne Semiconductor) in 1961. In 1967 Fairchild Semiconductor's general manager . . . left to rebuild a merger of Molectro (another Fairchild spinoff) into National Semiconductor as a major competitor. In 1968 . . . [two other Fairchild founders] . . . quit to form Intel. And the following year, a market manager . . . took seven other Fairchild employees with him to form Advanced Micro Devices. In 1971, 21 of the 23 semiconductor firms in the Bay Area could trace their ancestry to Fairchild . . . The Santa Clara Valley proved a fertile breeding ground for semiconductor companies, as well as other high technology ventures. Stanford University, by carrying out millions of dollars in electronics research for the Pentagon, creating the Stanford Research Institute, and leasing Stanford land to high technology companies, had consciously created a community of technical scholars.
>
> (Siegel, 1981, p. 4)

The result was the emergence of a remarkable geographical cluster of semiconductor and related industries in the Santa Clara Valley of southern California. Through the processes of localized territorial development discussed earlier in this book – notably the operation of traded and untraded interdependencies and the evolution of an innovative milieu in which localized learning processes predominated – Silicon Valley became the core of the United States semiconductor industry.[9] It was not the only localized concentration but it achieved the position of dominance. Initially, as we have seen, most of the new firms founded there were spin-offs from existing semiconductor producers. But as the area's dominance developed over the years, non-local firms, including Japanese, European and, more recently, other east Asian firms, have set up operations in the valley.

In fact, Silicon Valley has 'three faces':

> small entrepreneurial start-ups make up only one face of Silicon Valley. This has been true from the Valley's inception. Networks of alternately cooperating and competing small and large firms, manufacturing semiconductors and other electronics equipment in the manner of an industrial district, do exist in Silicon Valley and offer still another of its faces. A third face involves the deep interconnection between multiregional, often multinational American firms, Japanese and European corporations investing in the Valley, and such fundamentally outwardly directed, powerful institutions as Stanford University and the Pentagon.
>
> (Harrison, 1994, pp. 120–21)

The proliferation of new small firms in the US semiconductor industry as a whole slowed down as the barriers to entry increased during the 1970s. The dominance of large firms became especially marked in the manufacture of standard semiconductor devices which depend on mass-production technology. Even so, very high levels of new firm start-ups continued in the US semiconductor industry through the 1980s, especially among firms engaged in application-specific products (Angel, 1994, pp. 38, 39):

> While the germination of new products and production processes typically took place in the research facilities of large established firms, start-up companies were often at the forefront in developing applications for new technologies . . . During much of the first three decades of the industry, barriers to entry to semiconductor manufacturing remained relatively low. A large infrastructure of industrial support rapidly emerged to support start-up firms, including specialized venture-capital sources, equipment suppliers, and assembly subcontractors . . . Of greatest significance to the dynamic of new firm formation and technological development, however, has been the tendency toward a rapid circulation of technological knowledge and information among firms in the US semiconductor industry.

Outside the United States, the small entrepreneurial firm, starting its manufacturing life in the owner's garage, was never as important in the semiconductor industries of Japan, western Europe and the newly-emerging producers of South Korea and Taiwan. In these areas production tended to be within large, diversified electronics companies from the beginning.

Table 11.1   The world's leading semiconductor producers, 1995

Rank	Company	Country	Share of world market 1995 (%)	Share of world market 1989 (%)
1	Intel	United States	8.9	4.4
2	NEC	Japan	7.3	8.9
3	Toshiba	Japan	6.6	8.8
4	Hitachi	Japan	6.1	7.0
5	Motorola	United States	5.9	5.9
6	Samsung	South Korea	5.4	–
7	Texas Instruments	United States	5.2	5.0
8	Fujitsu	Japan	3.6	5.3
9	Mitsubishi	Japan	3.3	4.7
10	Hyundai	South Korea	2.8	–

Source: Press reports.

Table 11.1 lists the world's ten leading semiconductor firms (excluding IBM which, until 1994, was entirely an in-house producer of semiconductors). Together this small group produced 55 per cent of the world output in 1995. Five of the top ten semiconductor firms are Japanese, one fewer than in 1989, although the market shares of all the leading Japanese semiconductor firms have declined substantially since 1989. Three big changes have occurred in the top league since 1989. First, the US company, Intel, moved from eighth position to first as a result of the company doubling its share of the world semiconductor market. Secondly, the South Korean companies, Samsung and Hyundai, surged up the list to occupy sixth and tenth place respectively in 1995. Thirdly, there is no longer an indigenous European company in the top ten as Philips's share of the world semiconductor has fallen.

Such a simple listing of the leading semiconductor firms masks some important differences between them. Semiconductor producers can be classified into three broad types:

- *Vertically integrated captive producers*, which manufacture semiconductors entirely for their own in-house use.

- *Merchant producers*, specialist firms which manufacture semiconductors for sale to other firms.
- *Vertically integrated captive-merchant producers*, which manufacture semiconductors partly for their own use and partly for sale to others.

Most US semiconductor firms have tended to fall into the first two categories. IBM is the biggest vertically integrated captive producer but there are others such as AT&T, Delco (GM) and Hewlett-Packard. At the other extreme are the merchant producers which manufacture entirely for the open market. The leading US merchant firms are Intel, Motorola and Texas Instruments. In contrast, the leading Japanese and European semiconductor firms are primarily vertically integrated captive-merchant producers. They are, in general, parts of complex, diversified electronics companies operating across a very broad spectrum.

A general trend among the large semiconductor producers, especially in the United States, has been towards increased vertical integration. On the one hand, many large-scale users of semiconductors, particularly computer manufacturers but also automobile firms and others, have been integrating backwards into semiconductor production. On the other hand, the large specialist merchant producers began to integrate forwards into various user industries. Among the already highly integrated Japanese and European firms the strategy has been to strengthen such functional connections and to develop new applications in both industrial and consumer electronics. As in the United States, many of the major users of semiconductors have become increasingly involved in the manufacture of custom devices for their own applications.

A fourth category of semiconductor firm emerged especially in the United States during the 1980s. These were the 'fabless' semiconductor firms; design houses[10] which

> avoided the costs of building, equipping, and operating a fab. They developed core skills in product design and development, quality assurance, marketing, sales, and customer support. Besides design, they internalized the probe and final testing steps of the manufacturing chain, to safeguard product quality, and subcontracted for raw wafer manufacturing, wafer fabrication, and chip assembly. Beyond avoiding initial capital expenses and the associated overhead burden of owning a wafer fabrication facility, the fabless strategy sidestepped equipment obsolescence risks. Fabless companies held the flexibility to take advantage of new manufacturing process technologies as they became mainstream . . . [they] . . . were confident that others were willing to shoulder equipment obsolescence risks and make foundry capacity procurable. The latter included the Korean, Taiwanese, and Singaporean governments that heavily subsidized construction of new production facilities.
>
> (Ó hUallacháin, 1997, p. 222)

During the 1990s, however, shortages of production capacity forced the fabless firms to enter production themselves through a variety of mechanisms, including taking equity stakes in production facilities; entering joint ventures with established vertically integrated semiconductor firms; forging relational contracts with fabricators. These arrangements occurred both inside and outside the United States, notably in east and southeast Asia. In other words, the formerly disintegrated system of production developed by the fabless firms as a means of ensuring flexibility has been at least partially replaced by a reversion to vertically integrated production systems in which scale economies remain significant.

Table 11.2   International strategic alliances in the semiconductor industry

Alliance partners	Purposes of the alliance
Motorola (USA) – IBM (USA) – Siemens (Germany) – Toshiba (Japan)	To develop next generation of memory chips, including a 1 gigabit dynamic random access (D-RAM) device. Will build upon the existing alliance between IBM, Siemens, and Toshiba
Siemens (Germany) – Motorola (USA)	To build a new plant in the United States to make advanced memory chips
Mitsubishi (Japan) – Umax Data Systems (Taiwan) – Kanematsu (Japan)	To build a semiconductor facility in Taiwan to produce advanced memory chips
NEC (Japan) – Samsung (South Korea)	To collaborate in the production of memory chips for the European market
IBM (USA) – Toshiba (Japan)	To build a semiconductor facility in the United States to produce next-generation memory chips

Source: Press reports.

Acquisitions, mergers and strategic alliances

In striving to compete in global markets semiconductor firms from the United States, Japan and Europe have followed the two routes of acquisition and merger and of strategic alliances. During the 1970s and 1980s there was a substantial wave of mergers and acquisitions, both domestic and international. Within the United States, for example, Mostek was acquired by United Technologies, Synertek by Honeywell, Intersil by General Electric, American Microsystems by Gould, Signetics by Philips. Particularly vulnerable to take-over were the fast-growing smaller firms; by 1980 only seven of thirty-six post-1966 start-up companies were still independent. One of the biggest acquisitions of the late 1980s was the take-over of the British company ICL by the Japanese firm Fujitsu. Within Europe itself the biggest merger was that between the Italian firm SGS and the French firm Thomson to form SGS-Thomson, which is now the second largest semiconductor producer in Europe after Philips. In 1989 SGS-Thomson bought the sophisticated British producer INMOS, from Thorn-EMI.

International strategic alliances have become increasingly common in the semiconductor industry for all the reasons discussed in Chapter 7. In particular, the massive costs of R&D, the incredibly rapid pace of technological change and the escalating costs of installing new capacity all contribute towards the attractiveness of forming strategic alliances. Table 11.2 shows some of the more significant strategic alliances in the semiconductor industry which developed during the mid-1990s.

In addition to such international alliances, both US and European companies have been forming *local* alliances in an attempt to compete more effectively with the Japanese. In the United States a consortium of leading semiconductor firms formed Sematech to undertake joint research. Early in 1990 six firms (including IBM) joined together to purchase a leading US manufacturer of semiconductor equipment to prevent it being acquired by Japanese firms. In Europe an alliance of producers from different European companies formed European Silicon Structures (ES2) to manufacture custom chips.

Global organization of production

The most significant characteristic of the semiconductor industry is not simply that its markets are global but, rather, that its *production is organized globally*. The electronics industry was the first to which the label 'global factory' was applied because of its early use of offshore assembly. It is in the semiconductor industry that a *spatial hierarchy of production* at the global scale is most apparent, with clear geographical separation between different stages of the production process. It is overwhelmingly the *assembly* stages that have been partly relocated to certain developing countries. In general, the higher-level design, R&D and more complex stages of production have tended either to remain in the firm's home country or to be established in other developed countries where the necessary labour skills and physical infrastructure are more readily available. This pattern is changing, however. In the case of semiconductors the first two stages in the production sequence shown in Figure 11.6 have generally not been relocated overseas. Wafer fabrication, however, is carried out by American and Japanese firms in Europe and a great deal of the final assembly stage has been transferred to developing countries, especially in southeast Asia. In the past few years, some wafer fabrication and design facilities have developed in east and southeast Asia. Overall, the degree of offshore production varies somewhat between American, Japanese and European firms.

United States producers

Offshore production in the semiconductor industry first occurred in the early 1960s when a number of American firms began to seek out low labour-cost locations for their more routine assembly operations. In the semiconductor industry the initial stimulus was the intensifying competition within the United States itself, as new firms entered the industry and the need to reduce production costs accelerated. In the 1960s the differential between US labour costs and those in developing countries was especially great. Hence,

> during the 1960s, prompted by the particular needs of production, the semiconductor industry established a unique form of international production, the *integrated global assembly line*. Midway in the process of manufacturing transistors or integrated circuits, the producers shipped the unfinished components abroad for assembly – bonding – and they then shipped the assembled chips back to the US for testing. This global scheme, though much more complex today, still guides the industry.
>
> (Siegel, 1980, p. 7, emphasis added)

The first offshore assembly plant in the semiconductor industry was set up by Fairchild in Hong Kong in 1962. In 1964 General Instruments transferred some of its microelectronics assembly to Taiwan. In 1966 Fairchild opened a plant in South Korea. Around the same time, several US manufacturers set up semiconductor assembly plants in the Mexican border zone. In the later 1960s US firms moved into Singapore and subsequently into Malaysia. During the following decade semiconductor assembly grew very rapidly in these Asian locations and spread to Indonesia and the Philippines. Despite some developments elsewhere (for example in central America and the Caribbean) most of the growth of semiconductor assembly remained in east and southeast Asia.

By the early 1970s, therefore, every major American semiconductor producer had established offshore assembly facilities, a tendency greatly encouraged by the offshore assembly provisions operated by the US government. In 1978, 82 per cent of all imports under these provisions were of semiconductors from offshore locations. Figure 11.7 shows the global distribution of US semiconductor assembly plants, indicating clusters in Mexico and the Caribbean as well as in Europe but with by far the largest concentration in east and southeast Asia.

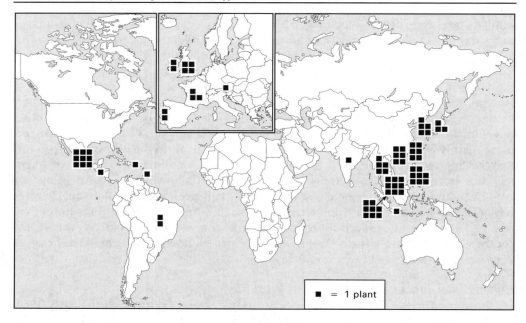

Figure 11.7 The global distribution of United States-owned semiconductor assembly plants
Source: Based on Scott and Angel (1988, Figure 5)

The structure of US semiconductor operations within east and southeast Asia has changed substantially since the early focus on low labour-cost assembly operations.[11] Gereffi (1996a) identifies three stages in the evolution of the Asian production network. In the first stage (late 1960s–70s), US firms simply sought cheap labour sites in the region as part of their international division of labour; the Asian assemblers constituted just one stage in this sequence. In the second stage (1980–85), the proportion of components sourced locally and the level of technological sophistication increased. In the third stage (1985–90s),

> the division of labor between US affiliates and local producers in East Asia deepened significantly. US electronics transnationals set up Asian networks based on a 'complementary' division of labor: US firms specialized in 'soft' competencies (the definition of standards, designs, and product architecture) and the Asian firms specialized in 'hard' competencies (the provision of components and basic manufacturing stages). The Asian affiliates of US firms developed extensive subcontracting relationships with local manufacturers, who in turn became increasingly skilled suppliers of components, subassemblies and, in some cases, entire electronics systems.
>
> (Gereffi, 1996a, p. 103)

Hence, a clear system of *intraregional specialization* has emerged, with higher-level functions being performed in the more advanced countries and the lower-level assembly functions being progressively relocated to later entrants. The other development has been the establishment of regional headquarters in the region by US semiconductor firms: for example, National Semiconductor in Singapore, Motorola, Sprague, Zilog in Hong Kong (Henderson, 1989). These latter developments reflect the growing importance of the region as a *market* for semiconductors and not merely as a low-cost assembly location.

Most of the semiconductor plants established by US semiconductor companies in Europe can be explained primarily by the 17 per cent tariff imposed on imports by the European Commission. However, the completion of the Single European Market in 1992 added a new stimulus. The increasingly stringent attitude by European governments towards full-scale local production (including the more advanced stages of design and manufacture) and the practice of using anti-dumping measures stimulated further direct investment in Europe by US semiconductor firms.

Within Europe there is a long-established US semiconductor presence with a particularly heavy concentration in Scotland and, to a much lesser extent, Ireland. United States firms dominate the Scottish semiconductor sector. Scotland has a long tradition of American manufacturing investment, a good supply of both high-skilled labour from the well developed higher-education sector and of female assembly workers. In addition, the country's investment promotion agency has been a strenuous, and successful, seeker of electronics companies to an area with a substantial package of investment incentives. Ireland, too, has strong links with the United States (one has only to think of the number of American presidents with Irish ancestry) and it, too, has adopted an aggressive strategy to attract foreign electronics firms on the basis of generous financial and tax incentives and a good labour supply. The need to operate behind the EC tariff wall, therefore, attracted large numbers of US semiconductor firms into Europe.

Japanese producers

As Figure 11.8 shows, Japanese semiconductor companies have also developed extensive offshore assembly operations, with an even greater emphasis on east and

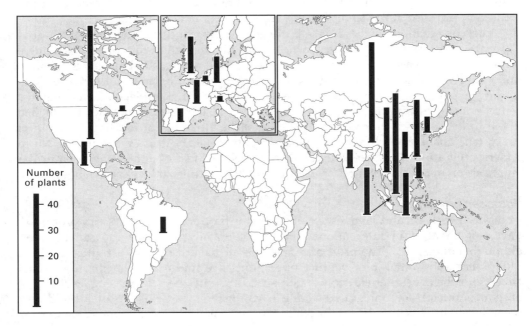

Figure 11.8   The global distribution of Japanese electronic component and devices plants
Source: Based on Electronic Industries Association of Japan (1996, *Facts and Figures on the Japanese Electronics Industry, 1996*, p. 91)

southeast Asia. Of the 636 electronic components and devices plants located outside Japan in 1995, 71 per cent were in Asia, 17 per cent in North America (including Mexico), and 10 per cent were located in Europe.

Of course, proximity to the Japanese domestic production base facilitated such developments in east and southeast Asia and helped to create a complex intraregional division of labour. The major locations within the region are Malaysia, Taiwan, Singapore, South Korea and, increasingly, China. Offshore production in the region by Japanese companies accelerated in the second half of the 1980s, after the major revaluation of the Japanese yen against the US dollar in 1985. The search for lower-cost production locations led to greatly increased offshore production in Asian countries. But whereas much American offshore production of semiconductors has been for eventual reimport to the United States, most Japanese offshore assembly plants in Asia supply Japanese plants outside Japan, particularly the growing consumer electronics operations set up by Japanese firms in both the developing and, more recently, the developed countries of the United States and Europe.

The establishment of semiconductor production facilities by Japanese firms in both Europe and North America is very much more recent than that of American firms in Europe. As we have already seen, Japanese firms have tended to concentrate their overseas plants in developing countries, particularly in Asia, mainly to reduce production costs. Until the 1970s there was virtually no direct Japanese investment in electronics in the developed countries. As Figure 11.8 shows, however, there are now substantial numbers of Japanese electronic component firms operating plants in both the United States and Europe. In the latter case more than two-thirds of the total are in the United Kingdom and Germany. Major Japanese semiconductor plants in the United Kingdom include the large-scale NEC fabrication plant in Scotland and the Shin-Etsu Handotai silicon manufacturing plant, also in Scotland, both of which arrived in the mid-1980s.

The combination of the Single European Market and the revalued yen – but especially the former – stimulated a new, and highly significant, wave of Japanese semiconductor investment in Europe. This new wave was led by Fujitsu's decision to build a $100 million chip fabrication plant in northeast England in 1989. The other leading Japanese companies followed this lead. Hitachi, for example, built an integrated semiconductor manufacturing plant in Germany. Japanese semiconductor production in the United States is much more firmly established than in Europe. In fact six out of every ten Japanese electronic component plants outside Asia are located in North America (Figure 11.8). Again, the same basic locational factors apply, with a particular emphasis on market access and proximity, and on scientific and technical labour.

European producers

Compared with US and Japanese semiconductor producers, the European producers have not engaged in international production to anything like the same extent, although Philips, SGS-Thomson and Siemens all have assembly activities in Asia. Philips has substantial joint venture operations in China manufacturing integrated circuits for the consumer electronics industry. It also has been increasing its sourcing of semiconductors and integrated circuits from the region. SGS-Thomson is building a second wafer fabrication plant in Singapore to make advanced wafers, mainly for export. But none of the leading European firms operates the kinds of global production network which have been developed by the Japanese and, especially, the US producers which are so heavily dependent on east and southeast Asia.

Within Europe itself, both Philips and Siemens have substantial production facilities in the United Kingdom. Indeed, Siemens is currently building what is claimed to be the world's most advanced semiconductor plant in northeast England at a cost of $1.8 billion, and is also establishing a $380 million memory chip plant in Portugal.

South Korean producers

The newest entrants into the global semiconductor industry – South Korean firms – are very new indeed. We noted earlier the targeted policies of the Korean government towards electronics. The speed with which the four leading Korean conglomerates – Samsung, Goldstar, Hyundai and Daewoo – have developed as extremely advanced semiconductor producers has been little short of spectacular. None of the Korean electronics firms was involved in semiconductors until Samsung entered the business in the mid-1970s. Goldstar followed in 1979, Hyundai and Daewoo (which were not then electronics companies) entered in 1983. Initially the Korean firms were totally dependent on technology acquired from the United States and Japan. They focused on simple types of semiconductor of low capacity: the products which were being abandoned by the American and Japanese companies. In the mid-1970s, the technology gap between Samsung and the industry standard was around thirty years; today it has virtually disappeared. Indeed, Samsung has totally caught up technologically with the industry leaders. By 1995, as Table 11.1 shows, not only had Samsung's aim to be in the world's top-ten semiconductor producers been achieved but also Hyundai had entered the top league as well.

Each of the Korean firms has established production facilities in the United States. The objective of these plants is to be close to the technological heart of the semiconductor industry – to absorb state-of-the-art technology. They are now moving into Europe, primarily to circumvent protectionist barriers as well as to be close to their growing markets. In the mid-1990s, Samsung, Hyundai and LG all announced plans to build massive semiconductor plants in the UK.

Corporate strategies in the consumer electronics industries

There are clearly very close links between semiconductor production and consumer electronics, both in general terms and also within those electronics companies which are in both businesses. As we saw earlier, such vertically integrated operations are especially common among Japanese and European electronics firms and less common among US firms. Some of the developments we have discussed in the case of the semiconductor industry apply also to colour television production and related consumer electronics products like VCRs. At the same time, there are important differences. Production of colour television receivers is even more highly concentrated than semiconductors. It is also almost totally dominated by Asian producers, notably Japanese, South Korean and Taiwanese. There is no longer a domestically owned TV manufacturer in the United States. On the other hand, despite the very heavy presence of Asian consumer electronics manufacturers in Europe, there remains substantial indigenous production by the Dutch firm Philips and by the French firm Thomson.[12]

Acquisitions, mergers and strategic alliances

The colour television industry has undergone major reorganization through acquisitions and mergers. The number of colour TV manufacturers in Europe has declined dramatically since 1980, when there were more than thirty:

In Britain the last major British-owned manufacturer – Thorn-EMI – sold its Ferguson division to the French Thomson company in 1987, at the same time as the latter acquired the RCA division of US General Electric . . . The German industry – the biggest in Europe – now has only one major German-owned producer, Bosch, which has a majority share holding in the medium-sized CTV producer, Blaupunkt. Managerial control of Grundig has passed to Philips, and Thomson owns Telefunken, Saba and Nordmende. Sony acquired Wega, and the ex-ITT subsidiary SEL is now owned by Nokia of Finland. This leaves Loewe-Opta, a small up-market manufacturer, as the only other CTV producer in German ownership.

(Cawson, 1989, p. 61)

Even greater changes have occurred in the ownership of the US colour TV industry. Acquisitions and mergers have been rife in the television sector, as many American firms quit an increasingly difficult market. For example,

Admiral sold out to Rockwell in 1973. In the next year, Motorola sold out to Matsushita, Magnavox to Philips, and Ford Motor sold its Philco consumer electronics business to GTE-Sylvania, which, in turn, sold its TV business to Philips in 1980. In 1976, Sanyo bought out Warwick, which was primarily a supplier to retailers like Sears.

(Turner, 1982, p. 57)

In 1966 there were sixteen US-owned television manufacturers; by 1980 this number had fallen to three. Today there is none. General Electric absorbed RCA and then sold out to the French company Thomson. Zenith disposed of its computer business (in which it was a highly successful producer of portable and lap-top machines) to concentrate on television production but was then acquired by the Korean firm, LG Electronics.

Strategic alliances have been less common in colour television production; the preference has been for outright merger or acquisition. In the light of the battle for the high-definition television standard, however, Philips and Thomson formed an alliance to develop a European technical standard. In related product areas, however, strategic alliances have been more common. Philips and Sony jointly developed the compact disc technology which had been invented by Philips. In VCRs a joint venture was created in Europe between the Japanese firm JVC, the British company Thorn EMI, the German company Telefunken and the French company Thomson. With Thomson's aggressive acquisition strategy this ultimately became a joint venture between just JVC and Thomson. On the other hand, there have been important failures in strategic alliances; both the GEC–Hitachi and Rank–Toshiba colour television ventures in Britain were abandoned and became wholly Japanese owned.

Global organization of production

United States producers

As in the case of semiconductors, internationalization of production is a fundamental competitive strategy of all the major consumer electronics companies. In both cases, it was US producers which initially moved some of their activities offshore in the early 1960s. But the precise stimulus for such a strategy was rather different from that in semiconductors. In that sector the major push factor towards offshore assembly was intensifying competition within the United States itself as new firms entered the industry and as domestic production costs accelerated. In consumer electronics, specifically television manufacture, the major stimulus for US companies to move offshore was intensifying competition from low-cost overseas competitors, notably the Japanese. Among American television manufacturers offshore production increased

further in the 1970s as Japanese import penetration of the domestic market accelerated. But, as we have seen, these offshore strategies were not sufficient to maintain a
US presence in the industry.

Japanese producers

Japanese consumer electronics manufacturers also developed substantial offshore assembly, notably in the east and southeast Asian countries, during the 1970s. The
motivation for Japanese offshore plants in the Asian NIEs was not only one of seeking
low labour costs, however, although this was probably the primary stimulus. An
additional consideration was the desire to get around the growing US import restrictions by producing in countries not covered by these restrictions. This helps to explain
the establishment of television plants by Matsushita, Mitsubishi, Sony and others in
Taiwan, Singapore, Malaysia and South Korea during the early 1970s.

Figure 11.9 shows the global distribution of Japanese consumer electronics plants
in 1995. Almost two-thirds of the total are located in Asia, with especially heavy
concentrations in Malaysia, China, Thailand and Taiwan. Significantly, the number of
Japanese consumer electronics plants in South Korea is relatively low, particularly
when compared with the components sector. This reflects the restrictive policies of the
South Korean government towards foreign consumer electronics firms. As in the case
of semiconductors, Japanese firms have created an intraregional division of labour
within east and southeast Asia which has become increasingly extensive and functionally differentiated since the mid-1980s' revaluation of the yen. Although much
Japanese electronics production has been shifted to neighbouring Asian countries
there is still a strong tendency for the high-end products and functions to remain
concentrated in Japan itself.

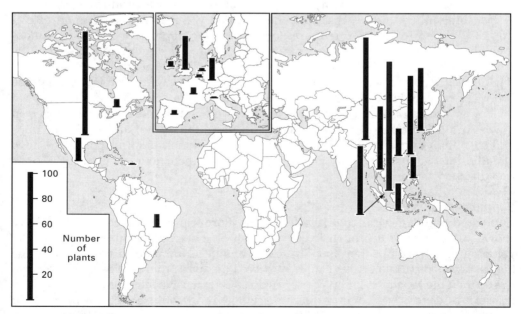

Figure 11.9   The global distribution of Japanese consumer electronics plants
Source: Based on Electronic Industries Association of Japan (1996, *Facts and Figures on the
Japanese Electronics Industry 1996*, p. 91)

Japanese consumer electronics firms have also greatly developed their North American and European production networks in response to both protectionist measures and an increasing desire to be close to the final consumer market. Almost 20 per cent of Japanese overseas consumer electronics plants are located in North America and 14 per cent in Europe. Given the substantial cost advantages of domestic Japanese production in the 1970s there was no case on cost grounds for Japanese firms to set up production facilities in either North America or Europe. As in the case of the automobile industry (see Chapter 10) Japanese consumer electronics producers could serve these markets through exports. However, the introduction of protective measures by American and European governments severely affected these advantages by imposing quantitative restrictions on the number of sets which could be sold in these markets. In the classic manner, therefore, Japanese firms 'jumped over' the trade barriers and invested heavily in production facilities in the United States and western Europe.

The influx of Japanese companies has been especially evident in television manufacture and, more recently, in the related field of video recorders. The establishment of Japanese plants in the United States totally changed the character of the American television industry, as all the major Japanese firms either acquired or built production facilities there. The same trend occurred in western Europe. All the leading Japanese television manufacturers now have production plants in Europe, notably in the United Kingdom, which has the highest concentration of Japanese television production in Europe. Apart from Sony, which set up the first Japanese television plant in Britain in 1974, and Matsushita in 1976, all the other ventures have involved the acquisition of existing plants. Following on from the surge of television plants in the 1970s, the 1980s saw a rush of openings of video recorder factories, mostly in the United Kingdom and in Germany, the leading markets for this product.

A combination of trade barriers, the revalued yen and also a growing realization of the need to have a direct presence in key geographical markets has fundamentally changed the international nature of Japanese consumer electronics activity. All the leading companies have develop explicitly *glocal* strategies though to varying degrees. All have set targets to increase the proportion of their total output to be produced outside Japan; the rate of growth of exports in several products has been declining. Production of low-value products is more and more being relocated in lower-cost Asian countries. Undoubtedly, some of the design and R&D functions will be shifted to the advanced markets of North America and Europe – some have already done so, notably Sony and Matsushita. But the extent of such shifts so far remains very limited.

European producers

Compared with American and Japanese firms, European electronics producers have shown rather less of a propensity to establish offshore production facilities in developing countries, but they have by no means been immune to the practice. The bigger firms, in particular, have gone offshore but, apart from Philips and Thomson, less so than the Japanese. Philips has a particularly extensive international production network in developing countries across a whole range of electronics activities. For example, Philips has an especially important manufacturing complex in Singapore as well as plants in Taiwan, Hong Kong, South Korea and Mexico. Indeed, a major part of its current strategy to compete with the Japanese and other east Asian producers is to shift a greater proportion of its labour-intensive activities to developing country locations. For example, between 1992 and 1993 alone, the number of Philips' plants in Europe (including The Netherlands) fell from 141 to 135; the number in North America declined from 58 to 45. Conversely, Philips increased its number of plants in Asia from 37 to 52 (Grotjohann, Sterkenburg and van Grunsven, 1996).

The French consumer electronics company, Thomson Consumer Electronics (TCE), has also developed an aggressive internationalization strategy:

> Since 1975, Thomson has begun to relocate components and low-cost products manufacturing to far east-Asian countries. That policy had intensified in the late 1980s and the 1990s. Shifting production from Europe and the USA to developing countries, southeast Asian countries and Mexico has been a main strategic issue for TCE. All European operations of low-cost products, such as audio products, black-and-white TVs and small-screen colour TVs, have been shifted from west Europe to Asian countries, such as Singapore, Thailand and Malaysia, in order to attain cost competitiveness. That huge shift was, in recent years, the only way for TCE to be able to continue to sell these products in Europe under its own brands, and to offer all the range of consumer electronics products . . . Today, TCE manages more than 13 plants in Singapore, Thailand, Malaysia, Taiwan, Indonesia and China . . . with more than 18,100 employees, and five plants in Mexico with more than 7,700 employees.
>
> (Savary, 1996, pp. 101, 102)

Despite these extensive international operations, both Philips and Thomson still have their major bases in Europe. Both have been engaged in large-scale rationalization and reorganization of their European production networks in the context not only of changing circumstances within Europe itself but also in the global context. Philips has been trying to rationalize its television production network in Europe since the 1980s. Because it began manufacturing TV sets before the emergence of the European Community, Philips developed TV production plants in every one of the major Community countries. More generally, Philips had long operated a strategy of giving a high degree of autonomy to its national subsidiaries. This strategy has been transformed, though painfully. In the case of Thomson, the rationalization process is being conducted at a time when its main shareholder, the French government, is endeavouring to privatize it.

East Asian NIE producers [13]

Finally, to an even greater extent than in semiconductors, a number of consumer electronics firms from east Asia (notably from South Korea and Taiwan) have become significant global players. In virtually every case, these firms – like Hyundai, LG, Samsung, Daewoo, for example – began as subcontractors for United States and Japanese firms. They were 'original equipment manufacturers' (OEM) whereby their customer provided all the detailed technical specifications and sold the product under its own brand name. Only in the last few years have the major Korean electronics firms become widely known through their own brands as they have made the transition from OEM to OBM (own brand manufacture). Today, 'Samsung' and 'Goldstar' are names which appear in the mass advertising slots at international sporting events and on airport baggage trolleys, along with longer-established Japanese, American and European brands. But even as recently as 1990 the leading company, Samsung, still manufactured about 40 per cent of its output for firms such as Sony and GE under OEM arrangements.

The links with Japanese firms were especially significant:

Samsung Electronics, Korea's largest electronics producer, began as a joint venture with Sanyo of Japan in 1969. That year it sent 106 employees to Sanyo and NEC in Japan to learn production know-how for radios, TV sets and components. Under joint ventures with Sanyo, NEC and Sumitomo of Japan, the firm acquired a variety of consumer and components technologies . . . Samsung was able to offer Japanese (and US) firms the advantage of low-cost labour, management and overheads. Production tended to be in high-volume standardized mature goods. Under OEM Japanese firms benefited by using Samsung to rapidly expand their capacity, while Samsung benefited from its first experiences in the international electronics market. Samsung received training in assembly methods and engineering, and began its own programme to reverse engineer products such as colour TVs and microwave ovens, even before a local market had emerged in Korea. OEM deals were often linked to licensing or joint venture arrangements. Later OEM provided an alternative to jointly owned ventures when these were discouraged by the Korean government.

(Hobday, 1994, pp. 344, 345)

The strategies of the leading Korean consumer electronics manufacturers increasingly involve them in establishing production plants overseas, not only in other Asian countries but also in North America and Europe. Both Samsung and Goldstar have built colour TV plants in the United States (in New Jersey and Alabama respectively). Samsung has manufacturing plants in the United Kingdom, Spain and Portugal while Goldstar has two plants in western Germany. A similar pattern is being followed by Tatung, the major television manufacturer in Taiwan. Tatung has plants in Japan and Singapore as well as in the United States and the United Kingdom. Indeed, during the 1990s, there was a massive wave of consumer electronics investment by South Korean and, to a lesser extent, Taiwanese firms in Europe. The French government, in particular, expressed its concern most overtly through its refusal to allow Daewoo to acquire the Thomson consumer electronics business. Press headlines heralding a 'Korean invasion' of Europe were reminiscent of reports of earlier 'invasions' by United States and Japanese firms.

Summary of global trends in the electronics industries

In a number of ways, therefore, the global pattern of electronics production shifts and changes as a result of the evolving corporate strategies of the major firms. Fierce competition forces firms to increase their degree of functional integration, to diversify into new product lines, to relocate production in more favourable locations in terms of markets or costs, and generally to rationalize their operations on a global basis. There has been considerable geographical shift to some developing countries, mostly in east and southeast Asia. In the case of television manufacture, such global shifts can be interpreted in terms of the product life-cycle. Television is a technologically mature product in which scale economies and the minimization of production costs are of key importance. But, like motor vehicles, the industry has experienced some technological rejuvenation which, together with government protectionist pressures, has kept a good deal of production within the developed countries. In both the United States and western Europe, however, there has been a substantial change in the ownership of production as Japanese firms in particular have established production bases there.

Semiconductors, integrated circuits and microprocessors, on the other hand, are by no means mature products. However, one stage in the production process – assembly – has been amenable to geographical separation and to location in low labour-cost areas. It is a clear example of the global combination of highly capital-intensive technology with low-cost, labour-intensive production. But the simplistic global division of labour characteristic of the 1960s and early 1970s no longer applies. Today's global map of semiconductor production is far more complex. Ernst (1985) identifies four basic trends:

- Locational shifts between the three leading global production regions – the United States, Japan, western Europe – involving:
  (a) the initial wave of investment by US semiconductor firms in western Europe
  (b) subsequent direct investment by Japanese and western European semiconductor firms in the United States and by Japanese firms in Europe
  (c) more recent investments by US semiconductor firms in Japan.
- Locational shifts from the centre of US and Japanese firms in Scotland, Ireland and Wales.
- An increased sophistication of semiconductor functions being carried out in the 'first wave' of offshore assembly platforms. Hong Kong, South Korea, Taiwan, Singapore (and to some extent Malaysia) now perform fabrication and testing (and even design in some cases).
- A consequent relocation of simple assembly operations from these original offshore locations to new sites in countries such as Thailand, the Philippines, Indonesia, China and the Caribbean basin.

To these four trends we should add a fifth:

- The development of international production by some leading firms from South Korea and Taiwan which have been establishing production bases in the United States and Europe as well as in Asia.

Jobs in the electronics industries

At first sight, the employment situation in the electronics industry would seem to be very different from that in the industries we have examined in the preceding two chapters. In both of those industries the story was one of substantial and long-standing employment decline in the older industrialized countries. There is no doubt that electronics is a growth industry. In semiconductors, for example, production has increased at a truly phenomenal rate since the early 1960s. But, for a number of reasons, this growth in output has not been accompanied by a comparable growth in employment. The consumer electronics sector of North America and western Europe has been hit hard by import penetration from lower-cost producers, particularly in Asia (including, of course, Japan). There is no doubt that some of the job loss in the consumer electronics industries has been caused by import penetration. But just how much is very difficult to assess and, in any case, import penetration is not the whole story. Two related factors are particularly important in any explanation – technological change and corporate rationalization – both of which have reduced the need for labour.

Technological change has been extremely rapid in the electronics industry. The result has been increasing labour productivity. All the major firms have been investing heavily in automated production and assembly and, thus, greatly reducing the number of workers required for a given level of production in both semiconductor and consumer electronics. In virtually all the major electronics companies, therefore, productivity has soared whilst employment has either not kept pace or has actually fallen.

At the same time, the strategies of rationalization and reorganization on a global scale have adversely affected employment, especially in North America and Europe. In the electronics industry there is, indeed, an element of truth in the 'runaway plant' charge: that firms close a plant in one country and transfer its production (and jobs) to another country. But the semiconductor industry, in particular, is not a classic runaway industry to the extent that consumer electronics and clothing may have been. In fact, geographical shifts in employment do not necessarily follow a simple path of reduction in countries where labour costs are high and growth in countries where they are low. The geographical complexity of employment change within global electronics corporations is far greater than this.

Some distinctive geographical concentrations of electronics employment have evolved in both developed and developing countries. In the United States, the Silicon Valley complex is the biggest and best-known but by no means the only important area of production. Of similar vintage is the agglomeration of electronics and high-technology firms along Route 128 around Boston in New England, while newer electronics clusters have been developing rapidly in Colorado, northern California, Oregon and North Carolina. Within Europe substantial geographical clusters of electronics production have emerged in the Central Valley of Scotland, the M4 corridor in southeast England and in the Grenoble area of France, amongst others. Within the developing countries most of the electronics production is concentrated in the export-processing zones which have proliferated during the past two decades. Roughly half of all the employment in the Asian EPZs is in the electronics industry.

But these geographical concentrations of electronics production are highly heterogeneous. They occupy different positions, and perform different roles, in the spatial hierarchy of production. Most importantly, they offer different kinds of job opportunity. Perhaps more significant than the simple geographical concentration of electronics employment is the pronounced *spatial differentiation of skills and occupations*. This is a direct consequence of the spatial and functional separation of the stages in the production process, especially in the semiconductor industry.

There is a clear *polarization* of skills in the industry between highly trained professional and technical workers on the one hand and low-skilled production workers on the other. Increased automation has led to a much steeper relative decline in the number of production workers employed, particularly the less skilled. But there is also a clear *spatial* pattern of skills in the electronics industry. Overall, professional and technical occupations are far more important in the developed countries than in the developing countries, a reflection of the fact that it is mainly the more routine assembly tasks which have been located in developing countries while the more sophisticated design, R&D and wafer fabrication stages remain mostly in developed countries. Hence, most, though by no means all, of the semiconductor employment in Silicon Valley and the other production clusters in the United States and Japan is in the higher-skill categories; conversely, virtually all the electronics employment in the EPZs of the Asian countries and in the Mexican border plants is of production workers. However, the increasingly complex intraregional division of labour within the semiconductor industry in east and southeast Asia is removing some of this simplicity.

In the European electronics clusters there is also a degree of spatial segmentation of labour skills, though in a less extreme form. For example, the Scottish electronics plants include both assembly and fabrication activities and, therefore, have a more even mix of skills than the developing country plants. Even so, the relative absence of higher-level design and of R&D activities in the Scottish branch plants creates a relative shortage of job opportunities for particular segments of the labour force. In contrast, the electronics cluster in southeast England is predominantly one of high-level administrative, design and R&D units within large firms together with highly innovative and specialized small electronics firms.

A further fundamental aspect of employment in the electronics industry is the predominance of female workers in the assembly stages.[14] This is apparent throughout the world and is not confined to the developing countries alone. An overwhelming majority of electronics assembly jobs are occupied by young female workers on relatively low wages. In this respect, there are clear parallels with the situation in the textiles and clothing industries. Female workers are the norm in the assembly processes wherever the plants are located:

> Most do assembly, the bonding of hair-thin wires to semiconductor chips, and the associated packaging. Though the work requires good eyesight and dexterity, little training is required. Women generally achieve peak productivity within a few months . . . In most countries, companies only hire unmarried women, of specific ages. The workforce, therefore, ranges in age from about 16 to 26. By avoiding the employment of married women except during times of severe labour shortages, companies avoid paying maternity benefits and ensure that an employee's primary loyalty will be to the company, not to her family or household. More important, by hiring young women, employers can get away with low wages and poor working conditions.
>
> (Siegel, 1980, p. 14)

However, more recent survey work of the electronics labour force in Malaysia and Singapore shows that changes are occurring (Lin, 1987). For example, the proportion of married women in the workforce has increased; neither marriage nor childbirth now automatically lead to a woman worker leaving her job; education levels of entrants are higher.

Although labour regulations are generally more stringent in developed countries, some undesirable characteristics may still apply. According to a study of the participation of women in the 'global factory',

> In Silicon Valley . . . 75 percent of the assembly line workers are women. The pattern is repeated along Route 128, outside Boston, and in North Carolina, an anti-union, 'right to work' state now favoured by the electronics industry. As in the US garment industry, immigrant women comprise a significant chunk of those workers: 40 percent. On the west coast, Filipinas, Thais, Samoans, Mexicans and Vietnamese have made the electronics assembly line a microcosm of the global production process. Management exploits their lack of familiarity with English and US labour law. Often, companies divide the assembly line according to race and nationality – one line may be all Vietnamese while another is all Mexican – to encourage competition and discourage cross-nationality alliances. Among the non-immigrant workers, many are married women, assuming paid employment after years of being homeworkers. Wages for semiconductor assembly in the US, while vastly superior to those overseas, are among the lowest in all of US industry . . . Since there is little upward mobility for women, especially Third World women . . . only 19 percent of all technicians and less than 9 percent of managers are women.
>   The illness rate in the electronics industry is one of the highest. Women are regularly exposed to solvents which may cause menstrual and fertility problems, liver and kidney damage, cancer, and chemical hypersensitisation.
>
> (Fuentes and Ehrenreich, 1983, pp. 53–54, 55)

The gleaming plate-glass and metal-clad plants of the modern electronics industry look, at first sight, to be the complete opposite of the dark, satanic mills of the
nineteenth century and the squalid sweatshops of the garment industry. In some
senses, of course, the contrast is undoubtedly what it seems. For many of the workers
employed in the industry wages are high, the work is stimulating and conditions are
superb. But for others the story is rather different. For the female workers of the
export-processing zones and for some in developed country plants as well, the contrast is less evident. At worst, conditions are just as bad, with long working hours, an
unpleasant and noxious working environment and little or no job security. It is indeed
paradoxical that an industry which epitomizes all that is new and up to date at the
same time harbours some of the oldest and least desirable attributes of work in
manufacturing industry.

Notes for further reading

1. See *The Financial Times* (13 November 1996).
2. Saxenian (1994) provides a recent analysis of Silicon Valley in comparison with the electronics cluster focused on Route 128 near Boston. Angel (1994) also devotes considerable
   attention to Silicon Valley in his analysis of the US semiconductor industry as does Harrison
   (1994) in his critique of Silicon Valley as an Italian-type industrial district.
3. Henderson (1989), Turok (1993) analyse the development of the electronics industry in
   Scotland while Morgan and Sayer (1988) trace the industry's development in Wales.
4. The changing structure of semiconductor production in east and southeast Asia is discussed
   by Henderson (1989; 1994), Hobday (1994; 1995), Gereffi (1996a).
5. See Angel (1994), Ó hUallacháin (1997).
6. Government policies in the semiconductor industry are discussed by Wade (1990b), Tyson
   and Yoffie (1993), Angel (1994, Chapter 6), Hobday (1994).
7. See Cawson (1989).
8. Henderson (1989), Mody (1990), Hobday (1994; 1995) discuss government policies towards
   the consumer electronics industries in Asia.
9. See the references in Note 2 above.
10. Ó hUallacháin (1997) provides a detailed analysis of these firms and their changing
    strategies.
11. Scott and Angel (1988), Henderson (1989; 1994), Gereffi (1996a) discuss the evolving intra-
    regional division of labour in the east and southeast Asian semiconductor industry.
12. Grotjohann, Sterkenburg and van Grunsven (1996) and Savary (1996) provide detailed
    analyses of the consumer electronics businesses of Philips and Thomson respectively.
13. The development of east Asian consumer electronics firms is discussed by McDermott
    (1991), Henderson (1989; 1994), Hobday (1994; 1995), Gereffi (1996a).
14. See Siegel (1980), Fuentes and Ehrenreich (1983), Lin (1987), Harrison (1994).

CHAPTER 12

'Making the world go round': the internationalization of services

Introduction: the growing importance of services in the global economy

One of the most significant developments of the last few decades in the global economy has been the rapid growth of the service industries. As we saw in Chapter 2, services account for the largest share of gross domestic product in all but the lowest-income countries (Table 2.4). They are increasingly the major source of employment in all the developed market economies and in many developing countries as well. International trade in commercial services has accelerated to become an important driving force in its own right in the global trading system. The services sector has also been attracting an increasing share of world foreign direct investment. Services now account for a larger share of foreign direct investment, for the leading industrialized countries as a whole, than manufacturing industry. Without doubt, therefore, services have become increasingly internationalized.

In the process, they also become a major focus of political friction within the framework of international trade negotiations of the Uruguay Round of the GATT. Indeed, one of the main reasons for the protracted length of the Uruguay Round was the failure to agree on the treatment of services. The developed economies, led by the United States, pressed for the inclusion of services in a new GATT agreement. Their objectives were to ensure unhindered access to developing country markets for services. Conversely, the developing countries were especially concerned to prevent their own service industries from being swamped either by trade or by direct investment. Some service industries are regarded as being extremely sensitive, either culturally or strategically, and are closely controlled and regulated by national governments.

In the event, a General Agreement on Trade in Services (GATS) was agreed. But although it is based upon the usual principles of non-discrimination of the GATT, their application to services is far more limited than in the case of manufacturing.[1] Individual countries can specify those services which they wish to be exempt from the non-discriminatory principles. These exemptions remain in force for ten years from the date of the agreement but must be subject to renegotiation in subsequent international trade negotiations. The major problems occurred in specific service sectors, notably financial services, transportation services and audio-visual services. Under a compromise agreement, negotiations in these sectors were to continue after the conclusion of the main Uruguay Round.

In fact, the whole debate over the 'liberalization of trade in services' and, indeed, of services in general, is highly confused. Certainly it is often very confusing. The reasons lie in the rather complicated nature of services themselves. The aims of this final case study chapter, therefore, are twofold. First, we need to disentangle the confusion over exactly what services are: what functions they perform, how they relate to the production of goods, how they operate at an international scale. Secondly, in line with our approach in the preceding three chapters, we need to look more closely at specific cases. Hence, one particularly important example – the *financial service industries* – will be examined in detail.

The nature of services: disentangling the confusion

Everybody knows what manufactured goods are: they can be seen; they can be handled; they are tangible. Some have at least a degree of permanence, although they may, of course, wear out or become obsolete. Others, such as food products, are consumed within a short period of time, although all goods can be stored in some way or other for future use long after they have been produced. Goods can also be acquired or consumed far away from their place of production: they can be traded at all

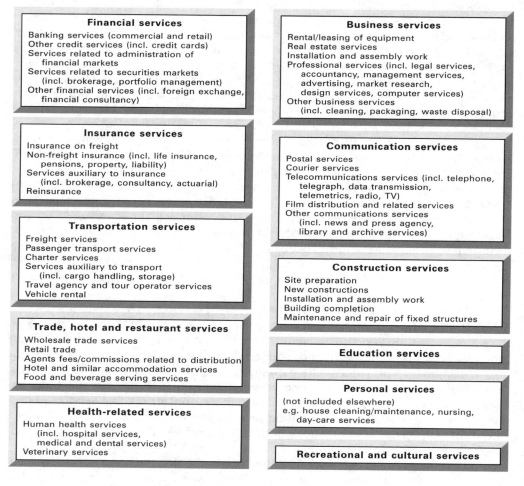

Figure 12.1   A typology of service activities
Source: Based on GATT (1989, Appendix II)

geographical scales, including the global scale. But what, exactly, is a service? To many it is essentially the opposite of a manufactured good. Services are commonly regarded as being intangible, perishable, requiring their consumption at the same time and in the same place as their production. Most services, by this definition, are not tradable. In the classifications used by government statistical agencies, services are often seen as a residual category: activities which are neither dug out of the ground nor manufactured.

The services sector, as Figure 12.1 shows, is extremely diverse; far more so, in many respects, than the manufacturing sector. It ranges from the highly sophisticated, knowledge- and information-intensive activities, performed in both private and public sector organizations, to the very basic services of cleaning and simple maintenance. It includes retail and wholesale distribution and entertainment as well as health care and education services. It encompasses construction activities, transport activities, financial activities, communications activities, professional services. The list is (almost) endless.

The traditional way of classifying this incredibly heterogeneous group of activities is to see them as part of a sequence of economic activities – primary, secondary, tertiary, quaternary – in which each element is further removed from direct involvement with the earth's physical resources. A major difference between these different sectors is seen to be the particular nature of the transactions they perform, especially their material or non-material nature. Thus, the primary and secondary sectors are supposedly characterized by flows of materials and material products through the transport system. Similarly, the tertiary sector also generates material flows through buying and service trips. The quaternary sector, on the other hand, is regarded as transmitting, receiving and processing information rather than materials; its component parts are linked together by flows of information, rather than materials, through the communications media, as well as by direct face-to-face contact. This kind of classification of economic activities encourages a linear view of change and development: as societies move up the development ladder the balance of their economic activities shifts further and further towards the quaternary sector. But, the world is not so simple. One of the major weaknesses of this kind of sectoral classification is that it implies a degree of separation between goods and services which does not really exist; it does not capture the *interdependencies* of the real world.

As the services sector has become an increasingly significant – even the dominant – element in many economies, particularly in terms of its contribution to GDP and employment, a rather different approach to classifying services has become popular. Services, it is suggested, can be separated into two categories:

- consumer services
- producer (or business) services.

This classification is based on what is seen to be the final output or customer of the particular service. It is derived from the distinction commonly made in the manufactured goods sector in which producer/intermediate goods are distinguished from consumer/final-demand goods. This is quite a clear and reasonable distinction to make in the goods-producing sector. There is relatively little overlap between intermediate and final-demand goods. But even a cursory glance at Figure 12.1 shows that this is emphatically not true in the case of services. Although there are some services which are clearly either 'producer' services on the one hand or 'consumer' services on the other, many services apply to both categories at the same time. In other words, most services are 'mixed' in that they are *both* producer *and* consumer services. Obvious examples are transportation and communication services, hotels, many financial services, insurance, legal and accountancy services. Thus, several of the activities listed in Figure 12.1 under 'business' services are also used by private consumers.

Services within the production chain and in the sphere of circulation

We need to think about services in a different way either from seeing them as being totally distinct from goods or, within the services sector itself, seeing them as being clearly divisible into producer and consumer services.[2] In Chapter 1 we introduced the concept of the *production chain* (Figure 1.1) to emphasize the interconnected nature of the production system. If we think of services in relation to the production chain then we can begin to see the kinds of role services play in the economy:

> Service activities not only provide linkages between the segments of production within a [production chain] and linkages between overlapping [production chains], but they also bind together the spheres of production and circulation. Services have come to play a critical role in [production chains] because they not only provide geographical and transactional connections, but they *integrate* and *coordinate* the atomized and globalized production process.
>
> (Rabach and Kim, 1994, p.123)

Figure 12.2 shows that service inputs are needed at each stage of the production process. It also shows an increasingly significant aspect of today's competitive environment: the fact that 'services have become a major source of value added. Downstream services in particular are both a factor contributing to competitive strength and a source of value-added' (UNCTAD, 1988, p. 178). But the value of services to the competitiveness of products is more general than this and applies throughout the production chain: 'In a number of product markets it is the "services" . . . design,

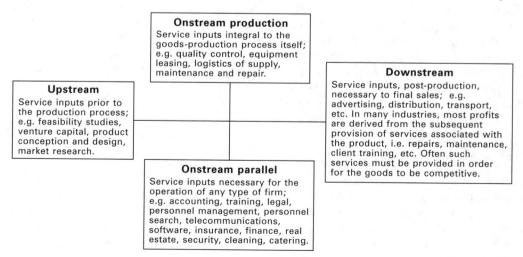

Figure 12.2　The interconnection between services and production in the production chain
Source: UNCTC (1988, p. 177)

styling, research, marketing, delivery, packaging, consumer credit – which determine the competitiveness of agricultural and manufacturing investment. As the length of production chains increases, so services are responsible for a greater share of value added to products' (Britton, 1990, p. 538). Many services are deeply embodied in goods.

The provision of such service inputs to production may be carried out within the producing firm itself (i.e. it can be *internalized*) or some or all of them may be put out to independent, specialist service firms (i.e. provision may be *externalized*). As we saw in Chapter 7, the boundary between the internalization and externalization of functions is continuously changing. Functions may be hived off or, alternatively, specialist service firms may be taken over or they may operate on a subcontracting basis. The development of dynamic organizational networks invariably involves the changing of the organizational – and often the geographical – location of the firm's various functions and altering the balance between production and service activities.

In one sense, therefore, services are not readily separable from the production of goods but, rather, are an *integral* part of such production. On the other hand,

> a substantial volume of transactions are generated within and between service industries themselves: the output of many producer service industries go to other services . . . one of the fastest-growing segments of the service sector is those services that either distribute the product of, service the input needs of, or act as market intermediaries for, other service industries.
>
> (Britton, 1990, p. 534)

There is no doubt, therefore, that labelling services as either 'producer services' or 'consumer services' can be misleading. It works in some cases but not others:

> While advertising, marketing and R and D are readily identified as producer services . . . the commercial and financial services which mediate and abbreviate the exchange processes within the economy are neither producer nor consumer services. They are circulation services, services produced within the process of circulation and for circulation, and not intermediate services produced primarily for other branches of industry or final services produced for consumers. *Circulation services . . . are concerned primarily with the velocity of turnover, whether of commodities, money or money capital . . . Whereas goods production is restricted to the sphere of production, service production occurs within both the sphere of production and the sphere of circulation.*
>
> (Allen, 1988, pp. 18, 19, emphasis added)

Hence the title of this chapter. In many respects it is service activities that 'make the world go round', which lubricate the wheels of production, distribution and exchange.

The internationalization of services

The internationalization of most economic activity generally occurs via two means. One way is through international *trade*; the other is through some kind of *direct presence* in a particular overseas location or market (e.g. via foreign direct investment, licensing, joint venture, subcontracting and the like, as shown in Figure 7.16). How far do services fit this pattern? They clearly fit the second of the two forms of internationalization: service firms may establish a presence in a foreign market to provide their services to local or locally based customers. But to what extent are services *tradable*? Precisely what is meant by trade in services is a complex issue, bound up with the ways in which balance of payments statistics are calculated by governments. Such technical matters need not concern us here.[3] What is clear, however, is that trade statistics vastly understate the extent to which services have become internationalized. This is because they do not fully capture some of the most important mechanisms involved: the embodiment of services in goods which are themselves traded, the trade in internalized services which occurs within transnational corporations.

In fact, many services are not tradable in the strict sense of the term because they need to be consumed at the point of production: they are not storable. It is claimed that technological innovations, especially in information technology, have made it possible – at least in principle – for some kinds of service transactions to be conducted with geographical separation between the producer and consumer of the service. For example, the development of sophisticated information technologies permits the international – or transborder – flow of data and information. In that respect, they allow 'information-based services' to be provided even where the producer and consumer are geographically separated, that is, they can be traded internationally. We shall see later in our examination of financial services that this is a particularly significant feature of current activity. But even in those cases, there is generally a need for the supplying firm to have an actual presence in the foreign market to deliver the service more efficiently and effectively.

Hence, the real issue in the current debate on the liberalization of international trade in services is not trade in itself but, rather, the conditions under which providers of services are permitted to establish an actual direct or indirect *presence* in a specific national market. In other words, it is really about foreign direct investment and the other modes of international involvement which firms may use (see Figure 7.16). As Gibbs (1985, p. 221) accurately observes, to the service TNC 'the distinction between trade and investment is academic, what they require is a presence in the market, the ability to compete free of regulations which place them at a disadvantage *vis-à-vis* domestic suppliers, the only trade which is vital is that in information, transborder data flows constituting their "lifeblood"'. However, it is the very capacity of TNCs to control information vital to many modern service industries which places developing countries at a disadvantage. Clearly, therefore, two of the major issues at the heart of both the processes and the politics of the internationalization of services are

- information technology
- government regulation.

Both of these will be examined more closely when we look at the case study of financial service industries.

We will look also at the specific ways in which such internationalization has occurred in those industries. For many service industries, particularly those which are primarily intermediate inputs into the production chain and those which are circulation activities, the initial stimulus to their internationalization was the rapid growth and global spread of TNCs in manufacturing industries. As manufacturing TNCs have proliferated globally so, too, have the major banks, advertising agencies, legal firms, property companies, insurance companies, freight corporations, travel and hotel chains, car rental firms and credit card enterprises. In many respects, the internationalization both of manufacturing and of these business service activities has become *mutually reinforcing*. In some cases (for example the Japanese general trading companies or *sogo shosha*) a network of service functions preceded overseas manufacturing investment. In other cases, the major business service corporations have tended to follow their manufacturing clients abroad. Conversely, the existence of an extensive global network of the familiar business service corporations helps to guide the further evolution of transnational manufacturing activities. A truly *symbiotic* relationship has developed. As these services have themselves become internationalized they have also acted as a stimulus for the internationalization of other service activities so that the process no longer is simply one of services following manufacturing.

The emergence of transnational service conglomerates

The global package deal

Most service companies are single-product organizations, specializing in a specific service function and supplying it in many different countries through a network of subsidiaries, branches, joint ventures and other forms of partnership. However, some services are strongly *complementary* to other services. For example, there is said to be a 'natural' relationship between airlines and international hotel chains; between different kinds of financial services; between accountancy and management consultancy; between advertising, public relations and the communications media. Hence, there has been a strong development of transnational service conglomerates: firms which operate internationally in a number of related service industries and aim to offer a 'global package deal'.

A good example is the recent development of the global advertising industry.[4] Through a whole series of aggressive acquisitions and mergers, since the early 1980s, the leading advertising agencies have attempted to become true global players with a vast network of overseas offices. Some of them have moved beyond simply offering advertising services; they have deliberately adopted a strategy of becoming much broader *marketing-service conglomerates*. They offer a global package which includes advertising, marketing, promotion, media services, public relations, management consultancy and similar functions. The rationale underlying this strategy is that it increases the firm's potential to serve global clients by offering an integrated package of services within a single co-ordinated operation. The argument is the old one – now somewhat discredited in manufacturing industries – of synergy (the whole as more than the sum of the parts). A client for one of the marketing-service conglomerates, it is argued, can be crossreferred to other parts of the same business.

Currently there is a small number of especially large marketing-service conglomerates, each of which is based upon a nucleus of global advertising companies and other operations. The leading players are the British company, WPP; two United States companies (Omnicom and Interpublic); and the Japanese firm, Dentsu.

The WPP Group, headquartered in London, is the biggest of these groups, employing 20,000 people in 780 offices in 83 countries. It grew very rapidly during the 1980s through a series of mergers and acquisitions. WPP owns two of the world's major advertising agencies – J. Walter Thompson and Ogilvy and Mather – together with major public relations, market research and strategic marketing consultancy firms. In its own words, the company is

> the leading marketing services company in the world . . . providing clients – local, national and multinational – with an unparalleled resource of specialized skills that include: advertising; public affairs and public relations; market research; direct marketing; consulting; identity and design; and sales promotion . . . WPP's central objective is the provision of outstanding service to clients. One of the great and growing advantages enjoyed by WPP companies, each trading individually but all belonging to the same Group, is the opportunity for the formation of client-directed partnerships.

The Japanese advertising firm, Dentsu, offers a variant on a similar theme. Dentsu is the fourth largest advertising agency in the world and has recently restructured its international operations to increase its emphasis on Asia, especially China. A key part of Dentsu's international network is its long-standing partnership with the American firm, Young and Rubicam, established in 1981. Dentsu operates eighty offices worldwide. Like WPP, Dentsu stresses the holistic nature of its operations: 'Dentsu is a Total Communications Service Company, embracing not just advertising but the entire range of communications activities, including sales promotion, public relations, event production and sports marketing.'

Despite the hype, the advertising/communications conglomerates have experienced some difficulties both in actually integrating their diverse activities into a coherent functional unit and also in selling the concept of the 'one-stop shop'. Although some major TNCs have opted for such a service package, others have preferred to use a number of different advertising agencies, rather than just one, for their global marketing activities. We shall look at diversified service companies again in the case study of financial services.

The Japanese sogo shosha

The most highly developed form of the transnational services conglomerate – involved in a vast array of mining, agricultural, manufacturing as well as service activities – is by no means a recent phenomenon. The giant Japanese general trading companies (the *sogo shosha*) are undoubtedly the most remarkable examples of this form of organization, with a long history in the development of the Japanese economy.[5] They have also been especially influential in the development of Japanese overseas direct investment as a whole. The common translation of the term *sogo shosha* is 'general trading company'. But they are very much more than this. The six leading *sogo shosha* – Mitsubishi, Mitsui, Itochu, Marubeni, Sumitomo, Nissho-Iwai – are gargantuan commercial, financial and industrial conglomerates. They operate a massive network of subsidiaries and thousands of related companies across the globe. Figure 12.3 shows the global distribution of their offices in 1996. Their huge size and the extent of their geographical operations have given them an enormous importance both in Japan's economic affairs and in the global economy as a whole. Each of the leading *sogo shosha* handles tens of thousands of different products. In the early 1990s, they were responsible for roughly 70 per cent of total Japanese imports and for 40 per cent of Japanese exports. This is the true Japanese general trading oligopoly, each member of which has a major co-ordinating role within one of the Japanese *keiretsu* (see Figure 7.11).

The *sogo shosha* were the first Japanese companies to invest on a large scale outside Japan. Such investments were to set up a global marketing and economic intelligence network. Once in place, this network, with all its supporting facilities, not only facilitated the growth of Japanese trade but also enabled a whole range of Japanese firms to venture overseas. Indeed, a good deal of the early overseas investment by Japanese manufacturing firms was organized by the *sogo shosha*. They perform four specific functions:

- *Trading and transactional intermediation*: matching buyers and sellers in a long-term contractual relationship.
- *Financial intermediation*: serving as a risk-buffer between suppliers and purchasers.

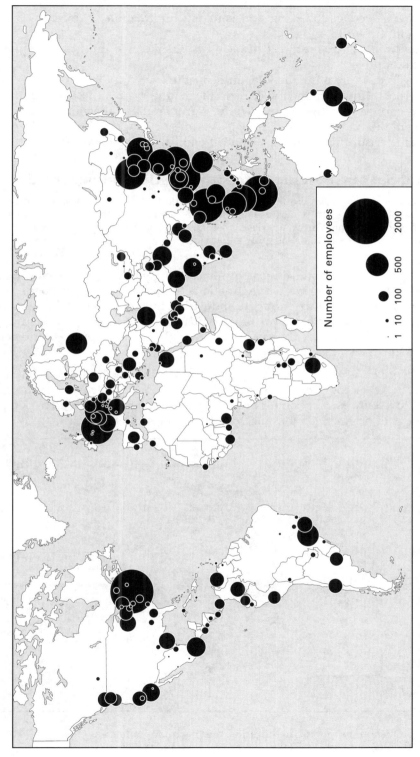

Figure 12.3   Global distribution of *sogo shosha* offices, 1996

- *Information gathering*: collecting and collating information on market conditions throughout the world.
- *Organization and co-ordination of complex business systems*: for example, major infrastructural projects.

The *sogo shosha* are a uniquely Japanese institution with deep historical, cultural and social roots. However, their success in facilitating the internationalization of the Japanese economy after the Second World War stimulated their imitation in South Korea, where the trading arms of the giant domestic conglomerates or *chaebol* have similar characteristics although on a smaller, and less diversified, scale.

International outsourcing in the information service industries[6]

> From a cruise ship off Bridgetown, the first view of Barbados is not sea and sand but an industrial estate where companies develop software, process health insurance claims and maintain databases for North American customers. Almost 3,000 people now work in such back-office jobs – as many as grow sugar cane.
>
> (*The Economist*, 29 March 1997)

This quotation hints at what has become an increasingly common characteristic of the information service industries: the international outsourcing of some parts of their production chain. To that extent, these industries are following in the footsteps of manufacturing firms which, as we saw in Chapter 7, have long engaged in such practices. For the most part the evidence is anecdotal rather than comprehensive. Figure 12.4 shows two cases:

- In the *airline industry*, it is now common for the processing of tickets and some other forms of ticket accounting to be carried out offshore. The materials are simply shipped to a low-cost location for processing and the results shipped back again to the country of origin. Examples include: American Airlines which

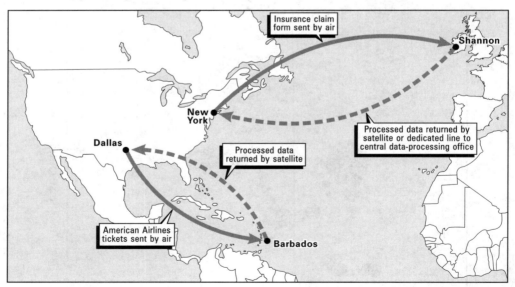

Figure 12.4 Offshore processing in the airline and insurance industries
Source: Based on material in Warf (1995, Figure 6); UNCTAD (1996, p. 107)

> assembles accounting material and ticket coupons in Dallas, Texas, for transport on its scheduled flights to Barbados for processing by its offshore subsidiary, AMR Information Services/Caribbean Data Services. In Barbados, details of 800,000 American Airlines tickets are entered daily on a computer screen, and the data are returned by satellite to its data centre in the United States.
>
> > (UNCTAD, 1996b, p. 107)

Similarly, Swissair has these functions performed in a newly established centre in Bombay, India.

- In the *insurance industry*, claim forms are processed in cheaper, offshore locations like Ireland. For example,

> New York Life insurance clients mail their health insurance claim forms to an address at Kennedy Airport in New York. The claims are sent overnight to Shannon Airport in Ireland, and then by courier to the firm's processing centre in Castleisland, about 60 miles from Shannon. The processing affiliate is linked via transatlantic telecommunications lines to the parent company. After processing, the claims are returned by dedicated line to the insurance firm's data-processing centre in New Jersey, and checks or responses are printed out and mailed to the clients.
>
> > (UNCTAD, 1996b, p. 107)

A far better documented case is that of the *computer software industry*, in which United States firms in particular have developed extensive outsourcing links.[7] The most significant software outsourcing location is India, which has attracted by far the largest share of this kind of work. Between 1980–81 and 1993–94, software exports from India grew at an annual average rate of 40 per cent per year although, as Heeks points out, Indian exports amount to less than 0.15 per cent of the world computer services and software market. Nevertheless, it is a highly significant development:

> By 1995, around 200 Indian software firms were active in global subcontracting but the field was highly concentrated with the top two firms – Tata Consultancy Services (TCS) and Tata Unysys Ltd (TUL) – responsible for about 30 per cent of all work, and the top eight players exporting roughly half of all software . . . In all, about 15,000 Indian software developers were involved in exports in 1994.
>
> > (Heeks, 1996, p. 371)

United States firms are by far the dominant clients of the Indian software industry, accounting for almost three-quarters of all outsourcing arrangements followed, a very long way behind, by some British and a few European firms. 'With the exception of Japanese firms, almost all major hardware multinationals are now outsourcing software development to India. The software multinationals have been slower to get involved and, instead, have concentrated on setting up agreements with Indian firms which act as their local distributors' (Heeks, 1996, p. 372).

Certain preconditions had to be met before software outsourcing could occur. First, software had to exist as a separate, 'outsourcable' commodity with a distrinctive market of its own. Secondly, the software production process had to be separable into a series of discrete operations with standardized and easily transportable technologies and skills. This process was aided by the change in programming languages from being manufacturer specific to being generic. As Figure 12.5 shows, the software production process can be divided into eight distinctive stages. It is stages 5 and 6 (coding and testing) which account for most of the Indian software work (65 per cent of total contracts). These stages are relatively less skill intensive than stages 3 and 4. In effect,

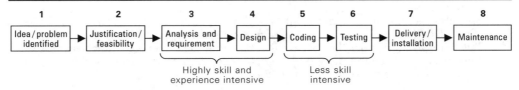

Figure 12.5   Stages in the production of computer software
Source: Based on material in Heeks (1996)

'Indian workers are far more often used as programmers rather than as systems analysts or designers. They work to requirements and design specifications set by Western software developers' (Heeks, 1996, p. 371). The kinds of customer–supplier relationship involved are of several different kinds, ranging from simple short-term trading relationships to longer-term close collaborations and alliances. Most of the arrangements begin with one-off orders and subsequently develop into more stable relationhips.

The question arises, of course, of why India in particular has developed such a dominant position in the software outsourcing business? Several contributory factors can be identified:

- Indian subcontractors offer large cost savings over United States' levels, even allowing for some differences in productivity. The wage levels for US software workers during the 1980s and 1990s were between six and eight times higher than those in India. Consequently, 'Indian subcontractors typically charge about 70 per cent of Western contract rates for onsite work and 40 per cent for work carried out offshore' (Heeks, 1996, pp. 379–80).
- There is a huge pool of well educated, English-speaking software workers in India who have the necessary technical skills. In contrast, the shortage of software workers in the United States is growing by the year at an estimated rate of 20,000. This pool of Indian workers can be drawn upon flexibly to meet fluctuations in staff requirements and skill shortages in contracting firms.
- The use of Indian subcontractors facilitates access to the Indian market itself which is growing very rapidly as the country strives to upgrade its technological base.
- Indian companies are very keen to enter into close collaborations and alliances with western firms in order to gain access to overseas markets and technologies.
- During the 1980s and 1990s, the Indian government policy became increasingly liberal in its attitude towards foreign involvement and also began to offer investment incentives to foreign firms.

The global software market continues to grow, providing both opportunities and threats to India's currently dominant position. Competition to attract software outsourcing is intensifying from such countries as Ireland, Singapore, the Philippines, China, Mexico, Israel and Hungary. The need for software services and development derives from a number of factors: the software maintenance of so-called 'legacy systems'; the need for software conversion for clients changing their systems; increasing programme modularity; the wider use of open systems. In addition, 'changes in software technology will also create new skills requirements and new markets. For the rest of the 1990s, networking, multimedia and client/server environments will be likely demand drivers' (Heeks, 1996, p. 389).

The internationalization of financial services[8]

> What is going on now is a revolution: a revolution in the way finance is organized, a revolution in the structure of banks and financial institutions and a revolution in the speed and manner in which money flows around the world.
>
> (Hamilton, 1986, p. 13)

International financial flows and foreign currency exchanges now dwarf the value of international trade in goods. Massive changes have been, and are, occurring both in the nature and composition of financial services themselves and also in the extent to which they have become internationalized. The global financial system has become extremely complex and extremely volatile. Its ramifications for the entire global economy are so enormous basically for two reasons:

- Financial services are *circulation* services: they are totally fundamental to the operation of every aspect of the economic system. Each element in the production chain depends upon necessary levels of finance to keep the chain in operation. This is true not only of manufacturing itself but also of all other intermediate and consumer services in the system.
- Many of its activities are *speculative*: financial investments aimed at making short- or long-term profits as ends in themselves. Hence, changes in the level and the geography of the flows of international portfolio investments (investments in stocks and bonds) greatly affect the financial well-being of entire national economies. For example, a heavy dependence on such 'hot' money was a major factor in the Mexican financial crisis of 1994 and in that of east and south east Asia in 1997. Much of the huge US current account and budget deficit is also financed by the inflow of foreign portfolio investment. Any major shift in such investment would have enormous repercussions on the US economy and, therefore, on households and individuals as well as on businesses. Although, in general terms, the speed and geographical complexion of international portfolio flows are determined by geographical differences in interest rates, exchange rates and political risk, the scale of international portfolio flows is now so great that they can, in themselves, help to determine exchange rates.

The changing pattern of demand for financial services[9]

The international financial system is made up of a great variety of different kinds of institution, each having a specific set of *core* functions (Table 12.1). In fact, the boundaries

Table 12.1   Major types of financial services

Type	Primary functions
Commercial bank	Administers financial transactions for clients (e.g. making payments, clearing cheques) Takes in deposits and makes commercial loans, acting as intermediary between lender and borrower
Investment bank/securities house	Buys and sells securities (i.e. stocks, bonds) on behalf of corporate or individual investors Arranges flotation of new securities issues
Credit card company	Operates international network of credit card facilities in conjunction with banks and other financial institutions
Insurance company	Indemnifies a whole range of risks, on payment of a premium, in association with other insurers/reinsurers
Accountancy firm	Certifies the accuracy of financial accounts, particularly via the corporate audit

between them have become increasingly blurred. One reason for this has been the changing pattern of demand for financial services as a whole. 'Ever since the mid- and late-1970s, financial services around the world have been characterized by a sharp intensification of competition and a rapid transformation of markets. These changes are leading to the emergence of a new, more dynamic market environment, quite unlike that which characterized the industry during the previous decades' (Bertrand and Noyelle, 1988, p. 16).

This 'new competition' consists of four major elements:

- *Market saturation*. From the late 1970s, it became apparent that the traditional financial services markets were reaching saturation. There were fewer and fewer new clients to add to the list; most were already being served, particularly in the commercial banking sector but also in the retail sector in the more affluent economies. A similar high level of market saturation was becoming apparent in insurance services.
- *Disintermediation*. This is the process whereby corporate borrowers in particular make their investments or raise their needed capital without going through the 'intermediary' channels of the traditional financial institutions, particularly the bank. Instead, they have increasingly sought capital from non-bank institutions, for example through securities. A similar trend has been occurring in the retail markets, where private investors have switched funds from traditional savings deposit accounts to higher-yielding funds such as investment trusts and mutual funds.
- *Deregulation of financial markets*. Financial services markets have traditionally been extremely closely regulated by national governments. One of the most important developments of the last few years has been the increasing deregulation or liberalization of financial markets. We shall examine the regulatory environment of financial services, and the trend towards deregulation, in some detail in a later section. At this point we should note that deregulation has been aimed primarily at three areas:
  - the opening of new geographical markets
  - the provision of new financial products
  - changes in the way in which prices of financial services are set.
- *Internationalization of financial markets*. Demand for financial services is no longer restricted to the domestic context: financial markets have become international and, in some cases, global. Three types of demand are especially significant in this context:
  - the massive growth in international trade has created a much increased demand for commercial financial services at an international scale
  - the global spread of transnational corporations has created a demand for financial services way beyond those corporations' home countries
  - the vastly increased institutionalization of savings – the channelling of savings into the pension funds and other institutions – has created an enormous pool of professionally administered investible capital seeking the best return, wherever that might be achieved.

Each of these forces for change in the demand for financial services is inter-related. Taken together they create a *new competitive environment* in which financial services companies now operate. Before looking specifically at the strategic responses of the major institutions, however, we need to consider two issues in more detail: first, technological change and, secondly, government regulation (and deregulation) of financial services.

Financial services and changing technologies[10]

In Chapter 5 we discussed technological developments in communications technologies and in the broader sphere of information technology. We have seen in the last three chapters that these technologies have been extremely important in the global evolution of the textiles and clothing, automobile and electronics industries. But they are even more fundamental to the service industries in general and to financial services in particular. This is because, to a far greater extent than in our manufacturing case studies, *information is both the process and the product of financial services*. Their raw materials are information: about markets, risks, exchange rates, returns on investment, creditworthiness. Their products are also information: the result of adding value to these informational inputs. In the words of one financial services executive: 'We don't have warehouses full of cash. We have *information* about cash – *that* is our product' (*The Financial Times*, 16 March 1994).

A particularly significant piece of added value is embodied in the *speed* with which financial service firms can perform transactions and the global extent over which such transactions can be made. 'Subject to the process of digitization, information and capital become two sides of the same coin' (Warf, 1995, p. 365). Not surprisingly, therefore, all the major financial services firms have invested extremely heavily in information transmission facilities. In 1995, for example, the top-ten United States banks spent more than $10 billion on information technology; the top-ten European banks spent almost $12 billion in the same year (*The Economist*, 26 October 1996).

Warf (1989) summarizes the major effects of the information technologies on financial services as follows:

- They have vastly increased productivity in financial services.
- They have altered the patterns of relationships or linkages both within financial firms and also between financial firms and their clients.
- They have greatly increased the velocity, or turnover, of investment capital. For example, the ability to transfer funds electronically – and, therefore, instantaneously – has saved billions of dollars in interest payments which were formerly incurred by the delay in making transfers.
- At the international scale, they have enabled financial institutions both to increase their loan activities and also to respond immediately to fluctuations in exchange rates in international currency markets.

From a technological viewpoint, therefore, it is now possible for financial services firms to engage in global twenty-four-hours-a-day trading, whether this be in securities, foreign exchange, financial and commodities futures or any other financial service. The ability to transmit data electronically over vast geographical distances creates the potential for continuous financial transactions worldwide, whatever the time of the day or the night because, as Figure 12.6 shows, the trading hours of the world's major financial centres overlap. True twenty-four-hour trading is currently limited to certain kinds of transaction partly because, although the technology is available, either the organizational structure or the national regulatory environment remain an obstacle. But there is no doubt at all that global twenty-four-hour trading is becoming standard in many kinds of financial service.

To the extent that such electronic transactions do not require direct physical proximity between seller and buyer they are a form of 'invisible' international trade.

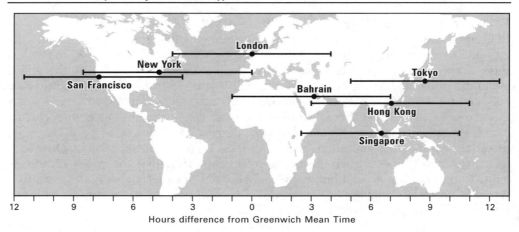

Figure 12.6 Twenty-four financial trading: the trading hours of the major world financial centres
Source: Warf (1989, Figure 5)

In that sense, therefore, financial services are one form of service activity which is *tradable*. Electronic communications have also contributed greatly to the bypassing of the commercial banks and the trend towards the greater *securitization* of financial transactions. Securitization, in the broad sense, is simply the conversion of all kinds of loans and borrowings into 'paper' securities which can be bought and sold on the market. Such transactions may be performed directly by buyers and sellers without necessarily going through the intermediary channels of the commercial banks.

The *global integration of financial markets* brings many benefits to its participants: in speed and accuracy of information flows and rapidity and directness of transactions, even though the participants may be separated by many thousands of miles and by several time zones. But such global integration and instantaneous financial trading also have their costs. 'Shocks' which occur in one geographical market now spread in-stantaneously around the globe, creating the potential for global financial instability.

The influence of telecommunications technologies on the speed and geographical extent of financial transactions between sellers and buyers of financial services is one expression of the effect of technological change on the financial services industries. Another is the effect of information technology on the *internal operations* of financial services firms. For example, banks and insurance companies were among the earliest adopters of computer technology to automate the internal processing of financial transactions; the so-called 'back-office' functions which are otherwise highly labour intensive. Subsequently, computerization and related technologies were applied to the firms' 'front offices': to the direct interface between firms and their customers. In retail banking, for example, counter clerks or tellers now operate online computer terminals; automated teller machines and cash dispensers have become the norm, giving customers access to certain services outside normal banking hours; direct banking by telephone has become the fastest-growing sector of retailing banking business. The equivalent of such services for the large corporate customers is the development of electronic cash management systems in which the corporate cus-tomer's computers are linked directly to the bank.

Table 12.2 Examples of product innovations in financial markets

Type of financial instrument	Basic characteristics
Floating-rate notes (FRNs)	Medium- to long-term securities with interest rates adjusted from time to time in accordance with an agreed reference rate, e.g. the London InterBank Offered Rate (LIBOR) or the New York banks' prime rate
Note issuance facilities (NIFs)	Short- to medium-term issues of paper which allow borrowers to raise loans on a revolving basis directly on the securities markets or with a group of underwriting banks
Eurocommercial paper	Non-underwritten notes sold in London for same-day settlement in US dollars in New York. More flexible than longer-term Euronotes of one, three or six months' duration
Loan sales	The sale of a loan to a third party with or without the knowledge of the original borrower
Interest rate swaps	A contract between two borrowers to exchange interest rate liabilities on a particular loan, e.g. the exchange of fixed-rate and floating-rate interest liabilities
Currency swaps	Financial transactions in which the principal denominations are in different currencies

Source: Based on Lewis and Davis (1987, pp. 415–31); UNCTC (1988, Box VII.2).

Innovations in telecommunications and in process technologies, therefore, have helped to transform the operations of financial services firms. But there has also been a variety of product innovations, especially the so-called *derivatives*.[11] A whole new array of financial instruments has appeared on the scene which can be categorized into two broad types:

● those which provide new methods of lending and borrowing
● those which facilitate greater spreading of risk.

Table 12.2 gives some examples of those financial innovations which have greatly increased the product diversity of financial markets.

Without question, therefore, technological developments, particularly in telecommunications and information technology, as well as in product innovations, have transformed the financial services industries. The global integration of financial markets has become possible, collapsing space and time and creating the potential for virtually instantaneous financial transactions in loans, securities and a whole variety of financial instruments. However, completely borderless financial trading does not actually exist, for the simple reason that most financial services remain very heavily supervised and regulated by individual national governments. Let us now see how the regulatory system operates and how it is changing.

The influence of governments: regulation and deregulation in financial services[12]

A tightly regulated system

Before the 1960s there was really no such thing as a 'world' financial market. The IMF, together with the leading industrialized nations, acted to ensure a broadly efficient global mechanism for monetary management based, initially, on the postwar Bretton Woods agreement. At the national level, financial markets and institutions were very closely supervised primarily because of concerns over the vulnerability of the financial system to periodic crisis and the centrality of finance to the operation of a country's economic system at both the macro- and the microscales.

Financial services, therefore, have been the most tightly regulated of all economic activities, certainly more so than manufacturing industries. The forms of such regulation can be divided into two major types:

- Those which govern the *relationships* between different financial activities. Most countries have restricted, or even prohibited, firms from participating in a range of financial service activities. In other words, each national financial services market has been *segmented* by regulation: banks operated in specified activities, securities houses in other areas of activity; the two were not allowed to mix across the boundaries. Neither was allowed to perform the functions of the other.
- Those which govern the *entry* of firms (whether domestic or foreign) into the financial sector. Restricted entry into the different financial services markets has been virtually universal, although the precise regulations differ from country to country. Most, however, have restrictions governing the entry of foreign firms into financial services. At the very least, national governments retain the discretionary power to restrict inward investments and all countries use their anti-trust/ competition laws to regulate foreign acquisitions of domestic financial activities. Governments have been especially wary of a too-ready expansion of the *branches* of foreign banks and insurance companies. Unlike subsidiary companies, which have to be separately incorporated, branches are far more difficult to supervise because they form an integral part of a foreign company's activities. In almost every case, there are limits on the degree of foreign ownership permitted.

'The crumbling of the walls'[13]

Although many of these restrictions are still operative, there is not the slightest doubt that the regulatory walls have been crumbling. The process was relatively slow at first but has accelerated rapidly since the late 1980s. The pressure towards deregulation has come from several sources, most notably the increasing abilities of international firms to take advantage of 'gaps' in the regulatory system and to operate outside national regulatory boundaries.

In Susan Stange's view, the emergence of the Eurodollar markets in the 1960s was the great technological breakthrough of international finance in the mid-twentieth century. Initially, Eurodollars were simply dollars held outside the US banking system largely by countries, like the USSR and China, which did not want their dollar holdings to be subject to US political control. 'Once it was appreciated that Eurodollars . . . were free of US political control, it did not take bankers long to recognise that the dollar balances were also free of US banking laws governing the holding of required reserves and controls upon the payment of interest' (Lewis and Davis, 1987, pp. 225, 228).

The rapid growth of this new currency market outside national regulatory control was certainly one of the major stimuli towards an international financial system. It was reinforced by the revolutions in telecommunications and information technologies, discussed earlier, which made possible the internationalization of financial transactions. Pressures built up, too, from the desires of banks and other financial services institutions to operate in a less constrained and segmented manner, both domestically and internationally. The internationalization of financial services and the deregulation of national financial services markets are virtually two sides of the same coin. Forces of internationalization were one of the pressures stimulating deregulation; deregulation is a necessary process to facilitate further internationalization.

Major deregulation has occurred, or is occurring, in all the major economies. A series of changes in the United States since the 1970s has both eased the entry of foreign banks into the domestic market and facilitated the expansion of US banks overseas. In 1981, the United States allowed the establishment of International Banking Facilities (IBFs), which created 'onshore offshore' centres able to offer specific facilities to foreign customers. Earlier, in 1975, the New York Stock Exchange had abolished fixed commissions on securities transactions. Recently, the federal government introduced a major reform of the US financial system aimed at allowing banks to become involved in a whole variety of financial services and to operate nationwide branching networks. One of the paradoxes of the US system has been that the 'non-bank banks', such as General Motors' Acceptance Corporation, could operate without many of the restrictions which applied to the 'proper' banks.

In the United Kingdom the so-called 'Big Bang' of October 1986 removed the barriers which previously existed between banks and securities houses and allowed the entry of foreign firms into the Stock Exchange. In France the 'Little Bang' of 1987 gradually opened up the French Stock Exchange to outsiders and to foreign and domestic banks. In Germany foreign-owned banks are now allowed to lead-manage foreign DM issues, subject to reciprocity agreements. In Japan the restrictions on the entry of foreign securities houses have been relaxed (though not removed) and Japanese banks are now allowed to open international banking facilities. But the Japanese financial system has remained tightly regulated. In 1996, the Japanese government announced its intention to undertake a wide-ranging deregulation of the country's financial system by 2001.

Increasing deregulation of financial services is an important component of both major regional economic blocs: the European Union and the NAFTA. Reforms within the EU have been directed at removing the individual national financial regulatory structures which have inhibited the creation of an EU-wide financial system so that financial services flow freely throughout the EU and financial services firms can establish a presence anywhere within the single market. The creation of a single European currency (discussed in Chapter 3) will obviously have enormous repercussions on the structure of the financial services sector in the EU. A financial services agreement was part of the Canada–United States Free Trade Agreement. The NAFTA provides for Mexico gradually to open up its financial sector to US and Canadian firms with all barriers to be eliminated by 2007. In addition, as we saw earlier, negotiations within the WTO to complete an international agreement on financial services continued beyond the conclusion of the Uruguay Round. The essence of those negotiations is to move towards a *multilateral*, rather than a bilateral, regulatary structure for financial services. An interim financial services agreement, involving some thirty countries, was reached by the WTO in 1995 which aimed to guarantee access to banking, securities, insurance markets. But the United States did not fully accept the agreement and reserved the right to restrict access to its own financial markets on the basis of reciprocity (although existing arrangements stayed in place).

The accelerating deregulation of financial services is the most important current development in the internationalization of the financial system. Even so, there remain substantial differences in the extent and nature of financial services regulation in different countries. The international regulatory environment is highly asymmetrical. Such differences, of course, have a powerful influence on the locational strategies of international financial services firms.

Corporate strategies in financial services

Shifting patterns in the demands for financial services, technological innovations which affect how these services can be delivered, and the changing regulatory framework interact together to form the environment in which financial services companies must operate. But, of course, such firms are not simply responding to environmental changes over which they have no influence. The processes are dynamic and interactive. Indeed, the strategies and actions of the major financial services firms – the banks, securities houses, insurance companies and the like – have themselves been highly influential in changing that environment. There is no simple cause–effect relationship but, rather, a complex interplay between actors and processes. In that respect, financial services are no different from any of the other industries we have been examining.

In this section we focus on the specialist financial services companies and exclude the in-house financial activities of TNCs in other industries. All TNCs, of course, operate large-scale and sophisticated financial operations. The corporate treasury departments of major companies are significant entities in their own right and are larger than many specialist financial services companies. However, the primary functions of these in-house operations are to contribute towards the efficiency of their own corporate systems. They do this by, for example, optimizing internal flows of funds, investing surplus capital effectively on the financial markets and hedging against the foreign currency fluctuations which are endemic in the operations of any TNC, whatever its line of business.

As far as the specialist financial services firms are concerned, two major strategic trends will be examined:

- the *internationalization* of their operations
- the *diversification* of companies into new product markets.

These two sets of strategies are very closely related.

Internationalization

> Banks have always engaged in international business. They have dealt in foreign exchange, extended credit in connection with foreign trade, traded and held foreign assets and provided travellers with letters of credit. Historically, the banks have carried out all this and some other types of business from their domestic locations. There was no need for a physical presence abroad. Business that could not be carried out by mail or telecommunications was handled by correspondent banks abroad.
>
> (Grubel, 1989, p. 61)

A small number of banks certainly set up a few overseas operations towards the end of the nineteenth century. Even in the early part of the twentieth century, however, the international network was very limited. Almost all international banking operations were 'colonial' – part of the imperial spread of British, Dutch, French and German business activities. In 1913 the four major US banks had only six overseas branches between them. By 1920 the number of branches had grown to roughly a hundred but there was little further change until the 1960s. During that period, Citibank had by far the most extensive international network of any US bank.

As with TNCs in manufacturing, the most spectacular expansion of transnational banking occurred in the 1960s and 1970s. Again, the initial surge was dominated by US firms, a reflection of both the focal role of the United States in the postwar

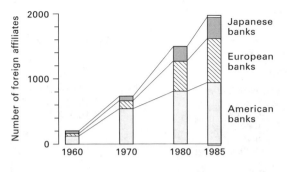

Figure 12.7 Growth in the number of overseas bank affiliates, 1960–85
Source: Based on *OECD Observer* (1989, Vol. 160, p. 36)

international financial and trading system and also the rapid proliferation of United States TNCs. But European, and later Japanese, banks increasingly internationalized their operations. Figure 12.7 shows that the number of foreign affiliates of banks increased from 202 in 1960 to 1,928 in 1985. Growth in this international network was especially rapid in the 1960s and 1970s. At the same time the *geographical composition* of the international banking network has changed. United States banks became less dominant than in the 1960s and 1970s, whilst European and Japanese banks increased the size of their international branch networks. In 1975, there were only two Japanese banks in the world's top ten (ranked seventh and tenth). By the late 1980s, Japanese banks occupied no less than seven of the top-ten places.

By 1996, the position had changed dramatically for a number of reasons. First, the Japanese banks suffered very badly in the collapse of the Japanese 'bubble economy' in the early 1990s. Having lent heavily (and, with hindsight, misguidedly) during the boom years of the 1980s, many of the major Japanese banks had to make large provisions in their balance sheets to cover losses on these big loans. Secondly, a number of very large bank mergers occurred in the mid-1990s. For example, six of the world's biggest banks were reduced in number to three through merger (*The Economist*, 6 April 1996). In Japan itself, Mitsubishi Bank and the Bank of Tokyo (sixth and tenth in Japan respectively) merged to form the world's biggest bank. In the United States, Chemical Bank and Chase Manhattan merged to form the biggest bank in the country whilst, at the same time, California Wells Fargo took over First Interstate Bank:

> These deals are the latest in a series of big banking mergers and acquisitions around the world that are being driven by a potent mixture of fear and ambition . . . too many banks with too much capital are chasing too few borrowers . . . Banks are afraid of competition from outside the industry as well as from within it. For years now, they have been trapped in a pincer. On the one hand, mutual funds (unit trusts) . . . have eaten away at banks' cheap deposits . . . On the other hand, bankers are facing stiff competition for lending business from the financing arms of industrial firms . . . Worse, their traditional role as middlemen has been decimated by the growth of the capital markets, which match borrowers directly to lenders . . . But their motives for merging are not purely defensive. In some cases, link-ups are aimed at uniting banks' common strengths in particular areas and at producing huge cost savings.
>
> (*The Economist*, 6 April 1996)

Table 12.3 shows the position in 1996 (just before the big mergers occurred). The measure used is 'Tier One capital' which is a better measure of a bank's strength than simple assets. It shows that, while still very strong – and the newly merged Tokyo-Mitsubishi Bank will be the number one – Japanese banks have lost some of their 1980s dominance. In fact, even in 1989, Japanese banks were rather less significant – in terms of both employment size and their international spread – than their asset values

Table 12.3   The world league table of commercial banks, 1996

Rank	Bank	Home country	Tier One capital ($m)
1	HSBC Holdings	United Kingdom	21,445
2	Crédit Agricole	France	20,386
3	Union Bank of Switzerland	Switzerland	19,903
4	Citicorp	United States	19,239
5	Dai-Ichi Kangyo Bank	Japan	19,172
6	Deutsche Bank	Germany	18,937
7	Sumitomo Bank	Japan	18,605
8	Sanwa Bank	Japan	17,676
9	Mitsubishi Bank	Japan	16,667
10	Sakura Bank	Japan	15,961
11	Fuji Bank	Japan	15,443
12	BankAmerica Corp.	United States	14,820
13	CS Holding	Switzerland	13,751
14	ABN-Amro Bank	The Netherlands	13,372
15	Groupe Caisse d'Epargne	France	12,667
16	Industrial Bank of Japan	Japan	12,497
17	Swiss Bank Corp	Switzerland	11,733
18	National Westminster Bank	United Kingdom	11,501
19	Banque National de Paris	France	11,453
20	Chemical Banking Corp	United States	11,436
21	Rabobank Nederland	The Netherlands	11,310
22	Bank of Tokyo	Japan	11,169
23	NationsBank	United States	11,074
24	Barclays Bank	United Kingdom	11,068
25	Compagnie Financière de Paribas	France	10,980

Source: The Banker (July 1996).

suggested. One reason is that their roles are primarily to serve the major Japanese business corporations. Their retail business is far less extensive than that of the American and European banks. In the late 1980s, the average number of foreign operations per bank was far lower among the leading Japanese banks than in banks from the other leading source countries, although the German figure was also low. Thus, although the Japanese banks have undoubtedly internationalized very rapidly in the last two decades, the really extensive global networks are still those operated by some of the American and European banks.

Although the specific reasons for, and the pace of, the internationalization of banks and other financial services may well vary from one case to another, there is an overall general pattern. In the immediate post-1945 period, the major international function of the small number of US transnational banks was the provision of finance for trade. But as the overseas operations of United States TNCs in manufacturing and other activities accelerated in the early 1960s, the functions of the US transnational banks evolved to meet their particular demands. Thus, the staple business of US banks in London became that of servicing the financial needs of their major industrial clients who were rapidly extending their operations outside the United States.

In the second half of the 1960s transnational banking functions took on a significant additional dimension with the growth of the Eurodollar market. This, as we have seen, was a market outside the regulatory control of the US government. As a consequence, US banks could raise money there for relending domestically as well as overseas. The attractions of international banking widened progressively as both the global capital market evolved and as local capital markets overseas developed. As banks from the other industrialized economies also internationalized, the process became *self-reinforcing*. All large banks had to operate internationally; they had to have a presence in all the leading markets:

> The internationalization of financial markets received fresh impetus in the 1970s from two sources. First, the oil-exporting countries emerged from the price increases of 1973–1974 with huge surplus savings which they sought to lend abroad, and much was recycled by the TNBs [transnational banks] to the deficit developing countries. Secondly in the mid-1970s, the Federal Republic of Germany and the United States dismantled their exchange controls on capital movements, and they were followed towards the end of the decade by Japan and the United Kingdom. The events of the 1970s catapulted the TNBs into prominence as the institutions dominating the world financial markets; but with the advent of the debt crisis in 1982, their pre-eminence as suppliers of international finance began to wane. This was in part because the flow of private loan capital going through the banks to the developing countries dropped to a trickle. But it was also because the dismantling of exchange controls on capital movements and other financial deregulation led to a large and vigorous growth in the international securities market, mainly among the developed market economies.
>
> (UNCTC, 1988, p. 103)

Figure 12.8 summarizes the major phases of development of international banking. As with all such models it should be seen as a broad general framework which captures the main features of the process, although the detail may well vary from case to case. However, it certainly fits the general experience of both US and Japanese transnational banks. These broad developments, intensified by the further deregulation of financial markets in many countries and by technological innovation, have pulled more and more securities firms into international operations:

> Up to 1979/80, the US multinational investment bank had little more than a large office in London and perhaps some much smaller ones in other European countries, perhaps an Arab country and possibly (though less likely) Japan. From 1980 onward, the development of the US investment bank as a multinational changed qualitatively.
>
> (Scott-Quinn, 1990, p. 281)

The same process applied to the leading Japanese *securities houses*, such as Nomura, Daiwa, Nikko and Yamaichi, although with differences of detail. Finally, as both

	Phase I National banking	Phase II International banking	Phase III International full- service banking	Phase IV World full- service banking
Internationalization of customer companies	Export – import.	Active direct overseas investment.	Multinational corporation.	
International operations in banking	Mainly foreign exchange operations connected with foreign trade. Capital transactions are mainly short-term ones.	Overseas loans and investments become important, as do medium- and longer-run capital transactions.	Non-banking fringe activities such as merchant banking, leasing, consulting and others are conducted. Retail banking.	
Methods of internationalization	Correspondence contracts with foreign bank.	To strengthen own overseas branches and offices.	By strengthening own branches and offices, capital participation, affiliation in business, establishing non-bank fringe business firms, the most profitable ways of fund-raising and lending are sought on a global basis.	
Customers of international operations	Mainly domestic customers.	Mainly domestic customers.	Customers are of various nationalities	

Figure 12.8   The major phases in the development of international banking
Source: Based on Fujita and Ishigaki (1986, Table 7.6)

industrial and service companies spread globally a further boost was given to the increased internationalization of *accountancy* companies.[14]

As a result, all the transnational financial services firms are now basing their strategy on a direct presence in each of the major geographical markets and on providing a local service based on global resources. They are, in fact, selling an *international brand image*, with the clear message that a global company can cope most easily and effectively with every possible financial problem which can possibly face any customer wherever they are located. For example, the Swiss Bank, UBS, now advertises itself as 'One bank worldwide. One brand worldwide'. The biggest United States bank, Citicorp, aims to establish itself as a global brand and has recently announced a global strategy to integrate its extensive, but so far very loosely connected, international operations (*The Financial Times*, 24 October 1996). The model for such a global strategy is, in fact, that of a consumer products manufacturer, like Gillette, which launches a new product simultaneously across the world. Citicorp's aim is to achieve a mix of consumer and corporate banking and of industrialized and emerging markets. It is placing a major emphasis on consumer banking and credit card business.

The drive for diversification

It is a short, and supposedly logical, step from this kind of global strategic orientation to the notion that global financial services should not only operate globally in their own core area of expertise but also that they should supply a *complete package of related services*. The financial conglomerate or the financial supermarket has arrived on the scene. Deregulation has been a major stimulus for such change. It is becoming increasingly possible for banks to act as securities houses, for securities houses to act as banks, and for both to offer a bewildering array of financial services way beyond their original operations. At the same time, entirely new non-bank financial services companies have emerged. Amongst all the leading accountancy firms, for example, there has been a strong move into management consultancy, including information systems. A leading UK bank's current portfolio of offerings includes: clearing banking, corporate finance, insurance broking, commercial lending, life assurance, mortgages, unit trusts, travellers' cheques, treasury services, credit cards, stockbroking, fund management, development capital, personal pensions and merchant banking.

Although some of this diversification has taken place by new greenfield investment, the majority has occurred through the process of acquisition and merger. This has been particularly the case in the rush of the securities houses to gain a foothold in newly deregulated markets, like the City of London in 1986, and in the creation of totally new financial services conglomerates. The rationale for diversification, both into new products and new geographical markets, is the familiar one of economies of scale and economies of scope:

> In financial services, the arguments are that there are large fixed costs in running a network of branches or agents. These can be spread over a lot of customers in large operations . . . Another supposed economy derives from the benefits of scale in participating in capital markets. Unit transaction costs tend to be lower, the larger you are. Consumer recognition is another advantage to the large company who might expect to attract customers most easily . . . a large diversified company can afford to cross-subsidize price wars in one sector with the profits made in another . . . As well as economies of scale, it is generally supposed that there are economies of scope in financial services. The bank doing business in all major European cities will find it easier to serve

clients than a similar size bank that is locally concentrated. The international bank would be better placed to serve international clients, and would gain strategically valuable information about a range of markets from its dealings abroad.

(Davis and Smales, 1989, p. 99)

The arguments for a strategy of internationally diversified financial service operations, then, are that it enables such a company to offer an entire package of services – a 'one-stop shop' – to customers. Their supply by a large, internationally recognized brand name – backed up, of course, by lavish advertising expenditure – is supposed to give a reassurance to potential customers that they will receive the highest quality of service. Whether or not economies of scale and scope really do exist to a significant extent is a matter over which there is considerable disagreement. The large financial companies themselves certainly seem to think so. Although there have been divestments as some financial services firms have shed parts of their diversified portfolio of activities, the diversification and internationalization trend continues. On the other hand, there are those who argue that there are distinct benefits in geographical specialization. Davis and Smales (1989), for example, suggest that

- the administrative benefits of size can be exaggerated
- local knowledge may be a major advantage in assessing local risks and opportunities
- the benefits of recognition through size may be less in national markets where names have already established themselves.

In other words, 'localization might be a marketing advantage'. This, of course, has been recognized by the major financial services companies in their emphasis on providing a locally sensitive service within a global organization.

Geographical structure of financial services activities[15]

At first sight, technological developments in communications systems would appear to release financial services companies from the spatial constraints on the location of their activities. Financial services firms, in particular, would appear to be especially footloose. They are not tied to specific raw material locations, whilst at least some of their transactions can be carried out over vast geographical distances, using telecommunications facilities. Such considerations led Richard O'Brien to entitle his 1992 book on global financial integration *The End of Geography*. That title undoubtedly reflects the views of many people, both inside and outside the financial services industries.[16] The aim of this section is to argue that such a view is misplaced. To be fair to O'Brien, however, it should be pointed out that his position was not totally unequivocal:

> The end of geography, as a concept applied to international financial relationships, refers to a state of economic development where geographical location no longer matters in finance, or matters much less than hitherto . . . In theory, the end of geography should mean that location no longer matters. Yet, 'everybody has to be somewhere', . . . Location still exists, even though electronic markets make that location more difficult to identify in traditional geographical ways. Firms, individuals, markets, even products, have to have a sense of place.

(O'Brien, 1992, pp. 1, 73)

There is no doubt that the revolutionary developments in the space- and time-shrinking technologies permit financial transactions to whizz around the world electronically, and that deregulation has reduced the impermeability of national boundaries to financial flows. But, far from heralding the 'end of geography', this has, in fact, made geography more – not less – important. In this section, we focus on two aspects of this issue: the relationship between geography and the nature of financial products, and the geographical structure of financial services.

Geography and the nature of financial products

Clark and O'Connor (1997) claim that the informational content of financial products themselves is inextricably related to the spatial structure of the global finance industry. They argue that

> the geography of places shapes the operation of finance markets and the global economy primarily because of the geography of information that is embedded in the provision of specific financial products. There is, in effect, a robust territoriality to the global financial industry . . . financial products often have a distinct spatial configuration of information embedded in their design.
>
> (Clark and O'Connor, 1997, pp. 90, 95)

They identify three types of financial product each of which has distinctive, geographically derived, properties (Table 12.4). The specific informational nature of these three types of financial product, and the fact that such information is, itself, locationally specific is reflected in a particular 'topography' of financial markets:

- *transparent products* are produced

> where the trading volume is large enough to make economical the repetitive small markets that are associated with this form of trading. These turnover efficiencies are only met in the larger global centers, so that the design and production of transparent products provides the apex of the world's financial system.
>
> (Clark and O'Connor, 1997, p. 101)

- *translucent products* (such as balanced equity products)

> draw upon a different set of factors and create opportunities for a second (national) level of financial centers . . . balanced equity products [are] typically differentiated by country of origin. They all require skills in design to structure them so as to maximize returns, spread risks, and maximize product differentiation in order to give such a

Table 12.4   Financial products and their geographical characteristics

Financial product type	Probable market scope and type of financial centre	Information intensity	Specialist expertise required	Perceived risk-adjusted return
Transparent	Global	Ubiquitous	Low (e.g. equity products like stock market indexes). Traded internationally	Low
Translucent	National	Third party-market specific	Significant. Require detailed knowledge of national/local companies. May be traded internationally	Medium
Opaque	Local	Transaction specific	Vital. Applies to products based upon trust and long-term relationships	High

Source: Based on Clark and O'Connor (1997, Table 4.1 and pp. 99–104).

product an edge in the competition for other funds. Hence there are opportunities for local operators and local markets. However, the costs of assembling and maintaining the information systems necessary to identify and monitor the various investments within a fund, and the skilled personnel required to market them, rely upon significant scale economies. These scale economies limit the effective diffusion of these products to smaller, third-level financial centers and provide the larger national centers with an important source of business . . . transaction costs can be minimized by concentrating the administration and organization and flow of funds, providing a further reason for the role of the first- and second-level centers; market share can also be maximized by location in these places, as companies can use their globally distributed networks of offices . . . to funnel funds to the large markets in these cities. The costs of the global networks become another source of scale economies available to the major markets.

> (Clark and O'Connor, 1997, pp. 101, 102)

- *opaque products* are those

> where the design and production is shrouded in some mystery to the outsider, and local knowledge is essential for confident trading. These are opaque products like the REIT or property trust, where a set of properties are packaged into a product that can be sold in units or shares . . . Property trusts will often be produced by firms with access to local markets as the information about them is so specialized, thereby providing the third layer in global financial geography. But, of course, these products may be consumed at distant locations.

> (Clark and O'Connor, 1997, p. 102)

Clark and O'Connor's (1997, p. 108) conclusion is that 'different types of products produce geography, as different geographies imply different products . . . local customs, market behavior, and conventions sustain the persistence of spatial differentiation'. We should not be surprised, therefore, to find that, at both the global and national scales, the major financial services activities continue to be extremely *strongly concentrated geographically*. They are, in fact, more highly concentrated than virtually any other kind of economic activity except those based on highly localized raw materials. However, there are some subtle variations according to the particular function involved; there is a *division of labour* within financial services firms, parts of which may show a greater degree of geographical decentralization.

Centralization: the hierarchy of international financial centres

Figure 12.9 maps the hierarchy of international financial centres. They are classified on the basis of sixteen statistical variables (Reed, 1989), of which the most important are

- the volume of international currency clearings
- the size of the Eurocurrency market
- the volume of foreign financial assets
- the number of headquarters of the large international banks.

Reed suggests that the hierarchy of international financial centres has three major levels. Interestingly, he places Tokyo in the second tier, although the structure of the global financial system is increasingly becoming articulated along the *tripolar axis* of New York–London–Tokyo.

Thrift (1994) identifies four interlocking types of locational determinant of international financial centres:

- The characteristics of the *business organizations* involved in international financial centres: 1) much of the production of financial products and services occurs at the

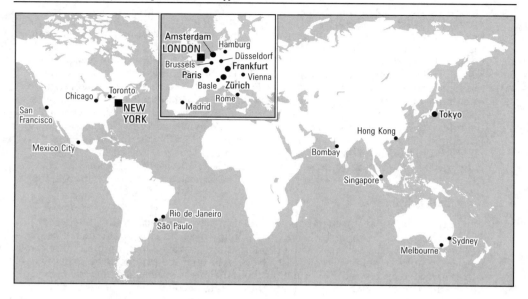

Figure 12.9   The hierarchy of international financial centres
Source: Based on Reed (1989, Figure 1)

boundaries of firms and there is a strong reliance on repeat business; 2) the firms tend to be 'flattened and non-hierarchical' and based around 'small teams of relationship and product specialists'; 3) firms need to co-operate as well as to compete, as in the case of syndicated lending; 4) firms need to compare themselves with one another to judge their performance; 5) there is a need for a constant search for new business and for rapid response:

> These shared characteristics point to two important correlates. First, in general, these firms must be sociable. Contacts are crucially important in generating and maintaining a flow of business and information about business. 'Who you know' is, in this sense, part of what you know . . . 'relationship management' is a vital task for both employees and firms. Second, this hunger for contacts is easier to satisfy if contacts are concentrated, are proximate. When contacts are bunched together they are easier to gain access to, and swift access at that.'

> (Thrift, 1994, p. 333)

- The *diversity of markets* in international financial centres: 1) their large size which makes them both flexible in terms of entry and exit and also socially differentiated: 'more likely to consist of social "micro-networks" of buyers and sellers, whose effect on price-setting can sometimes be marked'; 2) their basis in rapid dissemination of information which may lead to major market movements; 3) their speculative and highly volatile nature. 'Again, as with the case of organizations, there are the two obvious corollaries to these characteristics: the twin needs for sociability and proximity' (Thrift, 1994, p. 334).
- The *culture* of international financial centres: 1) such centres receive, send and interpret increasing amounts of information; 2) they are the focus of increasing amounts of expertise which arise from a complex division of labour involving workforce skills and machinery; 3) they depend on contacts and such contacts have become increasingly reflexive because of their basis in trust founded on relationships. Thrift

(1994, p. 334) argues that 'these cultural aspects of international financial centres are actually increasing in importance'.

- The *dynamic external economies of scale* which arise from the sheer size and concentration of financial and related services firms in such centres. Such economies include: 1) the sharing of the fixed costs of operating financial markets (e.g. settlement systems, document transport systems) between a large number of firms; 2) the attraction of greater information turnover and liquidity; 3) the enhanced probability of product innovations in such clusters (the 'sparking of mind against mind'; 4) the increased probability of making contacts which rises with the number of possible contacts; 5) the attraction of linked services such as accounting, legal, computer services, which reduces the cost to the firm of acquiring such services; 6) the development of a pool of skilled labour; 7) the enhanced reputation of a centre which, in a cumulative way, increases that reputation and attracts new firms. In other words, the constellation of traded and untraded interdependencies, described in Chapter 1, tend to be especially strongly developed in international financial centres.

The twenty-five cities shown in Figure 12.9 effectively control almost all the world's financial transactions. It is a remarkable level of geographical concentration. But they are more than just financial centres. There is clearly a close relationship with the distribution of the corporate and regional headquarters of transnational corporations which we examined in Chapter 7 (see Figure 7.4). These global cities are, indeed, the *control points of the global economic system*. As far as financial services firms are concerned, the international financial centres – especially those at or near the top of the hierarchy – are their 'natural habitat' (Thrift, 1987). Financial companies agglomerate together in these centres for the reasons outlined above.

But this does not mean that all international financial centres – even those at the top of the hierarchy – are identical. Far from it. Each has distinctive characteristics which reflect their specific historical and geographical embeddedness.[17] On those criteria which measure both the breadth and depth of global financial activity, New York and London occupy the apex of the international financial centre hierarchy. London is the more international of these two leading centres in terms of both foreign exchange and Eurocurrency transactions. The daily turnover on the London foreign exchange market is almost as large as the turnover in New York and Tokyo put together. Lewis and Davis (1987) attribute London's current great significance as a global financial centre to the following:

- The historical evolution of the City as a world centre has created both a large pool of relevant skills and also an almost unparalleled concentration of linked institutions within a very small geographical area. 'Unlike the position in some other countries, insurance, commodities trading, futures markets, stock broking, bond trading and legal services in the UK are all concentrated around the City' (Lewis and Davis, 1987, p. 236).
- Its geographical position in a time zone between New York and Tokyo.
- The regulatory environment, which has encouraged the growth of international banking and is favourably disposed towards international bankers. Foreign banks in London can operate as 'universal' banks, a particularly important feature for US and Japanese banks. They can combine both banking and securities businesses there in ways which are prohibited so far in their domestic operations.
- London is the centre of the Eurocurrency and Eurobond markets.

There is some concern in the London financial community that its pre-eminence may be threatened by competition from rapidly growing European financial centres, especially Paris and Frankfurt. This concern has been intensified by the potential development of a single European currency and the likely location of a European Central Bank in Frankfurt. However, the view of the accountancy firm, Arthur Anderson, is that 'London with the UK in or out of the euro will still be the primary centre in this time zone. In foreign exchange dealing, corporate finance, fund management, swaps and options and the international new issue market, London is likely to retain its strength' (quoted in *The Financial Times*, 18 November 1996). London still has the largest concentration of foreign banks in the world, with more than 550 foreign branches, subsidiaries or representative offices located there in 1997. In comparison, Paris has 280 and Frankfurt 250. The number located in London has been growing rapidly. There are now around 250 European banks in London, and a further 50 each from the United States and Japan.

London's strength as a financial centre rests on the scale of its foreign exchange and Eurocurrency business and its deregulated securities markets. New York, in comparison, is by far the world's largest securities market. But it also has a huge concentration of international banks and other financial activities. More than this,

> New York is the nerve centre of worldwide operations of those US banks with the largest presence in the Eurocurrency markets . . . Second, US banks use their overseas branches to switch US dollar funds between the Euro and domestic deposit markets according to the relative cost of funding. These operations help to tie Eurodollar interest rates to domestic money markets which revolve around New York. Finally, New York is the location for the majority of IBFs (International Banking Facilities).
>
> (Lewis and Davis, 1987, p. 239)

London and New York still stand apart from Tokyo as truly global financial centres. The international significance of Tokyo rests primarily on the strength of the Japanese economy itself:

> Whereas London and New York have a history of international financing, Tokyo has had a mainly domestic orientation. It is only since the 1970s that Japanese banks have made a concerted move into international banking. Yet this internationalization of banking was not accompanied by the internationalization of the yen nor of Japanese money and capital markets . . . Tokyo is now the world's second largest stock market . . . but at the end of 1985 only 20 listed companies, out of a total of 1,497 listings, were foreign . . . Japan stands out . . . in comparison with . . . especially the UK and USA, in terms of the relatively small number of foreign banking institutions operating in Japan.
>
> (Lewis and Davis, 1987, pp. 243–44)

Thus, New York, London and, increasingly, Tokyo sit at the apex of the global system of financial centres. This is not to say that the other centres shown in Figure 12.9 are not important. They certainly are, particularly those in the second level, some of which clearly have aspirations to gain promotion to the superleague. Centres in the third level serve mainly as regional financial centres which, again, is not to belittle their international significance but merely to put it in perspective. Each of these levels in the hierarchy has a particular role in developing different kinds of financial product, as Clark and O'Connor show.

Offshore financial centres[18]

> scattered across the globe, a series of little places – islands and micro-states – have been transformed by exploiting niches in the circuits of fictitious capital. These places have set

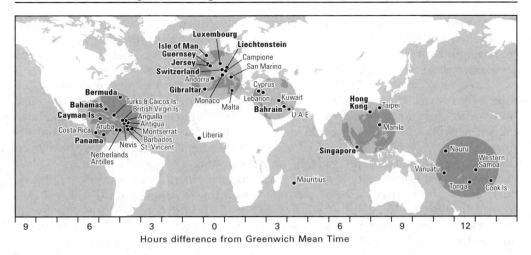

Hours difference from Greenwich Mean Time

Figure 12.10   Offshore financial centres
Source: Based on Roberts (1994, Figure 5.1)

themselves up as offshore financial centres; as places where the circuits of fictitious capital meet the circuits of 'furtive money' in a murky concoction of risk and opportunity. Furtive money is 'hot' money that seeks to avoid regulatory attention and taxes.

(Roberts, 1994, p. 92)

The speculative nature of financial transactions and flows and the desire to evade regulatory systems have led to the development of a number of offshore financial centres. With few exceptions, the sole rationale for such centres is to provide these kinds of services outside the regulatory reach of national jurisdictions. As Roberts points out, although the Cayman Islands has around 550 banks from all over the world, only 69 of these actually have a physical presence there. The vast majority are no more than 'a brass or plastic name plate in the lobby of another bank, as a folder in a filing cabinet or an entry in a computer system' (Roberts, 1994, p. 92).

Figure 12.10 shows the geographical distribution of offshore financial centres. Each tends to fill a specific niche which it exploits in competition with other centres in the same spatial cluster and with similar niche centres elsewhere in the world. Much of the growth of such centres occurred in the 1970s, in places which were already operating as tax havens, to act as banks' 'booking centres' for their Eurocurrency transactions:

> By operating offshore booking centres international banks could act free of reserve requirements and other regulations. Offshore branches could also be used as profit centres (from which profits may be repatriated at the most suitable moment for tax minimization) and as bases from which to serve the needs of multinational corporate clients.
>
> (Roberts, 1994, p. 99)

The location of these offshore centres and, especially their spatial clustering, is partly determined by time zones and the need for twenty-four-hour financial trading.

Decentralization? The geographical rearrangement of 'back-office' functions

All the discussion of international financial centres would seem to suggest that the potential for other cities, outside the favoured few shown in Figure 12.9, to develop as significant centres of finance and related activities is likely to be very limited. In the

United Kingdom, for example, the sheer overwhelming dominance of London makes it extremely difficult for provincial cities to develop more than a very restricted financial function. London, in that sense, is akin to the notorious upas tree, 'a fabulous Javanese tree so poisonous as to destroy all life for many miles around'.

It is, of course, the 'higher order' financial and service functions which are especially heavily concentrated in the major international financial centres. So-called 'front-office' functions, by definition, must be close to the customer – hence the huge branch networks of the retail banks and other financial services supplying final demand. The essence of all the financial services, indeed of all the services we have been discussing in this chapter, however, is the transformation of massive volumes of *information*. Much of that activity is routine data processing performed by clerical workers. Such 'back-office' activity can be separated from the front-office functions and performed in different locations, as we saw in the case of offshore sourcing of software services. The early adoption of large-scale computing by banks, insurance companies and the like from the late 1950s initially led many of them to set up huge *centralized* data-processing units. To escape the high costs (both land and labour) in the major financial centres such units were often relocated in less expensive centres or in the suburbs. Access to large pools of appropriate (often female) labour was a key requirement.

The introduction of microcomputers and networked computer terminals made such centralized processing units unnecessary and the tendency in recent years has been to *decentralize* back-office functions more widely in much smaller units. In effect,

> telecommunications accelerate a spatial bifurcation within many large finance firms by enhancing the attractiveness of downtown areas for skilled managerial activities while simultaneously facilitating the exodus of low wage, back office sectors. This process mimics the separation between headquarters and branch plant functions widely noted in manufacturing. In both cases, a vertical disintegration of production takes place, accompanied by the dispersal of standardized, capital-intensive functions and the concomitant reorganization of skilled labour-intensive functions around large, densely populated urban areas.
>
> (Warf, 1989, p. 267)

At the same time, however, the distinction between back-office and front-office functions is becoming less clear as distributed computer technology has been developed further. In fact, it is not only routine back-office activities that have been decentralized. It has become increasingly common for some of the higher-skilled functions to be relocated away from head office into dispersed locations, both nationally and, in some cases, internationally. But centralization of back-office functions by financial service companies is by no means an obsolete practice. For example, Citicorp, which as we saw earlier has embarked on an aggressive globalization strategy, plans to bring together all its back offices from around the world and to centralize their functions in large and more efficient centres to achieve economies of scale. Within this framework, all credit-card processing for Europe is carried out in South Dakota and all statement processing for the Caribbean region is concentrated in Maryland (*The Financial Times*, 24 October 1996). Even in the case of back offices, therefore, geographical centralization is far from dead.

Jobs in the financial services industries

In all the major industrialized economies, as well as in major financial centres such as Hong Kong and Singapore, the rapid growth and internationalization of financial services created impressive employment growth during the 1970s and 1980s. Whilst employment in manufacturing industries in many of the industrialized countries has been growing slowly, or even falling in many cases, jobs proliferated in financial services (as well as in other service sectors). Of course, it has to be remembered that such job growth is inevitably strongly concentrated within urban areas in general and in the big metropolitan areas in particular. In financial services, the biggest employment increases have been in the international financial centres themselves, as the experiences of both New York and London demonstrate. In both cases, deregulation of financial services, and the boost this gave to internationally oriented transactions, resulted in very substantial employment increases.

In New York, for example, employment in financial services increased by more than 30 per cent between 1979 and 1987, from 345,600 to 450,900. Growth was particularly spectacular in the securities industries (+106 per cent). Employment growth in financial services was apparent in virtually all major cities. 'Many of Britain's provincial financial centres enjoyed quite phenomenal rates of growth . . . for example, between 1974 and 1981, . . . purely financial service employment expanded by at least 25% in . . . 11 centres' (Leyshon, Thrift and Tommey, 1988, pp. 170, 172). In Britain as a whole, employment in financial services grew by 32 per cent between 1981 and 1989 (Tickell, 1997).

However, the story is not simply one of progressive employment expansion in the financial services industries. Technological and organizational developments are

Table 12.5   The changing nature of skills in banks and insurance companies

Old competencies	New competencies
Common emerging competencies	
1) Ability to operate in well defined and stable environment	Ability to operate in ill-defined and ever-changing environment
2) Capacity to deal with repetitive, straightforward and concrete work process	Capacity to deal with non-routine and abstract work process
3) Ability to operate in a supervised work environment	Ability to handle decisions and responsibilities
4) Isolated work	Group work, interactive work
5) Ability to operate within narrow geographical and time horizons	System-wide understanding; ability to operate with expanding geographical and time horizons
Specific emerging competencies	
Among upper-tier workers	
1) *Generalist competencies.* Broad, largely unspecialized knowledge; focus on operating managerial skills	*The new expertise.* Growing need for high-level specialized knowledge in well defined areas needed to develop and distribute complex products
2) *Administrative competencies.* Old leadership skills; routine administration; top-down, carrot-and-stick personnel management approach; ability to carry out orders from senior management	*The new entrepreneurship.* Capacity not only to manage but also set strategic goals; to share information with subordinates and to listen to them; to motivate individuals to develop new business opportunities
Among middle-tier workers	
1) *Procedural competencies.* Specialized skills focused on applying established clerical procedural techniques assuming a capacity to receive and execute orders	*Customer assistance and sales competencies.* Broader and less specialized skills focused on assisting customers and selling, capacity to define and solve problems
Among lower-tier workers	
1) *Specialized skills* focused on data entry and data processing	Disappearance of low-skill jobs

Source: Bertrand and Noyelle (1988, Table 4.1).

drastically changing the nature of work at all levels: redefining skills and increasing flexibility. Table 12.5 summarizes some of the major skills changes. In particular, there has been a

> widespread tendency for a dramatic decline in the volume of clerical processing work performed, until recently, manually by lower-tier personnel (with some assistance of mainframe computers for data crunching). This remarkable contraction in old-fashioned clerical processing work is the result not only of automation but also of the transformation of work done by personnel in the middle and upper rungs of a firm's occupational structure . . . Paralleling this transformation in data processing and data handling, increasing competition is generating new demands for both sales and assistance personnel and for specialists able to identify new markets, conceive new products, develop new systems and sell the new, often complex services (swaps, futures, etc.) . . . The outcome of this profound process of skill transformation is the emergence of a new matrix of competencies that may be viewed in terms of new skills that are being substituted progressively for older ones. Some of these new competencies are common to both middle- and upper-level workers; others are specific to various groups within the occupational hierarchy.
>
> (Bertrand and Noyelle, 1988, pp. 40–41)

Hence, there have been substantial changes in the demand for particular types of labour associated with

- technological change in both processes and products (including, for example, the growth of direct telephone and computer provisions of banking, insurance and other services)
- organizational changes in how new and existing service products are produced.

Such changes, together with the effects of the recession of the late 1980s/early 1990s, have meant that the labour market in financial services is no longer the 'job growth machine' it once appeared to be. Even in accountancy, the claim that 'like funeral parlours and the law business, the majors have discovered that accountancy is immune to economic downswings' (Clairmonte and Cavanagh, 1984, p. 268) no longer seems quite so valid. Particularly during the early 1990s, the 'golden hellos' of the burgeoning securities markets in the City of London and Wall Street were replaced by less-than-golden goodbyes to brokers and dealers. The former 'Masters of the Universe'[19] seemed to lose their positions of immense wealth and power. With economic recovery, however, some of that glitter has reappeared so that, once again, we are witnessing the obscenely high salaries and bonuses paid to a very small number of financial workers involved in speculative financial transactions on a twenty-four basis.

However, virtually all the major banks and financial service companies have undertaken major rationalization and restructuring programmes which have resulted in substantial job losses. Some of these have gone hand-in-hand with large-scale acquisitions and mergers. For example, following the merger of two United States banks, Chemical Bank and Manufacturers Hanover in 1991, the number of European staff was reduced from 3,400 to 2,000. The more recent merger between Chase and Chemical will certainly also result in job losses because one of the motivations for the merger is to increase efficiency (i.e. reduce costs).

The loss of financial jobs is especially marked in the retail banking sector where the number of retail branches is falling dramatically as the banks concentrate their services in a smaller number of centres and provide more of their services online. For example, in the UK, the National Westminster Bank shed around 8,000 jobs between

1994 and 1996 and announced plans to reduce its workforce by a further 10,000 jobs by the end of the decade. Barclays Bank reduced its workforce from 84,500 to 66,000 between 1991 and 1996. American Express reduced its worldwide labour force from 72,000 to 68,000 in 1997. In Britain, as a whole, there was a decline of 7.4 per cent in financial sector jobs between 1989 and 1993, in contrast to the growth of 32 per cent during the 1980s (Tickell, 1997). In the United States and Europe it is being predicted that there could be labour reductions in the financial services industries of up to 50 per cent by the year 2005 (*The Financial Times*, 28 March 1996).

More general employment reductions may well result from the continued deregulation of financial markets. The regulatory protection of national markets has allowed 'inefficient' practices to exist. As the regulatory walls come tumbling down in one country after another, and as the full impact of the internationalization of the industry takes hold, the terms 'overcapacity' and 'overbanking' have become commonplace. Such structural changes almost certainly signal the continuation of 'jobless growth' in the financial service industries, even though there may well be localized exceptions.

Notes for further reading

1. Hoekman and Kostecki (1995, Chapter 5) discuss this issue in some detail.
2. Allen (1988), Dunning (1989), Britton (1990) explore some of these conceptual issues.
3. The IMF definition of trade is that it constitutes a transaction between residents and non-residents of a country. Hence, it covers far more than the 'commonsense' definition of international trade as the movement of goods from one country to another. For example, the movement of tourists from, say, the United States to Britain is classified as British exports of tourism, even though the consumers actually move to the point of production of the tourist service.
4. Harris (1984), UNCTC (1988), Perry (1990), Leslie (1995) examine the globalization of the advertising industry.
5. Dicken and Miyamachi (1998) provide an up-to-date analysis of the *sogo shosha*.
6. There is a small, but growing, literature on this topic. See, for example, Wilson (1992), Warf (1995), Heeks (1996).
7. This section is based upon the excellent study by Heeks (1996).
8. General discussions of developments within the international financial system are provided by Strange (1986), Hamilton (1986), O'Brien (1992), Corbridge, Martin and Thrift (1994), Clark and O'Connor (1997). International banking is examined by Lewis and Davis (1987), Grubel (1989).
9. Bertrand and Noyelle (1988, Chapter 2), provide a concise summary of the major developments in the markets for financial services. This section draws extensively on that source.
10. See Hamilton (1986), Bertrand and Noyelle (1988), UNCTC (1988), Warf (1989; 1995).
11. Derivatives are 'financial tools derived from other financial products, such as equities and currencies. The most common of these are futures, swaps, and options . . . The derivatives market aims to enable participants to manage their exposure to the risk of movements in interest rates, equities, and currencies' (Kelly, 1995, p. 229). See also Hamilton (1986), Lewis and Davis (1987), UNCTC (1988).
12. See Hamilton (1986), Lewis and Davis (1987), UNCTC (1988), Corbridge, Martin and Thrift (1994), WTO (1996) for a discussion of the changing regulatory environment.
13. The term is borrowed from Clairmonte and Cavanagh (1984).
14. See Thrift (1987), Daniels, Thrift and Leyshon (1989).
15. The various chapters in Corbridge, Martin and Thrift (1994) represent an up-to-date perspective on the geography of finance. See also Clark and O'Connor (1997).
16. Martin (1994) and Clark and O'Connor (1997) take issue with O'Brien's thesis.
17. Thrift (1994) discusses the social and cultural structure of the City of London.
18. This section is based upon Roberts (1994).
19. See Tom Wolfe's novel, *The Bonfire of the Vanities,* where this term is used.

PART IV

Stresses and strains of adjustment to global shift

INTRODUCTION

A summary perspective

Putting the pieces together

The focus throughout the preceding twelve chapters has been on the patterns and processes of global shift; on the *forms* being produced by the increasing globalization of economic activities and on the *forces* producing those forms. By necessity, the approach has been that of concentrating on each individual element. However, when we try to understand the *impact* of globalizing processes on particular places and on the businesses, governments and people in such places we need to have

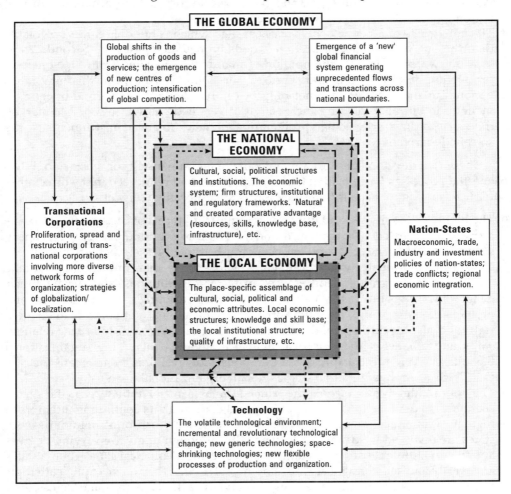

Figure IV.1   Globalizing processes as a system of interconnected elements and scales

some understanding of *how the individual parts fit together*. Figure IV.1 attempts to make these connections in a simplified, diagrammatic manner.

The upper section of the diagram identifies the two major sets of developments which constitute the contemporary globalizing world economy:

- The substantial global shifts in the *sphere of production* of goods and services which we examined in detail in Chapter 2 and in the industry case studies of Chapters 9–12. These are manifested in the emergence of new centres of production outside the formerly dominant core areas of the industrial world: notably (though not exclusively) the so-called newly industrializing economies of east and southeast Asia. The global economic map has increased vastly in complexity and is now a multipolar, multiscale structure; a 'mosaic of unevenness in a continuous of flux'. The geographical centre of gravity of the global production system has undoubtedly shifted from its long-established position 'in' the North Atlantic (focused on northwest Europe and the eastern seaboard of the United States) towards the Pacific Basin (focused on Japan and the newly industrializing economies of that region).

- The emergence of a new global *financial* system which is not only far more volatile than the one put in place at Bretton Woods immediately after the Second World War but which is now, in the opinion of many, disarticulated from the sphere of production of 'real' goods and services. The volume of flows in this new international financial system are not only of unprecedented – and mind-boggling – magnitude but also they are increasingly diversified. As we saw in Chapter 12, there has been a spectacular proliferation of new 'financial instruments' – the so-called 'derivatives'.

These two developments are the outcome of an immensely complex set of interconnected processes. The position taken in this book, and reinforced in the three other outer boxes in Figure IV.1, is that three forces are especially relevant:

- The highly differentiated and dynamic activities and forms of *transnational corporate activity*, as revealed in Chapters 6 and 7. TNCs are, undoubtedly, the primary movers and shapers of the global economy because of their potential ability to control or co-ordinate production chains across more than one country; to take advantage of geographical differences in factor distributions; and through their geographical flexibility to switch and reswitch resources internationally and perhaps globally. TNCs are both intricate organizational networks in their own right and also deeply embedded within dynamic networks of interfirm relationships and alliances. The empirical evidence suggests increasing organizational flexibility as TNCs restructure their operations. However, it is by no means the case that TNCs are converging to a single 'global' organizational form.

- The continuing significance of the *nation-state* as a major influence in the global economy, as demonstrated in Chapters 3 and 4, through its continuing attempts to regulate economic transactions within and across its territorial boundaries. All states are involved in such regulatory activity, although to greatly varying degrees and in very different ways according to their specific ideological stance. Some states are overtly and self-consciously 'developmental' in that they make explicit attempts to influence the shape and direction of economic activity within and across their borders. Without question, all the successful newly industrializing economies of the post-1945 period – from Japan onwards – can be regarded as developmental states although, as we showed in Chapter 4, this does not mean that they have all operated in the same way. More broadly, two sets of political forces have been especially significant in the last few years. One is the spread of 'deregulatory' forces as access to national markets has been opened up, initially to trade flows but, more recently, to foreign investment flows. The other is the proliferation of regional trade agreements which, in effect, shift the regulatory processes to a different scale.

- The role of *technology* as a fundamental enabling force in the internationalization and globalization of economic activities. As we demonstrated in Chapter 5, technological change is at the dynamic heart of all economic growth and development. It is, essentially, an evolutionary, learning process which occurs very unevenly through time and through space. The cumulative influence of small, incremental changes tends to be overshadowed by the massive radical changes involved in the periodic creation of entirely new technoeconomic paradigms which drastically shape and reshape both economy and society. The consensus view is that the current transformation is based upon the convergent streams of communications and computer technologies into a single stream of information technology. There is less consensus over the broader organizational effects of such technologies in an 'after-Fordist' world other than that various forms of flexibility are increasingly apparent. But the underlying message of Chapter 5 is the *social* nature of technological change.

Figure IV.1 'maps' the major *interconnections* between the broad patterns of change within the global economy, the major processes creating such changing patterns and, most importantly, the dynamic interaction between different geographical scales – notably the global, the national and the local – as indicated by the double-headed arrows connecting the individual boxes. The central message of Figure IV.1 is that the processes of globalization are not simply unidirectional, for example from the global to the local, but that all globalization processes are deeply embedded, produced and reproduced in particular contexts.[1] As Thrift (1990a, p. 181) argues, the global is not some *deus ex machina*. Rather, the 'local and the global intermesh, running into one another in all manner of ways'.

Hence, the specific assemblage of characteristics of individual nations and of local communities will not only influence *how* globalizing processes are experienced but also will influence the *nature* of those processes themselves. We must never forget that all 'global' processes originate in specific places. This is a theme which has been woven through each of the chapters in this book. We have repeatedly emphasized the strongly localized nature of economic activity, including that of technological change and the continuing significance of 'place' to the nature and behaviour of transnational corporations. Both nation-states and local communities are 'containers' of distinctive cultural, social, and political institutions and practices.

In both developed and developing countries, of course, the real *effects* of globalizing processes are felt not at the aggregate level of the national economy but at the *local* scale: the communities within which real people live out their daily lives. It is at this scale that the physical investments in economic activities are actually put in place, restructured or closed down. It is at this scale that most people make their living and create their own family or household communities. But, as Figure IV.1 shows, although the effects of globalizing processes on local communities may be direct they are, more commonly, 'refracted' through the medium of the national context within which the particular local community is embedded. The process is analogous to the way in which light rays become refracted or bent as they pass from one medium to another (just as a stick partly immersed in water appears bent rather than straight).

Problems of adjustment

The highly interconnected and uneven nature of the processes of globalization need to be borne in mind continuously as we consider how places adjust to the forces of change. In Chapter 13 we examine the problems which face countries and communities at different geographical scales and at different levels of economic development in their attempts to adjust to global change. A particular emphasis is placed upon the *employment* implications of global shifts, because it is largely through the incomes earned from employment (including self-employment) that levels of material well-being are determined. The question *'where will the jobs come from?'* is a crucial one throughout the world and one which has come to occupy the centre stage of much political debate at international, national and local levels.

In attempting to unravel this question in terms of the kinds of changes discussed throughout this book we find a very complex picture. The major employment changes that have been occurring in both developed and developing economies are the result of an intricate interaction of processes. Job losses in the developed market economies, for example, cannot be attributed simply to the relocation of production to developing countries. Although this is undoubtedly a factor, there is far more to it than this. What is clear, however, is that the industrialized economies face major problems of adjusting to the decline in manufacturing jobs. Nevertheless, the problems facing developing countries in a global economy are infinitely more acute. The spectacular success of a small number of NIEs should not blind us to the fact that the majority of developing countries face enormous problems of economic survival in an increasingly globalizing economic system.

Finally, in Chapter 14, we address, albeit very briefly, the question of international or global economic governance. Within the general governance context we focus on two specific issues, both of which are fundamentally important to the concerns of this book. The first addresses the problems posed by the nature of the international financial system which, as we see in Chapter 12, is both highly volatile in its operations and also unpredictable in its effects. The second governance issue discussed in Chapter 14 relates to international trade and, especially, to the contentious questions of the relationship between free trade and labour standards and between free trade and the environment.

Note for further reading

1. See Sayer (1989), Thrift (1990a), Dicken, Peck and Tickell (1997).

CHAPTER 13

Making a living in the global economy

Introduction: where will the jobs come from?

In this chapter our emphasis moves from describing the patterns of global shifts in economic activity and from explaining the underlying causes of those shifts to their *effects* on people. Such effects are manifold. The primary focus of this chapter is on the impact of the internationalization and globalization of economic activity on *employment opportunities*. There is a very good reason for adopting this specific focus.

The key to an individual's or a family's material well-being is *income*. The major source of income (for all but the exceptionally wealthy) is employment, or self-employment. In turn, there is no doubt that the complex, interlocking processes of internationalization and globalization exert a major influence on the structure of employment opportunities across the world. The position adopted in this book is that 'global shift' (used as a shorthand term to encompass the whole set of internationalization and globalization processes) is a *structural* phenomenon: a deep-seated *secular* trend. But the world economy, and its constituent parts, is also affected by *cyclical* forces: the roller-coaster of booms and slumps; prosperity and recession. The peaks and troughs of such business cycles, through their effects on investment (and disinvestment) decisions, tend to amplify structural change. This makes it very difficult to distinguish unambiguously between structural and cyclical forces.

Since the end of the Second World War, as we saw in Chapter 2, the world economy has experienced enormous cyclical variation in economic activity: the unparalleled growth of the long boom which lasted from the early 1950s to the mid-1970s; the deep world recession of the second half of the 1970s and the early 1980s; the impressive economic recovery of the later 1980s; the renewed recession of the early 1990s; the uncertainty of the prospects for the new millennium.

Underlying global cyclical trends, therefore, are *global structural changes* associated with the increasing internationalization and globalization of economic activity. The world economic map has become much more complicated than it was only forty or fifty years ago. Although world production, trade and investment are still dominated by the developed market economies, the position of individual industrial nations has changed dramatically. The United States is no longer the clear and undisputed industrial leader as it was in the immediate postwar period. Its hegemonic position was eroded initially by the resurgent European economies (notably Germany) but more recently, and comprehensively, by the spectacular rise of Japan as a world industrial power. The global economy is now *multipolar*.

At the same time, new centres of production have emerged in what had been, historically, the periphery of the world economy. The emergence of some newly industrializing economies is clear evidence that there has been a substantial geographical shift of economic activity away from the core, even though the extent of the shift is less than popular opinion tends to believe. The complexity of global change has rendered simplistic notions of 'core' and 'periphery' less useful or capable of capturing the nature of today's global economy. The world is more a 'mosaic of unevenness in a continual state of flux' than a simple dichotomous structure of core and periphery.

All parts of the world face major problems of adjustment to these far-reaching changes. But the nature of the problem, and certainly its perception, varies according to each country's position in the global system. In this respect, the view from the older industrialized countries is very different from the view from the newly industrializing economies and different again from the least industrialized countries. But position in the global economy is only part of the picture. It is far too simplistic to 'read off' a country's or a region's problems (and solutions) solely from its place in a global division of labour. Internal circumstances – cultural, social, political as well as economic – are of enormous importance. Nevertheless, there are problems which affect older industrialized countries, newly industrializing countries and least developed countries in different ways as *groups* of countries as they grapple with the repercussions of global economic change and attempt to adjust to its employment impact.

The question 'Where will the jobs come from?' faces all countries, whatever their position in the global economy. We face a truly desperate employment crisis at the global scale. In the mid-1990s, for example, the ILO estimated that there were roughly 34 million workers unemployed in the industrialized world, together with a further 15 million workers in involuntary, part-time employment. The number of unemployed in the seven leading industrialized economies virtually doubled between 1979 and 1995, from 13 million to 24 million.

Serious as the unemployment position is in the industrialized nations it pales into insignificance compared with the problems of most developing countries, particularly the least industrialized countries. At least in older industrialized countries the growth of the labour force is now easing. Only 1.1 per cent of the projected growth of the global labour force between 1995 and 2025 will be in the high-income countries. In most developing economies, on the other hand, extremely high rates of population growth mean that the number of young people seeking jobs will continue to accelerate for the foreseeable future. As Figure 13.1 shows, the low-income countries account for

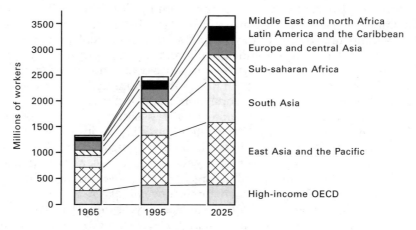

Figure 13.1   The distribution of the global labour force
Source: Based on World Bank (1995, Table 1.1)

a growing share of the global labour force. Indeed, more than two-thirds of the projected growth in the global labour force will occur in such countries:

> In the global economy, with a workforce of 2.4 billion, 125 million workers are formally unemployed, and a total of 750 million people are either unemployed or underemployed. Yet 200 million children are forced to work. 400 million new jobs need to be created in the next ten years merely to absorb newcomers to the labour market. *Unemployment is at the heart of the challenge facing the global economy.*
>
> (FIET, 1996, p. 16)

Not surprisingly, therefore, there has been a major resurgence of international concern about the employment issue. For example, the G7 group of industrialized countries now holds 'job summits' (the first took place in Detroit in 1994). The World Bank devoted its *1995 World Development Report* to the topic 'Workers in an integrating world'. The European Commission publishes regular surveys of employment in Europe. Most comprehensively of all, the International Labour Organization produces an annual *World Employment Report*. The whole question of employment and unemployment – its dimensions, the theories put forward to explain it, the solutions suggested for its alleviation – is far too complex and large a subject to be covered adequately here. No attempt is made to do this. The more limited aim is to show how the processes of global shift have contributed towards the employment and unemployment problems facing countries and communities occupying different positions in the global division of labour.

Employment problems of the older industrialized economies

Trends in employment, unemployment and income distribution

General trends in employment and unemployment

The major general trend, since at least the 1960s, has been for employment in services to grow far more rapidly than employment in manufacturing. Even so, it was only during the 1970s that manufacturing employment actually declined in absolute terms in the major European economies other than the United Kingdom, where manufacturing employment began its steep fall after 1966. It was this kind of experience which stimulated the intense – but often highly confusing – debate over what came to be called *deindustrialization*.[1]

In terms of total *employment*, the experience of the United States, the European economies and Japan has been very different, as the figures for the 1980–93 period show (European Commission, 1996). Total employment grew by almost 1.5 per cent per year in the United States, by a little over 1 per cent per year in Japan but by less than 0.5 per cent per year in the European Union. This translates into a very much higher level of employment growth in the United States than in Japan and, especially, Europe. In all three cases, the major employment growth occurred in services:

> In Europe, just over 18 million additional jobs were gained in services over the 13 years, not so many less than in the US, where almost 22 million extra jobs were created, despite the much slower rate of employment growth overall (in proportionate terms, the increase was 25 per cent in Europe, 33 per cent in the US). Similarly, the expansion of jobs in services was only slightly less than in Japan (where they increased by 28 per cent).
>
> (European Commission, 1996, pp. 101–02)

In both Europe and the United States, the number of people employed in manufacturing declined: by almost 0.4 per cent per year in Europe and by almost 0.2 per

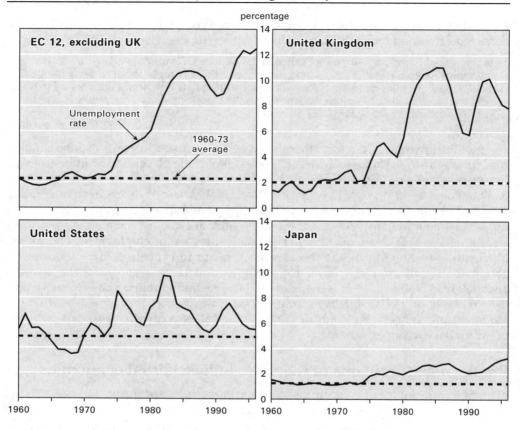

Figure 13.2   Unemployment rates in 1960–96 compared to the average rate for 1960–73: the United States, Europe, Japan
Source: Based on ILO (1997, Figure 3.1)

cent per year in the United States. In contrast, manufacturing employment growth continued in Japan (by a little more than 0.2 per cent per year). Overall, Europe's employment performance was very significantly worse than that of either the United States or Japan:

> In industry and agriculture combined . . . the number employed in Europe fell by just over 13 million over these 13 years, an average loss of 1 million jobs a year. In the US, the number declined by only just under 2 million and in Japan there was an increase of just over ½ million. In quantitative terms, therefore, the poor overall performance of the European economies in expanding employment relative to the US and Japan owes much more to the scale of job losses in the primary and secondary sectors than to the low rate of net job gains in services.
>
> (European Commission, 1996, p. 103)

Compared with the 1960s and early 1970s, *unemployment rates* in the industrialized countries have increased dramatically. Figure 13.2 shows the trends for the 1960–96 period. The graph shows as a horizontal line the average unemployment rate for the 1960–73 period (the so-called 'golden age of growth' referred to in Chapter 2). During that period, unemployment rates were highest in the US and lowest in Japan, with the

EC12 and the UK falling in between. But the post-1973 experience was very different. Figure 13.2 shows in stark terms the very unfavourable unemployment performance of Europe. Whereas both Japan and the US experienced an increase in the general level of their unemployment – more so in the case of the US – the European experience was appalling, with average unemployment rates in the EC12 rising to above 12 per cent in the early 1990s compared with less than 4 per cent in Japan and a little over 5 per cent in the US. Within these aggregate figures, there is a particular problem for those people who have been out of work for long periods of time. Long-term unemployment continues to be an especially serious problem in Europe.

The selective nature of unemployment

Although the general level of unemployment, including long-term unemployment, remains very high in most industrialized countries, its actual incidence is extremely uneven. Job loss is a *selective* process:

- different *social groups* experience different levels of unemployment
- unemployment tends to be *geographically uneven* within countries.

The socially differentiated nature of unemployment has several dimensions. For example, males aged between 25 and 54 years, with a good education and training, are far less likely to be unemployed, on average, than women, younger people, older workers and minorities. Most of these latter categories tend to be unskilled or semi-skilled workers. The vulnerability of women and young people to unemployment reflects two major features of the labour markets of the older industrialized countries. First, the participation of *women* in the labour force – particularly married women – has increased dramatically. A large proportion of these are employed as part-time workers in both manufacturing and services, especially the latter. Second, *youth unemployment* during the 1980s partly arose from the entry on to the labour market of vast numbers of 1960s 'baby-boom' teenagers. In most industrialized countries, therefore, unemployment rates among the young (under twenty-five years) are roughly twice as high as that for the over twenty-fives. In some cases youth unemployment is three times higher than adult unemployment.

Unemployment tends to be especially high among *minority groups* within the population. In the United States in the 1980s, for example, unemployment among black youths was 150 per cent higher than among white youths. Similarly, unemployment rates among Hispanic youths were at least 50 per cent higher than among white youths. In western Europe the problem of minority group unemployment reflects the large-scale immigration of labour in the boom years of the 1960s. Relatively easily absorbed – indeed welcomed as 'guestworkers' – in the good times, the migrant workers now face enormous problems both in times of economic recession and also because of longer-term decline in the demand for certain kinds of worker. In continental Europe most of the migrant labour came from the Mediterranean rim – north Africa, Turkey, Greece, Portugal, Spain and southern Italy. In the United Kingdom, where the nature of the immigration was different because of Commonwealth obligations, most migrants came from south Asia (India, Pakistan, Bangladesh) and the Caribbean. In the United States the major sources of new migrants were Mexico and parts of the Caribbean.

As we shall see later, such migration has been of great importance for the countries of origin. For the European host countries, too, the migrants have performed an extremely significant role. In the 1960s there were severe labour shortages as the European economies grew very rapidly. One response was to recruit migrant labour on temporary contracts. The migrant workers were overwhelmingly young, male, unaccompanied – and unskilled. Their numbers grew spectacularly:

> foreign workers in West Germany rose from about 300,000 in 1960 to a peak of about 2.6 million at the beginning of the energy crisis in September 1973 . . . The number then in north-west Europe as a whole (excluding Britain) was about 6 million, and they comprised about 30 per cent of the labour force in Switzerland and Luxembourg, 10 per cent in France and West Germany, 7 per cent in Belgium and 3 per cent in the Netherlands.
>
> (Jones, 1990, p. 246)

Thus, unemployment tends to fall especially heavily on certain sensitive groups within the population of the older industrialized countries. In addition, unemployment tends to be *geographically differentiated*, a reflection of the locational trends in economic activity within individual countries. We have noted examples of such trends at various points in the preceding chapters. In Chapter 7, for example, we discussed the tendency for the specialized units of large TNCs, to display specific locational preferences and to create a particular map of employment opportunity. In the industry case studies we identified numerous instances of highly localized employment loss associated with the closure or contraction of large plants.

Within the older industrialized countries, three broad types of geographical change in employment patterns – and, therefore, in unemployment – have been apparent in recent years:

- *Broad inter-regional shifts* in employment opportunities, as exemplified by the relative shift of investment from 'Snowbelt' to 'Sunbelt' in the United States, from north to south within the United Kingdom.
- The *relative decline of the large urban-metropolitan areas* as centres of manufacturing activity and the growth of new manufacturing investment in non-metropolitan and rural areas. Such a trend is apparent in both North America and many parts of western Europe as part of a quite powerful decentralization tendency.
- The decline of employment, especially manufacturing employment, in the *inner cities of the older industrialized countries*. In virtually every case, the inner urban cores have experienced massive employment loss as the focus of economic activity shifted first from central city to suburb and subsequently to less urbanized areas. This inner-city dimension is not confined solely to declining regions, though it is more prevalent there. But even cities in the growth regions of the United States Sunbelt or southeast England (notably inner London) have suffered. Some of the highest unemployment rates of all, therefore, are to be found in the older inner-city areas. For particular social groups within the inner cities the rate of unemployment is substantially higher. Among young, minority group members of the population of inner cities unemployment rates of 60–70 per cent are not unknown.

Thus, the phenomenon of deindustrialization is most dramatically experienced in the older industrial cities of all the developed market economies as well as in those broad regions in which the decline of specific industries has been especially heavy. The physical expression of this deindustrialization is the mile upon mile of industrial wasteland; the human expression is the despair of whole communities, families and individuals whose means of livelihood have disappeared. One outcome of these cataclysmic changes has been the growth of an *informal* or *hidden economy*, a world of interpersonal cash transactions or payments in kind for services rendered, a world much of which borders on the illegal and some of which is transparently criminal.

Of the leading industrialized countries only Japan appears not to have experienced the kinds of unemployment problems we have been discussing, or at least not to the same degree. Indeed, while virtually all the other industrialized countries suffered from labour surplus, Japan was experiencing labour shortages. This reflected Japan's marked success as both a domestic and an export economy. It also reflected some particular characteristics of the Japanese labour market, including the practice of lifetime employment in the larger companies (though not in the myriad of small firms). There is a considerable degree of 'hidden' unemployment in the Japanese economy.

However, Japan's older industrial regions have certainly been affected by job losses. By the end of the 1980s, too, it seemed that Japan might well begin to catch the 'western disease' of higher unemployment in the 1990s. The Japanese unemployment rate, though minuscule compared with that in North America and, especially, western Europe, is now more than twice as high as in the early 1970s. Demographic changes of the 1970s – the Japanese 'baby-boom' – will bring more young workers on to the labour market in the 1990s. As a result of these developments, we are now beginning to see an unravelling of the lifetime employment system in Japan's large firms. Although big Japanese firms are still reluctant to make workers redundant directly there is clear evidence of informal pressure on older 'salarymen' to leave.

The widening income gap in the older industrialized countries

A third significant trend in the jobs picture in the older industrialized countries concerns changes in the ratio of earnings of the highest- and the lowest-paid segments of the labour force. For the first 25 years or so after the Second World War, the general trend was for the earnings gap between the top and the bottom segments of the labour force to narrow whilst, at the same time, the overall level of per capita income increased substantially. In other words, most people became better off. During recent years, however, this trend towards reducing inequality has been reversed, especially in the United States and the United Kingdom but also in some other countries as well, as Figure 13.3 shows. In 1995, the ratio of the earnings of the highest segment of the labour force to that of the lowest segment rose in the United States from 3.2 to 4.4 and in the United Kingdom from 2.4 to 3.4. The average income of the top 5 per cent of US households was roughly seven times that of the bottom 40 per cent of households in the early 1970s. In the mid-1990s, the top 5 per cent earned on average ten times more than the bottom 40 per cent.

The pattern is more mixed across other industrialized countries. It is apparent, for example, that the same degree of increasing income dispersion within the labour force did not

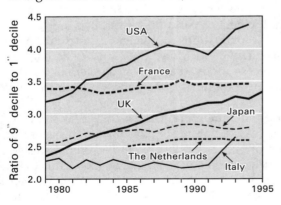

Figure 13.3 Trends in the dispersion of earnings for male workers, 1979–95
Source: Based on ILO (1997, Figure 3.5)

occur in many of the continental European countries. In some cases, indeed, the gap narrowed rather than widened. On the other hand, these countries have experienced much higher levels of unemployment than the United States in particular and even the United Kingdom. This suggests that labour market adjustments are occurring in different ways in different countries. In the United States and the United Kingdom adjustment has been primarily in the form of a relative lowering of wages at the bottom of the scale (i.e. of mainly unskilled workers); in other cases, such wage levels may have been maintained at the expense of jobs.

Causal factors: a recapitulation

The profound problem of unemployment which has gripped the older industrialized countries has no simple explanation (and, therefore, no simple, magic cure). The fact that, within the generally high level of unemployment, there are very substantial differences both between individual countries and also between parts of the same country suggests the operation of both general and specific forces. The most general explanation of an overall high level of unemployment in the older industrialized countries between the early 1970s and mid-1980s and, again, in the early 1990s was the effect of world recession. Recession, whatever its causes, drastically reduces levels of demand for goods and services. Thus, the ILO claims that the bulk of unemployment in the older industrialized countries as a whole is *demand-deficient* unemployment. But the general force of recession does not explain the spatial variation in unemployment between and within countries. In fact, a whole set of interconnected processes operates simultaneously to produce the changing map of employment and of its reverse image, unemployment. In this chapter we are concerned with the structural effect of 'global shift', the processes of internationalization and globalization, on employment. We cannot provide a quantitative assessment of this influence; we can merely suggest the kinds of influence involved. Let us now recapitulate the ways in which processes may help to explain the current situation in the older industrialized countries given the specific circumstances (social, cultural, institutional) which exist in individual countries.

Technological change

Technological developments in products and processes are widely regarded as being a major factor in changing both the number and the type of jobs available. In general, product innovations tend to increase employment opportunities overall as they create new demands. On the other hand, process innovations are generally introduced to reduce production costs and increase productive efficiency. They tend to be labour-saving rather than job-creating. Such process innovations are characteristic of the mature phase of product cycles and became a dominant phenomenon from the late 1960s onwards.

The general effect of process innovations, therefore, is to increase labour productivity and to permit the same, or an increased, volume of output from the same, or even a smaller, number of workers. Each of our industry case studies demonstrated this tendency. But, again, the impact of such technological change on jobs tends to be uneven. In most cases, it has been the semi- and unskilled workers who have been displaced in the largest numbers. It is manual workers rather than professional, technical and supervisory workers whose numbers have been reduced most of all, although the spread of the new information technologies is changing this situation as we saw in Chapter 12.

There can be little doubt that changes in process technology have adversely affected the employment opportunities of less skilled members of the population. The most vulnerable have been the young, the old and the members of minority groups in the older industrialized countries. However, there has been much debate about the overall contribution of technological change to unemployment. Some recent writers, such as Rifkin (1995), have argued that the 'end of work' is nigh and that much of this is due to the job-displacing effects of technological change. On the basis of his calculation that three out of four US jobs (including both white- and blue-collar jobs) could be automated, and that thousands of jobs have already disappeared from major US corporations, Rifkin predicts that hundreds of millions of workers will be without jobs by the middle of the next century.

But, the ILO is highly critical of the apocalytic view expressed in this kind of literature:

> The credibility of such sensational predictions is very limited since they are based on an invalid double generalization. On the one hand, the case of particular large corporations is proposed as representative of the whole economy; on the other, the direct labour-saving impact in production processes is proposed as the only consequence of technical change . . . the analyses make no allowance for the indirect effects of technical change and the jobs that can be created by the growth of new products and industries. From a static point of view it is simple to point out the adverse impact of labour-saving innovation, but from a proper dynamic perspective all the indirect effects of innovation have to be taken into account.
>
> (ILO, 1997, pp. 17–18)

However, although technological change may continue to create, in net terms, more jobs than it destroys, the problem is the actual *distribution* of such new jobs in relation to those destroyed. Although, as Tobin (1984, p. 83) points out, 'human labour in general has never yet become obsolete' he also notes that 'new technologies often displace particular workers and work hardships upon them, and upon whole industries and regions'.

Global restructuring and the transnationalization of production

Fundamentally, the highly uneven pattern of employment and unemployment change reflects the outcome of the complex industrial restructuring processes which have been occurring at the global scale since the mid-1970s. However, the problems created for particular social groups and particular geographical areas were greatly exacerbated by the deep economic recession and by demographic trends in the size and composition of the labour force. Restructuring of operations by business enterprises, especially the large TNCs, is a continuous process, as we argued in Chapter 7. But its effects are more keenly felt in times of recession than in times of growth. At the same time, recession intensifies the efforts of enterprises to rationalize and restructure their activities to sustain their profitability.

As we have seen, a key feature of the postwar period has been the development and intensification of global competition in virtually every industry, both old and new. In such circumstances, business enterprises adopt a variety of strategies to ensure their survival and to grow. Geographical and product diversification have greatly increased as have efforts to cut back production costs through processes of intensification, investment in new, labour saving technology and rationalization. New, more flexible work practices are being adopted on an extensive scale. Such flexibility takes two major forms:

- *Numerical* flexibility, whereby firms frequently adjust the number of workers in their labour force to meet changes in market conditions. Often this involves the more extensive use of part-time workers.
- *Functional* flexibility, in which the kind of work, and the number of tasks, are changed. Rigid divisions between tasks are broken down; a degree of multiskilling is created.

Within many major corporations in the west, especially in the United States and the United Kingdom, the late 1980s and early 1990s witnessed the phenomena of 'downsizing' and 'corporate re-engineering'. In both cases, the aim was drastically to reduce the numbers of workers employed within companies.

Most significant, however, is the fact that many of these restructuring processes have themselves become increasingly international, or even global, in their extent. Virtually all large firms, and many medium-sized ones, have become *transnational* enterprises. They engage in international, rather than merely domestic, production, whether directly through the establishment of overseas branch plants or indirectly through international alliances and subcontracting. As such, their strategies impinge upon both their home country and also the host countries in which they operate. But the process itself is very complex.

Both the reasons for engaging in international production and also the effects of such operations vary greatly between different types of firm in different industries and according to the characteristics of home and host countries. But there is no doubt that a good deal of the changing level and distribution of employment opportunities in the older industrialized countries is associated with the actions of TNCs. As we have seen, the large TNCs in particular have been altering the relative balance of their activities between home and overseas locations. Some of this involves the location and relocation of production units in developing countries, but the majority is still within the older industrialized countries themselves. In general, international production networks have become far more extensive. In one sense, therefore, TNCs can indeed be blamed for some of the loss of manufacturing jobs in their home countries. But as we argued in Chapter 8, the question is very complex. What would have been the realistic alternative to such overseas investment? Could the specific investment have been made at home? In an increasingly global economy is it possible for firms to opt out of international production? Would the absence of overseas investment have increased employment at home or would it have resulted in even greater employment loss?

Sweeping generalizations – pro or con – cannot be made. Nevertheless, where the volume of outward investment is very high in relation to domestic production, there are bound to be adverse effects on the domestic economy and, especially, on employment. The job losses associated with overseas investment have been both geographically and occupationally uneven. The older the plants the more likely they are to be replaced by new plants in different locations. Even relatively new plants may be closed if their efficiency falls below the target set by comparison with other plants in the transnational network. Generally, it has been the basic production activities, especially semi-skilled and unskilled operations, which have shown the greatest propensity to be relocated overseas. Where the displaced plants are either very large individually or where there are substantial localized concentrations the effects are especially severe. Where, as in the case of the automobile industry, linkages with other industries are especially strong the 'knock-on' effects of transnational restructuring can be very severe indeed.

Effects of import penetration from developing countries

One of the most contentious issues in the debate over unemployment and de-industrialization in the older industrialized countries is the contribution of imports of manufactured goods from developing countries, notably the NIEs. The rapid development of manufacturing production in a small number of NIEs, and their accelerating involvement in world trade, has been a major theme of this book. It is one of the most striking manifestations of global shift in the world economy. In some cases, notably Singapore and Taiwan, a major driving force has been the involvement of foreign TNCs. In other NIEs, such as South Korea and Hong Kong, the direct participation of TNCs has been relatively small.

The common element in virtually all the NIEs, however, is a very strong government involvement in guiding or directing the economy. The adoption and pursuit of vigorous export-oriented industrial strategies were discussed in detail in Chapters 3 and 4 and in the industry case studies of Part III. By enhancing their initial comparative advantage of a large and cheap labour force through the provision of basic physical infrastructure (including EPZs), by substantial financial and tax incentives and, most of all, by their 'guarantee' of a malleable labour force, the NIE governments created a powerful new force on the world economic scene.

The basic question is: how far has the industrialization of these fast-growing economies – as expressed through *trade* – contributed towards the deindustrialization of the older industrialized countries, to the increased levels of unemployment and to the pauperization of workers at the bottom end of the labour market?[2] As we showed in Chapter 2, imports from developing countries represent only about 20 per cent of total imports into developed countries, although they tend to be concentrated into specific sectors (for example, textiles, clothing, footwear, toys, electronics).

Wood (1994, pp. 8, 11, 13) argues that trade with developing countries has had a considerable impact, especially in widening the gap between skilled and unskilled workers:

> Countries in the South have increased their production of labour-intensive goods (both for export and domestic use) and their imports of skill-intensive goods, raising the demand for unskilled but literate labour, relative to more skilled workers. In the North, the skill composition of labour demand has been twisted the other way. Production of skill-intensive goods for export has increased, while production of labour-intensive goods has been replaced by imports, reducing the demand for unskilled relative to skilled workers . . . up to 1990 the changes in trade with the South had reduced the demand for unskilled relative to skilled labour in the North as a whole by something like 20 per cent . . . Thus expansion of trade with the South was an important cause of the deindustrialization of employment in the North over the past few decades. However, it does not appear to have been the sole cause . . .

However, the general conclusion of the ILO is that the results of the many economic studies of the relationship between trade and wage and income inequality in the older industrialized countries are 'inconclusive':

> although international trade has contributed to income inequality trends to some extent, it has not played a major role in pushing down the relative wage of less-skilled workers . . . [in the case of the United States] . . . employment patterns in industries least affected by trade moved in the same direction as those in trade-affected manufacturing industry, increasing the share of high-wage employment. This pattern of change in the employment structure is not well explained by the argument relying on the trade effect.
>
> (ILO, 1997, pp. 71, 73)

The basic problem in all the individual factor explanations – technology, transnational corporate restructuring, import penetration – is the fact that they are treated

Table 13.1   The positive and negative effects of globalization processes on employment in older industrialized countries

Positive effects	Negative effects
Cheaper imports of relatively labour-intensive manufactures promote greater economic efficiency through the demand side while releasing labour for higher productivity sectors	Particularly in relatively labour-intensive industries, the rising imports from developing countries, together with competition-driven changes in technology and other factors, lead to inevitable losses in employment and/or quality of jobs, including real wages. This increases inequality between skilled and unskilled workers, and causes extreme redeployment difficulties
Growth in developing countries through industry relocation and export-generated income leads to 1) increased demand for industrialized country exports; and 2) shifts in production in industrialized countries from lower to higher-valued consumer goods, to more capital- and/or skill-intensive manufacturing and services	Employment gains from rising industrialized country exports are unlikely to compensate fully for the job losses, especially if 1) industrialized country wages remain well above those of the NIEs and other emerging developing countries; and 2) the rates of world economic growth are relatively low, and/or excessively concentrated in east and southeast Asia
Employment growth and job quality improvement for skilled workers are likely to be significant in the short and medium term, even though in the long run the effects are unclear	The employment growth and job quality improvement for skilled workers will dwindle in the long run, as a result of relatively cheaper and more productive skilled labour in the NIEs
Relocation of production and/or imports causes negative short-term effects on workers but promotes labour market flexibility and efficiency through greater mobility of workers within countries (and, to a lesser extent, within regional economic spaces) to economic activities and areas with relative scarcities of labour	Increased trade will further reduce demand for unskilled labour. This exacerbates unemployment because, in a world of mobile capital, the industrialized countries no longer retain a capital-based comparative advantage

Source: Based on ILO (1996b, Table Int. 1).

independently of one another. It is as though changes in one of the variables are unrelated to the others. But this is clearly not the case. For example, although the *direct* effects of trade may be relatively small, the *indirect* effects may be larger because of the ways in which firms respond to the threat of increased international competition. They may, for instance, invest in labour-saving technologies to raise labour productivity and to reduce costs. This would appear as a 'technology effect' whereas the underlying reason for such technological change may be quite different. In fact, the decline in manufacturing employment in the older industrialized countries is primarily the result of increased productivity. But this has affected the labour force differentially with the greatest relative losses of jobs and of income falling on the least-skilled, least-educated workers.

Overall, then, the processes of globalization which are, themselves, complex as we have tried to show in previous chapters, produce complex employment effects. Table 13.1 summarizes the major positive and negative interpretations of the effects of globalization on the older industrialized countries.

Policy responses

Removing obstacles to adjustment

The period of low unemployment during the 1960s and early 1970s in the industrialized countries was associated with growing demand for manufactured products and also, in most cases, with a particular system of labour regulation. 'Based on the principle of full (male) employment, this system sought to maintain a balance between the normalization of aggregate demand, containment of class conflict, expansion of social welfare, and regulation of social reproduction' (Peck, 1996, p. 240).[3] However, the intensification of globalizing processes, the relative shift of production away from the older industrialized economies and intensified global competition have dramatically changed the labour policy environment. The over-riding concern is now with the problems of *adjustment* to intensified global competition and global shift.

Domestic policies vary from one country to another. For example, there has been a basic difference of approach between the United States, which has financed economic growth – and employment growth – through running a huge budget deficit, and most European countries, which have pursued strongly deflationary policies. In both cases, however, increasingly the focus has become that of *removing obstacles* to the efficient production or supply of goods and services in the context of keeping down inflation. In Europe, especially, cutting back on public expenditure has been a major strategy. This approach has been carried to its greatest extreme in the United Kingdom but it is one which now affects virtually all EU member states in their attempts to meet the qualifying criteria for the single European currency (see Chapter 3).

On the whole, labour is occupationally and geographically less mobile in Europe than in North America, for reasons which lie deep in social and economic history and in cultural attitudes. But there are many real obstacles to the geographical mobility of labour. Not only must deep community and family ties be broken but also the nature of the housing market may make relocation extremely difficult. For example, selling a house in an area of industrial decline may be virtually impossible; getting accommodation at a realistic price in an area of growth may be equally difficult. Rigidities in the public housing sector may also inhibit geographical mobility. The whole question of the geographical mobility of labour in the older industrialized countries is more complex than is often supposed, especially by politicians.

Nevertheless, a new conventional wisdom has emerged which is very different from the old. Its essence is that of removing what are seen to be *rigidities* in the labour markets of the older industrialized countries. Its aim is to make labour markets more *flexible*, in tune with what are seen to be the dominant characteristics of a 'post-Fordist' world (see the discussion in Chapter 5). The 'flexibilization' of labour markets through deregulation, including greatly increased pressures and restrictions on labour organizations, the drastic cutting back of welfare provisions and the move away from welfare towards 'workfare', has gone furthest in the United States. Its apparent success in continuing to create large numbers of jobs (albeit with the widening of income gaps) stimulated the United Kingdom government to move along the same path.

As yet, the countries of continental Europe have not moved as far, or as fast, down the flexibilization path. Most European governments are concerned that the social costs of reducing unemployment using the United States model may be politically unacceptable in a system in which the social dimension of the labour market is very strongly entrenched. But there are clear signs of change as governments become increasingly concerned about the financial costs of sustaining existing practices and the continuing loss of competitive edge. As a result, a variety of labour market measures, employed in various combinations in different European countries, has emerged.[4] These include

- the use of more temporary and fixed-term contracts
- the introduction of different forms of flexible working time
- moves to encourage greater wage flexibility by getting the long-term unemployed and the young to take low-paid jobs
- increased vocational training to provide more transferable skills
- reforms in state employment services
- incentives to employers to take on workers
- measures to encourage workers to leave the labour market
- reductions in the non-wage labour cost burdens on employers
- specific schemes to target the long-term unemployed

> Such policies are being pursued by governments with varying degrees of enthusiasm. The trends indicate most of the EU member states are moving towards a deregulated, flexible labour market a little closer to that of the US, although there is still no firm evidence this will prove any more effective.
> (Taylor, in *The Financial Times*, 22 August 1995)

Quite apart from efforts by governments to remove what they see as obstacles to economic expansion without increasing inflation, four types of policy have received a great deal of attention in the older industrialized countries:

- the promotion of small firms
- the development of new technologies
- the attraction of foreign investment
- the protection of certain industries against imports.

Each of these merits a chapter in itself; here our concern is simply to outline very briefly their implications for employment creation.

Promoting small firms

There has been a remarkable swing of the pendulum in attitudes towards firms of different sizes. In the 1960s the key to economic (and employment) growth was seen to be the very large firm and the very large plant. Only in such organizations, it was argued, could economies of scale be achieved to enhance competitiveness. Many governments encouraged mergers between enterprises towards this end. By the 1970s, in complete contrast, disillusionment with the large enterprise had set in and the employment panacea was seen to be the small firm with its supposed dynamism and lack of rigidity. This view was strongly reinforced by claims that the vast majority of all net new jobs created in the United States during the 1980s had been in firms employing fewer than twenty workers.

A number of writers make the connection between the development of flexible production methods, new ways of organizing production (which we discussed in Chapter 5) and the dynamism of small firms. Such a combination, it is argued, is leading to the development of 'new industrial districts' and 'new industrial spaces'. However, although such new industrial spaces are, indeed, an important development, their significance has been exaggerated and overgeneralized on the basis of a small number of cases.

Reservations have also been expressed about the job-creating significance of small firms.[5] It is certainly true that a large number of jobs are located in small firms in most industrialized countries. However, it would take many thousands of small firms to replace the jobs lost through the rationalization processes of even a few large firms, let alone to create additional jobs. There is no doubt that small firms are an important source of growth in an economy. But it should not be forgotten that the majority of small firms are far from dynamic, that most depend upon large firms for their markets and that the failure rate of small firms is very high. Thus, the 'small firm fix' is likely to be limited in its impact on unemployment: 'small, per se, is neither unusually bountiful nor especially beautiful, at least when it comes to job creation in the age of flexibility' (Harrison, 1994, p. 52).

Developing new technologies

A similar caution is necessary towards the 'technological fix' often advocated as the salvation of areas in economic distress. Certainly innovation is the life-blood of any economy. Not only must large firms 'innovate or die' but so must entire nations. As we have seen at several points in this book, governments have become universally involved in attempting to stimulate technological developments within their own national territories, either directly or indirectly. But controversy has always raged over the employment implications of introducing new technology. Does new technology create or destroy jobs? If it creates new jobs are they different from the old jobs either in terms of skills required or in their geographical location? As noted already, historical experience seems to show that, *in aggregate*, new technologies create more jobs than they destroy, at least over the longer term. This seems to occur because they create new demands for goods and services, many of which could not have been foreseen.

Petrella (1984, p. 353), however, warns against too ready an adoption of what he terms the technological fallacy,

> that simply by injecting a large dose of technological innovation into the economies of a region or country, one will inevitably recover a high growth rate and, consequently, a fresh growth in employment, increase in purchasing power and new rising standards of living. There are three reasons why this is a fallacy. First, . . . the interpretations made of the relationships between technology and employment are vague and insubstantial in terms of theory and analysis. The real world is not quite as simple and linear as many would like it to be. Second – and this applies particularly to Western Europe – more technology does not always mean more growth and more growth does not necessarily mean more employment . . . Third, the current technological and industrial changes are taking place in Europe against a background of national 're-industrialisation' processes activated by countries in competition with one another, the result being that even the modest benefits that technology can bring in terms of employment are being eroded and lost.

Thus, the new wave of information technologies may well not create the number of jobs some of its advocates suggest. Much of the growth associated with the new technologies may well be 'jobless growth'. The new information technologies are also having a major effect on many service industries (see Chapter 12) and may well reduce the capacity of the services sector to absorb employees displaced from manufacturing as they have done in the older industrialized countries since the 1960s. What is certain is that new technologies redefine the nature of the jobs performed, the skills required and the training and qualifications needed. They alter the balance of the labour force between different types of worker. To some writers the outcome is the deskilling of the labour force but this is by no means a universal outcome. Reskilling and multiskilling are also significant outcomes of technological change.

A further complication is that the new technologies, and the industries based upon them, will not necessarily emerge in the same geographical locations as the old industries. The terms 'sunrise' and 'sunset' industries themselves imply a geographical distinction (the sun does not rise and set in the same place!). This geographical dimension of technological change is apparent at both the international and the intranational scales. At the international scale, for example, western Europe is lagging behind both the United States and Japan in a number of key sectors, not least in the new information technologies. Within Europe, too, there are considerable technological gaps between individual nations, although these tend to vary by industry. In all cases, however, it is clear that much of the new economic activity based on new technologies has a different geographical profile from that of the obsolescent industries which are being displaced (and relocated overseas). The 'anatomy of job creation' is rather different from the 'anatomy of job loss'.

Attracting foreign investment

A third kind of adjustment policy pursued by the older industrialized countries in an attempt to cope with employment decline has been one of attracting foreign investment. As we saw in Chapter 8, rivalry for the investment favours of TNCs has become intense and often pursued at the highest governmental and state level. Prime ministers and presidents exert their influence either overtly or covertly to persuade the major TNCs to locate new, job-creating investment in their particular countries. Such efforts have been especially notable in industries such as automobiles, because of the large scale of the investments, and electronics, because of its high-tech nature. In recent years Japanese firms have become the prime target for both the US and western European enticement efforts. Whether such a policy is beneficial in the long term is a matter of debate, as we discussed in Chapter 8. From a political point of view, of course, the 'foreign investment fix' has the advantages of having a high profile and being relatively quick. Undoubtedly new jobs are created by inward investment but whether there is a gain in *net* terms depends on its impact on existing firms and on the effect of foreign investment on the country's technological development (see Chapter 8). In any case, there is a limited amount of internationally mobile investment to go round and since much of it will be in higher-technology industries anyway the number of jobs created may well be far less than in the past.

Protecting domestic industries

Running in parallel with attempts to attract foreign investment have been increasing measures to restrict imports of manufactured goods by most older industrialized countries. This 'protectionist fix' is especially contentious not only in terms of its likely effects on the older industrialized countries themselves but also globally because of its implications for the economic well-being of developing countries. The kinds of trade protection measures which governments can pursue were discussed in some detail in Chapter 3. Trade frictions have undoubtedly increased across the entire spectrum of countries. 'Managed' trade or 'strategic' trade policies have become more widespread in most of the older industrialized countries. Quite apart from the frictions between the United States and the EU in certain industries, and even between individual members of the EU, there have been two major targets for this new wave of protectionism: Japan and the Asian NIEs.

The crux of the issue is this: how far should those industries in which the older industrialized countries no longer have a comparative advantage be allowed to run down and be allowed to develop elsewhere? The neoliberal view is that the answer is self-evident. The older industrialized countries should simply move out of what are, for them, obsolete activities which can be manufactured more cheaply in developing countries and move to higher-technology products and into the more sophisticated service industries. But it is not as simple as this. The basic argument for protection against imports is that it is necessary to give domestic industry time to adjust: it provides a breathing space. There can be no doubt of the justification of such temporary measures in those sectors where international competition has intensified very rapidly. But if the breathing space is used, as it should be, to restore an industry's competitiveness or to shift into new activities, it will not necessarily preserve employment. As we have seen, new investment is likely to be labour-saving; new activities may be located in different places. It may also be the case that protection against imports in the sectors most sharply affected by NIE growth will not prevent an inevitable decline in employment in the older industrialized countries.

Clearly, the older industrialized countries continue to face considerable difficulties in adjusting to the intensified competitive environment of a global economy. Preserving, let alone creating, jobs for their active populations has become infinitely more complex in today's highly interconnected world. The imperatives which drive nation-states to strive to enhance their international competitiveness will not always be job-creating. In particular, all the industrialized economies face the task of alleviating the adjustment problems of declining industries and declining areas. However, the problems facing the affluent industrialized economies pale into insignificance when set beside the problems facing the world's developing countries.

Problems of developing countries in a globalizing economy

Heterogeneity of the developing world

In large part, though by no means entirely, the economic progress and well-being of developing countries are linked to what happens in the developed market economies. A continuation of buoyant economic conditions in the industrialized economies, with a general expansion of demand for both primary and manufactured products, would undoubtedly help developing countries. But the notion that 'a rising tide will lift all boats', while containing some truth, ignores the enormous variations that exist between countries. The shape of the 'economic coastline' is highly irregular; some economies are beached and stranded way above the present water level. For such countries there is no automatic guarantee that a rising tide of economic activity would, on its own, do very much to refloat them.

For the developing world as a whole, the basic problem is one of *poverty* together with a *lack of adequate employment opportunities*. More than twenty years ago, the problem was described in the following terms:

> More than 700 million people live in acute poverty and are destitute. At least 460 million persons were estimated to suffer from a severe degree of protein-energy malnutrition even before the recent food crisis. Scores of millions live constantly under a threat of starvation.
>
>   Countless millions suffer from debilitating diseases of various sorts and lack access to the most basic medical services. The squalor of urban slums is too well known to need further emphasis. The number of illiterate adults has been estimated to have grown from 700 million in 1962 to 760 million towards 1970. The tragic waste of human resources in the Third World is symbolised by nearly 300 million persons unemployed or underemployed in the mid-1970s.
>
> (ILO, 1976, p. 3)

Table 13.2 Variations in income and other indicators within the developing world

Country group		Per capita income ($)	Life expectancy at birth (yrs)	Infant mortality rate (per 1,000 live births)
Low income (51 countries)	mean	380	56	58
	range	80–720	38–73	25–163
Lower middle income (40 countries)	mean	1,590	67	36
	range	770–2,820	57–77	14–71
Upper middle income (17 countries)	mean	4,640	69	36
	range	2,970–8,260	54–78	6–89
High income countries (25 countries)	mean	23,420	77	7
	range	9,320–37,930	75–79	4–11

Source: Based on material in World Bank (1996, *World Bank Development Report, 1996*, Tables 1 and 3).

Twenty years later, the precise numbers may have changed but the basic dimensions of the problem surely have not. Indeed, the income gap between the rich and the poor has widened:

> In 1960, the richest 20% of the world's population had incomes 30 times greater than the poorest 20%. By 1990, the richest 20% were getting 60 times more. And this comparison is based on the distribution between rich and poor *countries*. Adding the maldistribution within countries, the richest 20% of the world's *people* get at least 150 times more than the poorest 20%.

(UNDP, 1992, p. 1)

Apart from the yawning gap between developed and developing countries as a whole, however, there are enormous disparities within the developing world itself. Table 13.2 follows the World Bank in distinguishing between three groups of developing countries based on income level. The weighted averages of three development indicators – per capita income, life expectancy and infant mortality – give some impression of the heterogeneity within the developing world as well as a stark indication of the gap between these countries and the high-income countries of the world. The income disparities are especially marked. The average per capita income of the fifty-one poorest countries in the mid-1990s was a mere $380 compared with $4640 in the upper middle-income group of developing countries and $23,420 for the high-income countries. This income gradient is, not surprisingly, reflected in the data for life expectancy and infant mortality.

Not only are the variations in well-being between developing countries much greater than those between industrialized countries but also variations between *different parts of the same developing country* tend to be much greater. In particular, the differential between urban and rural areas is especially great. United Nations data show that in Africa as a whole, 29 per cent of the urban population live in 'absolute' poverty compared with 58 per cent of the rural population. In Asia and Latin America the urban–rural differential is less but still substantial. In Asia, 34 per cent of the urban population and 47 per cent of the rural population live in absolute poverty; in Latin America the figures are 32 per cent and 45 per cent respectively (United Nations, 1996, p. 113).

Gilbert and Gugler (1982, pp. 23, 25) summarize the situation very concisely. On the one hand,

Urban poverty in the Third World is on a scale quite different to that in the developed countries . . . In the Third World city the relative poverty of the black Baltimore slum dweller is accentuated by absolute material deprivation. Some poor people in the United States suffer from malnutrition. Most of the poor in Indian cities fall into this category. Overcrowded tenement slums and too few jobs are abhorrent, but the lack of fresh water, medical services, drainage, and unemployment compensation adds to this problem in most Third World cities.

On the other hand, in the developing world,

cities are centres of power and privilege . . . Certainly, many urban dwellers live in desperate conditions . . . [but] . . . even those in the poorest trades reported that they were better off than they had been in the rural areas . . . The urban areas, and especially the major cities, invariably offer more and better facilities than their rural hinterlands.

(Gilbert and Gugler, 1982, pp. 50, 52)

Employment, unemployment and underemployment in developing countries

The basic problem: labour force growth outstrips the growth of jobs

Although the employment structure of Third World countries has undergone marked change (Table 13.3) the fact remains that most developing countries are predominantly agricultural economies. As Table 13.3 shows, an average of 69 per cent of the labour force in the lowest-income countries was employed in agriculture in 1990 (in some countries the figure was above 90 per cent) compared with only 5 per cent in the industrial market economies. Even in the upper middle-income group (in which most industrial development has occurred) agriculture employed around one-fifth of the labour force. In each category the relative importance of agriculture has declined even though in absolute terms the numbers employed in agriculture continued to grow. The balance of employment has shifted towards the other sectors in the economy: industry and services.

These broad sectoral changes in employment in developing countries have to be seen within the broader context of growth in the size of the labour force. The contrast with the experience of the industrialized countries in the nineteenth century is especially sharp. During that earlier period the European labour force increased by less than 1 per cent per year on average; in today's developing countries the labour

Table 13.3 The structure of the labour force in developing countries

Country group		Percentage of the labour force in					
		Agriculture		Industry		Services	
		1960	1990	1960	1990	1960	1990
Low income	mean	77	69	9	15	14	16
	range	54–95	18–94	1–18	2–31	3–37	–
Lower middle income	mean	71	36	11	27	18	37
	range	39–93	13–79	2–26	7–48	5–39	–
Upper middle income	mean	49	21	20	27	31	52
	range	8–67	6–51	9–52	16–46	19–69	–
High income	mean	18	5	38	31	44	64
	range	4–42	1–18	25–50	25–38	27–57	–

Note:
'Industry' includes mining, manufacturing, construction, electricity, water, gas.

Source: Based on World Bank (*World Development Report*, 1983, 1996).

force is growing at more than 2 per cent every year. Thus, the labour force in the developing world doubles roughly every thirty years compared with the ninety years taken in the nineteenth century for the European labour force to double. Hence, it is very much more difficult to absorb the exceptionally rapid growth of the labour force into the economy. The problem is not likely to ease in the near future because labour force growth is determined mainly by past population growth with a lag of about fifteen years. Virtually all the world's population growth – more than 90 per cent of it – since around 1950 has occurred in the developing countries.

There is, therefore, an enormous difference in labour force growth between the older industrialized countries on the one hand and the developing countries on the other. But the scale of the problem also differs markedly between different parts of the developing world itself. By far the greatest problem exists in low-income Asian countries, where the projected increase of 250 million in the labour force between 1975 and 2000 is twice that of the region's labour force growth rate between 1950 and 1975. Of course, pressure on the labour market is lessened where lower population growth rates occur. The basic dilemma facing most developing countries, therefore, is that the growth of the labour force vastly exceeds the growth in the number of employment opportunities available.

Formal and informal sectors in developing country labour markets

It is extremely difficult to quantify the actual size of the unemployment problem in developing countries. There are three main reasons for this. One is the simple lack of accurate statistics. A second is the nature of the unemployment itself, which tends to be somewhat different from that in the developed economies. A third reason is the structure of most developing economies, particularly their division into two distinctive, though closely linked, sectors: *formal* and *informal*.[6] Published figures in developing countries tend to show a very low level of unemployment, in some cases lower than those recorded in the industrial countries. But the two sets of figures are not comparable. Unemployment in developing countries is not the same as unemployment in industrial economies. To understand this we need to appreciate the strongly segmented nature of the labour market in developing countries:

- The *formal sector* is the sector in which employment is in the form of wage labour, where jobs are (relatively) secure and hours and conditions of work clearly established. It is the kind of employment which characterizes the majority of the workforce in the developed market economies. But in most developing countries the formal sector is not the dominant employer, even though it is the sector in which the modern forms of industry are found.
- The *informal sector* encompasses both legal and illegal activities, but it is not totally separate from the formal sector: the two are inter-related in a variety of complex ways. The informal sector is especially important in urban areas; some estimates suggest that between 40 and 70 per cent of the urban labour force may work in this sector (Gilbert and Gugler, 1982). But measuring its size accurately is virtually impossible. By its very nature, the informal sector is a floating, kaleidoscopic phenomenon, continually changing in response to shifting circumstances and opportunities.

In a situation where only a minority of the population of working age are 'employed' in the sense of working for wages or salaries, defining unemployment is thus a very different issue from that in the developed economies. In fact, the major problem in developing countries is *underemployment*, whereby people may be able to find work of varying kinds on a transitory basis, for example, in seasonal agriculture, as casual labour in workshops or in services.

The urban–rural dimension

Underemployment and a general lack of employment opportunities are widespread in both rural and urban areas in developing countries. There is a massive under-employment and poverty crisis in rural areas arising from the inability of the agri-cultural sector to provide an adequate livelihood for the rapidly growing population and from the very limited development of the formal sector in rural areas. Some industrial development has occurred in rural areas, notably in those countries with a well developed transport network. Mostly this is subcontracting work to small work-shops and households in industries such as garment manufacture. But the bulk of the modern industries are overwhelmingly concentrated in the major cities or in the export processing zones.

It is in the big cities that the locational needs of manufacturing firms are most easily satisfied. Yet despite the considerable growth of manufacturing and service industries in the cities the supply of jobs in no way keeps pace with the growth of the urban labour force. Not only is natural population increase very high in many de-veloping country cities but also migration from rural areas has reached gigantic dimensions. The pull of the city for rural dwellers is directly related to the fact that urban employment opportunities, scarce as they are, are much greater than those in rural areas.

In complete contrast to the older industrialized countries, therefore, where a growing *counterurbanization* trend was evident for some years, urban growth in most developing countries has continued to accelerate.[7] The highest rates of urban growth are now in developing countries where the number of very large cities has increased enormously. The sprawling shanty towns are the physical expression of this explosive growth. In the older industrialized countries, most industrial growth now occurs away from the major urban centres. In the developing countries, the reverse is the case: virtually all industrial growth is in the big cities. Like labour force growth in general, there is a stark contrast with the experience of the growing cities of the nineteenth-century industrial revolution:

> Whereas urbanisation in the industrialised countries took many decades, permitting a gradual emergence of economic, social and political institutions to deal with the problems of transformation, the process in developing countries is occurring far more rapidly, against a background of higher population growth, lower incomes, and fewer oppor-tunities for international migration. The transformation involves enormous numbers of people: between 1950 and 1975, the urban areas of developing countries absorbed some 400 million people; between 1975 and 2000, the increase will be close to one billion people . . . The rate of urban population growth in these countries is likely to decline after 1975, but it is expected to remain three to four times as high as the urban growth rates of the industrialised countries in this period.
>
> (World Bank, 1979, p. 72)

Labour migration as a 'solution'

Despite its considerable growth in at least some developing countries, manufacturing industry has made barely a dent in the unemployment and underemployment prob-lem of most developing countries. Only in the very small NIEs, such as Hong Kong

Table 13.4   Migrant workers' remittances as a percentage of home country exports, 1993

Home country	Remittances as per cent of exports
Morocco	48.7
Bangladesh	41.5
Greece	29.7
Portugal	24.9
Pakistan	23.5
Sri Lanka	21.8
Turkey	19.0

Source: Based on World Bank (1995, Tables 13, 17).

and Singapore – essentially city states with a minuscule agricultural population – has manufacturing growth absorbed large numbers of people. Indeed, Singapore has experienced a labour shortage and has had to resort to controlled in-migration while Hong Kong firms have increasingly located manufacturing production across the border in southern China. In all other cases, however, the problem is not so much that large numbers of people have not been absorbed into employment – they have – but that the *rate of absorption* cannot keep pace with the growth of the labour force.

One commonly adopted solution has been to 'export' labour to foreign countries. Although in general terms, labour is relatively immobile (certainly in comparison with capital) massive flows of international labour migration occur, often encouraged by developing country governments. In some cases, for example across the Mexico–United States border, much of this migration is illegal. The effects of such out-migration on the countries of origin can be very substantial, although they are not necessarily all beneficial. It is certainly true that out-migration helps to reduce pressures in local labour markets. It is also true that the remittances sent home by migrant workers make a very important contribution to the home country's balance of payments position and to its foreign exchange situation (as well as to the individual recipients and their local communities). Indeed, in many cases the value of foreign remittances is equivalent to a very large share of the country's export earnings as Table 13.4 shows. Other supposed benefits of out-migration include the learning of skills which, when the migrant returns, will help to upgrade the home country's economic and technological base.

The other side of the coin is less attractive for the labour-exporting countries. The migrants are often the young and most active members of the population. Further, as Jones (1990, p. 250) points out:

> growing familiarization with foreign consumption styles leads to disdain for domestic products and a growing dependence on expensive foreign imports . . . returning migrants are rarely bearers of initiative and generators of employment. Only a small number acquire appropriate vocational training – most are trapped in dead-end jobs – and their prime interest on return is to enhance their social status. This they attempt to achieve by disdaining manual employment, by early retirement, by the construction of a new house, by the purchase of land, a car and other consumer durables, or by taking over a small service establishment like a bar or taxi business; there is also a tendency for formerly rural dwellers to settle in urban centres. There is thus a reinforcement of the very conditions that promoted emigration in the first place. It is ironic that those migrants who are potentially most valuable for stimulating development in their home area – the minority who have acquired valuable skills abroad – are the very ones who, because of successful adaptation abroad, are least likely to return. There are also problems of demographic imbalance stemming from the selective nature of emigration. Many villages in Southern Europe have been denuded of young men, with consequences not only for family formation and maintenance but also for agricultural production.

Table 13.5   The positive and negative effects of globalization processes on employment in developing countries

Positive effects	Negative effects
Higher export-generated income promotes investment in productive capacity with a potentially positive local development impact, depending on intersectoral and interfirm linkages, the ability to maintain competitiveness, etc.	The increases in employment and/or earnings are (in contradiction to the supposed positive effects) unlikely to be sufficiently large and widespread to reduce inequality. On the contrary, in most countries, inequality is likely to grow because unequal controls over profits and earnings will cause profits to grow faster
Employment growth in relatively labour-intensive manufacturing of tradable goods causes 1) an increase in overall employment; and/or 2) a reduction of employment in lower wage sectors. Either of these outcomes tends to drive up wages, to a point which depends on the relative international mobility of each particular industry, labour supply–demand pressure and national wage-setting/bargaining practices	Relocations of relatively mobile, labour-intensive manufacturing from industrialized to developing countries, in some conditions, can have disruptive social effects if – in the absence of effective planning and negotiations between international companies and the government and/or companies of the host country – the relocated activity promotes urban-bound migration and its length of stay is short. Especially in cases of export assembly operations with very limited participation and development of local industry and limited improvement of skills, the short-term benefits of employment creation may not offset those negative social effects
These increases in employment and/or wages – if substantial and widespread – have the potential effect of reducing social inequality if the social structure, political institutions and social policies play a favourable role	Pressures to create local employment, and international competition in bidding for it, often put international firms in a powerful position to impose or negotiate labour standards and labour management practices that are inferior to those of industrialized countries and, as in the case of some EPZs, even inferior to the prevailing ones in the host country
Exposure to new technology and, in some industries, a considerable absorption of technological capacity leads to improvements in skills and labour productivity, which facilitate the upgrading of industry into more value-added output, while either enabling further wage growth or relaxing the downward pressure	

Source: Based on ILO (1996b, Table Int. 1).

There is no question, therefore, that the magnitude of the employment and unemployment problem in developing countries is infinitely greater than that facing the older industrialized countries, serious as their problem undoubtedly is. The biggest problem in developing countries is *underemployment* and its associated *poverty*. The high rate of labour force growth in many developing countries continues to exert enormous pressures on the labour markets of both rural and urban areas. Such pressures are unlikely to be alleviated very much by the development of manufacturing industry alone. With one or two exceptions among the NIEs, industrial growth has done little to reduce the severe problems of unemployment and underemployment – with their resulting poverty – in developing countries. Globalizing processes, whilst offering some considerable employment benefits to some developing countries, are, again, a double-edged sword as Table 13.5 shows.

Sustaining growth and ensuring equity in the newly industrializing economies

The spectacular industrial growth of a small group of developing countries – the NIEs – has been one of the most significant developments in the world economy in recent years. Indeed, in many respects the four leading Asian NIEs – the four 'dragons' or 'tigers' – should perhaps no longer be regarded as developing countries at all. Certainly, there can be no doubting the remarkable industrial progress of this group of countries although these 'industrial miracles' are not without their serious internal difficulties. Three kinds of problem can be identified here:

- sustaining economic growth
- ensuring that such growth is achieved with equity for the countries' own people
- the problem of foreign debt which faces some, though not all, newly industrializing economies.

Sustaining economic growth
Measured in terms of increased per capita income, larger shares of world production and trade the east and southeast Asian NIEs, in particular, have been phenomenally successful. But can such spectacular growth rates be maintained in the future? Although each of the four leading NIEs has managed to sustain very high rates of growth for a very long period there were suggestions in the mid-1990s that perhaps the 'miracle' was coming to an end. Press headlines such as 'Asia's precarious miracle'; 'is it over?'; 'roaring tiger is running out of breath' began to appear. Even though it is difficult to avoid a feeling that the western media were engaging in a little wish fulfilment the question of the sustainability of NIE growth is a significant one.[8]

Economic growth in the NIEs has been based primarily upon an aggressive export-oriented strategy. It is no coincidence that the take-off of the first wave of Asian NIEs occurred during the so-called 'golden age of growth' in the 1960s and early 1970s or that it was made possible by the relative openness of industrialized country markets. The growth and openness of such markets is, therefore, vital for the continued economic growth and development of the NIEs. During the 1960s the conditions were indeed favourable; future prospects look far less propitious as the older industrialized countries have reduced their demands for NIE exports, partly through the deliberate operation of protectionist trade measures. From the NIEs' viewpoint, therefore, the macroeconomic expansion of the industrialized economies is vital. But this will be effective only if trade barriers – especially non-tariff barriers – are also removed or at least lowered. The present political climate in the older industrialized countries makes both possibilities somewhat remote.

Trade tensions, particularly between the United States, their biggest export market, and the leading Asian NIEs are palpable and show little sign of disappearing. Countries like South Korea and Taiwan are regarded by the United States as being less open to industrialized country imports than they might be. Consequently, the NIEs will need to develop further their own domestic markets although if this were to be done by raising trade barriers which are already high, the likely result would be to make the older industrialized countries even more reluctant to modify their own protectionist stance. There is also the problem that the smaller NIEs have very limited domestic markets; this was the major reason for adopting an export-oriented strategy in the first place. However, it is significant that regional markets, notably in Pacific

Asia, are becoming increasingly important. This fits the policy position of UNCTAD which strongly urges NIEs to develop markets outside the older industrialized countries.

A second problem facing the leading NIEs in sustaining economic growth arises from the *growing competition from other developing countries* – the 'proto-NIEs'. In Chapter 2 we focused mainly on the leading NIEs, but there are further 'tiers' of potential NIEs which have also been growing substantially as manufacturing centres. This 'next tier' includes such countries as Malaysia, China, the Philippines, Thailand, Pakistan, Indonesia, Colombia, Chile, Peru, Turkey, and also the transitional economies of eastern Europe. Competition from these lower-wage countries has intensified as labour costs in the first tier of NIEs have risen. The competition is obviously most severe in the lower-skill, labour-intensive activities on which NIE industrialization was originally based.

Thus, the development of competition from other developing countries, together with trends in the automation of some labour-intensive processes in the industrialized countries, has added to the pressures on the leading NIEs to shift to more skill-intensive and capital-intensive products and processes. As we have seen at various points in this book, they have been very successful in making this transition so far. But the competitive pressure continues to intensify as other newly industrializing economies not only take on the less skilled functions but also, themselves, strive to upgrade their economies even further. As a result a complex intraregional division of labour has developed in east and southeast Asia with a clear hierarchical structure composed of countries at different levels of industrialization.[9]

Prospects for the other NIEs outside Asia depend very much on their specific regional context. For Mexico, the key issue is its ability to prosper within the NAFTA. Whilst its access to the United States and Canadian markets is now guaranteed, at least over a period of time, its own economy will also feel the full force of external competition within its own borders. Similarly, the southern European NIEs (Spain, Portugal and Greece), which are full members of the EU, have unfettered access to the entire EU market. However, not only are their own economies open to the full force of competition from the industrialized economies of the EU but also they also now face increasing competition from the opening up of eastern Europe (as, of course, do the NIEs in general).

How far a third or fourth tier of NIEs really will emerge to threaten the 'super-league' is, however, a matter of considerable argument. In the early 1980s, Cline put forward the 'fallacy of composition' argument: that what is possible for a small number of cases is not possible for all, or even the large majority of cases. He argued that if the east Asian model of export-led development were to become characteristic of all developing countries 'it would result in untenable market penetration into industrialized countries' (Cline, 1982, p. 88). The World Bank argues against this viewpoint:

> First, the capacity of industrial nations to absorb new imports may be greater than supposed . . . Second, the idea that a large number of economies might suddenly achieve export-to-GDP ratios for manufactures like Hong Kong, Korea or Singapore is highly implausible . . . Third, export-oriented countries would produce different products, and intra-industry trade is likely to be important. Finally, the first wave of newly industrializing countries is already providing markets for the labour-intensive products of the countries that are following.

> (World Bank, 1987, p. 91)

Table 13.6   Distribution of income within selected developing countries

Country (year)	Lowest 20%	Highest 20%
Brazil (1989)	2.1	67.5
Chile (1994)	1.4	61.0
Mexico (1984)	4.1	55.9
Malaysia (1989)	4.6	53.7
Philippines (1988)	6.5	47.8
South Korea (1988)	7.4	42.2
Singapore (1982/3)	5.1	48.9
India (1989/90)	8.8	41.3
Average for high-income countries	6.2	40.7

Note:
The figures are estimates from a variety of sources and for different dates.
Source: Based on World Bank (1996, *World Development Report, 1996*, Table 5).

Ensuring economic growth with equity

Sustaining economic growth is only one of the difficulties facing the NIEs (both existing and potential) in today's less favourable global environment. Sustaining growth *with equity* for the populations of the NIEs themselves is also a major problem. Two aspects of this issue are especially important: income distribution and the socio-political climate within individual countries.

A widely voiced criticism of industrialization in developing countries has been that its material benefits have not been widely diffused to the majority of the population. There is indeed evidence of highly uneven income distribution within many developing countries, as Table 13.6 reveals. In countries such as Brazil, Chile, Mexico and Malaysia, for example, the share of total household income received by the top 20 per cent of households was very much higher than that in the industrial market economies. However, this pattern does not apply in all cases. For example, India and South Korea have household income distributions very similar to that of the industrial market economies (though at much lower levels). Of course, the question of income distribution is very much more complex than these simple figures suggest and is the subject of much disagreement among analysts. The fact remains, however, that in general the Asian NIEs have a more equitable income distribution than the Latin American countries. Without doubt this reflects the specific historical experiences of these countries and, especially, the different patterns of land ownership and reform.

Income distribution is one aspect of the 'growth with equity' question. Another is the broader social and political issue of democratic institutions, civil rights and labour freedom. Although the degree of repression and centralized control in NIEs may sometimes be exaggerated, the fact is that such conditions do exist in a number of cases.[10] The very strong state involvement in economic management in most NIEs has brought with it often draconian measures to control the labour force. Labour laws tend to be extremely stringent and restrictive; in many instances strikes are banned:

> With regard to the maintenance of low-wage labour reserves, the proletarianized segment of labour in all Asian NIEs has suffered from state intervention to depress workers' wages below market rates in order to make exports competitive on the international market . . . But internal conditions and state strategies varied . . . In both Korea and Singapore, the government has been actively involved in the creation of a 'hyperproletarian' segment of the labour market that has been largely filled by women and is characterised by a high turnover of labour, institutionalised job insecurity, and low wages . . . The absence of either state or community restraint on exploitation of workers in Korea is manifested by

extremely poor working conditions for this segment. Labour laws have been arbitrarily enforced and favour employers over workers, particularly in the case of heavy industry and automobile production, the sectors in which labour has become most militant.

(Douglass, 1994, p. 554)

How far the success of NIEs, particularly in attracting foreign investment, really depends on the use of strongly authoritarian measures is difficult to ascertain. But until such repressive behaviour is relaxed, the achievements of some of the NIEs in the strictly economic sphere must be regarded with some reservations. In this respect, it is significant that in the late 1980s both South Korea and Taiwan began to move along the democratization path. As the persistence of labour disturbances in South Korea shows, however, the transition is not proving to be easy.

Environmental degradation is also a major social problem facing most of the NIEs. In the cases of South Korea and Taiwan, for example, Bello and Rosenfeld write of the 'toxic trade-off' and 'the making of an environmental nightmare' respectively. Although extensive environmental damage is certainly not confined to the NIEs,

> their single-minded pursuit of rapid economic growth has caused particularly severe environmental degradation. Much of the countryside in both South Korea and Taiwan is severely and perhaps irreparably damaged. South Korean rural areas suffer from extensive deforestation, related problems of soil erosion and flooding, and widespread chemical contamination of ground water . . . Rising environmental costs in both urban and rural areas are materializing in poor health, physical damage, loss of amenities, and other problems that demand extensive remedial spending . . . In order to stimulate rapid growth, the NIEs have used up significant environmental capital that can only be restored, if at all, at considerable cost to future generations.
>
> (Brohman, 1996, pp. 126, 127)

Although we have treated the questions of sustaining growth and sustaining growth with equity as separate they are, in fact, closely related. It is an open question as to how far the various forms of the developmental state which are manifested in different NIEs can continue to provide the basis for future economic development. This is an issue addressed by Douglass (1994). He concludes his analysis (p. 563) as follows:

> The strong state model of societal guidance that was a key feature of all of the Asian NIEs' industrialization processes is being challenged on all fronts, and whether a new complex of institutions reflecting an expanded political community that includes more democratic methods of regulation will appear to be or be effective is one of the most important questions for the coming years. If Bello and Rosenfeld (1990) are accurate in claiming that the command economies of the Asian NIEs are obsolete and that legitimation crises can no longer be met with the stick but must instead be resolved through democratic practice, the message from the West may not be comforting. But if there is a process of 'late democratization' accompanying late industrialization, a nonauthoritarian alternative to the developmental state based on the extension of political community beyond the state itself will also be on the political agenda.

The debt problem

A third serious problem for the NIEs is the burden of *financial debt* which many have incurred, particularly since the late 1970s.[11] Much of the industrial growth of the NIEs has been financed by overseas borrowing (as was that of the industrializing countries in the nineteenth century). Before the 'second oil shock' of 1979 there were no major problems. Exports from NIEs continued to hold up well despite the onset of recession in the older industrialized countries. Capital was needed for investment in the NIEs;

after the 1973 oil crisis the huge volume of petrodollars had to be *recycled* by the commercial banks to prevent the world financial system from seizing up. The banks were only too ready to lend to the more successful developing countries in order to achieve this.

The international debt crisis broke with great suddenness in the early 1980s when the first, and most spectacular, incident was the financial collapse of Mexico in August 1982. It soon became clear that a number of developing countries were in deep financial difficulty. The problem was particularly concentrated in the middle-income group of developing countries which had come to depend most heavily on commercial lending. The fact that the low-income countries were not greatly involved in no way indicates their lack of financial difficulty – on the contrary. But commercial banks have generally been unwilling to lend to the very poor countries so that most of their borrowing is in the public sector aid programmes.

In the early 1980s, therefore, the problem of recycling funds was suddenly displaced by that of *rescheduling* the massive debts of some developing countries. An increasing number of countries found themselves unable to repay the interest on the sums borrowed let alone reduce the basic sum. Some were having difficulty paying the interest on the interest as they had to borrow more simply to avoid going under completely. The emergence of this very serious problem reflects a whole host of factors. Some are internal to the countries involved: undoubtedly there has been some profligate spending on unnecessary prestige projects. But this is not the fundamental cause.

The most important factors relate to developments within the global economy since the late 1970s. As we have seen, demand for manufactured goods and materials in the older industrialized countries declined as recession deepened and as protectionist measures intensified. The market for NIE exports slackened very substantially.

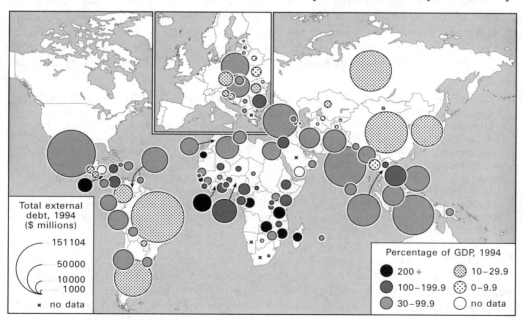

Figure 13.4 The map of developing country debt, 1994
Source: Based on World Bank (1996, *World Development Report, 1996*, Table 17)

At the same time, the governments of the industrialized countries began to pursue very tight financial and fiscal policies which, while reducing price inflation in their own economies, forced up interest rates. In particular, the need for the United States to finance its enormous budget deficit kept interest rates especially high. Floating interest rate debts subsequently cost far more to service than when they were initially incurred. New loans taken out became extremely expensive.

As Figure 13.4 shows, the largest debt problems are in Latin America. Brazil, Mexico and Argentina in particular have been continuously in the financial headlines since 1982 as each country seemed to be on the brink of financial collapse. Each of the major debtor nations has been forced to seek the co-operation of the international financial institutions – the IMF and the major commercial banks – in rescheduling their debts. So far, total breakdown has been averted but the very severe conditions the IMF places on the debtor countries as the price of rescue create serious social and political problems for the countries themselves. Whether the lid can be kept on such a volatile pot is by no means certain, as the collapse of the Mexican peso in December 1994 again illustrated. Prior to 1997, the east and southeast Asian NIEs had not experienced such problems. The relative buoyancy of the Asian NICs helped to keep them out of the grip of the moneylenders. For example, although South Korea has borrowed very heavily, its continuing high-level export performance enabled it to service the debt, although concerns surfaced in the mid-1990s over its rapid growth. However, the collapse of the Thai baht in the summer of 1997 and its knock-on effects in the region demonstrated the fragility of the situation.

Ensuring survival and reducing poverty in the least industrialized countries

Although the NIEs certainly face problems, they are not of the same magnitude or seriousness as those facing the least industrialized, low-income countries. As Table 13.2 demonstrated, the poorest fifty or so developing countries are poor not just in terms of income but also in virtually every other aspect of material well-being. They are the countries of the deepest poverty, several of which face mass starvation. For large numbers of people in the low-income countries (and in some of the higher-income countries, too) life is of the lowest material quality.

The causes of low income levels
Poverty is the most crushing burden of all. As Todaro (1989, p. 93) succinctly demonstrates: 'low levels of living (insufficient life-sustaining goods and inadequate or nonexistent education, health, and other social services) are all related in one form or another to low incomes. These low incomes result from the low average productivity of the entire labour force, not just those working.' Todaro argues that low labour force productivity can result from a variety of factors:

- On the *supply side*: poor health, nutrition and work attitudes; high population growth, and high unemployment and underemployment.
- On the *demand side*: inadequate skills, poor managerial talents, overall low levels of worker education; the importation of developed country labour-saving techniques of production which substitute capital for labour.
- The combination of low labour demand and large supplies results in the widespread underutilization of labour.

In addition,

- low incomes lead to low savings and investment which also restrict the total number of employment opportunities
- low incomes are also thought to be related to large family size and high fertility since children provide one of the few sources of economic and social security in old age for very poor families:

> The important point to remember . . . is that *low productivity, low incomes, and low levels of living are mutually reinforcing phenomena.* They constitute what Myrdal has called a process of 'circular and cumulative causation' in which low incomes lead to low levels of living (income plus poor health, education etc.) that keeps productivity low, which in turn perpetuates low incomes, and so on.
>
> (Todaro, 1989, p. 99)

Dependence on a narrow economic base

A most important contributory factor in the poverty of low-income countries (and of some of the lower middle-income countries too) is their dependence on a very narrow economic base together with the nature of the conditions of trade. We saw earlier (Table 13.3) that the overwhelming majority of the labour force in low-income countries is employed in agriculture. This, together with the extraction of other primary products, forms the basis of these countries' involvement in the world economy. Approximately four-fifths of the exports of developing countries are of primary products compared with less than one-quarter for the developed economies.

In the classical theories of international trade, based upon the comparative advantage of different factor endowments, it is totally logical for countries to specialize in the production of those goods for which they are well endowed by nature. Thus, it is argued, countries with an abundance of particular primary materials should concentrate on producing and exporting these and import those goods in which they have a comparative disadvantage. This was the rationale underlying the 'old' international division of labour in which the core countries produced and exported manufactured goods and the countries of the global periphery supplied the basic materials. According to traditional trade theory all countries benefit from such an arrangement. But such a neat sharing of the benefits of trade presupposes some degree of equality between trading partners, some stability in the relative prices of traded goods and an efficient mechanism – the market – which ensures that, over time, the benefits are indeed shared equitably.

In the real world – and especially in the trading relationships between the industrialized countries and the low-income, primary-producing countries – these conditions do not hold. In the first place, there is a long-term tendency for the composition of demand to change as incomes rise. Thus, growth in demand for manufactured goods is greater than the growth in demand for primary products. This immediately builds a bias into trade relationships between the two groups of countries, favouring the industrialized countries at the expense of the primary producers.

Over time, these inequalities tend to be reinforced through the operation of the *cumulative* processes of economic growth. The prices of manufactured goods tend to increase more rapidly than those of primary products and, therefore, the *terms of trade* for manufactured and primary products tend to diverge. (The terms of trade are simply the ratio of export prices to import prices for any particular country or group of countries.) As the price of manufactured goods increases relative to the price of primary products, the terms of trade move against the primary producers and in favour of the industrial producers. For the primary producers it becomes necessary to export a larger quantity of goods in order to buy the same, or even a smaller, quantity of manufactured goods. Although the terms of trade do indeed fluctuate over time, there is no doubt that they have generally deteriorated for the primary producing countries.

The severity of the situation facing the developing countries in general, and the low-income, least industrialized, countries in particular, led to demands for a radical change in the workings of the world economic system. The demand by a group of developing countries was for a *new international economic order* (NIEO), a demand set out formally in a United Nations Declaration in 1974. The United Nations Declaration 'called for the replacement of the existing international economic order, which was characterised by inequality, domination, dependence, narrow self-interest and segmentation, by a new order based on equity, sovereign equality, interdependence, common interest and co-operation among States irrespective of their economic and social systems' (United Nations, 1982, p. 3).

The NIEO declaration set out a very ambitious and detailed agenda. Overall, the demand was for better access to world markets for both primary products and manufactured goods from developing countries. In the case of primary products a key demand was for stabilization schemes to remove the serious fluctuations in both demand and prices for them, fluctuations which have an especially severe impact on many developing countries. In the case of manufactured goods, the demand was for a reduction in protection of developed country markets, for the older industrialized countries to adopt more positive adjustment policies in their home economies to permit the expansion of developing country exports and for measures to ease the transfer of technology.

The demand for a fairer distribution of world industrial production was formulated more precisely in the 1975 Lima Declaration and Plan of Action for Industrial Co-operation. The Lima Declaration set a target for 25 per cent of world industrial production to be located in the developing countries by the year 2000. As we saw in Chapter 2, developing countries now account for around 20 per cent of world manufacturing production. At first sight, therefore, it would seem that the Lima target is within grasp. Not so. Virtually all the manufacturing production ouside the core industrial economies is located in a very small number of developing countries. For the vast majority, as we have seen, levels of industrialization remain exceptionally low.

In fact, very little progress has been made on most of these demands. The very poor developing countries, those at the bottom of the well-being league table, have benefited least from the internationalization and globalization of economic activities. Both their present and their future are dire. Ways have to be found to solve the problems of poverty and deprivation: 'No task should command a higher priority for the world's policy makers than that of reducing global poverty. In the last decade of the twentieth century it remains a problem of staggering dimensions' (World Bank, 1990, p. 5).

In such a highly interconnected world as we now inhabit, and as our children will certainly inhabit, it is difficult to believe that anything less than global solutions can deal with such global problems. But how is such collaboration to be achieved? Can a new international economic order be created or is the likely future one of international

economic disorder? There can be no doubt that the present immense global inequalities are a moral outrage. The problem is one of reconciling what many perceive
to be conflicting interests. For example, one of the biggest problems facing the older
industrialized countries is unemployment. As we have seen, the causes of unemployment are complex but if one factor is perceived to be import penetration by developing country firms then the pressure to adopt restrictive trade policies becomes
considerable.

The alternative is to adopt policies which ease the adjustment for those groups
and areas adversely affected and to stimulate new sectors. But this requires a more
positive attitude than most western governments have been prepared to adopt. For
the poorer developing countries it is unlikely, however, that industrialization will
provide the solution to their massive problems. Despite its rapid growth in some
developing countries, manufacturing industry has made barely a small dent in the
unemployment and underemployment problems of developing countries as a whole.
For most, the answer must lie in other sectors, particularly agriculture, but the seriousness of these countries' difficulties necessitates concerted international action.
This, in turn, is part of a much larger debate about the overall governance of the world
economy.

Notes for further reading

1. The most recent contribution to this debate is by Rowthorn and Ramaswamy (1997). Important earlier contributions include Blackaby (1979), Bluestone and Harrison (1982), Cohen
   and Zysman (1987), Rowthorn and Wells (1987).
2. This debate intensified during the 1990s. For alternative views, see ILO (1997), together with
   the many articles in the press, notably in *The Economist, The Financial Times* and *The Wall
   Street Journal*.
3. Peck (1996; 1998) provides a stimulating discussion of changing labour market regulation in
   the United States and the United Kingdom.
4. See *The Financial Times* (22 August 1995).
5. For a stimulating polemic against the small firm phenomenon, see Harrison (1994).
6. The characteristics of the formal and informal sectors are discussed by Gilbert and Gugler
   (1982), Armstrong and McGee (1985), Roberts (1995).
7. The 1996 United Nations report, *An Urbanizing World*, discusses these processes in detail.
8. Bello and Rosenfeld (1990) are strongly of the view that the first tier of NIEs is in serious
   crisis. The financial crisis of 1997, which began in Thailand and spread to other east Asian
   countries (including Malaysia, Indonesia, and even Hong Kong), reinforced the
   doomsayers.
9. See Bernard and Ravenhill (1995), Child Hill and Fujita (1996) for a critical review of
   alternative theoretical perspectives on east Asian industrialization.
10. See Bello and Rosenfeld (1990), Douglass (1994), Brohman (1996).
11. Corbridge (1993) provides a substantive discussion of the debt problem in the context of
    development. See also Griffith-Jones and Stallings (1995).

CHAPTER 14

Issues of global governance

Confusion and contestation[1]

> While the world has become much more highly integrated economically, the mechanisms for managing the system in a stable, sustainable way have lagged behind.
> (Commission on Global Governance, 1995, pp. 135–36)

More than at any time in the last 50 years, virtually the entire world economy is now a *market economy*. The collapse of the state socialist systems at the end of the 1980s and their headlong rush to embrace the market, together with the more controlled opening up of the Chinese economy since 1979, has created a very different global system from that which emerged after the Second World War. Virtually all parts of the world are now, to a greater or lesser extent, connected into an increasingly integrated system in which the parameters of the market dominate.

The acceleration and intensification of technological change, and the emergence of transnational corporations, with their intricate internal and external networks, together ensure that what happens in one part of the world is very rapidly transmitted to other parts of the world although, as we have demonstrated, the processes of globalization are extremely uneven in both time and space. The massive international flows of goods, services and, especially, of finance in its increasingly bewildering variety, have created a real world whose rules of governance have not kept pace with such changes. As Strange (1996, p. 189) points out,

- power has shifted upwards, from weak states to stronger states with global or regional reach beyond their frontiers
- power has shifted sideways from states to markets and, hence, to non-state authorities which derive their power from their market shares
- some power has 'evaporated' in so far as no one exercises it.

Thus, although, as we argued in Chapter 3, the nation-state remains a highly significant actor in the world economy it is abundantly clear that its role has been changing and that it faces increasingly intractable problems in regulating its domestic economy in a flow-intensive, market-dominated, globalizing world.

There is no doubt that the market can be a highly effective mechanism for facilitating economic growth and development. But that does not mean that the market operates independently of a social context. On the contrary, all markets are socially embedded and constituted; all have to operate within socially defined rules. Totally unregulated markets are neither sustainable nor socially equitable; the unfettered market cannot be relied upon to create outcomes which maximize benefits for the many rather than just the few. To do so demands a regulatory or governance system which is legitimated by individual nation-states and by the communities and interest groups which constitute them.

The current international economic governance system is, in fact, made up of several levels operating at different, but interconnected, geographical scales:[2]

461

- International regulatory bodies established by agreement by nation-states to perform specific roles. Examples include the IMF and the WTO (formerly the GATT).
- International co-ordinating groups with a broader, but less formal, remit. Examples include the groups of leading industrialized countries (G3, G5, G7, etc.).
- Regional blocs, such as the EU or NAFTA.
- National regulatory bodies operating within individual nation-states.
- Local agencies operating at the level of the individual community:

> These five levels are interdependent to a considerable degree. Effective governance of economic activities requires that mechanisms be in place at all five levels, even though the types and methods of regulation are very different at each level . . . [However] . . . the different levels and functions of governance need to be tied together in a division of control that sustains the division of labour . . . The governing powers (international, national and regional) need to be 'sutured' together into a relatively well integrated system . . . The issue at stake is *whether* such a coherent system will develop . . .
>
> (Hirst and Thompson, 1996, pp. 122, 184)

In this discussion, we focus specifically on the *international* scales of governance.

Within a volatile global economy, there are many issues which pose very serious problems for all states and communities throughout the world and which need to be addressed at the international or global scale. In this section, we outline some of the problems associated with governance in two areas, both of which are central to our concerns throughout this book:

- international finance
- international trade, with specific reference to labour standards and the environment.

Governing the international financial system[3]

As we have seen, the regulatory basis of the postwar international financial system was established at Bretton Woods in the 1940s. However, through a whole series of developments (discussed in Chapter 12) the relatively stable basis of the Bretton Woods system was progressively undermined. In effect, we have moved from a 'government-led international monetary system (G-IMS) of the Bretton Woods era to the market-led international monetary system (M-IMS) of today' (Hirst and Thompson, 1996, p. 130). As a result,

> the international financial system has become, in effect, a separate state . . . we are left with a strange world, one in which money capital flows freely, and is becoming less and less regulated, while movement of goods in the 'productive' economy has become more and more negotiated and regulated. Yet, this productive economy is increasingly susceptible, in principle if not in practice, to the institutions of money capital.
>
> (Thrift, 1990b, p. 1136)

Susan Strange (1986, p. 1) coined the graphic term 'casino capitalism' to describe the international financial system in which

> every day games are played in this casino that involve sums of money so large that they cannot be imagined. At night the games go on at the other side of the world . . . (the players) . . . are just like the gamblers in casinos watching the clicking spin of a silver ball on a roulette wheel and putting their chips on red or black, odd numbers or even ones.

What we do not have, therefore, is a comprehensive and integrated global system of governance of the financial system. Instead, there are various areas of regulation performed by different bodies which are strongly *nationally* based (Hirst and Thompson, 1996, pp. 130–32):

- The G3 (the United States, Japan, EU) takes an overall view of the monetary, fiscal and exchange rate relationships between the G3 countries although, in practice, this has been confined primarily to attempts to determine the global money supply and to manipulate exchange rates. This has not resolved the basic problem of 'an institutional gap between the increasingly international nature of the financial system and the still predominantly "national" remits of the major central banks and the wider nationally located regulatory mechanisms for financial markets and institutions'. The problem is that the G3, as well as the broader G5 and G7, has no real institutional base. It is a largely informal arrangement structured around periodic summits of national leaders.

- The international payments system is operated through the national central banks rather than through an international central bank:

  > While this central banking function remains unfulfilled at the international level the risks of default increase and disturbances threaten to become magnified across the whole system . . . [there is] . . . a growing network of cooperative and coordinative institutionalized mechanisms for monitoring, codifying and regulating such transactions (centred on the BIS and headed by the Group of Experts on Payment Systems).

- The supervision of financial institutions themselves is carried out through the Bank for International Settlements (BIS), established in 1975 and now based upon the 1988 Basel Committee's Capital Accord:

  > Much like the monetary summits of the G3 to G7 the Basel Committee was initially designed as a forum for the exchange of ideas in an informal atmosphere with no set rules or procedures or decision making powers. But, although it maintains this original informal atmosphere, its evolution has been toward much more involvement in hard-headed rule-making and implementation monitoring.

Thus, there are several regulatory bodies in operation at the international level. Yet the fear remains that, in the absence of a more co-ordinated and institutionalized system, the international financial system could easily spiral out of control. What might be done to prevent this happening? Not surprisingly, there is no consensus. Martin (1994, pp. 274–75) sets out two broad alternatives:

> One route would be to reimpose national systems of control and regulation, along the lines of those now being dismantled. However, the very fact of global financial integration renders the feasibility and effectiveness of this response highly doubtful. An alternative response would be to introduce new forms of regulation based on international cooperation: if markets have gone global in their geography, so too should the institutions of regulation . . . This might be in the form of supranational regulation involving governing bodies outside the bounds of nations, or transnational regulation, involving the coordination of nation state policies through multilateral agreements. Such plans have their problems, of course, not least the difficulty of securing the international cooperation necessary for their implementation. But whatever the form, the case for re-regulation is strong: the power of global money over national economic space has already been allowed to extend too far.

Trade, labour standards and the environment

Compared with the international financial system, the governance of international trade in commodities and manufactured products is much clearer and well established. As we saw in Chapter 3, for the past fifty years the GATT and, since 1995, the WTO has constituted a trade regulatory regime based upon clearly defined, non-discriminatory, multilateral principles. Although significant trade friction continues to exist over specific issues and between certain groups of countries, the WTO system is generally accepted by virtually all countries in the world. Indeed, the eventual agreement on the Uruguay Round substantially widened the remit of the GATT/WTO. However, apart from the need still to reach agreement on certain sectors (such as services), there are two controversial issues which have moved to the centre of the trade (and development) debate: the relationship between free trade and labour standards and between free trade and the environment.

The basic question is: to what extent do international differences in labour standards and regulations (such as the use of child labour, poor health and safety conditions, repression of labour unions and workers' rights) and in environmental standards and regulations (such as industrial pollution, the unsafe use of toxic materials in production processes) distort the trading system and create unfair advantages? In both cases, the basic argument is that firms – as well as individual countries – may be able to undercut their competitors by capitalizing on cheap and exploited labour and lax environmental standards. Much of the focus of this concern is on the export processing zones which, as we saw in Chapter 4, have proliferated throughout the developing world.

These two issues were explicitly addressed in the negotiations for the NAFTA, in which the United States insisted on the signing of two side agreements to protect its domestic firms from low labour and environmental standards in Mexico. More recently, a group of countries led primarily by the United States but also including some European countries made a concerted attempt formally to incorporate the issue of *labour standards* into the WTO at its ministerial meeting in Singapore in December 1996. The attempt failed, partly because not all industrialized countries supported it but also because the developing countries were vehemently opposed. The argument of those opposed to its inclusion within the WTO's remit is that labour standards are the responsibility of the International Labour Organization; the counterargument is that the ILO lacks any powers of enforcement. It is also notable that the United States, despite its current position on including labour standards in trade agreements, has 'signed only one of the five core labour standards conventions issued by the International Labour Organization – and ratified only 12 of the total 176 ILO conventions. It says that though it respects their spirit, the UN body's conventions do not mesh with its own laws' (*The Financial Times*, 20 June 1996).

There is no doubt that stark differences do exist in labour standards in different parts of the world. As we have seen at various points in this book, basic workers' rights are denied in many countries. Working conditions, especially in the export processing zones (including the Mexican *maquiladoras*) – but not only in these zones – are often appalling.[4] As far as child labour is concerned, the ILO calculates that around 73 million children aged between ten and fourteen years are employed throughout the world, approximately 13 per cent of that age group. In Africa, one-quarter of children aged ten to fourteen are working. If children under ten are included, as well as young girls working full time at home, the ILO estimates that there are probably 'hundreds of millions' of child workers in the world. However, the ILO also points out that 90 per cent of children work in agriculture or linked activities in rural areas and that most are employed within the family rather than for outside employers. Even so, there is substantial evidence that, in many cases, young children are employed by manufacturers (whether in factories or as outworkers) in such industries as garments, footwear, toys, sports goods, artificial flowers, plastic products and the like. Their wages are a pittance and their working conditions often abysmal.

From the viewpoint of many developing countries, however, there is a strong feeling that the labour standards stance of many developed countries is merely another form of protectionism against their exports and, as such, an obstacle to their much-needed economic development. There is a suspicion, for example, that at least some of the developed country lobbies are pressing for international agreements on a minimum wage in order to lessen the low labour cost advantages of developing countries. By incorporating such labour standards criteria into the WTO framework, it is believed, developed countries could use trade regulations to enforce such indirect protectionism. There is clearly a basic dilemma for the international community. On the one hand, ethical considerations must be a basic component of international trade agreements; on the other hand, there is a real danger of threatened developed country interest groups using labour standards issues as a device to protect their own commercial interests.

A similar dilemma is central to the other trade-related question: that of the *environment*. To what extent should variations in environmental standards be incorporated into international trade regulations? Again, there is no doubting the existence of huge differences in the nature, scope and enforcement of environmental regulations across the world. There is no doubt, either, that the highest incidence of low environmental standards is in developing countries. The existence of such an 'environmental gradient' certainly constitutes a stimulus for some firms at least to take advantage of low standards. These may be domestic firms or they may be foreign firms.

Although it does not necessarily follow that such firms have relocated to environmentally lax areas simply to avoid more stringent regulations (and higher costs) elsewhere, the fact that they do operate there is a major problem. Again, one of the most notorious, and best-documented, cases is that of the *maquiladoras* of the US–Mexico border zone:

> Matamoros, with more than 100 *maquiladoras*, has only three government health and safety inspectors, who normally give advance warning of factory visits. Air quality and industrial effluents are tested only once a year. Senator Richard Gephardt . . . [noted] . . . '21st century technology combined with 19th century living and working conditions. We drove by industrial parks where companies continue to dump their toxic wastes at night into rivers. We saw furniture plants using highly toxic solvents and finishes that once operated in California and throughout the US, and which had moved to Mexico because of lax environmental enforcement'.

(*The Financial Times*, 6 June 1997)

But there are also broader environmental issues which have been associated with international trade and its governance.[5] It is self-evident, for example, that many environmental problems are not confined within national boundaries but 'spill over'. For example, 'acid rain' produced by certain types of energy creation is carried by the wind way beyond its points of origin to create environmental damage. The damage to the ozone layer in the earth's stratosphere is caused by the use of certain chemicals (such as chlorofluorocarbons – CFCs) which retain their stability over long periods of time and move upwards into the stratosphere, expelling chlorine which destroys ozone molecules. Again, such chemicals have an effect way beyond their points of use.

This is not the place to explore the details of the environmental debate as a whole. Various international environmental agreements are in place (with a greater or lesser degree of effectiveness) which are aimed at broader environmental governance. But how does the environmental issue relate specifically to the governance of international trade? At one level, the problem is exactly the same as that of labour standards. If a country allows lax environmental standards, it is argued, then it should not be able to use what is, in effect, a subsidy on firms located there to be able to sell its products more cheaply on the international market. The question then becomes one of whether the solution lies in using international trade regulations or in some other forms of sanction. The proponents and opponents of the 'trade solution' are the same as those discussed above in the case of labour standards.

However, there is an even more extreme position adopted by some environmentalists which is that the pursuit of ever-increasing international trade – which is clearly encouraged by a free-trade regime like the GATT/WTO – should be totally abandoned, not merely regulated. The argument here is basically that *sustainable* development is incompatible with the pursuit of further economic growth and, especially, with an economic system which is based upon very high levels of geographical specialization, since such specialization inevitably depends upon, and generates, ever-increasing trade in materials and products.

Daly (1993, p. 24, emphasis added) expresses this viewpoint as follows:

> No policy prescription commands greater consensus among economists than that of free trade based on international specialization according to comparative advantage. Free trade has long been presumed good unless proved otherwise . . . Yet that presumption should be reversed. *The default position should favour domestic production for domestic markets*. When convenient, balanced international trade should be used, but it should not be allowed to govern a country's affairs at the risk of environmental and social disaster. The domestic economy should be the dog and international trade its tail. GATT seeks to tie all the dogs' tails together so tightly that the international knot would wag the separate national dogs.

One of the bases of Daly's argument is that the energy costs of transporting materials and goods across the world are not taken into account in setting the prices of traded goods and that, in effect, trade is being massively subsidized at a huge short-term and long-term environmental cost. Another is that free trade injects new inefficiencies into the system: 'more than half of all international trade involves the simultaneous import and export of essentially the same goods. For example, Americans import Danish sugar cookies, and Danes import American sugar cookies. Exchanging recipes would surely be more efficient' (Daly, 1993, p. 25).

The counterview from a rather different environmentalist perspective is clearly expressed by Pearce (1995, pp. 74, 77, 78):

> Unquestionably, there are environmental problems inherent in the existing trading system. But there is also extensive confusion in the environmentalist critique of free trade . . . Given the potentially large gains to be obtained from free trade, adopting restrictions on trade for environmental purposes is a policy that needs to be approached with caution. Most importantly, all other approaches to reducing environmental damage should be exhausted before trade policy measures are contemplated . . . the policy implication of a negative association between freer trade and environmental degradation is not that freer trade should be halted. What matters is the adoption of the most cost-effective policies to optimize the externality. Restricting trade is unlikely to be the most efficient way of controlling the problem . . . The losses can best be minimized by firm *domestic* environmental policy design to uncouple the environmental impacts from economic activity . . . The 'first' best approach to correcting externalities is to tackle them directly through implementation of the polluter pays principle (PPP), not through restrictions on the level

of trade. Where the PPP is not feasible (e.g. if the exporter is a poor developing country), it is likely to be preferable to engage in cooperative policies, e.g. making clean technology transfers, assisting with clean-up policies etc., rather than adopting import restrictions.

'Through a glass, darkly'

> Time present and time past
> Are both perhaps present in time future
> And time future contained in time past
> (T.S. Eliot, *Four Quartets*)

What of the future? Although 'time present' and 'time past' are indeed 'present in time future' we cannot simply extrapolate current patterns and processes into the future. The future is an amalgam of the probable and the unpredictable. Change occurs within an existing context but it also transforms that context, usually gradually, occasionally rapidly. The future, then, is a land of many questions to which there are no certain answers. What does seem certain is that the tendency towards an increasingly highly interconnected and interdependent global economy will intensify. The fortunes of nations, regions, cities, neighhourhoods, families and individuals will continue to be strongly influenced by their position in the global network but in complex and highly reflexive ways. The world economy is structured as a multi-layered, multiscale mosaic of activities.

In trying to understand the processes of global economic transformation in this book we have emphasized the importance of two major kinds of institution – transnational corporations and nation-states – operating in a complex and volatile technological environment. Through their strategies and interactions TNCs and states have reshaped the global economic map and contributed towards the increasing globalization of economic activities. There is every reason to believe that they will continue to be the primary forces even though their particular behaviour, and the inter-relationships between them, will certainly change. The key point is that each is both a political and an economic institution. Nation-states, whilst essentially political institutions, have become increasingly involved in economic matters, arguably as increasingly competitive economic actors. Transnational corporations, though fundamentally economic in function, have become increasingly political in their actions and impact.

The 'topography' of tomorrow's global economic map, like that of yesterday's and today's, will be the outcome of both economic and political forces. The two cannot properly be separated. In the end, however, the issues are not merely academic. The global economy and all its participants – from transnational corporations and national governments, to local communities and individual citizens – face a major global challenge: to meet the material needs of the world community as a whole in ways which reduce, rather than increase, inequality and which do so without destroying the environment.

Notes for further reading

1. General discussions of issues of global governance are provided by Camilleri and Falk (1992), McGrew and Lewis (1992), Commission on Global Governance (1995), Michie and Grieve Smith (1995), Hirst and Thompson (1996), Strange (1996).
2. Hirst and Thompson (1996, p. 121).

3. There is a large literature on the problems of governance of the international financial system. See, for example, Martin (1994), Michie and Grieve Smith (1995), Hirst and Thompson (1996).
4. See, ICFTU (1996), ILO (1996a) for examples of labour conditions in developing countries.
5. For two contrasting viewpoints, see Daly (1993), Pearce (1995).

Bibliography

Abo, T. (ed) (1994) *Hybrid Factory: The Japanese Production System in the United States*, Oxford University Press, New York.

Abo, T. (1996) The Japanese production system: the process of adaptation to national settings, in R. Boyer and D. Drache (eds) *States Against Markets: The Limits of Globalization*, Routledge, London, Chapter 5.

Agnew, J. and Corbridge, S. (1995) *Mastering Space*, Routledge, London.

Allen, J. (1988) Service industries: uneven development and uneven knowledge, *Area*, Vol. 20, pp. 15–22.

Amin, A. (1992) Big firms versus the regions in the single European Market, in M. Dunford and G. Kafkalas (eds) *Cities and Regions in the New Europe: The Global-Local Interplay and Spatial Development Strategies*, Belhaven, London, pp. 127–49.

Amin, A. (ed) (1994) *Post-Fordism: A Reader*, Blackwell, Oxford.

Amin, A. and Malmberg, A. (1992) Competing structural and institutional influences on the geography of production in Europe, *Environment and Planning A*, Vol. 24, pp. 401–16.

Amin, A. and Robins, K. (1990) The re-emergence of regional economies? The mythical geography of flexible accumulation, *Environment and Planning D: Society and Space*, Vol. 8, pp. 7–34.

Amin, A. and Thrift, N. (1992) Neo-Marshallian nodes in global networks, *International Journal of Urban and Regional Research*, Vol. 16, pp. 571–87.

Amin, A. and Thrift, N. (1994) Living in the global, in A. Amin and N. Thrift (eds) *Globalization, Institutions and Regional Development in Europe*, Oxford University Press, Oxford, Chapter 1.

Amsden, A. (1989) *Asia's Next Giant: South Korea and Late Industrialization*, Oxford University Press, Oxford.

Anderson, M. (1995) The role of collaborative integration in industrial organization: observations from the Canadian aerospace industry, *Economic Geography*, Vol. 71, pp. 55–78.

Angel, D. (1994) *Restructuring for Innovation: The Remaking of the US Semiconductor Industry*, Guilford Press, New York.

Aoki, A. and Tachiki, D. (1992) Overseas Japanese business operations: the emerging role of regional headquarters, *RIM Pacific Business and Industries*, Vol. 1, pp. 28–39.

Aoki, M. (1984) Aspects of the Japanese firm, in M. Aoki (ed) *The Economic Analysis of the Japanese Firm*, North Holland, Dordrecht, pp. 3–46.

Appelbaum, R.P., Smith, D. and Christerson, B. (1994) Commodity chains and industrial restructuring in the Pacific Rim: garment trade and manufacturing, in G. Gereffi and M. Korzeniewicz (eds) *Commodity Chains and Global Capitalism*, Praeger, Westport, Conn. Chapter 9.

Armstrong, W. and McGee, T.G. (1985) *Theatres of Accumulation: Studies in Asian and Latin American Urbanization*, Methuen, London.

Badaracco, J.L. Jr (1991) The boundaries of the firm, in A. Etzioni and P.R. Lawrence (eds) *Socio-Economics: Towards a New Synthesis*, M.E. Sharpe, Armonk, NJ. pp. 293–327.

Bailey, P., Parisotto, A. and Renshaw, G. (eds) (1993) *Multinationals and Employment: The Global Economy of the 1990s*, ILO, Geneva.

Bairoch, P. (1982) International industrialization levels from 1750 to 1980, *Journal of European Economic History*, Vol. 11, pp. 269–333.

Bairoch, P. (1993) *Economics and World History*, Wheatsheaf, Brighton.

Bakis, H. (1987) Telecommunications and the global firm, in F.E.I. Hamilton (ed) *Industrial Change in Advanced Economies*, Croom Helm, London, pp. 130–60.

Barnet, R.J. and Cavanagh, J. (1994) *Global Dreams: Imperial Corporations and the New World Order*, Simon & Schuster, New York.

Barnet, R.J. and Muller, R.E. (1975) *Global Reach: The Power of the Multinational Corporation*, Jonathan Cape, London.

Barr, K. (1981) On the capitalist enterprise, *Review of Radical Political Economics*, Vol. 12, pp. 60–70.

Bartlett, C. (1986) Building and managing the transnational, in M.E. Porter (ed) *Competition in Global Industries*, Harvard Business School Press, Boston, MA, Chapter 12.

Bartlett, C.A. and Ghoshal, S. (1989) *Managing Across Borders: The Transnational Solution*, Harvard Business School Press, Boston, MA.

Batty, M. and Barr, R. (1994) The electronic frontier: exploring and mapping cyberspace, *Futures*, Vol. 26, pp. 699–712.

Baylin, F. (1996) *World Satellite Yearbook* (4th edition), Baylin Publications, Boulder, CO.

Behrman, J.N. and Fischer, W.A. (1980) *Overseas R & D Activities of Transnational Companies*, Oelgeschlager, Gunn & Hain, Cambridge, MA.

Behrman, J.N. and Grosse, R.E. (1990) *International Business and Governments: Issues and Institutions*, University of North Carolina Press, Chapel Hill, NC.

Bello, W. and Rosenfeld, S. (1990) *Dragons in Distress: Asia's Miracle Economies in Crisis*, Institute for Food and Development Policy, San Francisco, CA.

Benewick, R. and Wingrove, P. (eds) (1995) *China in the 1990s*, Macmillan, London.

Bernard, M. and Ravenhill, J. (1995) Beyond product cycles and flying geese: regionalization, hierarchy and the industrialization of East Asia, *World Politics*, 47, pp. 171-209.

Bertrand, O. and Noyelle, T. (1988) *Human Resources and Corporate Strategy: Technological Change in Banks and Insurance Companies*, OECD, Paris.

Bessant, J. and Haywood, B. (1988) Islands, archipelagos and continents: progress on the road to computer-integrated manufacturing, *Research Policy*, Vol. 17, pp. 349–62.

Bhagwati, J. (1988) *Protectionism*, MIT Press, Cambridge, MA.

Bhaskar, K. (1980) *The Future of the World Motor Industry*, Kogan Page, London.

Blackaby, F. (ed) (1979) *Deindustrialization*, Heinemann, London.

Blackburn, P., Coombs, R. and Green, K. (1985) *Technology, Economic Growth and the Labour Process*, Macmillan, London.

Bloomfield, G.T. (1978) *The World Automotive Industry*, David & Charles, Newton Abbot.

Bluestone, B. and Harrison, B. (1982) *The Deindustrialization of America*, Basic Books, New York.

Boyer, R. and Drache, D. (eds) (1996) *States Against Markets: The Limits of Globalization*, Routledge, London.

Bradford, C.I. Jr (1987) Trade and structural change: NICs and next tier NICs as transitional economies, *World Development*, Vol. 15, pp. 299–316.

Braudel, F. (1984) *Civilization and Capitalism, 15th–18th Centuries* (3 vols), Collins, London.

Britton, J.N.H. and Gilmour, J.M. (1978) *The Weakest Link*, Science Council of Canada, Ottawa.

Britton, S. (1990) The role of services in production, *Progress in Human Geography*, Vol. 14, pp. 529–46.

Brohman, J. (1996) Postwar development in the Asian NICs: does the neoliberal model fit reality? *Economic Geography*, Vol. 72, pp. 107–31.

Brooks, H.E. and Guile, B.R. (1987) Overview, in B.R. Guile and H.E. Brooks (eds) *Technology and Global Industry: Companies and Nations in the World Economy*, National Academy Press, Washington, DC, pp. 1–15.

Brunn, S.D. and Leinbach, T.R. (eds) (1991) *Collapsing Space and Time: Geographic Aspects of Communication and Information*, HarperCollins, New York.

Buckley, P.J. and Casson, M. (1976) *The Future of the Multinational Enterprise*, Macmillan, London.

Business International (1987) *Competitive Alliances: How to Succeed at Cross-Regional Collaboration*, Business International Corporation, New York.

Bylinsky, G. (1983) The race to the automatic factory, *Fortune*, 21 February, pp. 52–64.

Cable, V. and Henderson, D. (eds) (1994) *Trade Blocs? The Future of Regional Integration*, Royal Institute of International Affairs, London.

Camilleri, J.A. and Falk, J. (1992) *The End of Sovereignty? The Politics of a Shrinking and Fragmenting World*, Edward Elgar, Aldershot.

Cannon, T. and Jenkins, A. (eds) (1990) *The Geography of Contemporary China: The Impact of Deng Xiaoping's Decade*, Routledge, London.

Carter, C. (ed) (1981) *Industrial Policy and Innovation*, Allen & Unwin, London.

Casson, M. (ed) (1983) *The Growth of International Business*, Allen & Unwin, London.

Castells, M. (1989) *The Informational City*, Blackwell, Oxford.

Castells, M. (1996) *The Rise of the Network Society. Volume 1*, Blackwell, Oxford.

Castells, M. and Hall, P. (1994) *Technopoles of the World: The Making of 21st Century Industrial Complexes*, Routledge, London.

Caves, R.E. (1971) International corporations: the industrial economics of foreign investment, *Economica* NS, Vol. XXXVIII, pp. 1–27.

Cawson, A. (1989) European consumer electronics: corporate strategies and public policy, in M. Sharp and P. Holmes (eds) *Strategies for New Technology*, Philip Allan, London, Chapter 3.

Cecchini, P. (1988) *The European Challenge 1992: The Benefits of a Single Market*, Wildwood House, Aldershot.

Cerny, P.G. (1991) The limits of deregulation: transnational interpenetrations and policy change, *European Journal of Political Research*, Vol. 19, pp. 173–96.

Chandler, A.D. Jr (1962) *Strategy and Structure: Chapters in the History of the Industrial Enterprise*, MIT Press, Cambridge, MA.

Chant, S. and McIlwaine, C. (1995) Gender and export manufacturing in the Philippines: continuity or change in female employment? The case of the Mactan Export Processing Zone, *Gender, Place and Culture*, Vol. 2, pp. 147–76.

Chen, C.H. (1986) Taiwan's foreign direct investment, *Journal of World Trade Law*, Vol. 20, pp. 639–64.

Cherry, C. (1978) *World Communication: Threat or Promise?* Wiley, Chichester.

Chesnais, F. (1986) Science, technology and competitiveness, *Science Technology Industry Review*, Vol. 1, pp. 85–129.

Child Hill, R. and Fujita, K. (1996) Flying geese, swarming sparrows or preying hawks? Perspectives on east Asian industrialization, *Competition and Change*, Vol. 1, pp. 285–97.

Choate, P. (1990) *Agents of Influence: How Japan's Lobbyists in the United States Manipulate America's Political and Economic System*, Alfred Knopf, New York.

Clairmonte, F.F. and Cavanagh, J.H. (1981) *The World in their Web: Dynamics of Textile Multinationals*, Zed Press, London.

Clairmonte, F.F. and Cavanagh, J.H. (1984) Transnational corporations and services: the final frontier, *Trade and Development*, Vol. 5, pp. 215–73.

Clark, G.L. (1994) Strategy and structure: corporate restructuring and the scope and characteristics of sunk costs, *Environmental and Planning A*, Vol. 26, pp. 9–32.

Clark, G.L. and O'Connor, K. (1997) The informational content of financial products and the spatial structure of the global finance industry, in K. Cox (ed) *Spaces of Globalization: Reasserting the Power of the Local*, Guilford Press, New York, Chapter 4.

Clark, G.L. and Wrigley, N. (1995) Sunk costs: a framework for economic geography, *Transactions, Institute of British Geographers*, Vol. 20, pp. 204–23.

Cline, W.R. (1982) Can the east Asian model of development be generalised? *World Development*, Vol. 10, pp. 81–90.

Cline, W.R. (1987) *The Future of World Trade in Textiles and Apparel*, Institute for International Economics, Washington, DC.

Coase, R. (1937) The nature of the firm, *Economica*, Vol. 4, 386–405.

Cohen, R.B. (1981) The new international division of labour, multinational corporations and the urban hierarchy, in M. Dear and A.J. Scott (eds) *Urbanization and Urban Planning in Capitalist Society*, Methuen, London, Chapter 12.

Cohen, S.S. and Zysman, J. (1987) *Manufacturing Matters: The Myth of the Post-Industrial Economy*, Basic Books, New York.

Commission on Global Governance (1995) *Our Global Neighbourhood*, Oxford University Press, New York.

Coombs, R. and Jones, B. (1989) Alternative successors to Fordism, in H. Ernste and C. Jaeger (eds) *Information Society and Spatial Structure*, Belhaven Press, London, Chapter 8.

Corbridge, S. (1993) *Debt and Development*, Blackwell, Oxford.

Corbridge, S., Martin, R. and Thrift, N. (eds) (1994) *Money Power and Space*, Blackwell, Oxford.

Cowling, K. and Sugden, R. (1987) Market exchange and the concept of a transnational corporation, *British Review of Economic Issues*, Vol. 9, pp. 57–68.

Craig, L.C. (1981) Office automation at Texas Instruments Inc., in M.L. Moss (ed) *Telecommunications and Productivity*, Addison-Wesley, Reading, MA, Chapter 7.

Crane, G.T. (1990) *The Political Economy of China's Special Economic Zones*, M.E. Sharpe, Armonk, NY.

Crichton, M. (1992) *Rising Sun*, Random House, New York.

Dahrendorf, R. (1968) *Essays in the Theory of Society*, Routledge and Kegan Paul, London.

Daly, H.E. (1993) The perils of free trade, *Scientific American*, November, pp. 24–29.

Daniels, P.W. (1993) *Service Industries in the World Economy*, Blackwell, Oxford.

Daniels, P.W., Thrift, N.J. and Leyshon, A. (1989) Internationalization of professional producer services: accountancy conglomerates, in P. Enderwick (ed) *Multinational Service Firms*, Routledge, London, Chapter 4.

Davis, E. and Smailes, C. (1989) The integration of European financial services, in E. Davis *et al.* (eds) *1992: Myths and Realities*, London Business School, London, Chapter 5.

De Sola Pool, I. (1981) International aspects of telecommunications policy, in M.L. Moss (ed) *Telecommunications and Productivity*, Addison-Wesley, Reading, MA, Chapter 7.

Deyo, F.C. (1992) The political economy of social policy formation: east Asia's newly industrialized countries, in R.P. Appelbaum and J. Henderson (eds) *States and Development in the Asian Pacific Rim*, Sage, London, Chapter 11.

Dicken, P. (1987) A tale of two NICs: Hong Kong and Singapore at the crossroads, *Geoforum*, Vol. 18, pp. 151–64.

Dicken, P. (1988) The changing geography of Japanese foreign direct investment in manufacturing industry: a global perspective, *Environment and Planning A*, Vol. 20, 633–53.

Dicken, P. (1992) Europe 1992 and strategic change in the international automobile industry, *Environment and Planning A*, Vol. 24, pp. 11–31.

Dicken, P. (1994) Global–local tensions: firms and states in the global space-economy, *Economic Geography*, Vol. 70, pp. 101–28.

Dicken, P., Forsgren, M. and Malmberg, A. (1994) The local embeddedness of transnational corporations, in A. Amin and N. Thrift (eds) *Globalization, Institutions, and Regional Development in Europe*, Oxford University Press, Oxford, Chapter 2.

Dicken, P. and Kirkpatrick, C. (1991) Services-led development in ASEAN: transnational regional headquarters in Singapore, *The Pacific Review*, Vol. 4, pp. 174–84.

Dicken, P. and Lloyd, P.E. (1981) *Modern Western Society*, Harper & Row, London.

Dicken, P. and Lloyd, P.E. (1990) *Location in Space: Theoretical Perspectives in Economic Geography* (3rd edn), Harper & Row, New York.

Dicken, P. and Miyamachi, Y. (1998) 'From noodles to satellites': the changing geography of the Japanese *sogo shosha*, *Transactions, Institute of British Geographers*, Vol. 23, 1.

Dicken, P., Peck, J.A. and Tickell, A. (1997) Unpacking the global, in R. Lee and J. Wills (eds) *Geographies of Economies*, Arnold, London, Chapter 12.

Dicken, P., Tickell, A. and Yeung, H.W.C. (1997) Putting Japanese investment in Europe in its place, *Area*, Vol. 29, pp. 200–12.

Dickerson, K.G. (1991) *Textiles and Apparel in the International Economy*. Macmillan, New York.

Diebold, W. Jr (1982) Past and future industrial policy in the United States, in J. Pinder (ed) *National Industrial Strategies and the World Economy*, Croom Helm, London, Chapter 6.

Dizard, W.P. (1966) *Television: A World View*, Syracuse University Press, Syracruse, NY.

Dodwell Marketing Consultants (1992) *Industrial Groupings in Japan*, Dodwell, Tokyo.

Donaghu, M.T. and Barff, R. (1990) Nike just did it: international subcontracting and flexibility in athletic footwear production, *Regional Studies*, Vol. 24, pp. 537–52.

Dore, R. (1986) *Flexible Rigidities: Industrial Policy and Structural Adjustment in the Japanese Economy, 1970–1980*, Stanford University Press, Stanford, CA.

Dosi, G. (1983) Semiconductors: Europe's precarious survival in high technology, in G. Shepherd, F. Duchene and C. Saunders (eds) *Europe's Industries: Public and Private Strategies for Change*, Pinter, London, Chapter 9.

Dosi, G., Freeman, C., Nelson, R., Silverberg, G. and Soete, L. (eds) (1988) *Technical Change and Economic Theory*, Pinter, London.

Douglass, M. (1994) The 'developmental state' and the newly industrialized economies of Asia, *Environment and Planning A*, Vol. 26, pp. 543–66.

Doz, Y. (1986a) Government polices and global industries, in M.E. Porter (ed) *Competition in Global Industries*, Harvard Business School, Boston, MA, Chapter 7.

Doz, Y. (1986b) *Strategic Management in Multinational Companies*, Pergamon, Oxford.

Drache, D. (1996) New work and employment relations: lean production in Japanese auto transplants in Canada, in R. Boyer and D. Drache (eds) *States Against Markets: The Limits of Globalization*, Routledge, London, Chapter 10.

Drucker, P. (1946) *The Concept of the Corporation*, John Day, New York.

Drucker, P. (1986) The changed world economy, *Foreign Affairs*, Vol. 64, pp. 768–91.

Dunford, M. (1990) Theories of regulation, *Environment and Planning D: Society and Space*, Vol. 8, pp. 297–322.

Dunford, M. and Kafkalas, G. (1992) The global–local interplay, corporate geographies and spatial development strategies in Europe, in M. Dunford and G. Kafkalas (eds) *Cities and Regions in the New Europe: The Global–Local Interplay and Spatial Development Strategies*, Belhaven Press, London, Chapter 1.

Dunning, J.H. (1973) The determinants of international production, *Oxford Economic Papers*, Vol. 25, pp. 289–336.

Dunning, J.H. (1977) Trade, location of economic activity and the MNE: a search for an eclectic approach, in B. Ohlin, P.O. Hesselborn and P.M. Wijkman (eds) *The International Allocation of Economic Activity*, Macmillan, London, Chapter 12.

Dunning, J.H. (1979) Explaining changing patterns of international production: in defence of the eclectic theory, *Oxford Bulletin of Economics and Statistics*, Vol. 41, pp. 269–96.

Dunning, J.H. (1980) Towards an eclectic theory of international production: some empirical tests, *Journal of International Business Studies*, Vol. 11, pp. 9–31.

Dunning, J.H. (1983) Changes in the level and structure of international production: the last 100 years, in M. Casson (ed) *The Growth of International Business*, Allen & Unwin, London, Chapter 5.

Dunning, J.H. (1989) Transnational corporations and the growth of services: some conceptual and theoretical issues, *UNCTC Current Studies*, 9.

Dunning, J.H. (1992) The competitive advantages of countries and the activities of transnational corporations, *Transnational Corporations*, Vol. 1, pp. 135–68.

Dunning, J.H. (1993) *Multinational Enterprises and the Global Economy*, Addison-Wesley, Reading, MA.

Dunning, J.H. and Rugman, A.M. (1985) The influence of Hymer's dissertation on the theory of foreign direct investment, *American Economic Review, Papers & Proceedings*, Vol. 75, pp. 228–32.

Dwyer, D. (ed) (1994) *China: The Next Decades*, Longman, Harlow.

Eden, L. and Molot, M.A. (1993) Insiders and outsiders: defining 'who is us' in the North American automobile industry, *Transnational Corporations*, Vol. 2, pp. 31–64.

Elson, D. (1988) Transnational corporations in the new international division of labour: a critique of 'cheap labour' hypotheses, *Manchester Papers in Development*, Vol. IV, pp. 352–76.

Elson, D. (1989) The cutting edge: multinationals in the EEC textiles and clothing industry, in D. Elson and R. Pearson (eds) *Women's Employment and Multinationals in Europe*, Macmillan, Basingstoke, Chapter 5.

Elson, D. (1990) Marketing factors affecting the globalization of textiles, *Textiles Outlook International*, March, pp. 51–61.

Encarnation, D.J. and Wells, L.T. Jr (1986) Competitive strategies in global industries: a view from host governments, in M.E. Porter (ed) *Competition in Global Industries*, Harvard Business School, Boston, MA, Chapter 8.

Enderwick, P. (1982) Labour and the theory of the multinational corporation, *Industrial Relations Journal*, Vol. 13, pp. 32–43.

Enderwick, P. (1989) Multinational corporate restructuring and international competitiveness, *California Management Review*, Vol. 32, pp. 44–58.

Ernst, D. (1985) Automation and the worldwide restructuring of the electronics industry: strategic implications for developing countries, *World Development*, Vol. 13, pp. 333–52.

Euh, J.D. and Min, S.H. (1986) Foreign direct investment from developing countries: the case of Korean firms. *The Developing Economies*, Vol. XXIV, pp. 149–68.

European Commission (1996) *Employment in Europe*, EC, Brussels.

Fernandez-Kelly, M.P. (1989) International development and industrial restructuring: the case of garment and electronics industries in southern California, in A. MacEwan and W.K. Tabb (eds) *Instability and Change in the World Economy*, Monthly Review Press, New York, pp. 147–65.

FIET (1996) *A Social Dimension to Globalization*, FIET, Geneva.

Forester, T. (ed) (1985) *The Information Technology Revolution*, Blackwell, Oxford.

Forester, T. (1987) *High-Tech Society: The Story of the Information Technology Revolution*, Blackwell, Oxford.

Fransman, M. (1990) *The Market and Beyond: Cooperation and Competition in Information Technology in the Japanese System*, Cambridge University Press, Cambridge.

Freeman, C. (1982) *The Economics of Industrial Innovation*, Pinter, London.

Freeman, C. (1987) The challenge of new technologies, in OECD, *Interdependence and Cooperation in Tomorrow's World*, OECD, Paris, pp. 123–56.

Freeman, C. (1988) Introduction, in G. Dosi, C. Freeman, R. Nelson, G. Silverberg and L. Soete (eds) *Technical Change and Economic Theory*, Pinter, London.

Freeman, C., Clark, J. and Soete, L. (1982) *Unemployment and Technical Change*, Pinter, London.

Freeman, C. and Perez, C. (1988) Structural crises of adjustment, business cycles and investment behaviour, in G. Dosi, C. Freeman, R. Nelson, G. Silverberg and L. Soete (eds) *Technical Change and Economic Theory*, Pinter, London, Chapter 3.

Friedmann, J. (1986) The world city hypothesis, *Development and Change*, Vol. 17, pp. 69–83.

Fröbel, F., Heinrichs, J. and Kreye, O. (1980) *The New International Division of Labour*, Cambridge University Press, Cambridge.

Fruin, W.M. (1992) *The Japanese Enterprise System*, Clarendon Press, Oxford.

Fuentes, A. and Ehrenreich, B. (1983) *Women in the Global Factory*, South End Press, Boston, MA.

Fuentes, N.A., Alegria, T., Brannon, J.T., James, D.D. and Lucker, G.W. (1993) Local sourcing and indirect employment: multinational enterprises in northern Mexico, in P. Bailey, A. Parisotto and G. Renshaw (eds) *Multinationals and Employment: The Global Economy of the 1990s*, ILO, Geneva, Chapter 6.

Fujita, M. and Ishigaki, K. (1986) The internationalization of commercial banking, in M.J. Taylor and N.J. Thrift (eds) *Multinationals and the Restructuring of the World Economy*, Croom Helm, London, Chapter 7.

Gabriel, P. (1966) The investment in the LDC: asset with a fixed maturity, *Columbia Journal of World Business*, Vol. 1, pp. 113–20.

Gamble, A. and Payne, A. (eds) (1996) *Regionalism and World Order*, Macmillan, London.

GATT (1989) Services in the domestic and global economy, in *International Trade 1988–1989*, GATT, Geneva, Part III.

Gereffi, G. (1990) Paths of industrialization: an overview, in G. Gereffi and D.L. Wyman (eds) *Manufacturing Miracles: Paths of Industrialization in Latin America and East Asia*, Princeton University Press, Princeton, NJ, Chapter 1.

Gereffi, G. (1994) The organization of buyer-driven global commodity chains: how US retailers shape overseas production networks, in G. Gereffi and M. Korzeniewicz (eds) *Commodity Chains and Global Capitalism*, Praeger, Westport, CT, Chapter 5.

Gereffi, G. (1995) Global production systems and Third World development, in B. Stallings (ed) *Global Change, Regional Response: The International Context of Development*, Cambridge University Press, Cambridge, Chapter 4.

Gereffi, G. (1996a) Commodity chains and regional divisions of labor in east Asia, *Journal of Asian Business*, Vol. 12, pp. 75–112.

Gereffi, G. (1996b) Global commodity chains: new forms of coordination and control among nations and firms in international industries, *Competition & Change*, Vol. 1, pp. 427–39.

Gereffi, G. and Korzeniewicz, M. (eds) (1994) *Commodity Chains and Global Capitalism*, Praeger, Westport, CT.

Gereffi, G. and Wyman, D.L. (eds) (1990) *Manufacturing Miracles: Paths of Industrialization in Latin America and East Asia*, Princeton University Press, Princeton, NJ.

Gerlach, M.L. (1992) *Alliance Capitalism: The Social Organization of Japanese Business*, University of California Press, Berkeley, CA.

Germidis, D. (ed) (1980) *International Subcontracting: A New Form of Investment*, OECD, Paris.

Gertler, M. (1988) The limits to flexibility: comments on the post-Fordist vision of production and its geography, *Transactions, Institute of British Geographers*, Vol. 13, pp. 419–32.

Gibb, R. and Michalak, W. (eds) (1994) *Continental Trading Blocs: The Growth of Regionalism in the World Economy*, Wiley, Chichester.

Gibbs, M. (1985) Continuing the international debate on services, *Journal of World Trade Law*, Vol. 19, pp. 199–218.

Giddy, I.H. (1978) The demise of the product cycle model in international business theory, *Columbia Journal of World Business*, Vol. 13, pp. 90–97.

Gilbert, A. and Gugler, J. (1982) *Cities, Poverty and Development: Urbanization in the Third World*, Oxford University Press, Oxford.

Glasmeier, A., Thompson, J.W. and Kays, A.J. (1993) The geography of trade policy: trade regimes and location decisions in the textile and apparel complex, *Transactions, Institute of British Geographers*, Vol. 18, pp. 19–35.

Glickman, N.J. and Woodward, D.P. (1989) *The New Competitors: How Foreign Investors are Changing the US Economy*, Basic Books, New York.

Glyn, A. and Sutcliffe, B. (1992) 'Global but leaderless'? The new capitalist order, in R. Miliband and L. Panitch (eds) *New World Order: The Socialist Register*, Merlin Press, London, pp. 76–95.

Gordon, D.M. (1988) The global economy: new edifice or crumbling foundations? *New Left Review*, Vol. 168, pp. 24–64.

Graham, E.M. and Krugman, P.R. (1989) *Foreign Direct Investment in the United States*, Institute for International Economics, Washington, DC.

Graham, S. and Marvin, S. (1996) *Telecommunications and the City: Electronic Spaces, Urban Places*, Routledge, London.

Granovetter, M. and Swedberg, R. (eds) (1992) *The Sociology of Economic Life*, Westview Press, Boulder, CO.

Gretschmann, K. (1994) Germany in the global economy of the 1990s: from player to pawn? In R. Stubbs and G.R.D. Underhill (eds) *Political Economy and the Changing Global Order*, Macmillan, London, Chapter 29.

Griffiths-Jones, S. and Stallings, B. (1995) New global financial trends: implications for development, in B. Stallings (ed) *Global Change, Regional Response: The New International Context of Development*, Cambridge University Press, Cambridge, Chapter 5.

Grosse, R. and Behrman, J.N. (1992) Theory in international business, *Transnational Corporations*, Vol. 1, pp. 93–126.

Grotjohann, J., Sterkenburg, T. and van Grunsven, L. (1996) Sourcing strategies and local embeddedness of MNCs in the Asian Pacific Rim: the case of Philips Electronics in Malaysia, *Geographical Studies of Development and Resource Use, University of Utrecht*.

Grubel, H.G. (1989) Multinational banking, in P. Enderwick (ed) *Multinational Service Firms*, Routledge, London, Chapter 3.

Grunsven, L. van, Egeraat, C. and Meijsen, S. (1995) New manufacturing establishments and regional economy in Johor: production linkages, employment and labour fields, *Department of Geography of Developing Countries, University of Utrecht Report Series*.

Guisinger, S. (1985) *Investment Incentives and Performance Requirements*, Praeger, New York.

Hachten, W.A. (1974) Mass media in Africa, in A. Wells (ed) *Mass Communications: A World View*, National Press Books, Palo Alto, CA, Chapter 9.

Hagedoorn, J. and Schakenraad, J. (1990) Inter-firm partnerships and cooperative strategies in core technologies, in B. Dankbaar, J. Groenewegen and H. Schenk (eds) *Perspectives in Industrial Economics*, Kluwer, Dordrecht, pp. 47–65.

Haggard, S. (1995) *Developing Nations and the Politics of Global Integration*, The Brookings Institution, Washington, DC.

Haig, R.M. (1926) Toward an understanding of the metropolis, *Quarterly Journal of Economics*, Vol. 40, pp. 421–33.

Hall, P. and Preston, P. (1988) *The Carrier Wave: New Information Technology and the Geography of Innovation, 1846–2003*, Unwin Hyman, London.

Hamill, J. (1984) Labour relations, decision-making within multinational corporations, *Industrial Relations Journal*, Vol. 15, pp. 30–35.

Hamill, J. (1993) Employment effects of the changing strategies of multinational enterprises, in P. Bailey, A. Parisotto and G. Renshaw (eds) *Multinationals and Employment: The Global Economy of the 1990s*, ILO, Geneva, Chapter 3.

Hamilton, A. (1986) *The Financial Revolution*, Penguin Books, Harmondsworth.

Hamilton, G. (ed) (1991) *Business Networks and Economic Development in East and South East Asia*, Centre of Asian Studies, University of Hong Kong.

Hampden-Turner, C. and Trompenaars, A. (1994) *The Seven Cultures of Capitalism*, Doubleday, New York.

Harris, G. (1984) The globalization of advertising, *International Journal of Advertising*, Vol. 3, pp. 223–34.

Harrison, B. (1994) *Lean and Mean: The Changing Landscape of Corporate Power in the Age of Flexibility*, Basic Books, New York.

Harvey, D. (1982) *The Limits to Capital*, Blackwell, Oxford.

Harvey, N. (ed) (1993) *Mexico: Dilemmas of Transition*, Institute of Latin American Studies, University of London, London.

Hawkins, R.G. (1972) Job displacement and the multinational firm: a methodological review, *Center for Multinational Studies, Occasional Paper 3*, Washington, DC.

Heeks, R. (1996) Global software outsourcing to India by multinational corporations, in P. Palvia, S.C. Palvia and E.M. Roche (eds) *Global Information Technology and Systems Management: Key Issues and Trends*, Ivy League Publishing, Nashua, NH, Chapter 17.

Heenan, D.A. and Perlmutter, H. (1979) *Multinational Organizational Development: A Social Architecture Perspective*, Addison-Wesley, Reading, MA.

Helleiner, G.K. and Lavergne, R. (1979) Intra-firm trade and industrial exports to the United States, *Oxford Bulletin of Economics and Statistics*, Vol. 41, pp. 297–312.

Helou, A. (1991) The nature and competitiveness of Japan's *keiretsu*, *Journal of World Trade*, Vol. 25, pp. 99–131.

Henderson, J. (1989) *The Globalization of High Technology Production*, Routledge, London.

Henderson, J. (1993) The role of the state in the economic trasnformation of east Asia, in C. Dixon and D. Drakakis-Smith (eds) *Growth Economies of Asia*, Routledge, London, Chapter 5.

Henderson, J. (1994) Electronics industries and the developing world: uneven contributions and uneven prospects, in L. Sklair (ed) *Capitalism and Development*, Routledge, London, Chapter 13.

Henderson, J. and Appelbaum, R.P. (1992) Situating the state in the east Asian development process, in R.P. Appelbaum and J. Henderson (eds) *States and Development in the Asian Pacific Rim*, Sage, London, Chapter 1.

Henderson, J. and Castells, M. (eds) (1987) *Global Restructuring and Territorial Development*, Sage, London.

Henry, N. (1992) The new industrial spaces: locational logic of a new production era? *International Journal of Urban and Regional Research*, Vol. 16, pp. 375–96.

Hepworth, M. (1989) *Geography of the Information Economy*, Belhaven, London.

Herod, A. (1997) From a geography of labor to a labor geography: rethinking conceptions of labor in economic geography, *Antipode*, Vol. 29, pp. 1–31.

Hesselman, L. (1983) Trends in European industrial intervention, *Cambridge Journal of Economics*, Vol. 7, pp. 197–208.

Higgott, R. (1993) Competing theoretical approaches to international cooperation: implications for the Asia-Pacific, in R. Higgott, R. Leaver and J. Ravenhill (eds) *Pacific Economic Relations in the 1990s*, Allen & Unwin, London, Chapter 14.

Hill, H. (1990) Foreign investment and east Asian economic development, *Asian-Pacific Economic Literature*, Vol. 4, pp. 21–58.

Hill, H. (1993) Employment and multinational enterprises in Indonesia, in P. Bailey, A. Parisotto and G. Renshaw (eds) *Multinationals and Employment: The Global Economy of the 1990s*, ILO, Geneva, Chapter 7.

Hirsch, S. (1967) *Location of Industry and International Competitiveness*, Clarendon Press, Oxford.

Hirsch, S. (1972) The United States electronics industry in international trade, in L.T. Wells Jr (ed) *The Product Life Cycle and International Trade*, Harvard Business School, Boston, MA, pp. 39–54.

Hirst, P. and Thompson, G. (1992) The problem of 'globalization': international economic relations, national economic management and the formation of trading blocs, *Economy and Society*, Vol. 24, pp. 408–42.

Hirst, P. and Thompson, G. (1996) *Globalization in Question*, Polity Press, Cambridge.

Hobday, M. (1994) Export-led development in the four dragons: the case of electronics, *Development and Change*, Vol. 25, pp. 333–61.

Hobday, M. (1995) East Asian latecomer firms: learning the technology of electronics, *World Development*, Vol. 23, pp. 1171–93.

Hobsbawm, E. (1979) The development of the world economy, *Cambridge Journal of Economics*, Vol. 3, pp. 305–18.

Hobsbawm, E. (1994) *Age of Extremes: The Short Twentieth Century, 1914–1991*, Michael Joseph, London.

Hodgson, G.M. (1993) *Economics and Evolution: Bringing Life Back into Economics*, Polity Press, Cambridge.

Hoekman, B. and Kostecki, M. (1995) *The Political Economy of the World Trading System: From GATT to WTO*, Oxford University Press, Oxford.

Hoffman, K. (1985) Clothing, chips and competitive advantage: the impact of microelectronics on trade and production in the garment industry, *World Development*, Vol. 13, pp. 371–92.

Hoffman, K. and Rush, H. (1988) *Microelectronics and Clothing: The Impact of Technical Change on a Global Industry*, Praeger, New York.

Hofstede, G. (1980) *Culture's Consequences*, Sage, London.

Holmes, J. (1986) The organization and locational structure of production subcontracting, in A.J. Scott and M. Storper (eds) *Production, Work and Territory: The Geographical Anatomy of Industrial Capitalism*, Allen & Unwin, London, Chapter 5.

Holmes, J. (1990) The globalization of production and the future of Canada's mature industries: the case of the automobile industry, in D. Drache and M. Gertler (eds) *The New Era of Global Competition: State Policy and Market Power*, McGill-Queens Press, Montreal, Chapter 7.

Holmes, J. (1992) The continental integration of the North American automobile industry; from the Auto Pact to the FTA, *Environment and Planning A*, Vol. 24, pp. 95–120.

Hood, N. and Young, S. (1982) US multinational R&D: corporate strategies and policy implications for the UK, *Multinational Business*, Vol. 2, pp. 10–23.

Hopkins, T.K. and Wallerstein, I. (1986) Commodity chains in the world-economy, *Review*, Vol. X, pp. 157–70.

Howells, J.R.L. (1990) The internationalization of R&D and the development of global research networks, *Regional Studies*, Vol. 24, pp. 495–512.

Hu, Y.-S. (1992) Global firms are national firms with international operations, *California Management Review*, Vol. 34, pp. 107–26.

Hu, Y.-S. (1995) The international transferability of the firm's advantages, *California Management Review*, Vol. 37, pp. 73–88.

Hudson, R. and Schamp, E. (eds) (1995) *Towards a New Map of Automobile Manufacturing in Europe?* Springer-Verlag, Berlin.

Humbert, M. (1994) Strategic industrial policies in a global industrial system, *Review of Intrnational Political Economy*, Vol. 1, pp. 445–64.

Hymer, S.H. (1972) The multinational corporation and the law of uneven development, in J.N. Bhagwati (ed) *Economics and World Order*, Macmillan, London, pp. 113–40.

Hymer, S.H. (1976) *The International Operations of National Firms: A Study of Direct Foreign Investment*, MIT Press, Cambridge, MA.

ICFTU (1996) *Behind the Wire*, ICFTU, Brussels.

ILO (1976) *Employment, Growth and Basic Needs: A One-World Problem*, ILO, Geneva.

ILO (1981a) *Employment Effects of Multinational Enterprises in Developing Countries*, ILO, Geneva.

ILO (1981b) *Employment Effects of Multinational Enterprises in Industrialized Countries*, ILO, Geneva.

ILO (1984) *Technology Choice and Employment Generation by Multinational Enterprises in Developing Countries*, ILO, Geneva.

ILO (1988) *Economic and Social Effects of Multinational Enterprises in Export Processing Zones*, ILO, Geneva.

ILO (1996a) *Child Labour: What is to be Done?*, ILO, Geneva.

ILO (1996b) *Globalization of the Footwear, Textiles and Clothing Industries*, ILO, Geneva.

ILO (1997) *World Employment, 1996/97: National Policies in a Global Context*, ILO, Geneva.

Inoguchi, T. and Okimoto, D.I. (1988) *The Political Economy of Japan*, Stanford University Press, Stanford, CA.

Isard, W. (1956) *Location and Space-Economy*, MIT Press, Cambridge, MA.

Jenkins, R. (1984) Divisions over the international division of labour, *Capital and Class*, Vol. 22, pp. 28–57.

Jessop, B. (1994) Post-Fordism and the state, in A. Amin (ed) *Post-Fordism: A Reader*, Blackwell, Oxford, Chapter 8.

Johnson, C. (1982) *MITI and the Japanese Economic Miracle: The Growth of Industrial Policy, 1925–1975*, Stanford University Press, Stanford, CA.

Johnson, C. (1985) The institutional foundations of Japanese industrial policy, *California Management Review*, Vol. XXVII, pp. 59–69.

Johnston, R. and Lawrence, P.R. (1988) Beyond vertical integration – the rise of the value-adding partnership, *Harvard Business Review*, July–August, pp. 94–101.

Jones, H.R. (1990) *A Population Geography* (2nd edition), Paul Chapman, London.

Julius, DeAnne (1990) *Global Companies and Public Policy: The Growing Challenge of Foreign Direct Investment*, Pinter, London.

Kao, J. (1993) The worldwide web of Chinese business, *Harvard Business Review*, March–April, pp. 24–36.

Kaplinsky, R. (1988) Restructuring the capitalist labour process: some lessons from the car industry, *Cambridge Journal of Economics*, Vol. 12, pp. 451–70.

Kelly, R. (1995) Derivatives: a growing threat to the international financial system, in J. Michie and J. Grieve-Smith (eds) *Managing the Global Economy*, Oxford University Press, Oxford, Chapter 9.

Kennedy, P. (1987) *The Rise and Fall of the Great Powers*, Random House, New York.

Kenney, M. and Florida, R. (1989) Japan's role in a post-Fordist age, *Futures*, Vol. 21, pp. 136–51.

Khanna, S.R. (1993) Structural changes in Asian textiles and clothing industries: the second migration of production, *Textile Outlook International*, September, pp. 11–32.

Kim, H.Y. and Lee, S.-H. (1994) Commodity chains and the Korean automobile industry, in G. Gereffi and M. Korzeniewicz (eds) *Commodity Chains and Global Capitalism*, Praeger, Westport, CT, Chapter 14.

Kindleberger, C.P. (1969) *American Business Abroad*, Yale University Press, New Haven, CT.

Kindleberger, C.P. (1988) The 'new' multinationalization of business, *Asean Economic Bulletin*, Vol. 5, pp. 113–24.

Kitson, M. and Michie, J. (1995) Trade and growth: a historical perspective, in J. Michie and J. Grieve-Smith (eds) *Managing the Global Economy*, Oxford University Press, Oxford, Chapter 1.

Knickerbocker, F.T. (1973) *Oligopolistic Reaction and Multinational Enterprises*, Harvard Business School Press, Boston, MA.

Kobrin, S.J. (1987) Testing the bargaining hypothesis in the manufacturing sector in developing countries, *International Organization*, Vol. 41, pp. 609–38.

Kobrin, S.J. (1988) Strategic integration in fragmented environments: social and political assessments by subsidiaries of multinational firms, in N. Hood and J.E. Vahlne (eds) *Strategies in Global Competition*, Croom Helm, London, Chapter 4.

Koo, H. and Kim, E.M. (1992) The developmental state and capital accumulation in South Korea, in R.P. Appelbaum and J. Henderson (eds) *States and Development in the Asian Pacific Rim*, Sage, London, Chapter 5.

Korzeniewicz, M. (1994) Commodity chains and marketing strategies: Nike and the global athletic footwear industry, in G. Gereffi and M. Korzeniewicz (eds) *Commodity Chains and Global Capitalism*, Praeger, Westport, CT, Chapter 12.

Kostecki, M. (1987) Export-restraint agreements and trade liberalization, *The World Economy*, Vol. 10, pp. 425–53.

Kozul-Wright, R. (1995) Transnational corporations and the nation-state, in J. Michie and J. Grieve-Smith (eds) *Managing the Global Economy*, Oxford University Press, Oxford, Chapter 6.

Krasner, S. (1994) International political economy: abiding discord, *Review of International Political Economy*, Vol. 1, pp. 13–19.

Krugman, P. (ed) (1986) *Strategic Trade Policy and the New International Economics*, MIT Press, Cambridge, MA.

Krugman, P. (1990) *Rethinking International Trade*, MIT Press, Cambridge, MA.

Krugman, P. (1991) *Geography and Trade*, Leuven University Press, Leuven.

Krugman, P. (1994) Competitiveness: a dangerous obsession, *Foreign Affairs*, March–April, pp. 28–44.

Krugman, P. (1995) *Development, Geography and Economic Theory*, MIT Press, Cambridge, MA.

Krugman, P. (1996) *Pop Internationalism*, MIT Press, Cambridge, MA.

Krugman, P. and Obstfeld, M. (1994) *International Economics: Theory and Policy*, (3rd edition), HarperCollins, New York.

KSA (1995) *Cost Comparison for 46 Countries*, KSA, Dusseldorf.

Laigle, L. (1996) New relationships between suppliers and car makers: towards development cooperation, *EUNIT Discussion Paper*, 2.

Lall, S. (1973) Transfer pricing by multinational manufacturing firms, *Oxford Bulletin of Economics and Statistics*, Vol. 35, pp. 173–95.

Lall, S. (1978) Transnationals, domestic enterprises and industrial structure in host LDCs: a survey, *Oxford Economic Papers*, Vol. 30, pp. 217–48.

Lall, S. (1979) Multinationals and market structure in an open developing economy: the case of Malaysia, *Weltwirtschaftliches Archiv*, Vol. 115, pp. 325–50.

Lall, S. (1984) Transnationals and the Third World: changing perceptions, *National Westminster Bank Quarterly Review*, May, pp. 2–16.

Lall, S. (1994) Industrial policy: the role of government in promoting industrial and technological development, *UNCTAD Review*, 1994, pp. 65–90.

Lall, S. and Streeten, P. (1977) *Foreign Investment, Transnationals and Developing Countries*, Macmillan, London.

Langdale, J.V. (1989) The geography of international business telecommunications: the role of leased networks, *Annals of the Association of American Geographers*, Vol. 79, pp. 501–22.

Lawrence, R.Z. (1996) *Regionalism, Multilateralism, and Deeper Integration*, The Brookings Institution, Washington, DC.

League of Nations (1945) *Industrialization and Foreign Trade*, League of Nations, New York.

Lee, N. and Cason, J. (1994) Automobile commodity chains in the NICs: a comparison of South Korea, Mexico, and Brazil, in G. Gereffi and M. Korzeniewicz (eds) *Commodity Chains and Global Capitalism*, Praeger, Westport, CT, Chapter 11.

Leonard, H.J. (1988) *Pollution and the Struggle for World Product: Multinational Corporations, Environment and International Comparative Advantage*, Cambridge University Press, Cambridge.

Leontiades, J. (1971) International sourcing in the LDCs, *Columbia Journal of World Business*, Vol. VI, pp. 19–28.

Leslie, D.A. (1995) Global scan: the globalization of advertising agencies, concepts, and campaigns, *Economic Geography*, Vol. 71, pp. 402–26.

Levitt, T. (1983) The globalization of markets, *Harvard Business Review*, May–June, pp. 92–102.

Lewis, M.K. and Davis, J.T. (1987) *Domestic and International Banking*, Philip Allan, Oxford.

Leyshon, A. (1992) The transformation of regulatory order: regulating the global economy and environment, *Geoforum*, Vol. 23, pp. 249–67.

Leyshon, A., Thrift, N.J. and Tommey, C. (1988) The rise of the British provincial financial centre, *Progress in Planning*, Vol. 31, pp. 151–229.

Lim, C.H. (ed) (1988) *Policy Options for the Singapore Economy*, McGraw-Hill, Singapore.

Lim, L.Y.C. and Pang, E.F. (1986) *Trade, Employment and Industrialization in Singapore*, International Labour Office, Geneva.

Lin, V. (1987) Women electronics workers in south east Asia: the emergence of a working class, in J. Henderson and M. Castells (eds) *Global Restructing and Territorial Development*, Sage, London, Chapter 6.

Lovering, J. (1990) Fordism's unknown successor: a comment on Scott's theory of flexible accumulation and the re-emergence of regional economies, *International Journal of Urban and Regional Research*, Vol. 14, pp. 159–74.

Lyons, D. and Salmon, S. (1995) World cities, multinational corporations, and urban hierarchy: the case of the United States, in P.L. Knox and P.J. Taylor (eds) *World Cities in a World-System*, Cambridge University Press, Cambridge, Chapter 6.

MacBean, A.I. and Snowden, P.N. (1981) *International Institutions in Trade and Finance*, Allen & Unwin, London.

McConnell, J. and Macpherson, A. (1994) The North American Free Trade Agreement: an overview of issues and prospects, in R. Gibb and W. Michalak (eds) *Continental Trading Blocs: The Growth of Regionalism in the World Economy*, John Wiley, Chichester, Chapter 6.

McDermott, M. (1991) Taiwan's electronic companies are targeting Europe, *European Management Journal*, Vol. 9, pp. 466–74.

McGrew, A.G. (1992) Conceptualizing global politics, in A.G. McGrew and P.G. Lewis (eds) *Global Politics: Globalization and the Nation-State*, Polity Press, Cambridge, Chapter 1.

McGrew, A.G. and Lewis, P.G. (eds) (1992) *Global Politics: Globalization and the Nation-State*, Polity Press, Cambridge.

McHale, J. (1969) *The Future of the Future*, George Braziller, New York.

McLuhan, M. (1960) *Understanding Media*, Routledge and Kegan Paul, London.

Magaziner, I.C. and Hout, T.M. (1980) *Japanese Industrial Policy*, Policy Studies Institute, London.

Mair, A., Florida, R. and Kenney, M. (1988) The new geography of automobile production: Japanese transplants in North America, *Economic Geography*, Vol. 64, pp. 352–73.

Majaro, S. (1982) *International Marketing: A Strategic Approach to World Markets*, Allen & Unwin, London.

Malmberg, A. (1996) Industrial geography: agglomeration and local milieu, *Progress in Human Geography*, Vol. 20, pp. 386–97.

Malmberg, A. and Maskell, P. (1997) Towards an explanation of regional specialization and industry agglomeration, *European Planning Studies*, Vol. 5, pp. 25–41.

Malmberg, A., Solvell, O. and Zander, I. (1996) Spatial clustering, local accumulation of knowledge and firm competitiveness, *Geografiska Annaler*, Vol. 76B, pp. 85–97.

Mandel, E. (1980) *Long Waves of Capitalist Development*, Cambridge University Press, Cambridge.

Martin, R. (1994) Stateless monies, global financial integration and national economic autonomy: the end of geography? In S. Corbridge, R. Martin and N. Thrift (eds) *Money, Power and Space*, Blackwell, Oxford, Chapter 11.

Martin, R. and Sunley, P. (1996) Paul Krugman's economics and its implications for regional development theory, *Economic Georgraphy*, Vol. 72, pp. 259–92.

Mason, M. (1994) Historical perspectives on Japanese direct investment in Europe, in M. Mason and D. Encarnation (eds) *Does Ownership Matter? Japanese Multinationals in Europe*, Clarendon Press, Oxford, Chapter 1.

Mensch, G. (1979) *Stalemate in Technology. Innovations Overcome the Depression*, Ballinger, New York.

Metcalfe, J.S. and Diliso, N. (1996) Innovation, capabilities and knowledge: the epistemic connection, in J. de la Mothe and G. Paquet (eds) *Evolutionary Economics and the New International Political Economy*, Pinter, London, Chapter 3.

Michalet, C-A. (1980) International subcontracting: a state of the art, in D. Germidis (ed) *International Subcontracting: A New Form of Investment*, OECD, Paris.

Michie, J. and Grieve Smith, J. (eds) (1995) *Managing the Global Economy*, Oxford University Press, Oxford.

Miles, R.E. and Snow, C.C. (1986) Organizations: new concepts for new forms, *California Management Review*, Vol. XXVIII, pp. 62–73.

Miller, A. (1987) *Timebends: A Life*, Methuen, London.

Mody, A. (1990) Institutions and dynamic comparative advantage: the electronics industry in South Korea and Taiwan, *Cambridge Journal of Economics*, Vol. 14, pp. 291–314.

Mody, A. and Wheeler, D. (1987) Towards a vanishing middle: competition in the world garment industry, *World Development*, Vol. 15, pp. 1269–84.

Morgan, K. and Sayer, A. (1988) *Microcircuits of Capital: 'Sunrise' Industry and Uneven Development*, Polity Press, Cambridge.

Morris, D. and Hergert, M. (1987) Trends in international collaborative agreements, *Columbia Journal of World Business*, Vol. XXII, pp. 15–21.

Mothe de la, J. and Paquet, G. (eds) (1996) *Evolutionary Economics and the New International Political Economy*, Pinter, London.

Musgrave, P.B. (1975) *Direct Investment Abroad and the Multinationals: Effects on the United States Economy*, USGPO, Washington, DC.

Miyashita, K. and Russell, D. (1994) *Keiretsu: Inside the Hidden Japanese Conglomerates*, McGraw-Hill, New York.

Negrine, R. and Papathanassopoulos, S. (1990) *The Internationalization of Television*, Pinter, London.

Nelson, R.R. (ed) (1993) *National Innovation Systems: A Comparative Study*, Oxford University Press, New York.

Newfarmer, R.S. (1983) Multinationals and marketplace magic in the 1980s, in C.P. Kindleberger and D.B. Andretsch (eds) *The Multinational Corporation in the 1980s*, MIT Press, Cambridge, MA, Chapter 8.

Nixson, F. (1988) The political economy of bargaining with transnational corporations: some preliminary observations, *Manchester Papers in Development*, Vol. IV, pp. 377–90.

O'Brien, R. (1992) *Global Financial Integration: The End of Geography*, Royal Institute of International Affairs, London.

Odagiri, H. (1992) *Growth Through Competition, Competition Through Growth: Strategic Management and the Economy in Japan*, Clarendon Press, Oxford.

OECD (1979) *The Impact of the Newly Industrializing Countries on Production and Trade in Manufactures*, OECD, Paris.

OECD (1996a) *International Direct Investment Statistics Yearbook, 1996*, OECD, Paris.

OECD (1996b) *Regionalism and its Place in the Multilateral Trading System*, OECD, Paris.

Office of Technology Assessment (1993) *Multinationals and the National Interest*, Office of Technology Assessment, Washington, DC.

Office of Technology Assessment (1994) *Multinationals and the US Technology Base*, Office of Technology Assessment, Washington, DC.

Ogden, M.R. (1994) Politics in a parallel universe: is there a future for cyberdemocracy? *Futures*, Vol. 26, pp. 713–29.

Ohlin, B. (1933) *Inter-regional and International Trade*, Harvard University Press, Cambridge, MA.

Ohmae, K. (1985) *Triad Power: The Coming Shape of Global Competition*, Free Press, New York.

Ohmae, K. (1990) *The Borderless World: Power and Strategy in the Interlinked Economy*, Free Press, New York.

Ohmae, K. (1995a) *The End of the Nation State: The Rise of Regional Economies*, Free Press, New York.

Ohmae, K. (ed) (1995b) *The Evolving Global Economy: Making Sense of the New World Order*, Harvard Business School, Boston, MA.

Ó hUallacháin, B. (1997) Restructuring the American semiconductor industry: Vertical integration of design houses and wafer fabricators, *Annals of the Association of American Geographers*, Vol. 87, pp. 217–37.

Okimoto, D.I. (1989) *Between MITI and the Market: Japanese Industrial Policy for High Technology*, Stanford University Press, Stanford, CA.

Oman, C. (1989) *New Forms of Investment in Developing Country Industries: Mining, Petroleum, Automobiles, Textiles, Food*, OECD, Paris.

Ostry, S. (1990) *Governments and Corporations in a Shrinking World*, Council on Foreign Relations Press, New York.

Ozawa, T. (1979) *Multinationalism, Japanese Style*, Princeton University Press, Princeton, NJ.

Paliwoda, S.J. (1986) *International Marketing*, Heinemann, London.

Palloix, C. (1975) The internationalization of capital and the circuit of social capital, in H. Radice (ed) *International Firms and Modern Imperialism*, Penguin Books, Harmondsworth, Chapter 3.

Palloix, C. (1977) The self-expansion of capital on a world scale, *Review of Radical Political Economics*, Vol. 9, pp. 1–28.

Patel, P. (1995) Localized production of technology for global markets, *Cambridge Journal of Economics*, Vol. 19, pp. 141–53.

Patel, P. and Pavitt, K. (1991) Large firms in the production of the world's technology: an important case of 'non-globalization', *Journal of International Business Studies*, Vol. 22, pp. 1–21.

Pauly, L.W. and Reich, S. (1997) National structures and multinational corporate behavior: enduring differences in the age of globalization, *International Organization*, Vol. 51, pp. 1–30.

Pearce, D.E. (1995) *Capturing Environmental Value*, Earthscan Publications, London.

Pearce, R. and Singh, S. (1992) *Globalizing Research and Development*, Macmillan, London.

Pearson, C.S. (ed) (1987) *Multinational Corporations, Environment and the Third World*, Duke University Press, Durham, NC.

Peck, J.A. (1990) Circuits of capital and industrial restructuring: adjustment in the Australian clothing industry, *Australian Geographer*, Vol. 21, pp. 33–52.

Peck, J.A. (1996) *Work-Place: The Social Regulation of Labour Markets*, Guilford, New York.

Peck, J.A. (1998) *Workfare States*, Guilford, New York.

Peck, J.A. and Dicken, P. (1996) Tootal: internationalization, corporate restructuring and 'hollowing out', in J.-E. Nilsson, P. Dicken and J.A. Peck (eds) *The Internationalization Process: European Firms in Global Competition*, Paul Chapman, London, Chapter 7.

Peck, J.A. and Miyamachi, Y. (1995) Regulating Japan? Regulation theory versus the Japanese experience, *Environment and Planning D: Society and Space*, Vol. 12, pp. 639–74.

Peet, R. (1983) Relations of production and the relocation of United States manufacturing industry since 1960, *Economic Geography*, Vol. 59, pp. 112–43.

Perez, C. (1985) Microelectronics, long waves and world structural change, *World Development*, Vol. 13, pp. 441–63.

Perry, M. (1990) The internationalization of advertising, *Geoforum*, Vol. 21, pp. 35–50.

Petrella, R. (1984) Technology and employment in Europe: problems and proofs, *Science and Public Policy*, Vol. 11, pp. 352–59.

Phillips, D.R. and Yeh, A.G.O. (1990) Foreign investment and trade: impact on spatial structure of the economy, in T. Cannon and A. Jenkins (eds) *The Geography of Contemporary China: The Impact of Deng Xiaoping's Decade*, Routledge, London, Chapter 9.

Picciotto, S. (1991) The internationalization of the state, *Capital and Class*, Vol. 43, pp. 43–63.

Piore, M.J. and Sabel, C.F. (1984) *The Second Industrial Divide: Possibilities for Prosperity*, Basic Books, New York.

Pitelis, C. (1991) Beyond the nation-state? The transnational firm and the nation-state, *Capital and Class*, Vol. 43, pp. 131–52.

Pitelis, C. and Sugden, R. (eds) (1991) *The Nature of the Transnational Firm*, Routledge, London.

Plant, R. (1981) *Industries in Trouble*, ILO, Geneva.

Porter, M.E. (ed) (1986) *Competition in Global Industries*, Harvard Business School Press, Boston, MA.

Porter, M.E. (1990) *The Competitive Advantage of Nations*, Macmillan, London.

Poynter, T.A. (1985) *Multinational Enterprises and Government Intervention*, Croom Helm, London.

Pralahad, C.K. and Doz, Y. (1987) *The Multinational Mission*, Free Press, New York.

Rabach, E. and Kim, E.M. (1994) Where is the chain in commodity chains? The service sector nexus, in G. Gereffi and M. Korzeniewicz (eds) *Commodity Chains and Global Capitalism*, Praeger, Westport, CT, Chapter 6.

Radice, H. (ed) (1975) *International Firms and Modern Imperialism*, Penguin Books, Harmondsworth.

Ramesh, M. (1995) Economic globalization and policy choices: Singapore, *Governance: An International Journal of Policy and Administration*, Vol. 8, pp. 243–60.

Ramstetter, E.D. (1993) Asian multinationals in the world economy, *International Economic Insights*, Vol. 4, pp. 19–22.

Rapkin, D.P. and Avery, W.P. (eds) (1995) *National Competitiveness in a Global Economy*, Lynne Rienner, Boulder, CO.

Rapkin, D.P. and Strand, J.R. (1995) Competitiveness: useful concept, political slogan, or dangerous obsession? In D.P. Rapkin and W.P. Avery (eds) *National Competitiveness in a Global Economy*, Lynne Rienner, Boulder, CO, Chapter 1.

Redding, S.G. (1991) Weak organizations and strong linkages: managerial ideology and Chinese family business networks, in G. Hamilton (ed) *Business Networks and Economic Development in East and South East Asia*, Centre for Asian Studies, University of Hong Kong, Chapter 3.

Reed, H.C. (1989) Financial centre hegemony, interest rates, and the global political economy, in Y.S. Park and N. Essayyad (eds) *International Banking and Financial Centres*, Kluwer Academic, Boston, MA, Chapter 16.

Reich, R.B. (1991) *The Work of Nations*, Alfred A. Knopf, New York.

Reich, R.B. and Mankin, E.D. (1986) Joint ventures with Japan give away our future, *Harvard Business Review*, March–April, pp. 78–86.

Reich, S. (1989) Roads to follow: regulating direct foreign investment, *International Organization*, Vol. 43, pp. 543–84.

Reich, S. (1996) 'Manufacturing' investment: national variations in the contribution of foreign direct investors to the US manufacturing base in the 1990s, *Review of International Political Economy*, Vol. 3, pp. 27–64.

Reid, N. (1990) Spatial patterns of Japanese investment in the US automobile industry, *Industrial Relations Journal*, Vol. 21, pp. 49–59.

Richardson, J.D. (1990) The political economy of strategic trade policy, *International Organization*, Vol. 44, pp. 107–35.

Rifkin, J. (1995) *The End of Work: The Decline of the Global Labour Force and the Dawn of the Post-Market Era*, Putnam, New York.

Robert, A. (1983) The effects of the international division of labour on female workers in the textile and clothing industries, *Development and Change*, Vol. 14, pp. 19–37.

Roberts, B.R. (1995) *The Making of Citizens: Cities of Peasants Revisited*, Edward Arnold, London.

Roberts, S. (1994) Fictitious capital, fictitious spaces: the geography of offshore financial flows, in S. Corbridge, R. Martin and N. Thrift (eds) *Money, Power and Space*, Blackwell, Oxford, Chapter 5.

Rodan, G. (1991) *The Political Economy of Singapore's Industrialization*, Forum Books, Petaling Jaya.

Root, F.R. (1990) *International Trade and Investment* (6th edition), South Western Publishing, Cincinnati, OH.

Rosenberg, N. (1982) *Inside the Black Box: Technology and Economics*, Cambridge University Press, New York.

Ross, R. and Trachte, K. (1983) Global cities and global classes: the peripheralization of labour in New York City, *Review*, Vol. VI, pp. 393–431.

Rowthorn, R. and Ramaswamy, R. (1997) Deindustrialization: causes and implications, *IMF Working Paper*.

Rowthorn, R. and Wells, J.R. (1987) *Deindustrialization and Foreign Trade: Britain's Decline in a Global Perspective*, Cambridge University Press, Cambridge.

Rugman, A.M. (1981) *Inside the Multinationals*, Croom Helm, London.

Rugman, A.M. and Verbeke, A. (1992) Multinational enterprise and national economic policy, in P.J. Buckley and M. Casson (eds) *Multinational Enterprises in the World Economy: Essays in Honour of John Dunning*, Edward Elgar, Aldershot, pp. 194–211.

Ruigrok, W. and van Tulder, R. (1995) *The Logic of International Restructuring*, Routledge, London.

Sabel, C.F. (1989) Flexible specialization and the re-emergence of regional economies, in P. Hirst and J. Zeitlin (eds) *Reversing Industrial Decline? Industrial Structure and Policy in Britain and her Competitors*, Berg, Oxford, Chapter 1.

Sadler, D. (1995) National and international regulatory framework: the politics of European automobile production and trade, in R. Hudson and E. Schamp (eds) *Towards a New Map of Automobile Manufacturing in Europe?* Springer-Verlag, Berlin, Chapter 2.

Savary, J. (1995) The rise of international co-operation in the European automobile industry: the Renault case, *European Urban and Regional Studies*, Vol. 2, pp. 3–20.

Savary, J. (1996) Thomson Consumer Electronics: from national champion to global contender, in J.-E. Nilsson, P. Dicken, and J.A. Peck (eds) *The Internationalization Process: European Firms in Global Competition*, Paul Chapman, London, Chapter 6.

Saxenian, A. (1994) *Regional Advantage: Culture and Competition in Silicon Valley and Route 128*, Harvard University Press, Cambridge, MA.

Sayer, A. (1986) New developments in manufacturing: the just-in-time system, *Capital and Class*, Vol. 30, pp. 43–72.

Sayer, A. (1989) The 'new' regional geography and the problems of narrative, *Environment and Planning D: Society and Space*, Vol. 7, pp. 253–76.

Sayer, A. and Walker, R. (1992) Beyond Fordism and flexibility, in A. Sayer and R. Walker, *The New Social Economy*, Blackwell, Oxford, Chapter 5.

Scammell, W.M. (1980) *The International Economy Since 1945*, Macmillan, London.

Scheffer, M. (1992) *Trading Places: Fashion, Retailers and the Changing Geography of Clothing Production*, Department of Geography, University of Utrecht, Utrecht.

Schiller, D. (1982) Business users and the telecommunications network, *Journal of Communication*, Vol. 32, pp. 84–96.

Schoenberger, E. (1988a) From Fordism to flexible accumulation: technology, competitive strategies and international location, *Environment and Planning D: Society and Space*, Vol. 6, pp. 245–62.

Schoenberger, E. (1988b) Multinational corporations and the new international division of labour: a critical appraisal, *International Regional Science Review*, Vol. 11, pp. 105–19.

Schoenberger, E. (1989) Thinking about flexibility: a response to Gertler, *Transactions, Institute of British Geographers*, Vol. 14, pp. 98–108.

Schoenberger, E. (1997) *The Cultural Crisis of the Firm*, Blackwell, Oxford.

Schonberger, R.J. (1982) *Japanese Manufacturing Techniques: Nine Hidden Lessons in Simplicity*, Free Press, New York.

Schroeder, T.G. (1989) *Operations Management: Decision Making in the Operations Function* (3rd edition), McGraw-Hill, New York.

Schumpeter, J. (1939) *Business Cycles: A Theoretical, Historical and Statistical Analysis of the Capitalist Process*, McGraw-Hill, New York.

Schumpeter, J. (1943) *Capitalism, Socialism and Democracy*, Allen & Unwin, London.

Scott, A.J. (1988a) Flexible production systems and regional development, *International Journal of Urban and Regional Research*, Vol. 12, pp. 171–85.

Scott, A.J. (1988b) *New Industrial Spaces: Flexible Production, Organization and Regional Development in North America and Western Europe*, Pion, London.

Scott, A.J. (1995) The geographic foundations of industrial performance, *Competition & Change*, 1, pp. 51–66.

Scott, A.J. and Angel, D. (1988) The global assembly operations of US semiconductor firms: a geographical analysis, *Environment and Planning A*, Vol. 20, pp. 1047–67.

Scott-Quinn, B. (1990) US investment banks as multinationals, in G. Jones (ed) *Banks as Multinationals*, Routledge, London, Chapter 5.

Servan-Schreiber, J.-J. (1968) *The American Challenge*, Hamish Hamilton, London.

Sharpston, M. (1975) International subcontracting, *Oxford Economic Papers*, Vol. 27, pp. 94–135.

Sheard, P. (1983) Auto production systems in Japan: organizational and locational features, *Australian Geographical Studies*, Vol. 21, pp. 49–68.

Shepherd, G. (1983) Textiles: new ways of surviving in an old industry, in G. Shepherd, F. Duchene and C. Saunders (eds) *Europe's Industries: Public and Private Strategies for Change*, Pinter, London, Chapter 2.

Shepherd, G., Duchene, F. and Saunders, C. (1983) *Europe's Industries: Public and Private Strategies for Change*, Pinter, London.

Siegel, L. (1980) Delicate bonds: the global semiconductor industry, *Pacific Research*, Vol. 11, pp. 1–26.

Singh, A. (1994) Growing independently of the world economy: Asian economic development since 1980, *UNCTAD Review*, 1994, pp. 91–106.

Singh, A. (1995) How did east Asia grow so fast? Slow progress towards an analytical consensus, *UNCTAD Bulletin*, Vol. 32, pp. 4–5, 8, 14.

Sklair, L. (1989) *Assembling for Development: The Maquila Industry in Mexico and the United States*, Unwin Hyman, London.

Sklair, L. (ed) (1995) *Sociology of the Global System* (2nd edition), Prentice-Hall/Harvester Wheatsheaf, Hemel Hempstead.

Smelser, N. and Swedberg, R. (eds) (1994) *The Handbook of Economic Sociology*, Princeton University Press, Princeton, NJ.

Smith, D.M. (1981) *Industrial Location: An Industrial-Geographical Analysis* (2nd edition), Wiley, New York.

Stallings, B. (ed) (1995) *Global Change, Regional Response: The New International Context of Development*, Cambridge University Press, Cambridge.

Stegemann, K. (1989) Policy rivalry among industrial states: what can we learn from models of strategic trade policy? *International Organization*, Vol. 43, pp. 73–100.

Stopford, J.M. and Strange, S. (1991) *Rival States, Rival Firms: Competition for World Market Shares*, Cambridge University Press, Cambridge.

Storper, M. (1985) Oligopoly and the product cycle: essentialism in economic geography, *Economic Geography*, Vol. 61, pp. 260–82.

Storper, M. (1992) The limits to globalization: technology districts and international trade, *Economic Geography*, Vol. 68, pp. 60–93.

Storper, M. (1995) The resurgence of regional economies, ten years later: the region as a nexus of untraded interdependencies, *European Urban and Regional Studies*, Vol. 2, pp. 191–221.

Storper, M. (1997) *The Regional World: Territorial Development in a Global Economy*, Guilford Press, New York.

Storper, M. and Walker, R. (1983) The theory of labour and the theory of location, *International Journal of Urban and Regional Research*, Vol. 7, pp. 1–41.

Storper, M. and Walker, R. (1984) The spatial division of labour: labour and the location of industries, in L. Sawers and W.K. Tabb (eds) *Sunbelt/Snowbelt: Urban Development and Regional Restructuring*, Oxford University Press, New York, Chapter 2.

Storper, M. and Walker, R. (1989) *The Capitalist Imperative: Territory, Technology, and Industrial Growth*, Blackwell, Oxford.

Strange, S. (1986) *Casino Capitalism*, Blackwell, Oxford.

Strange, S. (1994) Wake up, Krasner, the world *has* changed, *Review of International Political Economy*, Vol. 1, pp. 209–20.

Strange, S. (1996) *The Retreat of the State: The Diffusion of Power in the World Economy*, Cambridge University Press, Cambridge.

Stubbs, R. and Underhill, G.R.D. (eds) (1994) *Political Economy and the Changing Global Order*, Macmillan, London.

Taplin, I.M. (1994) Strategic reorientations of US apparel firms, in G. Gereffi and M. Korzeniewicz (eds) *Commodity Chains and Global Capitalism*, Praeger, Westport, CT, Chapter 10.

Taylor, M.J. (1986) The product cycle model: a critique, *Environment and Planning A*, Vol. 18, pp. 751–61.

Taylor, P.J. (1994) The state as container: territoriality in the modern world-system, *Progress in Human Geography*, Vol. 18, pp. 151–62.

Taylor, M.J. and Thrift, N. (eds) (1986) *Multinationals and the Restructuring of the World Economy*, Croom Helm, London.

Terpstra, V. and David, K. (1991) *The Cultural Environment of International Business*, South-Western Publishing, Cincinnati, OH.

Thoburn, J. and Howell, J. (1995) Trade and development: the political economy of China's open policy, in R. Benewick and P. Wingrove (eds) *China in the 1990s*, Macmillan, London, Chapter 14.

Thrift, N. (1987) The fixers: the urban geography of international commercial capital, in J. Henderson and M. Castells (eds) *Global Restructuring and Territorial Development*, Sage, London, Chapter 9.

Thrift, N. (1990a) Doing regional geography in a global system: the new international financial system, the City of London and the south east of England, 1984–1987, in R.J. Johnston, J. Hauer and G.A. Hoekveld (eds) *Regional Geography: Current Developments and Future Prospects*, Routledge, London, pp. 180–207.

Thrift, N. (1990b) The perils of the international financial system, *Environment and Planning A*, Vol. 22, pp. 1135–36.

Thrift, N. (1994) On the social and cultural determinants of international financial centres: the case of the City of London, in S. Corbridge, R. Martin and N. Thrift (eds) *Money, Power and Space*, Blackwell, Oxford, Chapter 14.

Tickell, A. (1997) Restructuring the financial system into the 21st century, *Capital and Class*, Vol. 62, pp. 13–19.

Tickell, A. and Peck, J.A. (1992) Accumulation, regulation and the geographies of post-Fordism: missing links in regulationist research, *Progress in Human Geography*, Vol. 16, pp. 190–218.

Tobin, J. (1984) Unemployment in the 1980s: macroeconomic diagnosis and prescription, in A.J. Pierre (ed) *Unemployment and Growth in the Western Economies*, Council on Foreign Relations, New York, pp. 79–112.

Todaro, M.P. (1989) *Economic Development in the Third World*, (4th edition), Longman, New York.

Toffler, A. (1971) *Future Shock*, Pan, London.

Tolchin, M. and Tolchin, S. (1988) *Buying into America: How Foreign Money is Changing the Face of our Nation*, Times Books, New York.

Tomaney, J. (1994) A new paradigm of work organization and technology? In A. Amin (ed) *Post-Fordism: A Reader*, Blackwell, Oxford, Chapter 5.

Toyne, B., Arpan, J.S., Barnett, A.H., Ricks, D.A. and Shimp, T.A. (1984) *The Global Textile Industry*, Allen & Unwin, London.

Turner, L. (1982) Consumer electronics: the colour television case, in L. Turner and N. McMullen (eds) *The Newly Industrializing Countries: Trade and Adjustment*, Allen & Unwin, London. Chapter 4.

Turok, I. (1993) Inward investment and local linkages: how deeply embedded is 'Silicon Glen'? *Regional Studies*, Vol. 27, pp. 401–17.

Tyson, L.D. (1993) *Who's Bashing Whom? Trade Conflict in High-Technology Industries*, Institute for International Economics, Washington, DC.

Tyson, L.D. and Yoffie, D.B. (1993) Semiconductors: from manipulated to managed trade, in D.B. Yoffie (ed) *Beyond Free Trade: Firms, Governments, and Global Competition*, Harvard Business School Press, Boston, MA, Chapter 2.

UNCTAD (1988) Services in the world economy, in *Trade and Development Report 1988*, United Nations, New York, Part 2.

UNCTAD (1993a) *Environmental Management in Transnational Corporations: Report on the Benchmark Corporate Environmental Survey*, United Nations, New York.

UNCTAD (1993b) *World Investment Report 1993: Transnational Corporations and Integrated International Production*, United Nations, New York.

UNCTAD (1994) *World Investment Report 1994: Transnational Corporations, Employment and the Workplace*, United Nations, New York.

UNCTAD (1995) *World Investment Report 1995: Transnational Corporations and Competitiveness*, United Nations, New York.

UNCTAD (1996a) *Trade and Development Report 1996*, United Nations, New York.

UNCTAD (1996b) *World Investment Report 1996: Investment, Trade and International Policy Arrangements*, United Nations, New York.

UNCTC (1983) *Transnational Corporations in World Development: Third Survey*, United Nations, New York.

UNCTC (1988) *Transnational Corporations in World Development: Trends and Prospects*, United Nations, New York.

Underhill, G.R.D. (1994) Conceptualizing the changing global order, in R. Stubbs and G.R.D. Underhill (eds) *Political Economy and the Changing Global Order*, Macmillan, London, pp. 17–44.

UNDP (1992) *Human Development Report*, United Nations, New York.

UNIDO (1980) Export Processing Zones in developing countries, *UNIDO Working Paper on Structural Change, 19*.

UNIDO (1981) *Restructuring World Industry in a Period of Crisis – The Role of Innovation; An Analysis of Recent Developments in the Semiconductor Industry*, UNIDO, Vienna.

UNIDO (1996) *Industrial Development: Global Report 1996*, Oxford University Press, Oxford.

United Nations (1982) *Towards the New International Economic Order*, United Nations, New York.

United Nations (1996) *An Urbanizing World: Global Report on Human Settlements, 1996*, Oxford University Press, New York.

van Duijn, J.J. (1983) *The Long Wave in Economic Life*, Allen & Unwin, London.

Vernon, R. (1966) International investment and international trade in the product cycle, *Quarterly Journal of Economics*, Vol. 80, pp. 190–207.

Vernon, R. (1971) *Sovereignty at Bay: The Multinational Spread of US Enterprises*, Basic Books, New York.

Vernon, R. (1974) The location of economic activity, in J.H. Dunning (ed) *Economic Analysis and the Multinational Enterprise*, Allen & Unwin, London, pp. 89–114.

Vernon, R. (1979) The product cycle hypothesis in a new international environment, *Oxford Bulletin of Economics and Statistics*, Vol. 41, pp. 255–68.

Villareal, R. (1990) The Latin American strategy of import substitution: failure or paradigm for the region? In G. Gereffi and D.L. Wyman (eds) *Manufacturing Miracles: Paths of Industrialization in Latin America and East Asia*, Princeton University Press, Princeton, NJ, Chapter 11.

Wade, R. (1990a) *Governing the Market: Economic Theory and the Role of Government in East Asian Industrialization*, Princeton University Press, Princeton, NJ.

Wade, R. (1990b) Industrial policy in east Asia: does it lead or follow the market? In G. Gereffi and D.L. Wyman (eds) *Manufacturing Miracles: Paths of Industrialization in Latin America and East Asia*, Princeton University Press, Princeton, NJ, Chapter 9.

Wade, R. (1996) Globalization and its limits: reports of the death of the national economy are greatly exaggerated, in S. Berger and R. Dore (eds) *National Diversity and Global Capitalism*, Cornell University Press, Ithaca, NY, Chapter 2.

Wade, R. and Evans, D. (1994) Comments on World Bank's Study: The East Asian Miracle, *Working Paper, Institute of Development Studies*, University of Sussex.

Walker, R. (1988) The geographical organization of production systems, *Environment and Planning A*, Vol. 6, pp. 377–408.

Wallerstein, I. (1979) *The Capitalist World Economy*, Cambridge University Press, Cambridge.

Warf, B. (1989) Telecommunications and the globalization of financial services, *The Professional Geographer*, Vol. 41, pp. 257–71.

Warf, B. (1995) Telecommunications and the changing geographies of knowledge transmission in the late 20th century, *Urban Studies*, Vol. 32, pp. 361–78.

Warr, P.G. (1987) Malaysia's industrial enclaves: benefits and costs, *The Developing Economies*, Vol. XXV, pp. 30–55.

Webber, M.J. and Rigby, D.L. (1996) *The Golden Age Illusion: Rethinking Postwar Capitalism*, Guilford, New York.

Wells, L.T. Jr (ed) (1972) *The Product Life Cycle and International Trade*, Harvard Business School, Boston, MA.

Whitley, R.D. (1992a) *Business Systems in East Asia: Firms, Markets and Societies*, Sage, London.

Whitley, R.D. (ed) (1992b) *European Business Systems*, London, Sage.

Whitley, R.D. (1994a) Dominant forms of economic organization in market economies, *Organization Studies*, Vol. 15, pp. 153–82.

Whitley, R.D. (1994b) The internationalization of firms and markets: its significance and institutional structuring, *Organization*, Vol. 1, pp. 101–24.

Whitley, R.D. (1997) Business systems and global commodity chains: competing or complementary forms of economic organization? *Competition & Change*, Vol. 2, pp. 411–26.

Williams, K. *et al.*, (1992) Against lean production, *Economy and Society*, Vol. 21, pp. 321–54.

Williamson, O.E. (1975) *Markets and Hierarchies*, Free Press, New York.

Wilson, M. (1992) The office farther back: business services, productivity, and the offshore back office, Michigan State University, mimeo.

Winters, L.A. (1985) *International Economics* (3rd edition), Allen & Unwin, London.

Wise, M. (1994) The European Community, in R. Gibb and W. Michalak (eds) *Continental Trading Blocs: The Growth of Regionalism in the World Economy*, Wiley, Chichester, Chapter 3.

Womack, J.R., Jones, D.T. and Roos, D. (1990) *The Machine that Changed the World*, Rawson Associates, New York.

Wong, K.Y. and Chu, D.K.Y. (eds) (1985) *Coordination in China: The Case of the Shenzhen Special Economic Zone*, Oxford University Press, Hong Kong.

Wong, K.Y., Lau, C.-C. and Li, E.B.C. (eds) (1988) *Perspectives on China's Modernization*, Chinese University of Hong Kong, Hong Kong.

Wood, A. (1994) *North-South Trade, Employment and Inequality: Changing Fortunes in a Skill-Driven World*, Clarendon Press, Oxford.

World Bank (annual) *World Development Report*, Oxford University Press, New York.

World Bank (1979) *World Development Report, 1979*, Oxford University Press, New York.

World Bank (1987) *World Development Report, 1987*, Oxford University Press, New York.

World Bank (1993) *The East Asian Miracle: Economic Growth and Public Policy*, Oxford University Press, New York.

World Bank (1995) *World Development Report, 1995: Workers in an Integrating World*, Oxford University Press, New York.

WTO (annual) *Annual Report*, World Trade Organization, Geneva.

Yeung, H.W.C. (1994a) Third World multinationals revisited: a research critique and future agenda, *Third World Quarterly*, Vol. 15, pp. 296–317.

Yeung, H.W.C. (1994b) Hong Kong firms in the ASEAN region: transnational corporations and foreign direct investment, *Environment and Planning A*, Vol. 26, pp. 1931–56.

Yeung, H.W.C. (1995a) Transnational corporations from Asian developing countries: their characteristics and competitive edge, *Journal of Asian Business*, Vol. 10, pp. 17–58.

Yeung, H.W.C. (1995b) The geography of Hong Kong transnational corporations in the ASEAN region: some empirical observations, *Area*, Vol. 27, pp. 318–34.

Yeung, H.W.C. (1997) Business networks and transnational corporations: a study of Hong Kong firms in the ASEAN region, *Economic Geography*, Vol. 73, pp. 1–24.

Yeung, H.W.C. (1998) The political economy of transnational corporations: a study of the regionalization of Singaporean firms, *Political Geography*.

Yeung, Y.M. and Lo, F.C. (1996) Global restructuring and emerging urban corridors in Pacific Asia, in Y.M. Yeung and F.C. Lo (eds) *Emerging World Cities in Pacific Asia*, United Nations University Press, Tokyo, Chapter 2.

Yoffie, D.B. (ed) (1993) *Beyond Free Trade: Firms, Governments, and Global Competition*, Harvard Business School Press, Boston, MA.

Yoffie, D.B. and Milner, H.V. (1989) An alternative to free trade or protectionism: why corporations seek strategic trade policy, *California Management Review*, Vol. 31, pp. 111–31.

Zhan, J.X. (1995) Transnationalization and outward investment: the case of Chinese firms, *Transnational Corporations*, Vol. 4, pp. 67–100.

Zukin, S. and DiMaggio, P. (eds) (1990) *Structures of Capital: The Social Organization of the Economy*, Cambridge University Press, Cambridge.

Index